AF538919

Pvt. Ltd.

**Dedicated to**

**Beloved *parents* and *My wife***

**Er. S. Arisutha, M.Tech, MBA, Ph.D**

**(S. Suresh)**

**Dedicated to**

***My wife*, Khushi and *Son*, Pranav**

**(Amit Keshav)**

**Separation system synthesis & design problems**

*"The problems in the separation tasks, techniques employed to solve these problems, sequencial steps to achieve desired separation, and design of the separation units"*

# *Textbook of* SEPARATION PROCESSES

**S. Suresh**
Department of Chemical Engineering,
Maulana Azad National Institute of Technology
Bhopal - 462 051 (Madhya Pradesh)

&

**Amit Keshav**
Department of Chemical Engineering,
National Institute of Technology
Raipur - 492 010 (Chhattisgarh)

**2012**

**Studium Press (India) Pvt. Ltd.**

# *Textbook of*
# SEPARATION PROCESSES



ISBN: 978-93-80012-32-2

*Published by*:

**Studium Press (India) Pvt. Ltd.**
*4735 / 22, 2nd Floor, Prakash Deep Building*
*(Near Delhi Medical Association),*
*Ansari Road, Darya Ganj, New Delhi-110 002*
*Tel.: +91-11-43240200-15 (15 lines), Fax: +91-11-43240215;*
*E-mail: studiumpress@gmail.com*

*Printed at*: Salasar Imaging Systems

# About the Authors

## Dr. S. Suresh

**Dr. S. Suresh** is an Assistant Professor in Department of Chemical Engineering, Maulana Azad National Institute of Technology Bhopal - 462 051, Madhya Pradesh, India. He received his Ph.D in Chemical Engineering from Indian Institute of Technology Roorkee, India. He has extensive research background in the field of separation process, reactor design and published many international and national research papers. Part of his work is also published in a textbook titled *"Comprehensive Biotechnology: Engineering Fundamentals in Biotechnology"*, edited by Murray Moo-Young in Elsevier Sciences. He has also worked in Pondicherry (Central) University and Indian Institute of Technology Kanpur in the area of Environmental pollution & its control and Membrane process. His research interests include process modeling and simulation, preparation of nano-catalyst/materials, Biochemical Engg. (Fermentation, Bioreactor design); Environmental Engineering and Energy Engineering (Multicomponent adsorption, phytoremediation, composting technology, wastewater treatment, membrane process, Biomass to biofuels/ bioproducts, carbon sequestration, desulphurization, wet air oxidation, sequential batch reactor, treatment of petroleum oily sludge). He has received many awards and honours for his research contribution. Uttarakhand State Council for Science & Technology, Department of Science & Technology, India has honoured him with *Young Scientist Award* in 2009. He has received *Visiting Research Fellowship* at International Centre for Materials Science, JNCASR, Bangalore, India; 2011-2012. During his last two years, he had been involved at various levels in organization of two national level short-term courses. He is actively involved in research and had been reviewer of many international journal papers. He has been keynote speaker and resource person at several international and national conferences and programmes.

## Dr. Amit Keshav

**Dr. Amit Keshav** is working as Assistant Professor in Department of Chemical Engineering, National Institute of Technology, Raipur - 492 010, Chhattisgarh, India. He has received Ph.D. from Indian Institute of Technology Roorkee (IITR),

India. He has published many papers in international/national research journals (more than 22) and attended international and national conferences (more than 40). He worked as lecturer at Seth Jai Parkash and Mukand Lal Instutute of Engineering and Technology (JMIT), Radaur for more than three years and presently working as Assistant Professor at National Institute of Technology, Raipur for last 2 years. He had worked as head of department at JMIT Radaur for a period of more than a year and had set up various labs at both the institutes he had worked. During the last five years he had been involved at various levels in organization of three national level seminars. He is actively involved in research and had been reviewer of many international journal papers. His working areas are Reactive extraction, Vapour liquid equilibria, Process intensification, Food science, Adsorption and CFD.

## HIGHLIGHTS OF THIS TEXTBOOK

The textbook of separation processes covers, the syllabus of B.Tech, M.Tech Chemical, Biochemical, Environment and Energy Engineering. The book illustrates the basic principles as well as advanced separation processes: Supercritical fluid extraction, Pervaporation, Application of membrane modules, Liquid Membrane particularly ionic liquids, Hybrid systems, Reactive extraction with process intensification & its modelling, Adsorption, Biosorption and its modelling, Phyto-sorption, Ion exchange, Chromotography, Catalytic processes. The book also deals with Case studies for better understanding. This book is certainly useful for research scholars who are working in the areas of Chemical, Biochemical, Environment and Energy Engineering.

# Preface

In this book our endeavour is to present a few texts on the unit operations and unit processes of Chemical Engineering. It is written for undergraduate and postgraduate students who have completed the usual course in mathematics, physics, chemistry and introduction to chemical engineering as well as promoting research based work. One semester or one quarter of undergraduate/ postgraduate courses may be based on any of below section or combination of them.

Separate chapters are devoted to each of the principle operations/processes, which are prepared in five sections: Chapter 1 cover details about some of unit operations/processes like: absorption/stripping of dilute solution, liquid-liquid extraction including Supercritical Fluid Extraction. Distillation includes especially, Bubble-cap column, divide wall column with an emphasis on graphical methods of solution, these separation achieved by phase addition or creation. Leaching of materials from a solid into a liquid solvent, crystallization from a liquid, Evaporation, Drying of solids with a section on psychrometry. Mass transfer operations play a dominant role in chemical processing and this is reflected in the continued attention given to the operations of distillation, gas absorption and liquid–liquid extraction. The remaining separations involve applications of mass-transfer processes, in the presence of solid particles in Leaching (solid–liquid extraction), Drying and Crystallisation. In Distillation, Gas Absorption and Liquid–Liquid Extraction, interactions occur between two fluid streams with mass transfer taking place across a phase boundary. Usually these operations are carried out as continuous counter-current flow processes, either stage wise (as in a plate-column) or with differential contacting (as in a packed-column). Rate-based models for membrane separations include basic processes, Applications of membrane modules, liquid membranes including ionic liquid, hybrid membrane systems and *case study*: Ammonia removal in polypropylene flat circular membrane module. These techniques are of particular interest in that they provide a means of separating molecular species which are difficult to separate by other techniques and which may be present in very low concentrations. Such species include large molecules, sub-micrometre size particles, stereoisomers and the products from bioreactors. The separations can be highly specific and may depend on molecular size and shape, and the configuration of the constituent chemical groups of the molecules. Chapter 2 covers reactive extraction which includes process intensification, various process parameters affects extraction, kinetic and linear solvation energy relationship modeling, and *case studies* of reactive extraction of organic acids. Chapter 3 is concerned with equilibrium-based and rate-based models of adsorption, biosorption, phyto-sorption, ion-exchange and chromatography which are solid or solid like sorbents and *case studies*: Adsorption of phenol into activated alumina and molecular sieve, Adsorption of colour in activated carbon packed column, Biosorption of As(III) into dried *Eichhonia crasspies,* Treatment of domestic wastewater using *Eichhonia crasspies* biomass. These operations are of growing importance, especially in pollution control. Then, in Chapter 4, covers catalytic processes include various homogenous and heterogeneous reaction studies. In the Chapter 5, a brief description of the various separation processes backgrounds, research scopes and objectives are presented. Some of solved examples are given in the end of the book. Some unsolved questions are also included. Students of biology, chemistry, pharmaceutical science/technology and biochemical engineering require the theoretical knowledge which can be adopted for laboratory as well as industrial scale operations. Related chromatographic techniques such as ion-exchange chromatography, gas chromatography, high performance liquid chromatography and other methods have been discussed in the Chapter 3.

It is a pleasant privilege to acknowledge of my teachers Professor S. Sundarmoorthy of the Department of Chemical Engineering, Pondicherry Engineering College (India), Professor S.A. Abbasi of the Centre for Pollution Control and Enviornmental Engineering, Pondicherry University (India), Professor P.K. Bhattacharya of the Department of Chemical Engineering, Indian Institute of Technology Kanpur (India), Professor I.M. Mishra and Dr. V.C. Srivastava of the Department of Chemical Engineering, Indian Institute of Technology Roorkee (India), for their constant advice and encouragement.

M.N. Gernal Rana, Ravi Kant Tripathi, Siddarth Jain, Pavi Gupta and Kshtij Prabhu are students who have helped us during the preparation of chapters and solved examples. We are beholder to professors and other staff members in the MANIT Bhopal (India) and NIT Raipur (India) for encouraging us in various ways, personal as well as professional. Our thanks are also to a few of my relatives and friends who constantly encouraged us in this endeavour. Last but not the least, we wish to appreciate the role of our family members for constant endurance during the period of writing this book.

It has always been our endeavour to keep the contents of the book crisp and easily comprehensible. We feel that the contents in the book are certain mistakes and omissions may have inadvertently crept in. Readers may kindly indicate them to us for future improvement. Any suggestions for improving this book are most welcome. Users of the book can contact us at sureshpecchem @gmail.com/dr.amitkeshav@gmail.com. It has been a pleasure to work with Studium Press (India) Pvt. Ltd and we are looking forward to associate with them in future.

**S. Suresh**
**Amit Keshav**

# Contents

# Nomenclature

### NOTATIONS

| | |
|---|---|
| $a$ | Specific interface surface area; surface area; interfacial area per stage; constant in equation of state |
| $a_h$ | Specific hydraulic area of packing |
| $a_R$ | Constant of redlich-peterson isotherm |
| $a_v$ | Surface area per unit volume |
| $a_{ij}$ | Competition coefficients of component $i$ by component $j$, dimensionless |
| $A$ | Absorbance; absorption factor, dimensionless |
| $a_R$ | Constant of redlich-peterson isotherm |
| Am | Projected area |
| $A.S$ | Activated complex |
| $B$ | Constant in equations of state, bottoms flow rate; number of binary azeotropes |
| $B_{DR}$ | Dubinin and radushkevich isotherm constant |
| $B_T$ | Temkin isotherm constant, related to the heat of adsorption |
| $b$ | Temkin isotherm constant |
| c | Concentration; mass fraction; mole fraction; or volume fraction; number of components; constant; constant in tray liquid holdup; rate of production of crystals |
| $C$ | Constant that gives idea about the thickness of boundary layer |
| $c^*$ | Liquid concentration in equilibrium with gas at its bulk partial pressure |
| c' | Concentration in liquid adjacent to a membrane surface |
| $^oC$ | Degree celsius |
| $C_0$ | Initial concentration of adsorbate in solution |
| $C_e$ | Equilibrium liquid phase concentration |
| $C_L$ | Heat capacity of a liquid |
| $C_P$ | Heat capacity of a gas at constant pressure |
| $C_S$ | Adsorbent concentration in the solution |
| $C_t$ | Equilibrium liquid phase concentration after time t |
| $C_0$ | Initial concentration of adsorbate in solution |
| $C_{0,i}$ | Initial concentration of each component in solution |
| $C_e$ | Unadsorbed concentration of the single-component at equilibrium |
| $C_{e,i}$ | Unadsorbed concentration of each component in the binary mixture at equilibrium |
| D | Dipole moment; diffusivity; distillate flow rate; amount of distillate; desorbent |
| $D_e$, $D_{eff}$ | Effective diffusivity |
| $D_i$ | Impeller diameter |
| $D_p$ | Effective packing diameter; particle diameter |
| $D_s$ | Surface diffusivity |
| $E$ | Activation energy; evaporation rate; mean free energy of sorption per molecule of the sorbate |

| | |
|---|---|
| $E_a$ | Activation energy |
| $f$ | Component flow rate in feed; residual; exchange of number of moles of water per mole of adsorbate at the adsorption site |
| $F(t)$ | Fractional uptake of adsorbate on adsorbent, $0 < F(t) < 1$ |
| $F_i(C)$ | Positive fractional deviation of the synergistic langmuir adsorption isotherm from the single component langmuir isotherm for each component |
| G | Gibbs free energy; mass velocity; volumetric holdup on a tray; rate of growth of crystal size |
| $\Delta G$ | Gibbs free energy for active complex formation |
| $\Delta G^0$ | Gibbs free energy of adsorption |
| g | Molar Gibbs free energy; acceleration due to gravity |
| $g_c$ | Universal |
| $h$ | Molar enthalpy; heat-transfer coefficient; specific enthalpy; liquid molar enthalpy; initial sorption rate |
| H | Henry's law coefficient; height or length of vessel; molar enthalpy |
| $H_j$ | Residual of energy balance equation; heat of adsorption; heat of condensation; heat of crystallization; heat of dilution; integral heat of solution at saturation; heat of solution at infinite dilution; molar enthalpy of vaporization |
| $H_G$ | Height of a transfer unit for the gas |
| $H_L$ | Height of transfer unit for the liquid |
| $H_{OG}$ | Height of an overall transfer unit based on the gas phase |
| $H_{OL}$ | Height of an overall transfer unit based on the liquid phase |
| HETP | Height equivalent to a theoretical plate |
| HTU | Height of a transfer unit |
| $\Delta H$ | Enthalpy for active complex formation |
| $\Delta H^0$ | Enthalpy of adsorption |
| $\Delta H_w$ | Heat of adsorption of water (normally assumed to be zero) |
| $\Delta H_{sol}$ | Heat of solution |
| $\Delta H_{st,0}$ | Isosteric heat of adsorption with zero coverage |
| $\Delta H_{st,a}$ | Apparent isosteric heat of adsorption |
| $\Delta H_{st,net}$ | Net isosteric heat of adsorption |
| $J_i$ | Molar flux of i by ordinary molecular diffusion relative to the molar average velocity of the mixture |
| $j_i$ | Mass flux of i by ordinary molecular diffusion relative to the molar average velocity of the mixture |
| $I$ | Constant that gives idea about the thickness of boundary layer |
| $i$ | Number of donor particles |
| $k$ | Mass-transfer coefficient that takes into account the bulk flow effect; Rate constant |
| $k_c$ | Mass-transfer coefficient based on a concentration; thermal conductivity of crystal layer |
| $K$ | Equilibrium ratio for vapor-liquid equilibria; equilibrium partition coefficient in and for a component distributed between a fluid and a membrane; overall mass transfer coefficient; adsorption equilibrium constant |
| $K_G$ | Overall mass-transfer coefficient based on the gas phase with a partial pressure driving force |
| $K_L$ | Overall mass-transfer coefficient based on the liquid phase with a partial pressure driving force |
| $k_{0_B}$ | Constant in bangham equation |
| $k_0$ | Pre-exponential factor in arrhenius equation |
| $k_A$ | Adsorption rate constant for the adsorption equilibrium |
| $k_B$ | Boltzmann constant |
| $k_D$ | Desorption rate constant for the adsorption equilibrium |
| $k_d$ | Distribution coefficient |
| $k_f$ | Rate constant of pseudo-first-order adsorption model |
| $k_{id}$ | Intra-particle diffusion rate constant |
| $k_S$ | Rate constant of pseudo-second-order adsorption model |

| | |
|---|---|
| $K_F$ | Constant of freundlich isotherm |
| $K_{EL,i}$ | Constant of extended-langmuir isotherm for each component |
| $K_L$ | Constant of langmuir isotherm |
| $K_R$ | Constant of redlich-peterson isotherm |
| $K_F$ | Mono-component (non-competitive) constant of freundlich isotherm of the single component |
| $K_{F,i}$ | Individual freundlich isotherm constant of each component |
| $K_L$ | Constant of langmuir isotherm |
| $K_{L,i}$ | Individual langmuir isotherm constant of each component |
| $K_T$ | Equilibrium binding constant corresponding to the maximum binding energy |
| L | Liquid; length; height; liquid flow rate; under flow rate; crystal size |
| $L_B$ | Length of adsorption bed |
| LES | Length of equilibrium (spent) of adsorption bed |
| LUB | Length of unused bed in adsorption |
| $m$ | Slope of equilibrium curve; mean flow rate; mass; mass of adsorbent per liter of solution |
| M | Molecular weight; molar liquid holdup |
| n | Molar flowrate; moles; constant in Freundlich equation; number of pores per cross sectional area of membrane; number of units |
| N | Number of phases; number of moles; molar flux; number of equilibrium (theoretical stages); rate of rotation; number of transfer units; cumulative number of crystals size; number of stable nodes; number of data points |
| $N_A$ | Number of additional variables; avogadro's number |
| $N_G$ | Number of gas phase transfer units |
| $N_L$ | Number of liquid phase transfer units |
| $N_{min}$ | Minimum number of stages |
| $N_{Nu}$ | Nusselt number = temperature gradient at wall or interface/temperature gradient across fluid |
| $N_{Re}$ | Reynolds number = intertial force/viscous force |
| $N_{Pr}$ | Prandtl number = momentum diffusivity/thermal diffusivity |
| $N_{Pe}$ | Peclet number = convective transport/molecular transfer |
| $N_{Sc}$ | Schmidt number momentum diffusivity/mass diffusivity |
| $N_{Sh}$ | Sherwood number = conventration gradient at wall or interface/concentration gradient across fluid |
| $N_{St}$ | Stanton number for mass transfer, heat transfer |
| $N_{OG}$ | Number of overall gas-phase transfer units |
| $N_{OL}$ | Number of overall liquid-phase transfer units |
| NTU | Number of transfer units |
| $\bar{n}$ | Average number of donor particle |
| $n_m$ | Number of measurements |
| $n_P$ | Number of parameters |
| $1/n$ | Mono-component (non-competitive) freundlich heterogeneity factor of the single component, dimensionless |
| $1/n_i$ | Individual freundlich heterogeneity factor of each component, dimensionless |
| $N_i(Q)$ | Number of sites having energy $Q$, dimensionless |
| P | Pressure; power |
| p | Partial pressure |
| $P$ | Percentage contribution |
| $p_c$ | Critical pressure |
| $P_M$ | Permeability; permeance |
| $P_r$ | Reduced pressure; vapor pressure |
| $pH_0$ | Initial pH of the solution |
| $pH_{PZC}$ | Point of zero charge |
| $q_e$ | Equilibrium single-component solid phase concentration |
| $q_{e,i}$ | Equilibrium solid phase concentration of each component in binary mixture |
| $q_{e,cal}$ | Calculated value of solid phase concentration of adsorbate at equilibrium |

| | |
|---|---|
| $q_{e,\,exp}$ | Experimental value of solid phase concentration of adsorbate at equilibrium, mg/g |
| $q_m$ | Maximum adsorption capacity of adsorbent |
| $q_{max}$ | Constant in extended langmuir isotherm |
| $q_t$ | Amount of adsorbate adsorbed by adsorbent at time t |
| $Q$ | Rate of heat transfer; volume of liquid; volumetric flow rate; adsorption energy |
| $Q_L$ | Volumetric liquid flow rate |
| R | Universal gas constant, 8.314 J/K mol |
| $R^2$ | Correlation coefficients |
| $R_a$ | Radius of the adsorbent particle assumed to be spherical |
| $R_{min}$ | Minimum reflux ratio |
| r | Radius; reaction rate |
| $r_H$ | Hydraulic radius = flow cross section/wetted perimeter |
| $r_p$ | Pore radius |
| $R_L$ | Liquid-phase withdrawal factor; separation factor, dimensionless |
| S | Solid; rate of entropy; total entropy; solubility; cross sectional area for flow; solvent flow rate; mass of adsorbent; stripping factor; surface area; inert solid flow rate; flow rate of crystals; supersaturation |
| $S_g$ | Surface area per unit volume of a porous particle |
| $SS_T$ | Total sum of squares |
| $S$ | Active site on the adsorbent |
| $\Delta S$ | Change in the entropy for active complex formation |
| $\Delta S^0$ | Change in the entropy for adsorption |
| $T$ | Overall mean of response |
| $t$ | Time; residence time; average residence time |
| $t_b$ | Time to breakthrough in adsorption |
| $t_E$ | Elution time in chromatography |
| $T$ | Absolute temperature |
| $T_c$ | Critical temperature |
| U | Superficial velocity; Overall heat-transfer coefficient; Liquid stream molar flow rate |
| $U_a$ | Superficial vapor velocity |
| u | Velocity; Interstitial velocity; Bulk average velocity; Flow average velocity |
| $u_c$ | Superficial liquid velocity |
| $u_s$ | Superficial velocity |
| $u_v$ | Gas velocity |
| $V$ | Vapour; volume; vapour flow rate; overflow rate; volume of the solution |
| $V_p$ | Pore volume per unit mass of particle |
| $w$ | Mass fraction; mass of the adsorbent |
| $W_i$ | Initial weight of the sample |
| $W\tau$ | Actual weight of the sample |
| $W_f$ | Final weight of the sample |
| $x_i, y_i, z_i$ | Multi-component (competitive) freundlich adsorption constants of each component, dimensionless |
| x | Mole fraction in liquid phase; mole fraction in any phase; distance; mass fraction in raffinate; mass fraction in underflow; mass fraction of particles |
| $X$ | Equilibrium moisture |
| $X_B$ | Bound moisture content |
| $X_C$ | Critical free moisture content |
| $X_T$ | Total moisture content |
| $X_{Ae}$ | Fraction of the adsorbate adsorbed on the adsorbent under equilibrium |
| $X_i$ | Mass of solute per volume of solid |
| y | Mole fraction in vapour phase; distance; mass fraction in extract; mass fraction in overflow |
| Y | Mole or mass ratio; mass ratio of soluble material to solvent in overflow; concentration of solute in solvent |
| z | Mole fraction in any phase; overall mole fraction in combined phases |

## GREEK LETTERS

| | |
|---|---|
| $\sigma$ | Stefan-Boltzmann constant = 5.671 × $10^{-8}$ W/$m^2$ $K^4$ |
| ±' | Degree of transformation, fractional pyrolysis of the adsorbent |
| $\alpha$ | Bangham constant |
| $\alpha_D$ | Ratio of the isosteric enthalpy at zero coverage to that at monolayer coverage |
| $\alpha_i$ | Constant in SRS model for each component, dimensionless |
| $\beta_i$ | Constant in multi-component redlich-peterson model for each component, dimensionless |
| $\beta$ | Constant of redlich-peterson isotherm |
| $\beta_D$ | Pattern parameter |
| $\lambda$ | Wavelength |
| $\lambda_{vap}$ | Heat of vaporization |
| $\eta$ | Index of refraction |
| $\eta_{L,i}$ | Multi-component (competitive) langmuir adsorption constant of each component, dimensionless |
| $\eta_{R,i}$ | Multi-component (competitive) redlich-peterson adsorption constant of each component, dimensionless |
| $\theta_i(Q)$ | Coverage of each component at energy level $Q$, dimensionless |
| $\mu$ | Viscosity; linear absorption coefficient; ionic strength |
| $\mu_D$ | Interaction energy of the adsorbed adsorbate molecules |
| $\rho$ | Density; resistivity |
| $\tau$ | Time constant; inhibition function |
| $\chi$ | Transmission coefficient |
| $2\theta$ | Scattering angle |
| $\pi$ | Pi;type of electron or bond |
| $\varepsilon$ | Molar absorptivity |
| $\Sigma$ | Summation symbol |
| $\omega$ | Angular frequency |

## ABBREVIATIONS

| | |
|---|---|
| Avg | Average |
| ACC | Activated carbon commercial grade |
| AAS | Atomic absorption spectrometry |
| AFS | Atomic fluorescence spectrometry |
| $\overset{o}{A}$ | Angstrom |
| ACs | Activated carbons |
| bbl | Barrel |
| bar | 0.9869 atmosphere or 100 kPa |
| Btu | British thermal unit |
| BET | Brunauer-emmett-teller |
| BFA | Bagasse fly ash |
| BJH | Barrett-joyner-halenda |
| BOD | Bio-chemical oxygen demand |
| C | Degrees celsius, K-273.3 |
| cal | Calorie |
| cfs | Cubic feet per second |
| cm | Centimeter |
| cmHg | Pressue in centimetres head of mercury |
| cP | Centipoise |
| COD | Chemical oxygen demand |
| CV | Coefficient of variation |
| CP | 4-Chlorophenol |
| d | Differential |
| DOE | Design of experiments |
| DTA | Differential thermal analysis |

| | |
|---|---|
| DOF | Degree of freedom |
| e | Exponential function |
| exp | Exponential function |
| eq | Equivalent |
| ECD | Electron capture detector |
| F | Degree fahrenheit, R 459.7 |
| f | Function |
| ft | Feet |
| FTIR | Fourier transform infrared spectroscopy |
| FID | Flame ionization detector |
| g | Gram |
| g-mol | Gram-mole |
| gpd | Gallons per day |
| GAC | Granular activated carbon |
| GC | Gas chromatography |
| GC-MS | Gas chromatography-mass spectrometry |
| h | Hour |
| HPLC | High pressure liquid chromatography |
| HQ | Hydroquinone |
| HSDB | Hazardous substances data bank |
| HYBRID | Hybrid fractional error function |
| in | Inch |
| ICP | Inductively coupled plasma |
| J | Joule |
| K | Degrees kelvin |
| IUPAC | International union of pure and applied chemistry |
| kg | Kilogram |
| kmol | Kilogram-mole |
| L | Liter |
| LHS | Left-hand side of an equation |
| lb | Pound |
| $lb_f$ | Pound-force |
| ln | Logarithm to the base e |
| log | Logarithm to the base 10 |
| M | Molar |
| m | Meter |
| meq | Milliequivalents |
| mg | Milligram |
| min | Minute; minimum |
| max | Maximum |
| mm | Milimeter |
| mmol | Millimole (0.001 mole) |
| mol | Gram-mole |
| mole | Gram-mole |
| mmHg | Pressure in mm head of mercury |
| MPSD | Marquardt's percent standard deviation |
| MW | Molecular weight |
| N | Newton; normal |
| NP | 4-Nitrophenol |
| P | Phenol |
| ppm | Parts per million |
| psi | Pounds force per square inch |
| psia | Pounds force per square inch absolute |
| PV | Pervaporation |
| $P_s$ | Phenols |
| R | Resorcinol |

| | |
|---|---|
| RHS | Right-hand side of an equation |
| RHA | Rice husk ash |
| RO | Reverse osmosis |
| R | Degrees Rankine |
| S.G. | Specific gravity |
| STP | Standard conditions of temperature and pressure (usually 1 atm and either °C or 60 F) |
| s | Second |
| stm | Steam |
| SEM | Scanning electron microscopy |
| SSE | Sum of squares of error |
| TCD | Thermal conductivity detector |
| TG | Thermal gravimetery |
| TGA | Thermo-gravimetric analysis |
| TLV | Threshold limit value |
| vs | Versus |
| VOC | Volatile organic compound |
| USEPA | United states environmental agency protection |
| wt | Weight |
| y | Year |
| yr | Year |
| µm | Micron = micrometer |

# Dimensions and Units

*Conversion Factors for American Engineering and CGS Units to SI Units*

| *To convert from* | *To* | *Multiply by* |
|---|---|---|
| **Area** | | |
| $ft^2$ | $m^2$ | 0.0929 |
| $in.^2$ | $m^2$ | $6.452 \times 10^{-4}$ |
| **Acceleration** | | |
| $Ft/h^2$ | $m/s^2$ | $2.325 \times 10^{-8}$ |
| **Density** | | |
| $Lb_m/ft^2$ | $Kg/m^3$ | 16.02 |
| $Lb_m$/gal (US) | $Kg/m^3$ | 119.8 |
| $g/cm^3$ | $Kg/m^3 = g/l$ | 1000 |
| **Diffusivity, Kinematic Viscosity** | | |
| $ft^2/h$ | $m^2/s$ | $2.581 \times 10^{-5}$ |
| $cm^2/s$ | $m^2/s$ | $1 \times 10^{-4}$ |
| **Energy, Work, Heat** | | |
| $ft.lb_f$ | J = N . m | 1.356 |
| Btu (IT) | J = N . m | 1055 |
| Cal (IT) | J = N . m | 4.187 |
| erg | J = N . m | $1 \times 10^{-7}$ |
| kW.h | J = N . m | $3.6 \times 10^6$ |
| **Enthalpy** | | |
| Btu (IT)/$lb_m$ | J/kg = N . m/kg | 2326 |
| Cal (IT)/g | J/kg = N . m/kg | 4187 |
| **Force** | | |
| $lb_f$ | N | 4.448 |
| dyne | N | $1 \times 10^{-5}$ |

| | | |
|---|---|---|
| **Heat-transfer Coefficient** | | |
| Btu (IT)/h. ft$^2$. °F | W/m$^2$ . K | 5.679 |
| cal (IT)/s. cm$^2$. °C | W/m$^2$ . K | $4.187 \times 10^{-4}$ |
| **Interfacial Tension** | | |
| $lb_f$/ft | N/m = kg/s$^2$ | 14.59 |
| dyne/cm | N/m = kg/s$^2$ | $1 \times 10^{-4}$ |
| **Length** | | |
| ft | m | 0.3048 |
| in. | m | 0.0254 |
| **Mass** | | |
| $lb_m$ | Kg | 0.4536 |
| ton | Kg | 907.2 |
| tonne (metric ton) | Kg | 1000 |
| **Mass Flow Rate** | | |
| $lb_m$/h | Kg/s | $1.26 \times 10^{-4}$ |
| $lb_m$/s | Kg/s | 0.4536 |
| **Mass Flux, Mass Velocity** | | |
| $lb_m$/h. ft$^2$ | Kg/s. m$^2$ | $1.356 \times 10^{-3}$ |
| **Power** | | |
| ft . $Lb_f$/h | W = J/s = N. m/s | $3.766 \times 10^{-4}$ |
| ft. $Lb_f$/s | W = J/s = N. m/s | 1.356 |
| hp | W = J/s = N. m/s | 745.7 |
| btu (IT)/h | W = J/s = N. m/s | 0.2931 |
| **Pressure** | | |
| $lb_f$/ft$^2$ | Pa = N/m$^2$ | 47.88 |
| $lb_f$/in$^2$ | Pa = N/m$^2$ | 6895 |
| atm | Pa = N/m$^2$ | $1.013 \times 10^5$ |
| Bar | Pa = N/m$^2$ | $1 \times 10^5$ |
| torr = mmHg | Pa = N/m$^2$ | 133.3 |
| In. Hg | Pa = N/m$^2$ | 3386 |
| In. $H_2O$ | Pa = N/m$^2$ | 249.1 |
| **Specific Heat** | | |
| Btu (IT)/$lb_m$ . °F | J/kg. K = N . m/kg . K | 4187 |
| cal/g . °C | J/kg . K = N . m/kg . K | 4187 |
| **Surface Tension** | | |
| $lb_f$/ft | N/m | 14.59 |
| dyne/cm | N/m | 0.001 |
| erg/cm$^2$ | N/m | 0.001 |
| **Thermal Conductivity** | | |
| Btu (IT). Ft/h . ft$^2$ . °F | W/m . K = J/s . m . K | 1.731 |
| cal (IT) . cm/s . cm$^2$ . °C | W/m . K = J/s . m . K | 418.7 |
| **Velocity** | | |
| ft/h | m/s | $8.467 \times 10^{-4}$ |
| ft/s | m/s | 0.3048 |

**Viscosity**

| | | |
|---|---|---|
| $lb_m$/ft .s | Kg/m . s | 1.488 |
| $lb_m$/ft .h | Kg/m . s | $4.134 \times 10^{-4}$ |
| cP | Kg/m . s | 0.001 |

**Volume**

| | | |
|---|---|---|
| $ft^3$ | $m^3$ | 0.02832 |
| l | $m^3$ | $1 \times 10^{-3}$ |
| gal (US) | $m^3$ | $3.785 \times 10^{-3}$ |

## Physical Constants

**Universal (ideal) gas law constant, R**

1,987 cal/mol . K or Btu/lbmol . °F
8315 J/kmol . K or Pa . $m^3$/kmol . K
8.315 kPa . $m^3$/kmol . K
0.08325 bar . L/mol . K
82.06 atm . $cm^3$/mol . K
0.7302 atm . $ft^3$/lbmol . °R
10.73 psia . $ft^3$/lbmol . °R
1544 ft . lbf/lbmol . R
62.36 mmHg . L/mol . K
21.9 in. Hg . $ft^3$/lbmol . °R

**Atmospheric pressure (sea level)**

101.3 kPa = 101300 Pa = 1.013 bar
760 torr = 29.92 in. Hg
1 atm = 14.696 psia

**Avogadro's number**

$6.022 \times 10^{23}$ molecules/mol

**Boltzmann constant**

$1.381 \times 10^{-23}$ J/K . molecules

**Faraday's constant**

96490 charge/g-equivalent

**Gravitational acceleration (sea level)**

9.807 $m/s^2$ = 32.174 $ft/s^2$

**Joule's constant (mechanical equivalent of heat)**

4.184 J/cal
778.2 ft. $lb_f$/Btu

**Planck's constant**

$6.626 \times 10^{-34}$ J. s/molecule

**Speed of light in vacuum**

$2.998 \times 10^{8}$ m/s

**Stefan- Boltzmann constant**

$5.671 \times 10^{-8}$ W/$m^2$ . $K^4$
$0.1712 \times 10^{-8}$ Btu/h . $ft^2$ . $°R^4$

# Chapter 1

# Various Separation Processes

## 1.1. GENERAL INTRODUCTION-DEFINITIONS

In all industries today, separation has now become an important part of the process. Different types of separation procedures have been introduced from time to time and each separation has its own significances and drawbacks. Some of the major separation techniques are listed below:

***Absorption*:** It is a physical or chemical phenomenon or a process in which atoms, molecules, or ions enter some bulk phase - gas, liquid or solid material. This is a different process from adsorption, since the molecules are taken up by the volume, not by surface.

*Disadvantages*: (1) Requires further separation. (2) Limitation of absorbent.

***Distillation*:** It is the process used for separation of mixtures of two components based on their relative volatility or boiling points.

*Disadvantages*: (1) More complex separation. (2) Requires high heat and energy.

***Extraction & Leaching*:** The basic principle of the above techniques is separation of two mixtures by introduction of a foreign substance/solvent. It involves separation of two liquid mixtures whereas Leaching involves separation of a solid-liquid mixture.

*Disadvantages*: (1) Due to introduction of foreign material purity is hindered. (2) Requires further separation.

***Evaporation*:** It is the process by which molecules of a liquid in a solid-liquid mixture spontaneously become gaseous without being heated to its boiling point and thus being separated.

*Disadvantages*: (1) Requires high amount of heat and energy. (2) Limited application.

***Crystallization*:** It is a solid-liquid separation technique, in which mass transfer of a solute from the liquid solution to a pure solid crystalline phase occurs.

*Disadvantages*: (1) Clusters need to attain critical size for separation. (2) No uniformity in shape and size of crystals.

## 1.2. ABSORPTION/STRIPPING

### 1.2.1. *Introduction*

***Gas absorption***: Two components of a gas are separated by contact with a liquid (in which one component is preferentially soluble).

***Stripping***: Two components of a liquid are separated by contact with a gas. An example of gas absorption is the removal of ammonia from air by contact with liquid water. Ammonia is very soluble in water whereas air is only slightly soluble.

Both gas absorption and stripping involve at least three components (see Fig. 1.2.1). Usually only one of these components crosses the phase boundary. In the example of amine and steam with $CO_2$, amine is the component whose molar flowrate changes by the largest percentage of the inlet value. Although some $CO_2$ will also dissolve in water, and some water will evaporate into the $CO_2$, the molar flowrates of $CO_2$ and steam change by negligible fractions: their flows can usually be considered constant. (1) Chemical absorption technology is one of very potential technology for $CO_2$ separation. Suitable absorption liquid and optimal operation parameters seem to realize high efficiency and low energy penalty in which ammonia and some mixing absorbent looks very promising. (2) Membrane absorption system looks superior to chemical absorption system unless serious membrane pore-wetting and pore-plugging happened. Research on how to reduce the membrane price and how to prolong the membrane lifetime is very important in the future. The $CO_2$ absorption process generally consists of (1) an absorption unit where $CO_2$ is recovered from a gas phase into a liquid solvent and (2) a heat exchange system where rich solution is heated by lean solution before being pumped into desorption unit, and lean solution is cooled to a specific temperature value before being pumped into absorption unit; (3) a

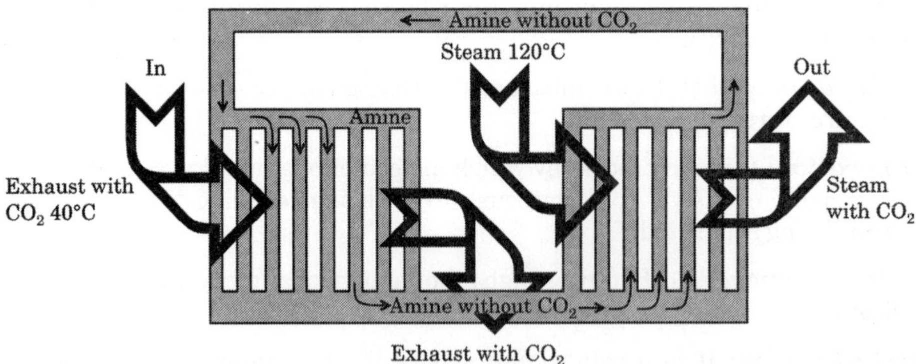


**Fig. 1.2.1:** Schematic of absorber/stripper system

desorption unit where the absorption capability of the used solvent is recovered before being reintroduced to the absorption unit. $CO_2$ separation by membrane absorption has advantage of operation flexibility, be easy to be scale up and economical, but membrane wetting is big challenge.

### 1.2.2. *Equipment for Absorption/Stripping Operation*

The gas absorption column is a unit designed for studying hydrodynamics and absorption processes in packed columns. The installation absorbs CO or ammonia from an air mixture into an aqueous solution flowing down the column. The Unit consists of the following parts: packed column, liquid (water) circuit, gas (air and CO) circuit and measurement and analysis elements. The flow rate is measured with a rotameter placed in the front panel. Once the column is crossed, the effluent liquid is sent back to the storage tank. The gas circuit is formed by an air blowing that supplies air and a pressure cylinder (not included), that supplies CO or ammonia. The flow rate of both gases are measured by means of two rota meters installed in line in the frontal panel. The two gas streams are mixed and introduced in the base of the column trough a lateral inlet. The effluent gas leaves the top of the column and is intended to be exhausted to the laboratory building. The measurement and analysis elements include two manometric tubes for determining the pressure drops across the column and equipment for the CO analysis (Fig. 1.2.2.).

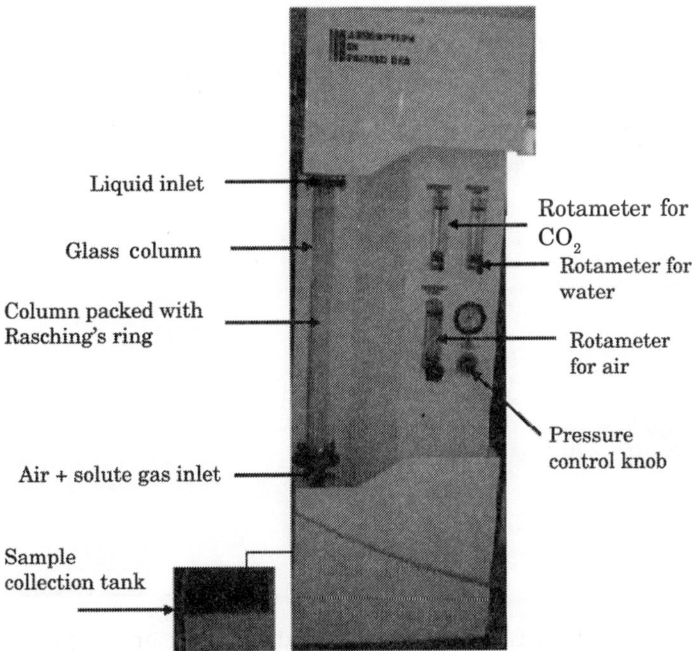


**Fig. 1.2.2:** Absorption in packed bed reactor (Suresh, unpublished figure)

In case of absorption in a packed bed column of cross sectional area, S, neglecting the change in molar feed rate of gas, V, the amount of solute absorbed in section dz is:

$$V'dy = K_Y \alpha (y - y^*) S dz \tag{1.2.1}$$

$$\frac{K_y \alpha S}{V'} \int dZ = \frac{K_y \alpha S Z_T}{V'} = \int_a^b dy / (y - y^*) \tag{1.2.2}$$

V'= Molal flow rate of gas, kgmol/h

also $Z_T = H_{0y} N_{0y}$ (1.2.3)

$$V = \frac{V'}{S} = \text{Superficial mass velocity kgmol/h m}^2 \tag{1. 2.4}$$

Where, $N_{ay} = \int_a^b dy / (y - y^*)$ (1.2.5)

$$H_{ay} = \frac{V'/S}{K_{ay} \alpha} \tag{1.2.6}$$

In case of absorption with chemical reaction, (*e.g.* $CO_2$ absorption in NaOH solution) the equilibrium P.P. is zero ($Y^* = 0$) (if reaction is irreversible)

$$N_{ay} = \int_a^b \frac{dy}{y} = \ln \frac{y_a}{y_b} \tag{1.2.7}$$

$$H_{ay} = \frac{Z_1}{N_{ay}} \tag{1.2.8}$$

But

$$H_{ay} = \frac{V'/S}{K_{ay} \alpha} = \frac{V}{K_y \alpha} \tag{1.2.9}$$

$$K_y \alpha = \frac{V'/S}{H_{ay} \alpha} = \frac{V}{H_{ay}} \text{ and } K_G \alpha = \frac{K_y \alpha}{P} \tag{1.2 .10}$$

Although liquid and gas streams for absorption or stripping could be contacted using a tray column (like that used in distillation), tray columns are seldom used. The reason is that tray efficiencies are generally much lower for absorption and stripping than for distillation (perhaps only 5% instead of 50%).

Because of this very low efficiency, very large numbers of trays are required-perhaps 100s or 1000s. Fabrication cost just becomes prohibitively expensive. Fortunately, a viable alternative exists: the packed tower.

A packed tower is simply a tube or pipe, which is filled with some sort of "packing". The packing typically consists of particles around an inch in diameter. In commercial packed towers, the usual choices are particles with one of three different shapes: (1) Raschig ring (which is just a piece of pipe which has been cut into segments, whose length and diameter are about the same). (2) Berl saddle. (3) Pall ring. The purpose of the packing is to promote good contact between the liquid and vapour streams which are being brought together to permit interfacial mass transfer. The liquid stream is usually fed into the top of the tower while the vapour is fed into the bottom. Thus we have countercurrent flow of the two streams, which has the same advantages for mass transfer as it did for heat transfer. The packing promotes good contact between the phases by dividing the two fed streams into many parallel interconnected paths. Ideally, you would like the liquid to flow downward as a thin film over the surface of the packing. This would give the maximum surface area of contact between the gas and liquid. If you just pour the liquid from the end of the pipe onto the top of the packing in tower having much larger diameter than the pipe, most of the packing will not even be wet. Only some of the channels will be carrying flow. *This is called*: channelling- maldistribution of liquid flow. So some sort of device to distribute the flow over the entire cross section of the tower is needed. This device is called a distributor. Even if the flow is evenly distributed at the top of tower, channelling might still develop as the fluid trickles down. When two thin films converge they tend to form a thick film and a dry patch, which results in a reduction in contact area. So redistributors are placed every 10-15 feet along the length of the tower.

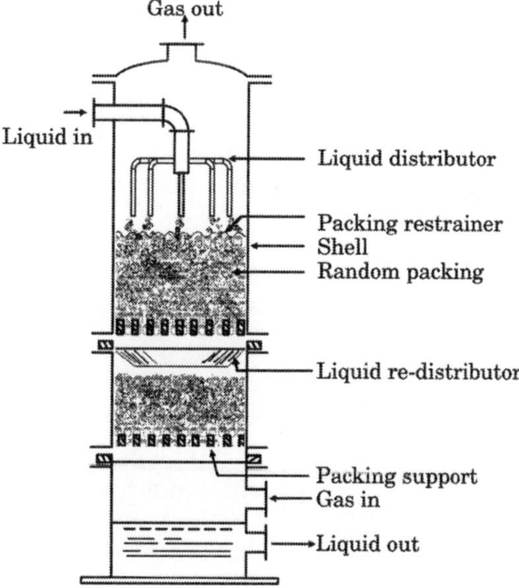


**Fig. 1.2.3:** Packed bed absorption tower (Suresh, unpublished figure)

### 1.2.3. *Packed – Column Efficiency, Capacity and Pressure Drop*

The parameters like mass – transfer coefficient and corresponding HTUs depends upon gas and/or liquid flow rates per unit inside cross – sectional area of the packed column. So, column diameter must be estimated before determining required height of packing. The estimation of a suitable column diameter for a given system, packing and operational conditions requires consideration of liquid holdup, flooding and pressure drop.

#### 1.2.3.1. *Liquid Holdup*

As the liquid flows down the packing at an increasing rate, gas phase pressure drop for a given gas velocity increasing. For a given liquid velocity, the upper lmit to the gas velocity for a constant liquid holdup is termed the loading point. Below this point, the gas phase is the continuous phase. Above this point, liquid begins to accumulate or load the bed, replacing gas holdup and causing a sharp increase in pressure drop. Finally, a gas velocity is reached at which the liquid surface is continuous across the top of the packing and the column is flooded. The region between the loading point and flooding point is the loading region, where significant liquid entrainment is observed, liquid holdup increases sharply, mass – transfer efficiency decreases, and column operation is unstable. Typically, the superficial gas velocity at the loading point is approximately 70% of that at the flooding point.

The specific liquid holdup in the preloading region has been found, from extensive experiments for a wide variety of random and structured packing and, for a number of gas – liquid systems, to depend on packing characteristics, and the viscosity, density, and superficial velocity of the liquid, $u_L$, according to the dimensionless expression

$$h_L = \left(12\frac{N_{Fr_L}}{N_{Re_L}}\right)^{1/3}\left(\frac{a_h}{a}\right)^{2/3} \tag{1.2.11}$$

Where $N_{Re_L}$ = liquid Reynolds number = $\dfrac{\text{inertial force}}{\text{viscous force}}$

$$N_{Re_L} = \frac{u_L\rho_L}{a\mu_L} = \left(\frac{u_L}{a\nu_L}\right) \tag{1.2.12}$$

Where

$V_L$ = kinematic viscosity

$N_{Fr_L}$ = liquid Froud number = $\dfrac{\text{inertial force}}{\text{gravitional force}}$

$$= \left(\frac{u_L^2 a}{g}\right) \tag{1.2.13}$$

and the ratio of the specific hygrulic area of packing, $a_{h,}$ to specific area of packing, a, is given by

$$\frac{a_h}{a} = C_h N_{Re_L}^{0.15} N_{Fr_L}^{0.1} \text{ for } N_{Re_L} < 5 \tag{1.2.14}$$

$$\frac{a_h}{a} = 0.85 C_h N_{Re_L}^{0.25} N_{Fr_L}^{0.1} \text{ for } N_{Fr_L} \geq 5 \tag{1.2.15}$$

Value of $a_h/a > 1$ are responsible for the creation of droplet and jet flow besides the rivulets that cover the packing surface. To ensure adequate wetting of packing, proven liquid distributors and redistributors should be used and supercritical liquid velocities should exceed the following value: Ceramic=0.00015 m/s; Metals=0.0003 – 0.0009 m/s; Plastic=0.0012 m/s.

1.2.3.2. *Column Diameter and Pressure Drop*

Most packed columns consist of cylindrical vertical vessels. The column diameter is determined so as to safely avoid flooding and operate in the preloading region with a pressure drop of no greater than 1.5 in. of water head per foot of packed height. Flooding data for packed columns with countercurrent flow of liquid and gas, used the following liquid – to – gas kinetic energy ratio,

$$F_{LV} = \left(\frac{LM_L}{VM_V}\right)\left(\frac{\rho_V}{\rho_L}\right)^{0.5} \tag{1.2.16}$$

Many experiments were conducted on ring and saddle packings to extend the earlier flooding correlation to include lines of constant pressure drop, with the resulting chart becoming known as the generalized pressure–drop correlation. For given fluid flow rates and properties and a given packing material, the pressure – drop correlation chart is used to compute $u_{vf}$, the superficial velocity at flooding. Then a fraction of flooding, if, is selected (usually from 0.5 to 0.7), followed by calculation of the tower diameter from an equation given below:

$$D_T = \left(\frac{4VM_V}{\rho_v f \pi u_V}\right)^{0.5} \tag{1.2.17}$$

Experimental flooding – point data for a variety of packing material are in reasonable agreement with upper curve of the pressure – drop correlation chart (refer perry handbook of chemical engineers). Such good agreement is not always the case for pressure, particularly for operation at superficial velocities above 50% of flooding, where pressure drop is greater than 0.5 in. of water head per foot of packed height. Studies conducted for both random and structured packings showed that the pressure drop at flooding is strongly dependent on the packing fraction, $F_p$, by the expression:

$$\Delta P_{flood} = 0.115 F_P^{0.7} \tag{1.2.18}$$

Where $\Delta P_{flood}$ has unit of inches of water head per foot of packed height and $F_P$ has unit of ft²/ft². Based on extensive experimental studies using more than 50 packing materials, including structured packings, Billet and Schulets [1991] developed a correlation for dry gas pressure drop, $\Delta P_o$. Their dimensionless consistent correlation is

$$\frac{\Delta P_o}{l_T} = \psi_o \frac{a u_V^2 \rho_V}{2\varepsilon^3 K_W} \tag{1.2.19}$$

Where, $l_T$ = height of packing, $K_w$ = a wall factor, $K_W$ is an important for columns with an inadequate ratio of effective packing diameter to inside column diameter, and is given by

$$\frac{1}{K_W} = 1 + \frac{2}{3}\left(\frac{1}{1-\varepsilon}\right)\frac{D_P}{D_T} \tag{1.2.20}$$

Where the effective packing diameter, $D_P$, is determined from

$$D_P = 6\left(\frac{1-\varepsilon}{a}\right) \tag{1.2.21}$$

The dry packing coefficient (a modified friction factor), $\Psi_o$, is given by the empirical expression

$$\psi_o = C_P\left(\frac{64}{N_{Re_V}} + \frac{1.8}{N_{Re_V}^{0.08}}\right) \tag{1.2.22}$$

Where

$$N_{Re_V} = K_W\left(\frac{u_V D_P \rho_V}{(1-\varepsilon)\mu_V}\right) \tag{1.2.23}$$

and $C_P$ is packing constant, determined for experimental data. The laminar flow region is characterized by, while the next term characterizes the more common turbulent flow regime. When the packed tower is irrigated with a downward – flowing liquid, the cross sectional area for gas flow is reduced by the liquid holdup and the surface structure exposed to the gas is changed as the result of the coating of the liquid film. The pressure now becomes dependent on the holdup and a two phase flow resistance, and was found by Billet amd Schulets [1991] to depend on the liquid flow and froud number as follows for flow rates up to the loading point:

$$\frac{\Delta P}{\Delta P_o} = \left(\frac{\varepsilon}{e - h_L}\right)^{3/2} \exp\left[\frac{13300}{a^{3/2}}\left(N_{Fr_L}\right)^{1/2}\right] \tag{1.2.24}$$

### 1.2.3.3. *Mass Tranfer Efficiency*

The mass transfer efficiency of a packed column incorporated in the HETP or the more theoretically based HTUs and volumetric mass transfer coefficients. Although the HETP concept lack a theoretical basis, its simplicity, coupled with the relative ease with which equilibrium stage can be made with computer-aided simulation programs, has made it a widely used method for estimating packing height. In the preloading region where distribution of liquid and vapor is initiated and maintained, values of HETP depend mainly on packing type and size, liquid viscosity and surface tension. In general, lower values of HETP are achieved with smaller size random packings, particularly in small-diameter columns, and with structured packing particularly with large value of a, the packing surface area per packed volume.

The two film theory defines an overall coefficient in terms of the individual volumetric mass–transfer coefficients. Most commonly, reference is made to the overall gas phase resistance

$$\frac{1}{K_y a} = \frac{1}{k_y a} + \frac{K}{k_x a} \tag{1.2.25}$$

For mass transfer rates expressed in terms of mole fration driving force

$$r = k_y a\left(y - y_1\right) = k_x a\left(x_1 - x\right) = K_y a\left(y - y^*\right) \tag{1.2.26}$$

Where K is vapor – liquid equilibrium ratio. Mass transfer rate can be expressed in terms of liquid phase concentration and gas phase partial pressure

$$r = k_P a\left(p - p_1\right) = k_L a\left(c_1 - c\right) = K_G a\left(p - p^*\right) \tag{1.2.27}$$

If we define a Henery's Law constant at the equilibrium interface between the two phases by $p_I$ = H', and let

$p^*$ = H'

Then

$$\frac{1}{K_G a} = \frac{1}{k_p a} + \frac{H'}{k_L a} \tag{1.2.28}$$

Alternatively expression can be derived for $K_x a$ and $K_L a$

### 1.2.3.4. *Packing Height: The Method of Transfer Units*

A concept in the analysis of packed column centred on the method of transfer units. This method is more appropriate because the changes in compositions of the liquid and vapour phases occur differentially in a packed column rather than in stepwise fashion as in tray column.

In this method, height of packings required can be evaluated either based on the gas-phase or the liquid-phase. The packed height (z) is calculated using the following formula:

$$z = N \times H \tag{1.2.29}$$

Where, N = number of transfer units (NTU)
H = height of transfer units (HTU)

The number of transfer units (NTU) required is a measure of the difficulty of the separation. A single transfer unit gives the change of composition of one of the phases equal to the average driving force producing the change. The NTU is similar to the number of theoretical trays required for tray column. Hence, a larger number of transfer units will be required for a very high purity product. The height of a transfer unit (HTU) is a measure of the separation effectiveness of the particular packing for a particular separation process. As such, it incorporates the mass transfer coefficient that we have seen earlier. The more efficient the mass transfer (*i.e.* larger mass transfer coefficient), the smaller the value of HTU. The values of HTU can be estimated from empirical correlations or pilot plant tests, but the applications are rather restricted.The applications of the equations rather than any derivation of them. Determination of the packed height can be based on either the gas-phase or the liquid-phase.

For the gas-phase, we have: z = NOG × HOG (1.2.30)

$$N_{OG} = \frac{y_1 - y_2}{\left(y - y^*\right)_{LM}} \tag{1.2.31}$$

$$(y - y^*)_{LM} = \frac{\left(y_1 - y_1^*\right) - \left(y_2 - y_2^*\right)}{\ln\left[\frac{\left(y_1 - y_1^*\right)}{\left(y_2 - y_2^*\right)}\right]} \tag{1.2.32}$$

$$H_{OG} = \frac{G}{K_Y a \left(1 - y\right)^*_{LM}} \tag{1.2.33}$$

$$\left(1 - y\right)^*_{LM} = \frac{\left(1 - y_1\right) - \left(1 - y_1^*\right)}{\ln\left[\frac{\left(1 - y_1\right)}{\left(1 - y_1^*\right)}\right]} \tag{1.2.34}$$

and $K_Y$ is the overall gas-phase mass transfer coefficient. 'a' is the packing parameter. *i.e.*, That characterizes the wetting characteristics of the packing material (area/volume). Normally, packing manufacturers report their data with both $K_Y$ and 'a' combined as a single parameter. Since $K_Y$ has a unit of mole/(area.time.driving force), and 'a' has a unit of (area/volume), the combined parameter $K_Y$ a will have the unit of mole/ (volume.time.driving force). As seen earlier, other than mole fraction, driving force can be expressed in partial pressure etc. $y_1^*$ is the mole fraction of solute in vapour that is in equilibrium with the liquid of mole fraction $x_1$ and $y_2^*$ is mole fraction of solute in vapour that is in equilibrium with the liquid of mole fraction $x_2$. The values of $y_1^*$ and $y_2^*$ can be obtained from the equilibrium line see the *figure* below (Fig. 1.2.4).

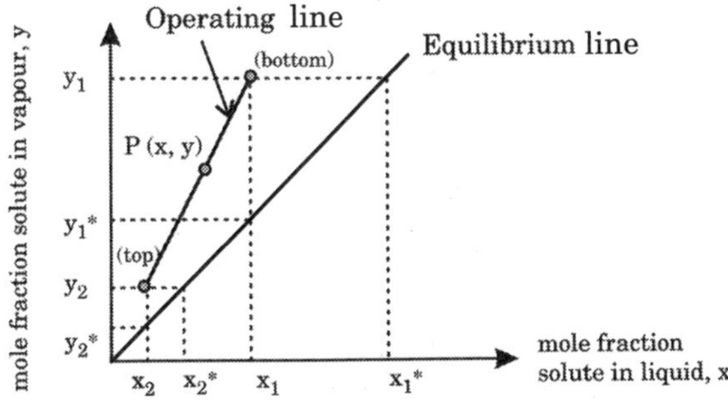


**Fig. 1.2.4:** Graph representing oprating line and equilibrium line for absorption

$(y_1 - y_1^*)$ is the concentration difference driving force for mass transfer in the gas phase at point 1 (bottom of column) and $(y_2 - y_2^*)$ is the concentration difference driving force for mass transfer in the gas phase at point 2 (top of column). Point P (x, y) as shown is any point in the column. The concentration difference driving force for mass transfer in the gas phase at point P is $(y - y^*)$. Equilibrium values $y_1^*$ and $y_2^*$ can also be calculated using Henry's Law ($y = m\,x$, where m is the gradient) which is used to represents the equilibrium relationship at dilute conditions.

*Note*: Both equilibrium line and operating line are straight lines under dilute conditions.

Thus, we have: $y_1^* = m\,x_1$; $y_2^* = m\,x_2$

Similarly for the liquid-phase we have: $z = N_{OL} \times H_{OL}$

$$N_{OL} = \frac{x_1 - x_2}{\left(x^* - x\right)_{LM}} \tag{1.2.35}$$

$$(x^* - x)_{LM} = \frac{\left(x_1^* - x_1\right) - \left(x_2^* - x_2\right)}{\ln\left[\frac{\left(x_1^* - x_1\right)}{\left(x_2^* - x_2\right)}\right]} \tag{1.2.36}$$

$$H_{OL} = \frac{L}{K_X a (1-x)^*_{LM}} \tag{1.2.37}$$

$$(1-x)^*_{LM} = \frac{(1-x_1)-(1-x_1^*)}{\ln\left[\frac{(1-x_1)}{(1-x_1^*)}\right]} \tag{1.2.38}$$

and $K_X$ is the overall liquid-phase mass transfer coefficient, and 'a' is the packing parameter seen earlier. Again, normally both $K_X$ and 'a' combined as a single parameter. Likewise, $x_1^*$ is the mole fraction of solute in liquid that is in equilibrium with the vapour of mole fraction $y_1$ and $x_2^*$ is mole fraction of solute in liquid that is in equilibrium with the vapour of mole fraction $y_2$. Alternatively, $x_1^* = y_1/m$ and $x_2^* = y_2/m$.

$(x_1^* - x_1)$ is the concentration difference driving force for mass transfer in the liquid phase at point 1 (bottom of column) and $(x_2^* - x_2)$ is the concentration difference driving force for mass transfer in the liquid phase at point 2 (top of column). Using either gas-phase or liquid-phase formula should yield the same required packing height Z = N x H.

$H_{OG} = N_{OL} \times H_{OL}$

Where, N = number of transfer units (NTU); H = height of transfer units (HTU)

### 1.2.3.5. *HETP (Height Equivalent to a Theoretical Plate)*

As we have noted, instead of a tray (plate) column, a packed column can be used for various unit operations such as continuous or batch distillation, or gas absorption. With a tray column, the vapours leaving an ideal plate will be richer in the more volatile component than the vapour entering the plate by one equilibrium "step". With plate columns, the vapour leaving an ideal plate is richer in the more volatile component than the vapour entering the plate, by one equilibrium step. Experiments suggest that that same enrichment of the vapour will occur in a certain height of packing, which is termed the height equivalent of a theoretical plate (HETP). When packings are used instead of trays, the same enrichment of the vapour will occur over a certain height of packings, and this height is termed the HETP. As all sections of the packings are physically the same, it is assumed that one equilibrium (theoretical) plate is represented by a given height of packings. Thus the required height of packings for any desired separation is given by (HETP x No. of ideal trays required).

HETP values are complex functions of temperature, pressure, composition, density, viscosity, diffusivity, pressure drop, vapour and/or liquid flowrates, packing characteristics, etc. Empirical correlations, though available to calculate the values of HETP, are restricted to limited applications. The main difficulty lies in the failure to account for the fundamentally different action of tray and packed columns. In industrial practice, the HETP concept is used to convert empirically the number of theoretical trays to packing height. Most data have been derived from small-scale operations and they do not provide a good guide to the values which will be obtained on full-scale plant. As all of the packing are physically the same, it is assumed that one equilibrium stage is represented by a given height of packing. Thus the required height of packing for any desired separation is given by HETP × (No. of ideal stages required). This is a simple method of representation which has been widely used as a method of design. Despite this fact, there have been few developments in the theory. Columns 50–750 mm diameter and packed over heights of 0.9–3.0 m with rings, saddles, and other packings have been considered. Most of the results were for conditions of total reflux, with a vapour rate of 0.18–2.5 kg/m²s which corresponded to 25–80 per cent of flooding. The relationship is:

$$HETP = C_1 G^{C_2} d_c^{C3} Z^{1/3} \left( \frac{\alpha \mu_L}{\rho_L} \right) \tag{1.2.39}$$

Where the values of $C_1$, $C_2$, $C_3$ varied with packings

It may also be noted that the mixtures considered were mainly hydrocarbons with values of relative volatilities only slightly in excess of 3. In equation G = mass velocity of vapour (kg/m² s of tower area), d = column diameter (m), Z = packed height (m), $\alpha$ = relative volatility, $\mu_L$ = liquid viscosity (N s/m²), and $\rho L$ = liquid density (kg/m³).The following dimensionally consistent equation for the HETP (Z) of packed columns using 25 and 50 mm Raschig rings

$$Z = 18d + 12m \left[ \frac{G}{L} - 1 \right] \tag{1.2.40}$$

Where: $d_r$ is the diameter of the rings, m is the average slope of equilibrium curve, G is the vapour flowrate, and L is the liquor flowrate. In practice, the HETP concept is used to convert empirically the number of theoretical stages to packing height. As most data in the literature have been derived from small-scale operations, these do not provide a good guide to the values which will be obtained on full-scale plant.

## 1.3. DISTILLATION

### 1.3.1. *Important Definitions*

***Distillation:*** Process of separating liquid phases on the basis of their volatility. The liquid phase get separated into distillate (light phase) and bottom product(heavy phase). Light phase is consisted of vapour with more volatile component whereas less volatile component constitute the heavy phase.

***Azeotrope:*** Is a mixture of two or more pure compounds (chemicals) in such a ratio that its composition cannot be changed by simple distillation. This is because when an azeotrope is boiled, the resulting vapor has the same ratio of constituents as the original mixture of liquids.

***Bottoms*:** The stream of liquid product collected from the reboiler at the bottom of a distillation tower.

***Distillate*:** The vapor from the top of a distillation column is usually condensed by a total or partial condenser. Part of the condensed fluid is recycled into the column (reflux) while the remaining fluid collected for further separation or as final product is known as distillate or overhead product.

***Bubble point*:** The temperature at constant pressure (or the pressure at constant temperature) at which the first vapor bubble forms when a liquid is heated (or decompressed).

***Dew point*:** The temperature at constant pressure (or the pressure at constant temperature) at which the first liquid droplet forms when a gas (vapor) is cooled (or compressed).

***Reboiler*:** Is a heat exchanger typically used to provide heat to the bottom of industrial distillation columns. They boil the liquid from the bottom of a distillation column to generate vapors which are returned to the column to drive the distillation separation.

***Condenser*:** Is a heat exchanger which condenses a substance from its gaseous to its liquid state.

***Equation of state*:** A relation between the pressure, volume and temperature of a system, from which other thermodynamic properties may be derived. The relation employs any number of 'constants' specific to the system. For example, for a pure component, the constants may be generalized functions of critical temperature, critical pressure and acentric factor, while for a mixture, mixing rules (which may be dependent on composition or density), are also used.

***Key component*:** A distillation column is assigned with two key components. The key components in the feed are the main components to be separated in that column. The volatility of the two key components must be in adjacent order when the volatilities of all the components in the feed are arranged in either ascending or descending order.

***K-value*:** Vapor-liquid equilibrium constant or distribution coefficient. It is used in nonideal (hydrocarbon) systems.

***Heavy key*:** The heavier (less volatile) of the two key components. Heavy key is collected at the bottoms. All non-key components heavier than the heavy key are known as the heavy components.

***Light key*:** The lighter (more volatile) of the two key components. Light key is collected at the distillate. All non-key components lighter than the light key are known as the light components.

***Reflux ratio*:** The ratio of the reflux stream to the distillate. The operating reflux ratio could affect the number of theoretical stages and the duties of reboiler and condenser.

***Relative volatility***: Relative volatility is defined as the ratio of the concentration of one component in the vapor over the concentration of that component in the liquid divided by the ratio of the concentration of a second component in the vapor over the concentration of that second component in the liquid. For an ideal system, relative volatility is the ratio of vapor pressures *i.e.* $\alpha = P_2/P_1$.

***Vapor-liquid equilibrium (VLE)***: Liquid and its vapor (gas phase) are in equilibrium with each other, a condition or state where the rate of evaporation (liquid changing to vapor) equals the rate of condensation (vapor changing to liquid) on a molecular level such that there is no net (overall) vapor-liquid interconversion.

***Vapor pressure***: The pressure exerted by the vapor phase that is in equilibrium with the liquid phase in a closed system. For moderate temperature ranges, the vapor pressure at a given temperature can be estimated using the Antoine equation.

### 1.3.2. *Distillation History*

*The history of distillation*: Forbes, a Chinese scientist has chronicled the history of distillation in 1948. It was later introduced to India, Arabia, Britain and the rest of the world. Early distillation consisted of simple batch stills to produce ethanol. Crude ethanol was placed in a still and heated, and the vapor drawn from the still was condensed for consumption. Lamp oil was later produced using the same method, with crude oil heated in batch stills. The next progression in the history of distillation was to continually feed the still and recover the light product. Further advancements include placing the stills in series and interchanging the vapor and liquid from each still to improve recovery. This was the first type of counter-current distillation column that we have today. Distillation columns are used for about 95% of liquid separations and the energy use from this process accounts for an estimated 3% of world's energy consumption. Motivated by this large energy requirement for distillation, researchers have developed various column arrangements that can bring in savings in both energy and capital cost. Any reduction of energy consumption will not only bring economical benefits but also environmental benefits in terms of reduced usage of fossil fuels and its associated emissions. Column configurations vary from simple to complex configurations. Simple column configurations refer to direct sequence, indirect sequence and distributed sequence, and are the conventional column systems operated worldwide today. Complex column arrangements refer to recycling the vapor and/or liquid, heat integration, etc. Thermally coupled arrangements are realized by setting up two-way vapor/liquid flows between different columns of the simple column configurations. The Dividing-Wall Column (DWC) has been known for several decades, introduced the thermal coupling for separating ternary mixtures and presented a fully thermally coupled configuration-Petyluk column. Literature reported studies reveal that Fully Thermally Coupled Distillation System provides the maximum energy reduction in columns. In most cases, this is implemented in the form of a DWC in which both columns are housed in a single shell. This reduces not only the energy consumption but also space and investment requirements compared to the conventional column system.

When considering the workhorse of process industry, a distillation column with one feed and two, *i.e.* overhead and bottom products, a better overall performance can be achieved, if appropriate, by employing better operation and control, by using more advanced column internals, or by employing side heat exchangers These options are particularly interesting because they allow capacity and/or efficiency increase (reduction of specific energy requirement) of existing columns at minimum investment. However substantial energy savings can be expected only from implementation of energy saving concepts in new designs. A problem with new designs, *i.e.* implementation of highly efficient energy saving technologies is that these are generally capital intensive. Typical example is employment of heat pump assisted distillation, which, technically speaking solves the problem at its source, by avoiding use of external heat source in reboiler. However, due to high capital expenditures required to purchase fitting compressor(s), it appeared to be industrially viable only in case of large capacity, end of the train (practically binary) separations of close boiling mixtures which require minimum compressor/compression effort related costs. This appeared to be the case mainly with separation of rather light hydrocarbons, like C2, C3 and C4 components, where adiabatic exponent of substances is large enough to enable a significant temperature rise by a relatively low compression effort. Costlier and more risk associated option in this respect is to implement theoretically and experimentally proven, but industrially not yet applied column configurations that offer potential for energy saving beyond that achievable with conventional heat pump assisted distillations. Typical example is so called internally heat integrated column which, by combining direct vapour recompression with uniform distribution of heat duty along integrated part of the column (diabatic operation) allows minimisation of energy requirement of a distillation column. However, being a heat pump assisted process requiring a heavy heat transfer infrastructure to enable a uniform distribution of required heat duty along the column, a internally heat integrated column represents a complex and capital highly intensive column configuration. This fact as well as the lack of necessary process and mechanical design know-how is a major technical barrier to be overcome to realize this, regarding the potential, ultimate energy saving concept in case of standalone columns. On the distillation system design level, various methods can be used to make continuous distillation process more energy efficient and thus more sustainable. When the feed is a multicomponent mixture and more than two products need to be obtained, the number of columns required to isolate pure products or specified fractions is equal to the number of components minus one. The problem is that the number of possible column sequences increases exponentially with the number of components to be separated. Therefore an effort toward finding the optimal separation sequence is necessary, and the best solution is usually that generating lowest total annualized cost.

Recent developments around sharp and non-sharp separations where heat coupling can be combined in a highly beneficial way with equipment compactness are done, simply by placing two columns into one shell. The application windows as well as potential gains in this respect could be further significantly increased by adopting recently introduced non-welded wall technology, which, most importantly, allows placing three or even more columns into one shell.

Distillation is by far the most important separation process in the petroleum and chemical industries. It is the separation of key components in a mixture by the difference in their relative volatility, or boiling points. It is also known as fractional distillation or fractionation. In most cases, distillation is the most economical separating method for liquid mixtures. However, it can be energy intensive. Distillation can consume more than 50% of a plant's operating energy cost. There are alternatives to distillation process such as solvent extraction, membrane separation or adsorption process. On the other hand, these processes often have higher investment costs. Therefore, distillation remains the main choice in the industry, especially in large-scale applications.

It is noted that the industrial acceptance and commercialization of DWC by organizations such as BASF AG, M.W. Kellogg (together with BP, later known as BP Amoco), and Sumitomo Heavy Industries Co. together with Kyowa Yuka. Linde AG constructed the world's largest DWC for Sasol, an estimated 107 m tall and 5 min diameter. Recently reported that BASF operates about 30 DWCs worldwide in their plants. It is established that the Petyluk system requires the least energy among all the options for a three-product system. This key advantage is also applicable for a DWC, which has the minimum vapor flow rate for the particular separation compared to the conventional system.

It has shown that, for a mixture of three components, the DWC reduces the total vapor flow by 10–50% compared to conventional systems using direct and indirect sequences. Reduction in vapor flow contributes to lower duties of reboiler and condenser, and consequently capital and operating costs. Furthermore, DWC uses only 1 reboiler and 1 condenser when compared to 2 reboilers and 2 condensers for a conventional 2-column (C2C) system. This would add to the savings in capital as well as operating cost. A search of the open literature had revealed that several applications of DWCs were successfully studied as new design. A few other applications involve a side draw stream in a single column; these can be tried in a DWC for efficient separation.

### 1.3.3. *Basic Distillation Equipment and Operation*

Distillation columns are made up of several components, each of which is used either to tranfer heat energy or enhance materail transfer. A typical distillation contains several major components: (1) a vertical shell where the separation of liquid components is carried out (2)column internals such as trays/plates and/or packings which are used to enhance component separations (3)a reboiler to provide the necessary vaporisation for the distillation process (4) a condenser to cool and condense the vapour leaving the top of the column (5) a reflux drum to hold the condensed vapour from the top of the column so that liquid (reflux) can be recycled back to the column The vertical shell houses the column internals and together with the condenser and reboiler, constitute a distillation column. A schematic of a typical distillation unit with a single feed and two product streams is shown in the Fig. 1.3.3.

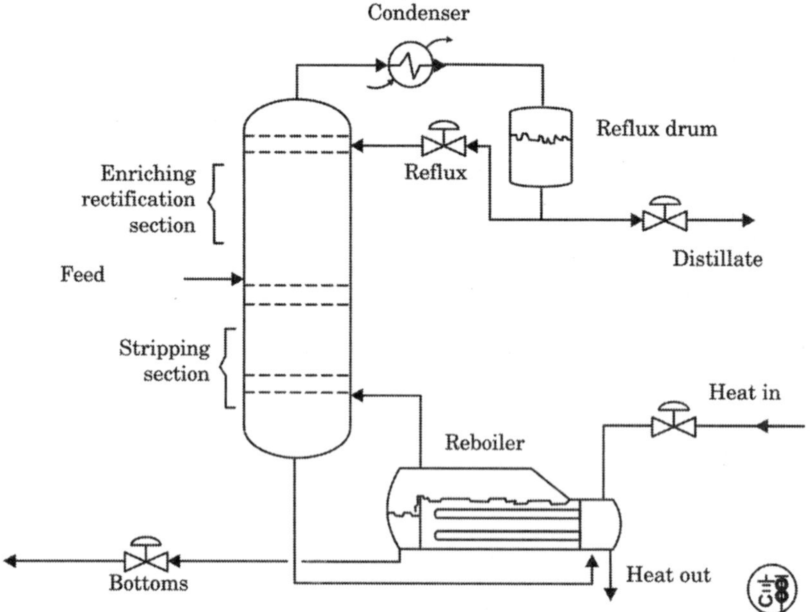


**Fig. 1.3.3:** A typical distillation unit

The liquid mixture that is to be processed is known as the feed and this is introduced usually somewhere near the middle of the column to a tray known as the feed tray. The feed tray divides the column into a top (enriching or rectification) section and a bottom (stripping) section. The feed flows down the column where it is collected at the bottom in the reboiler. Heat is supplied to the reboiler to generate vapour. The source of heat input can be any suitable fluid, although in most chemical plants this is normally steam. In refineries, the heating source may be the output streams of other columns. The vapour raised in the reboiler is re-introduced into the unit at the bottom of the column. The liquid removed from the reboiler is known as the bottoms product or simply, bottoms. The vapour moves up the column, and as it exits the top of the unit, it is cooled by a condenser. The condensed liquid is stored in a holding vessel known as the reflux drum. Some of this liquid is recycled back to the top of the column and this is called the reflux. The condensed liquid that is removed from the system is known as the distillate or top product. Thus, there are internal flows of vapour and liquid within the column as well as external flows of feeds and product streams, into and out of the column. There are a number of designs of reboilers. It is beyond the scope of this set of introductory notes to delve into their design principles. However, they can be regarded as heat-exchangers that are required to transfer enough energy to bring the liquid at the bottom of the column to boiling boint. The following are examples of typical reboiler types. A novel development in reboiler design is the self-cleaning shell-and-tube heat exchangers by Klarex Technology for applications where heat exchange surfaces are prone to fouling by the process fluid. Particles are introduced into the process stream and these produce a scouring action on the heat exchange surfaces.

## 1.3.4. *Methods of Distillation*

### 1.3.4.1. *Batch (Differential) Distillation*

In this distillation techniques, a known quantity of a liquid mixture is charged into a jacketed kettle or still. The jacket is provided for heating the mass in the still with the help of a heating media such as steam. The charge is boiled slowly, vapours formed are withdrawn and fed to a condenser where they are liquefied and collected in a receiver as distillate. In the early stage of distillation proceeds the liquid in the still becomes lean with respect to the more volatile component. The composition of the lesss volatile component thereby increases and hence the boiling point increases. The product (distillate) from such units can be collected in several receivers, called cuts, to give products of various purities over the length of distillation period. The distillation is continued till the boiling point of the liquid reaches a predetermined value and the contents of the still are finally removed as residual liquid containing majority of the less volatile component. In many instances processes are carried out in batches, and it is more convenient to distill each batch separately. In these cases the whole of a batch is run into the boiler of the still and, on heating, the vapour is passed into a fractionation column, as shown in Fig. 1.3.4.1. As with continuous distillation, the composition of the top product depends on the still composition, the number of plates in the column and on the reflux ratio used. When the still is operating, since the top product will be relatively rich in the more volatile component, the liquid remaining in the still will become steadily weaker in this component. As a result, the purity of the top product will steadily fall. Thus, the still may be charged with mols of a mixture containing a mole fraction of the more volatile component. Initially, with a reflux ratio, the top product has a composition. If after a certain interval of time the composition of the top product starts to fall, then, if the reflux ratio is increased to a new value, it will be possible to obtain the same composition at the top as before, although the composition in the still is weakened to final product. This method of operating a batch still requires a continuous increase in the reflux ratio to maintain a constant quality of the top product. An alternative method of operation is to work with a constant reflux ratio and allow the composition of the top product to fall. One of the added merits of batch distillation is that more than one product may be obtained. Thus, a binary mixture of alcohol and water may be distilled to obtain initially a high quality alcohol. As the composition in the still weakens with respect to alcohol, a second product may be removed from the top with a reduced concentration of alcohol. In this way it is possible to obtain not only two different quality products, but also to reduce the alcohol in the still to a minimum value. This method of operation is particularly useful for handling small quantities of multi-component organic mixtures, since it is possible to obtain the different components at reasonable degrees of purity. To obtain the maximum recovery of a valuable component, the charge remaining in the still after the first distillation may be added to the next batch.

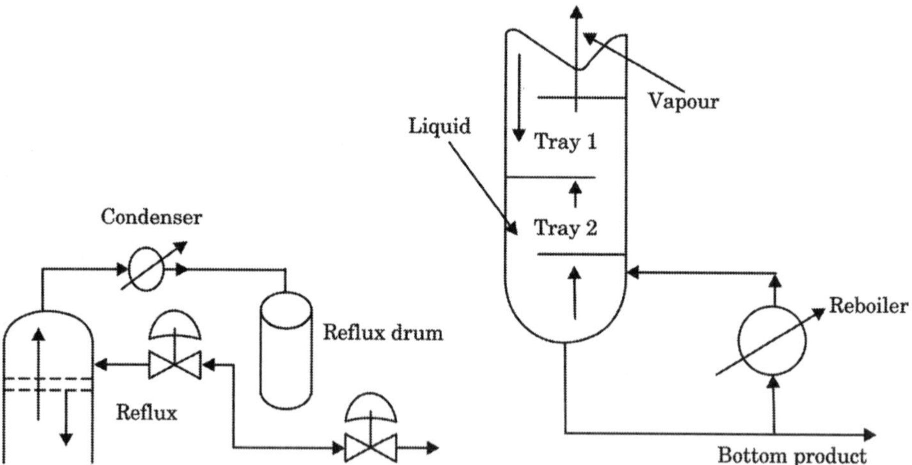


**Fig. 1.3.4.1:** Column for batch distillation

1.3.4.2. *Steam Distillation*

When a material to be distilled has a high boiling point, and particularly where decomposition might occur if direct distillation is employed, the process of steam distillation may be used. Steam is passed directly into the liquid in the still and the solubility of the steam in the liquid must be very low. Steam distillation is perhaps the most common example of differential distillation (Fig. 1.3.4.2). Two cases may be considered. The steam may be superheated and so provide sufficient heat to vaporise the material concerned, without it condensing. Alternatively, some of the steam may condense, producing a liquid water phase.

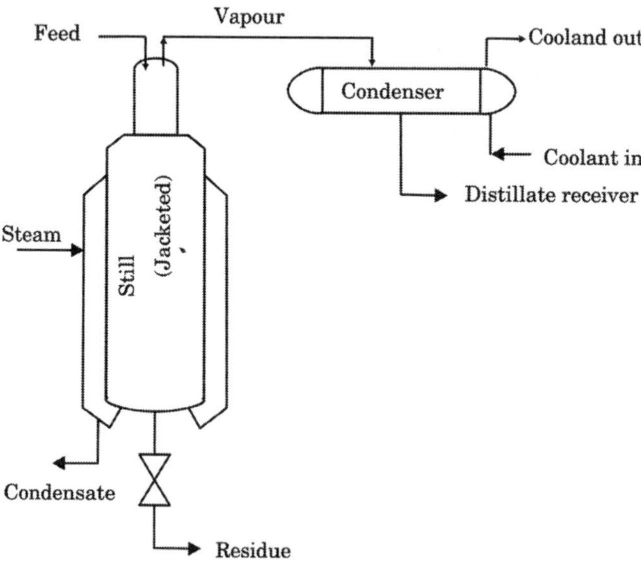


**Fig. 1.3.4.2:** Simple distillation unit

### 1.3.4.3. *Flash Distillation*

Flash or equilibrium distillation, frequently carried out as a continuous process, consists of vaporising a definite fraction of the liquid feed in such a way that the vapour evolved is in equilibrium with the residual liquid. The feed is usually pumped through a fired heater and enters the still through a valve where the pressure is reduced. The still is essentially a separator in which the liquid and vapour produced by the reduction in pressure have sufficient time to reach equilibrium. The vapour is removed from the top of the separator and is then usually condensed, while the liquid leaves from the bottom. Flash distillation is normally carried out in a continuous manner. In this method, a liquid mixture is partially vaporised, the vapour and liquid are allowed to attain equilibrium and finally withdrawn separately. Feed is heated in a tubular heat exchanger. The hot liquid mixture is then fed to a separator via a pressure reducing valve whereby the pressure is reduced and a vapour is formed at the expense of liquid adiabatically. The liquid is withdrawn from the bottom of the separator and the equilibrium vapour leaves the separator from the top which is then liquified in a condenser. Flash distillation is commonly used in petroleum industry, handling multicomponent system in the pipe stills (Fig. 1.3.4.3).

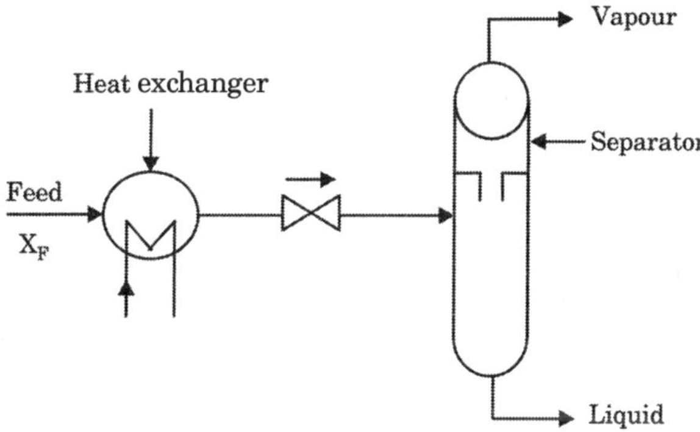


**Fig. 1.3.4.3:** Flash/Equilibrium distillation unit

Consider one mole of a liquid mixture, having $x_F$ mole fraction of the more volatile component, is fed to a flash distillation unit. Let f be the fraction of the feed that is vaporised and is of composition y. Then (1 - f) will be the mole of the residual liquid obtained. Let x be the mole fraction of the residual liquid obtained. Let x be the mole fraction of the more volatile component in the liquid.

Then, a material balance of the more volatile component gives

$$x_F = f.y + (1 - f)x \tag{1.3.1}$$

$$x_y = -(1 - f)x + x_F \tag{1.3.2}$$

$$y = -\frac{(1 - f)}{f}x + \frac{x_F}{f} \tag{1.3.3}$$

Equation (1.3.3) is the material balance/operating line for flash distillation with a slope equal to - (1 - f)/f and an intercept equal to $x_F/f$. The point of intersection of the operating line and the diagonal (x = y) is,

$$y = -\frac{(1-f)x}{f} + \frac{x_F}{f} \tag{1.3.4}$$

$$x = y \tag{1.3.5}$$

$$x = -\frac{(1-f)}{f}x + \frac{x_F}{f} \tag{1.3.6}$$

$$xf = -x + xf + x_F \tag{1.3.7}$$

$$x = y_F \quad and \quad y = x_F \tag{1.3.8}$$

for f = 1, feed totally vaporised (feed 100 mole per cent vaporised)

$$slope = -\frac{(1-f)}{f} = 0 \tag{1.3.9}$$

and hence the operating line is parallel to x- axis through the point ($x_F$, $x_F$) on the diagonal.

For f = 0, - no feed is vaporised.

Slope = ∞ and hence the operating line will be parallel to y - axis through the point ($x_F$, $x_F$) on the diagonal.

Overall material balance over

Liquid flow to plate m + 1 = Vapour flow from plate m + 1 + Bottom product withdrawn

$$L_m = V_{m+1} + W \tag{1.3.10}$$

Similarly, material balance of more volatile component (A):

$$L_m x_m = V_{m+1}\, y_{m+1} + W x_w \tag{1.3.11}$$

Rearranging the Equation (1.3.11),

$$y_{m+1} = \frac{L_m}{V_{m+1}} x_m - \frac{W x_W}{V_{m+1}} \tag{1.3.12}$$

We know from the Equation (1.3.10), that $V_{m+1} = L_m - W$

$$y_{m+1} = \frac{L_m}{L_m - W} x_m - \frac{W x_W}{L_m - W} \tag{1.3.13}$$

The Equation (1.3.13) is known as the material balance line of operating line of the stripping section.

### 1.3.4.4. *Bubble Cap Distillation*

The bubble cap distillation column is the oldest type of the distillation column. Being the oldest has no effect on its application in the process of distillation. The main feature of this distillation column is its tray containing the bubble-caped perforations. A bubble cap tray has two major parts – a bell-shaped cap known as bubble cap. The riser is inserted through the perforation on the tray floor and the bubble cap is bolted over it. A ring gasket is used to connect the cap with the riser. The riser is simply a tube with expanded bottom end for the passage of vapour and holds the cap. A bubble cap has a slot on his wall. The slot can be of any shape *i.e.* rectangular, triangular, trapezoidal etc. In the bubble cap tray, the cup is so mounted so that there is a space between riser and cap to allow the passage of vapour. Vapour rises through the chimney and is directed downward by the cap, finally discharging through slots in the cap, and finally bubbling through the liquid on the tray. The caps and the riser are made of low carbon steel, stainless steel or any other suitable material that can withstand the environment with the tower. The bubble cap distillation column got high turndown ratio, but are now mostly replaced by other tray columns like sieve tray column, value tray column etc. due to high cost of investment and comparatively low efficiency. The 'bubble-column reactors' (BCRs) are gaining an important place in both chemical and biochemical industries. Unlike mechanically agitated reactors, bubble columns are simple to construct and operate. They consist of vessels (usually cylindrical) in which gas is sparged into a liquid. They have no moving parts, as adequate levels of mixing can be achieved with the sparged gas. In BCRs, all the energy needed for agitation, as well as the oxygen required for the culture, is provided by sparged air. Internal structures that modify the flow characteristics within the reactor can be installed; one such modification, which is gaining increasing industrial acceptance, is the air-lift reactor. Bubble columns are cylindrical vessels wherein gas is sparged, via a distributor, in the form of bubbles, into liquid or liquid-solid suspension. Bubble columns serve as multiphase contactors and reactors in the chemical, petrochemical, biochemical, and metallurgical industries. As reactors, bubble columns are used for chemical processes involving oxidation, chlorination, alkylation, polymerization, and hydrogenation reactions .Other processes that employ bubble columns include hydrotreating and conversion of petroleum residues and direct and indirect liquefaction in the production of liquid fuels from coal. Bubble columns have been identified as a suitable type of reactor for a variety of gas conversion processes involving the production of liquid fuels from synthesis gas. Typical examples are the Fischer-Tropsch process, synthesis of methanol, and manufacture of other synthetic fuels that offer environmental advantages over petroleum-derived fuels. Technology for the liquid-phase methanol synthesis in slurry bubble columns has been recently developed. Bubble columns also have found wide application as fermentors, in wastewater treatment, and in a variety of metallurgical operations such as leaching of metal ores. Excellent heat- and mass-transfer characteristics, lack of moving parts, and therefore reduced wear and tear, higher durability of catalyst, ease of operation, compactness, and low operating and maintenance costs are the advantages that render bubble columns as an attractive reactor choice for various multiphase processes, especially for processes involving highly exothermic reactions.

Bubble columns usually operate with a length-to-diameter ratio, or aspect ratio, of at least five. They are operated in either semi batch (zero liquid throughput) or continuous mode (co current or countercurrent flow of gas and liquid), with liquid superficial velocities lower than the gas superficial velocity by at least an order of magnitude. As a result, it is the flow of gas that controls the hydrodynamics of the individual phases in these systems. In the presence of a solid (catalyst) phase, these reactors are commonly referred to as slurry bubble columns. The size of particles used in slurry bubble columns are typically in the range of 5 to 100 μm, with a solids loading of up to 50% by weight. Heat-exchanger tubes can be inserted into the reactor form cooling the system and maintaining isothermal conditions, especially for highly exothermic reactions. In addition, in some cases the column may be sectionalized using baffles to inhibit liquid back mixing. In general, the design and scale-up of bubble-column reactors depends on the quantification of three main phenomena: (1) heat and mass transfer, (2) mixing characteristics; and (3) chemical kinetics of the reacting system. The first two are system dependent and intimately linked with the fluid dynamics of bubble columns. Although simple in construction, due to the complex hydrodynamics and its influence on transport characteristics, the design and scale-up of bubble columns has involved considerable empiricism, and has been the subject of extensive ongoing research. Reported studies indicate that the accurate and successful design and scale-up of bubble-column reactors requires an improved understanding of the multiphase fluid dynamics and its influence on phase holdup distribution, mixing, and transport characteristics. In recent years considerable effort has been directed toward fundamental fluid-dynamic modeling of two-phase flows in bubble-columns with the aim of using these models as tools in the design of bubble-column reactors. This has resulted in a number of computational fluid dynamics (CFD) studies. However, the uncertainty regarding the various phase interaction terms and turbulence closure schemes, in addition to the computational difficulties associated with computation of large flow fields, have in practice delayed the full implementation of CFD codes. Simulation of the churn-turbulent flow regime, which is preferred for industrial bubble columns, poses further difficulties due to the complications arising from the multiple bubble sizes and bubble interactions.

It is widely recognized that the present fluid-dynamic models used in the numerical simulation of bubble columns require experimental data as input and for model verification. In general, the flow in a bubble column is highly turbulent, and the instantaneous flow patterns tend to lack symmetry about the central axis. Therefore, one would expect the need for a 3-D simulation to capture the dynamic features of the flow field. These computations are, however, costly and time-consuming to run. Hence, until recently, most reported CFD studies of bubble columns were based on the 2-D axis symmetric assumption, which results in the computed liquid-phase flow field that exhibits a single-cell recirculation pattern with upflow in the center and down flow at the walls.

Serving as a simplified means of simulating the full 3-D flow, the 2-D axisymmetric simulations can be, if properly validated, useful for engineering calculations of the overall velocity and gas holdup distribution. Moreover, the results also provide the basis for the development and improvement of 1-D models, as widely used in the design of bubble columns employed in F-T synthesis and other gas conversion processes. However, a 2-D simulation cannot capture the dynamic features of the flow, such as the spiral motions. It is important to examine the effect of neglecting these features on the prediction of the overall reactor flow field and performance. A satisfactory answer to these issues only can be obtained through a detailed comparison between the experimental data and simulations. Hence, the efforts in modeling and simulation have to be supplemented with reliable experimental data for the fluid-dynamic parameters describing phase velocities, phase interaction, and turbulence.

Numerous experimental studies have been conducted to provide insight into gas holdup and liquid-velocity distribution in bubble columns. Notable among these is the using pilot tubes, provided the radial profiles of liquid velocity at different gas superficial velocity. His measurements were among the first for local averaged liquid velocities and holdup in bubble columns. Other investigations used invasive andror local probes such as flywheel anemometer, heat-pulse anemometry, hot-film anemometry. The laser doppler anemometry, which is a noninvasive technique, had been used to obtain liquid velocities at designated locations of the columns. Fan and coworkers introduced particle image velocimetry, which is also a noninvasive technique, to measure the liquid velocity field and flow structure in, mainly 2-D, bubble columns. While all of these investigations provided valuable insight into the flow structure in bubble columns, the techniques employed have several shortcomings. They are cumbersome to use in large diameter columns; it is very difficult and extremely time-consuming to map the flow field in the entire column; they fail at large-volume fractions of the dispersed phase, and yet it is the operation in churn turbulent flow at high gas holdup that is of current industrial interest. In contrast, introduction of radioactive particle tracking for studies of liquid motion in bubble columns. This technique is capable of providing the flow field in the whole column in a reasonable period of time, can handle large-diameter columns, and most importantly is not hindered by and is fully operational in systems with high-volume fraction of the gas. Moreover the radioactive-particle tracking technique produces Lagrangian data for the whole column. Column is made of stainless steel with seven plates for mounting the bubble caps. Each plate is provided with one bubble cap. An electrically heated reboiler is installed at the bottom of the column. The bottom product is collected in the tank. The vapors from the top of the column are condensed in the shell and tube type condenser by circulating cooling water. A tank with pump and Rotameter is provided for cooling water. The condensate is feedback to column as reflux and part of it is collected as distillate in product receiving tank with the help of reflux divider. The complete column is insulated for minimizing the heat loss. The instrumentation is done for pressure and temperature measurement wherever is necessary (Fig. 1.3.4.4).

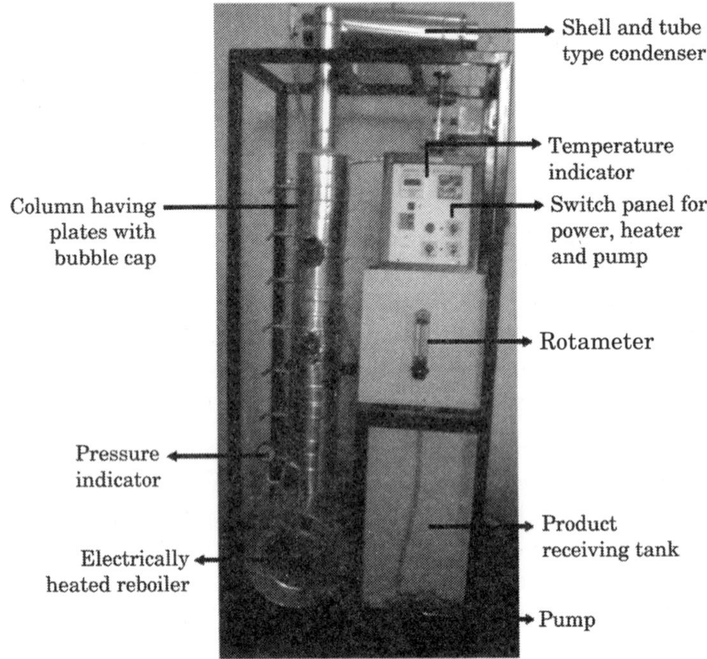


**Fig. 1.3.4.4:** Laboratory scale bubble cap distillation column (Suresh, unpublished work)

Batch distillation with only a single still does not give a good separation unless the relative volatility is very high. To obtain product with a narrow composition range, a rectifying batch still used that consist of a reboiler, a rectifying column, a condenser, some means of splitting of a portion of condensed vapour or distillate as reflux and the receiver. The operation of the batch still and column can be analyzed using the same operating line equation as for the rectifying section of the continuous distillation:

$$y_{n+1} = \frac{R}{(R+1)} x_n + \frac{x}{(R+1)} \tag{1.3.14}$$

for the binary system:

$$y_2 = 1 - y_1 \text{ and } x_2 = 1 - x_1 \tag{1.3.15}$$

Since the slope of the operating line is $\frac{R}{R+1}$, the slope increases as the reflux increases, until when reflux if infinite. Under total reflux slope is 1. The operating line then coincides with the diagonal. The number of plate is minimum at total reflux. Minimum number of plates required can be calculated from the terminal concentration of $x_b$ and $x_b$ based on the relative volatility of the component α, which is defined in terms of equilibrium concentrations:

$$\alpha = \frac{y_{1e}/x_{1e}}{y_{2e}/x_{2e}} \tag{1.3.16}$$

Minimum number of theoretical plates required,

$$N_{min} = \frac{\log[x_D(1 - x_B) / x_B(1 - x_D)]}{\log \alpha} - 1 \tag{1.3.17}$$

An ideal mixture follows Raults' law, and the relative volatility is the ratio of vapor pressure. Thus

$$p_1 = P_1 x_1 \tag{1.3.18}$$

$$p_2 = P_2 x_2 \tag{1.3.19}$$

$$Y_2 = \frac{P_2}{P} \tag{1.3.20}$$

$$Y_2 = \frac{P_2}{P} \tag{1.3.21}$$

$$\alpha = \frac{y_1/x_2}{y_2/x_2} = \frac{P_1}{P_2} \tag{1.3.22}$$

The final ratio does not change much over the range of temperature encountered in a typical column, so the relative volatility is taken as constant. The most important advantages of bubble columns result from their simplicity. As there is no need to introduce energy by mechanical means, the sealing of the stirrer-shaft assembly, which is the most difficult problem to overcome in the mechanical design of a bioreactor, is eliminated. In addition, the absence of a shaft in the head space of the vessel gives more room for entry ports, an important factor in vessels with relatively small dimensions. Mechanical simplicity is particularly important in biological processes, where sterility has to be maintained over extended periods. The absence of shafts eliminates a very expensive and equally vulnerable feature of mechanically stirred bioreactors, thereby increasing process reliability. With few exceptions, the discussion of flow regimes and heat and mass transfer in the published literature are generally applicable to liquids with a viscosity, or an effective viscosity, <50 m Pa.s. The behaviour of bubbly dispersions in highly viscous liquids is quite different. Bubbles in such liquids coalesce readily, and very large bubbles (some as large as the column diameter) form in tall columns. In addition, the coalescence of bubbles in the bulk volume of such liquids, and the break up of bubbles at the surface, generates very tiny bubbles, resulting in an apparently bimodal distribution of bubble size.

### 1.3.4.5. *Packed / Tray Column Distillation*

A packed bed distillation column is the type of distillation column which uses the packing instead of tray as the contacting medium for the vapor and liquid phases. It consists of a cylindrical shell filled with suitable packing material to provide a large interfacial area of contact between the phases (Fig. 1.3.4.5.1). A set of 'tower internals' such as packing support, liquid and gas distributor etc. are also used in the packing bed. The liquid is distributed at the top of the packing which wet the packing material and the gas is fed at the bottom. It flows through the void in the wetted material of packed bed and the mass transfer will take place here. The packed bed column provides more surfaces for the interfacial contact than the tray columns and is generally shorter than the tray distillation columns. The height of the column acts as the parameter for the designing of the packed bed columns. The height of the packed bed is taken as the equivalent to the number of tray in the tray distillation column.

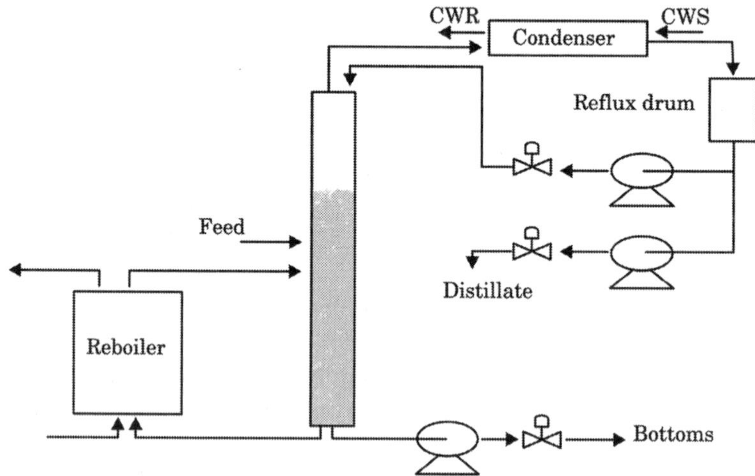


**Fig. 1.3.4.5.1:** Packed bed distillation column

The packed bed column is packed with glass rasching rings. An electrically heated reboiler is installed at the bottom of the column. The bottom product is collected in the tank. The vapors from the top of the column are condensed in the shell and tube type condenser by circulating cooling water. A tank with pump and Rotameter is provided for cooling water. The condensate is collected in reflux drum and feedback to column as reflux and part of it is collected as distillate in product receiving tank. The complete column is insulated for minimizing the heat loss (Fig. 1.3.4.5.2).

### 1.3.4.6. *Extractive Distillation*

One type of variation of the distillation processes is extractive distillation. In this type of process, an external solvent is added to the system to increase the separation. The external solvent changes the relative volatility between two

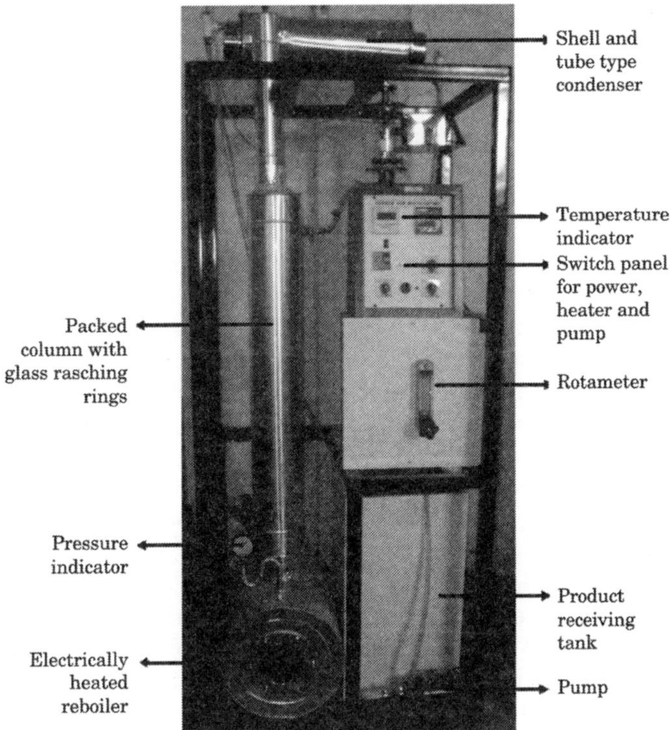


**Fig. 1.3.4.5.2:** Laboratory scale packed bed distillation column (Suresh, unpublished figure)

'close' components by extracting one of the components, forming a ternary mixture with different properties. The solvent is recycled into the system after the extracted component is separated from it. (Fig. 1.3.4.6) is applied when a binary mixture is difficult or impossible to separate by ordinary distillation because the relative volatility of the components is low. A third component, called solvent, is added, which alters the relative volatility of the orginial components. The solvent itself is of low volatility and is not appreciably vaporized in the column.

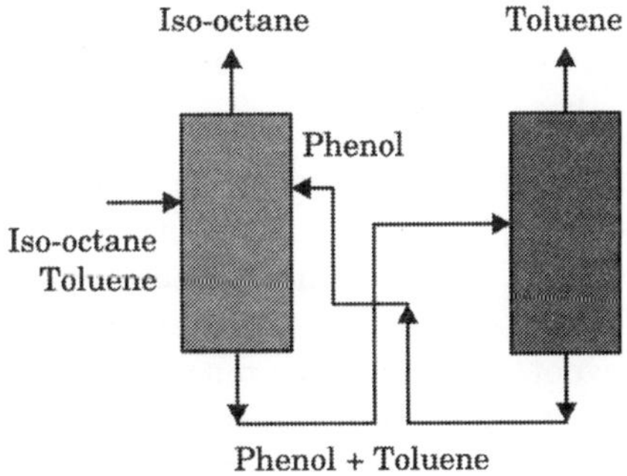


**Fig. 1.3.4.6:** Principle of extractive distillation

Extractive distillation is often applied for the separation of aromatics from alkanes that thave similar volatility. For example, separation of toluene (b.p. 383.8 K) and isooctane (b.p. 372.4 K) is achieved by adding phenol (b.p. 455 K) which strips toluene from the mixture so that iso-octane is obtained at the top. The separation of toluene and phenol is relatively easy.

1.3.4.7. *Azeotropic Distillation*

The course is aimed at process engineers who have experience of conventional distillation and wish to learn more about non-ideal distillation processes. Azeotropic distillation methods are required when separating mixtures that form azeotropes. Such separations are usually achieved by using a solvent to break the azeotrope. The course focuses on graphical design methods (including residue curve maps) for assessing feasibility, solvent selection and column and flowsheet design. The design of processes that exploit liquid-liquid phase splitting will also be addressed. Separation sequences are complicated by the presence of azeotropes, often involving mixtures of oxygenated organic compounds: (1) Alcohols, (2) Ketones, (3) Ethers, (4) Acids, (5) Water. In azeotropic distillation of two components that form an azeotrope or are difficult to separate otherwise, a third component (an azeotropic agent) is added to increase the relative volatility of the two components. An example of industrially applied azeotropic distillation is the dehydration of ethanol. Ethanol and water form a minimum boiling azeotrope at a temperature of 351 K, where the mixture contains 89 mole % ethanol. Starting with a mixture containting a lower proportion of ethanol, it is not possible to obtain a product richer in ethanol than 89%. If benzene is added to the azeotrope, a ternay azeotrope is formed with a boiling point of 338 K. The relatively non-polar benzene serves to volatilize water (a highly polar molecule) more than ethanol (a moderately polar molecule). A virtually pure ethanol product is obtained. The first tower gives nearly pure ethanol as the bottoms product and the ternary azeotrope at the top. Cooling results in the formation of two phase: a benzene-rich organic phase, which is used as reflux, and a second, aqueous phase, which also contains benzene and ethanol. The second column separates the ternary azeotrope, which is recycled to the first column, from water. The catalyst complex must be recovered with an extreme efficiency and returned to the reactor, because rhodium is a harmful metal and it is very expensive. Nearly complete recovery of rhodium is affected in the light ends column. Furthermore the volatile and toxic methyl iodide has to be retained in the system. This is achieved in the methanol scrubber. Overhead gases from all columns are sent to this scrubber.

An azeotrope that contains one liquid phase in contact with vapour is called a homogenous azeotrope. A homogenous azeotrope cannot be separated by conventional distillation. However, vacumn distillation may be used as the lower pressures can shift the azeotropic point. Alternatively, an additional substance may added to shift the azeotropic point to a more 'favourable' position. (1) When this additional component appears in appreciable amounts at the top of the column, the operation is called azeotropic distillation. (2) When the additional component appears mostly at the bottom of the column, the operation is called extractive distillation The VLE curve on the left is also generated by an azeotropic system, in this case a heterogenous azeotrope. Heterogenous azeotropes can be identified by the 'flat' portion on the equilibrium diagram. They may be separated in 2 distillation columns since these substances usually form two liquid phases with widely differing compositions. The phases may be separated using settling tanks under appropriate conditions.

***Homogeneous Azeotropes*:** For non-ideal mixtures, the activity coefficients are different from unity: If Y>1 the mixture has a minimum-boiling azeotrope; Example – Phase diagrams for Isopropyl ether-Isopropyl alcohol. For non-ideal mixtures, the activity coefficients are different from unity: If Y<1 the mixture has a maximum-boiling azeotrope; Example – Phase diagrams for Acetone-Chloroform (Fig. 1.3.4.7.1).

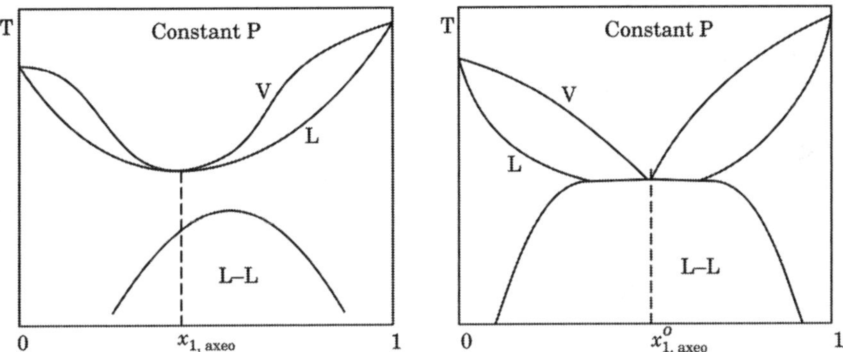


**Fig. 1.3.4.7.1:** Homogeneous azeotropic distillation

***Heterogeneous Azeotropes*:** For a minimum-boiling azeotrope with large deviation from Raoult's law (Y>>1), phase splitting may occur and a minimum-boiling heterogeneous azeotrope forms, having a vapor phase in equilibrium with two liquid phases. *Example*: Dehydration of Ethanol. Heterogeneous

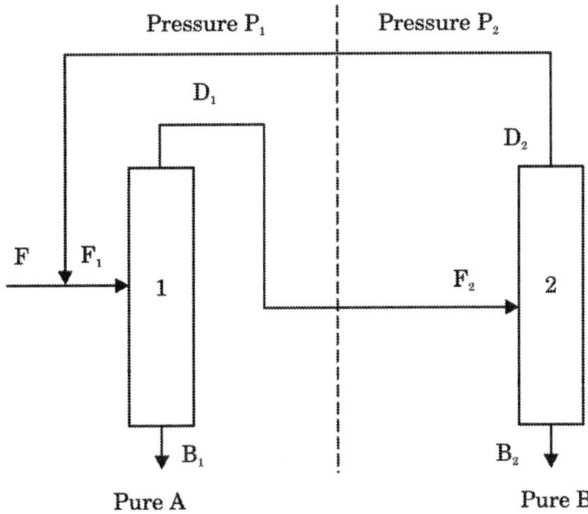


**Fig. 1.3.4.7.2:** Distillation sequence for minimum-boiling azeotrope (*Pressure-swing distillation*)

azeotropic distillation is a widely practiced process for the dehydration of a wide range of materials including acetic acid, chloroform, ethanol, and many higher alcohols. The technique involves separating close boiling components by adding a third component, called an entrainer, to form a minimum boiling, normally ternary azeotrope which carries the water overhead and leaves dry product in the bottom Fig. 1.3.4.7.2. The overhead is condensed to two liquid phases; the organic, 'entrainer rich' phase being refluxed while the aqueous phase is decanted. Given the non-idealities, the phase splitting, the distillation boundaries present, and the possible existence of multiple steady states in such a system, columns like these can be extremely difficult to simulate and to operate. A systematic approach to the problem can save you many headaches. Example: Dehydration of Tetrahydrofuran and water and Ethanol-water-n-pentane.

A distillation column may also have a catalyst bed and reaction occurring in it. This type of column is called a reactive distillation column. The targeted component reacts when it is in contact with the catalyst, thereby separated from the rest of the components in the mixture (Fig. 1.3.4.7.3).

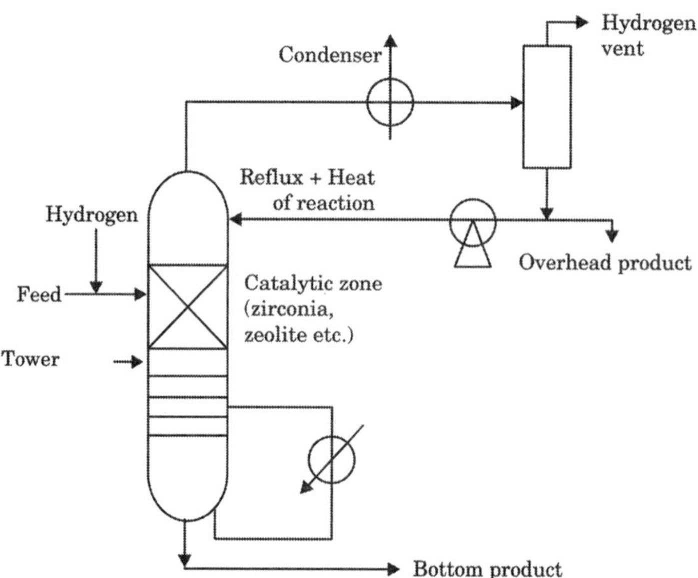


**Fig. 1.3.4.7.3:** Catalytic distillation unit

### 1.3.4.8. *Divide-wall Column*

The divided wall distillation column (DWDC) to separate three or more components in a single column is receiving interest in industry and academia due to potential for capital cost and energy savings. The design of a DWDC was proposed almost 50 years ago. Since then many authors have studied design, control, and most importantly energy saving aspects of DWDC. For separating a mixture of three components A, B, and C (in order of increasing boiling point) using simple distillation columns, at least two columns are needed which can be arranged in different ways leading to direct, indirect, or transition split of components. In a direct split, component A is the desired distillate obtained as top product of first column, while a second column is required to separate mixture of B and C obtained as the bottom product of first column. In an indirect split, component C is obtained as the bottom product of first column, while a second column is used to separate the desired product A from B. In a transition split, however, three columns are required as distillate and bottom products of first column (primary fractionator) are mixtures of A and B, and B and C, respectively. It is possible to merge two secondary fractionators by removing a condenser and reboilers to act as the main column and to remove reboiler and condenser of primary, the arrangement becomes a Petyluk column. Further, when both columns (prefractionator and main column) of the Petyluk arrangement are placed in the same shell, it becomes a dividing-wall distillation column (DWDC). The first half of DWDC separates the lightest component (A) from the heaviest component (C) while the middle component (B) remains distributed. The middle component B is separated from A in the second half of the column in trays above side stream product, and from C in trays below the side stream product. Thermodynamically a DWDC can be considered as equivalent to Petlyuk column (Petlyuk and Slavinski, 1965). Both the configuration allows removal of the condenser and reboiler from the prefractionator, with a consequent reduction in energy needs.

DWC is a single shell, fully thermally coupled distillation column capable of separating mixtures of three or more components into high purity products. Compared to conventional columns-in-series and/or in-parallel configurations a DWC requires much less energy, capital and space. This makes DWC to something that corresponds with the present day idea of sustainable process technology. Based on published papers and patent literature this paper aims to give a complete overview of the work done so far on the research and implementation of DWCs, from early ideas on thermal coupling of distillation columns to practical issues that needed to be solved for their successful implementation. Approaches to short-cut and rigorous simulation, optimization, and control are highlighted, with particular focus on column internals and dimensioning, which is only conceptually considered in academic publications. A survey of relevant patents is included providing information on equipment innovations and application areas of industrial interest. Finally authors look at what is needed on research and engineering side to enable maximization of potential gains by building DWCs for obtaining four or even more products containing two or more partition walls in parallel, which is something not yet attempted in industrial practice. Some of various application of divide-wall column distillation is shown in Table 1.3.4.8. The main advantages of divided wall columns include: Two columns combined in one shell *i.e.* lowering of the overall capital cost, Compact Design, lower space requirements, increased energy efficiency.

Distillation is still the bullwork of the separation technology in chemical process industry, even when important developments have taken places in new

**Table 1.3.4.8:** Some of various application of divide-wall column distillation

| |
|---|
| Separating mixtures of n-hexane, n-heptane and n-octane |
| Separation of close boiling mixture of $CH_4$-separate 1-butene from the feed mixture containing i-butane, 1-butene, n-butane, trans-2-butene and cis-2-butene |
| Separation of (1) acetaldehyde-methanol-water, (2) acetone-chloroform-benzene and (3) ethanol-water-ethylene glycol system |
| Separation of methanol, iso-propanol and butanol, air to produce argon |
| Separation of alkane mixtures containing 2-methylbutane, pentane, hexane and heptane |
| Separation of pure ethyl acetate from the the mixture of lower and higher boiling impurities |
| Separation of a mixture of ethanol, n-propanol and n-butanol |
| Separation of LPG from mixed pentanes and a heavy end in a light-end refinery distillation train |
| Separation of equilimolar mixture of benzene, toluene and o-xylene |
| Separation of a mixture with butanol, pentanol and hexanol |
| Separation of benzene, toluene, ethyl benzene mixture |
| Reduction of benzene in gasoline and separation of benzene, toluene and xylene mixture |
| Mixture of benzene, toluene and biphenyl, (2) mixture of alkanes, and (3) mixture of benzene, toluene and p-xylene |

separation technologies in the last two decades. Distillation continues to be the most important separation method although it is the most energy intensive unit operation. Distillation is well suited for the separation of high purity products from a liquid mixture since any degree of separation can be obtained by increasing the number of trays (equilibrium stages) in one column. For more than two components in the mixtures (or mixtures forming azeotropes) cascading of more than one distillation column is required. In the classical direct sequence arrangement for separating ternary mixtures, (Fig. 1.3.4.8.1), the feed is introduced to the first column where the low-boiling fraction A is recovered as the top product. The bottom product is fed to the second column where the medium-boiling fraction B is distilled off as the top product and the high-boiling fraction C remains as the bottom product (residue). Alternatively, the indirect sequence (Fig. 1.3.4.8.1) may be applied, where the high-boiling

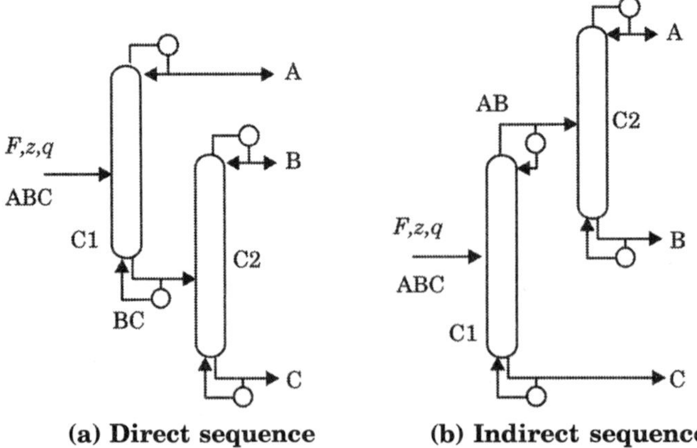


**Fig. 1.3.4.8.1:** Classical divide-wall column of direct and indirect sequence arrangement for separating ternary mixtures. (a) Direct sequences (b) Indirect sequence

fraction C is obtained as the residue of the first column. The optimum column sequence is selected after the consideration of the requirements to be met, *e.g.* the quantity of the individual fractions, the relative volatility and the thermal stability of the components, the available utilities etc., so that the investment and operating costs are minimized. In an alternative procedure, ternary mixtures can be separated in thermally coupled columns. The first thermally coupled system designed by Wright (1949), can also be integrated into a single column, known as a divided wall column. Detailed theoretical study of thermally coupled column was provided by Petlyuk *et al.* (1965). These investigators suggested thermally coupled columns to be a novel alternative to the two-column configuration (Fig. 1.3.4.8.2).

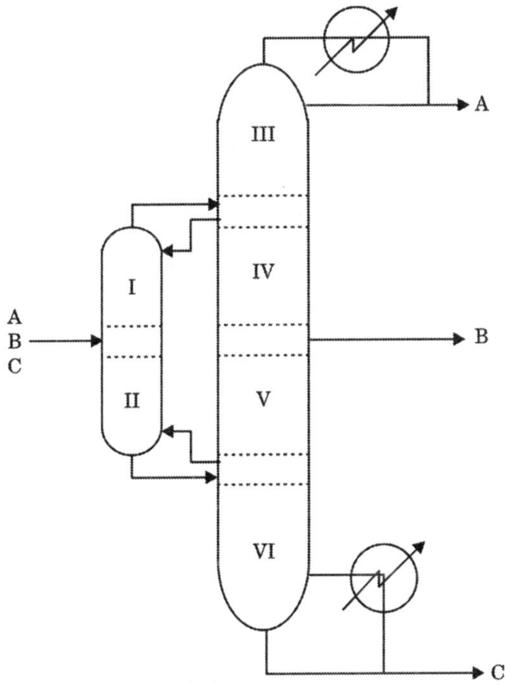


**Fig. 1.3.4.8.2:** Thermally coupled divide-wall columns

### 1.3.4.9. *Hybrid Separation Schemes*

Design of thermally coupled columns (I) The Petlyuk Column: Combines two column shells in one separation unit with one reboiler and one condenser, in a fully thermally coupled system, the work of separation with this configuration is the minimum (Petlyuk *et al.*, 1965). Example: s-Butanol-i-Butanol-n-Butanol. By simple comparison of driving forces for various separation techniques, an optimal hybrid separation scheme (Fig. 1.3.4.9) is visually generated (Mixture of MTBE and MeOH). The configuration exhibiting the largest driving force, is the combination of pervaporation and distillation. Detailed simulations have shown that this configuration is at least 35% more energy efficient than any other configuration considered.

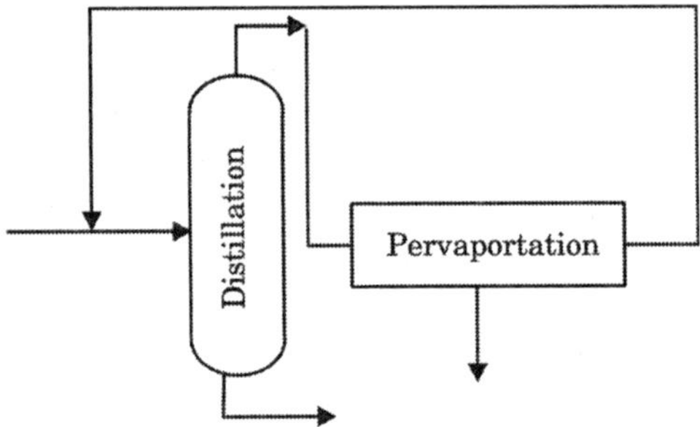


**Fig. 1.3.4.9:** Hybrid separation scheme

### 1.3.5. *Vapour Liquid Equilibria*

Distillation columns are designed based on the boiling point properties of the components in the mixtures being separated. Thus the sizes, particularly the height, of distillation columns are determined by the vapour liquid equilibrium (VLE) data for the mixtures. VLE curves at constant pressure, VLE data are obtained from boiling point diagrams shown in Fig. 1.3.5.1.

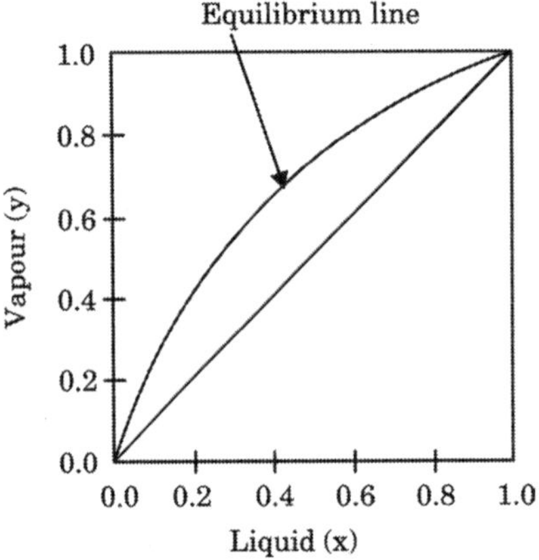


**Fig. 1.3.5.1:** Vapour-Liquid-Equilibrium curves at constant pressure

VLE data of binary mixtures are often presented as a plot, as shown in the figure on the right. The VLE plot expresses the bubble-point and the dew-point of a binary mixture at constant pressure. The curved line is called the equilibrium line and describes the compositions of the liquid and vapour in equilibrium at some fixed pressure. This particular VLE plot shows a binary mixture that has a uniform vapour-liquid equilibrium that is relatively easy to separate. The most intriguing VLE curves are generated by azeotropic systems. An azeotrope is a liquid mixture which when vaporised, produces the same composition as the liquid. In other words azeotropic systems give rise to VLE plots where the equilibrium curves crosses the diagonals. Note the shapes of the respective equilibrium lines in relation to the diagonal lines that bisect the VLE plots.

**1.3.5.1. *Driving Force*:** Different separation mechanisms exist, and the generation of a two-phase (composition dependent) separation can be achieved in different ways (Fig. 1.3.5.1).

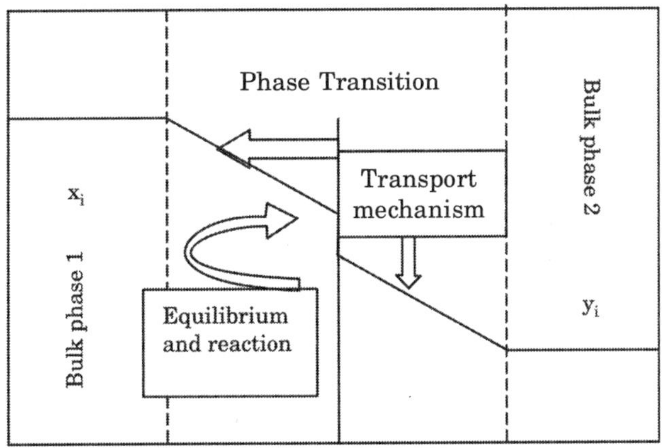


**Fig. 1.3.5.1:** Separation mechanism

Operation with the largest driving force leads to the easiest separation, and the optimal design (including energy consumption).

**1.3.5.2. *Raoult's Law*:** At boiling and higher temperatures the sum of the individual component partial pressures becomes equal to the overall pressure, which can symbolized as $P_{tot}$.

Under such conditions, Dalton's Law would be in effect as follows:

$$P_{tot} = P_1 + P_2 + \dots \tag{1.3.34}$$

Then for each component in the vapor phase:

$$y_1 = P_1/P_{tot},\ y_2 = P_2/P_{tot},\ \dots \text{ etc.} \tag{1.3.35}$$

where $P_1$ = partial pressure of component 1, $P_2$ = partial pressure of component 2, etc.

Raoult's Law is approximately valid for mixtures of components between which there is very little interaction other than the effect of dilution by the other components. Examples of such mixtures includes mixtures of alkanes, which are non-polar, relatively inert compounds in many ways, so there is little attraction or repulsion between the molecules. Raoult's Law states that for components 1, 2, etc. in a mixture:

$$P_1 = x_1\,P^0{}_1,\ P_2 = x_2\,P^0{}_2, \text{ etc.} \tag{1.3.36}$$

Where $P^0{}_1$, $P^0{}_2$, etc. are the vapor pressures of components 1, 2, etc. when they are pure, and $x_1$, $x_2$, etc. are mole fractions of the corresponding component in the liquid.

Vapor pressures of liquids are very dependent on temperature. Thus the $P^0$pure vapor pressures for each component are a function of temperature (T):

For example, commonly for a pure liquid component, the Clausius-Clapeyron equation (not shown here) may be used to approximate how the vapor pressure varies as a function of temperature. This makes each of the partial pressures dependent on temperature also regardless of whether Raoult's Law applies or not. When Raoult's Law is valid these expressions become:

$P_1(T) = x_1 P^0_1(T)$, $P_2(T) = x_2 P^0_2(T)$, etc.

At boiling temperatures if Raoult's Law applies, the total pressure becomes:

$$P_{tot} = x_1 P^0_1(T) + x_2 P^0_2(T) + \ldots \quad (1.3.37)$$

At a given $P_{tot}$ such as 1 atm and a given liquid composition, T can be solved for to give the liquid mixture's boiling point or bubble point, although the solution for T may not be mathematically analytical (may require a numerical solution or approximation). For a binary mixture at a given $P_{tot}$, bubble point T can become a function of $x_1$ (or $x_2$) and this function can be shown on a two-dimensional graph like a binary boiling point diagram.

At boiling temperatures if Raoult's Law applies, a number of the preceding equations in this section can be combined to give the following expressions for vapor mole fractions as a function of liquid mole fractions and temperature:

$$y_1 = x_1 P^0_1(T)/P_{tot}, \quad (1.3.38)$$

$$y_2 = x_2 P^0_2(T)/P_{tot}, \ldots \text{ etc.} \quad (1.3.39)$$

Once the bubble point T's as a function of liquid composition in terms of mole fractions have been determined, these values can be plugged into the above equations to obtain corresponding vapor composition in terms of mole fractions. When this is finished over a complete range of liquid mole fractions and their corresponding temperatures, one effectively obtains a temperature (T) function of vapor composition mole fractions. This function effectively acts as the dew point T function of vapor composition.

In the case of a binary mixture: $x_2 = 1 - x_1$ and the above equations can be expressed as:

$y_1 = x_1 P^0_1(T)/P_{tot}$ and

$y_2 = (1 - x_1) P^0_2(T)/P_{tot}$

For many kinds of mixtures, particularly where there is interaction between components beyond simply the effects of dilution, Raoult's Law do not work well for determining the shapes of the curves in the boiling point or VLE diagrams. Even in such mixtures, there are usually still differences in the vapor and liquid equilibrium concentrations at most points, and distillation is often still useful for separating components at least partially. For such mixtures, empirical data are typically used in determining such boiling point and VLE diagrams. Chemical engineers have done a significant amount of research trying to develop equations for correlating and/or predicting VLE data for various kinds of mixtures which do not obey Raoult's Law well.

**1.3.5.3. *Relative volatility*:** Relative volatility is a measure comparing the vapor pressures of the components in a liquid mixture of chemicals. This quantity is widely used in designing large industrial distillation processes. In effect, it indicates the ease or difficulty of using distillation to separate the more volatile components from the less volatile components in a mixture. By convention, relative volatility is usually denoted as α. Relative volatilities are used in the design of all types of distillation processes as well as other separation or absorption processes that involve the contacting of vapor and liquid phases in a series of equilibrium stages. Relative volatilities are not used in separation or absorption processes that involve components reacting with each other (for example, the absorption of gaseous carbon dioxide in aqueous solutions of sodium hydroxide). For a liquid mixture of two components (called a binary mixture) at a given temperature and pressure, the relative volatility is defined as

$$\alpha = \frac{(y_i/x_i)}{(y_j/x_j)} K_i/K_j \tag{1.3.40}$$

Where, α is the the relative volatility of the more volatile component i to the less volatile component j, $y_i$ is the the vapor-liquid equilibrium concentration of component i in the vapor phase, $x_i$ is the the vapor-liquid equilibrium concentration of component i in the liquid phase, $y_j$ is the the vapor-liquid equilibrium concentration of component j in the vapor phase, $x_j$ is the the vapor-liquid equilibrium concentration of component j in the liquid phase, (y/x) is the commonly called the K value or vapor-liquid distribution ratio of a component.

When their liquid concentrations are equal, more volatile components have higher vapor pressures than less volatile components. Thus, a K value (= y/x) for a more volatile component is larger than a K value for a less volatile component. That means that $\alpha \geq 1$ since the larger K value of the more volatile component is in the numerator and the smaller K of the less volatile component is in the denominator. α is a unitless quantity. When the volatilities of both key components are equal, $\alpha = 1$ and separation of the two by distillation would be impossible under the given conditions because the compositions of the liquid and the vapor phase are the same (azeotrope). As the value of α increase above 1, separation by distillation becomes progressively easier.

### 1.3.6. Design of distillation column

| | |
|---|---|
| Graphical design | *e.g.* frequently used to get a visual idea of the process behaviour |
| Binary | *e.g.* McCabe and Thiele, 1925; Ponchon and Savarit, 1921; 1922 |
| Ternary | *e.g.* Laroche *et al.*, 1992; Stichmair *et al.*, 1989 |
| Empirical design method | frequently used to obtain a design fast |
| Binary | *e.g.* Underwood, 1932-1948; Kirkbride, 1944 |
| Complex | *e.g.* Glinos and Malone, 1985; Meiers *et al.*, 1995 |
| Mathemamatical formulations | *e.g.* Dilnnebeir and Pantelides, 1999 |

**1.3.6.1. *Material balance-binary mixtures*:** As the composition of the vapour issuing from the distillation still and that of the liquid remaining in it changes during the course of operation, the mathematical approach should be differential. Let F be the kmol of a liquid mixture (A+B) containing $x_F$ mole fraction of A which is charged to a distillation still. Let D be the kmol of distillate and W be the kmol of residual liquid in the still which are obtained at the end of operation. Let $y_{Davg}$ and $x_W$ be the mole fractions of A in the distillate and the bottom residual liquid. Let L be the kmol of liquid in the still at any time during the course of distillation and let x be the mole fraction of A in the liquid. let a very small amount dD kmol of the distillate of the composition y in equilibrium with the liquid is vaporised. Then, the composition and the quantity of liquid decreases from x to x - dx and L to L-dL respectively.

Then

Overall material balance at any time is:

$$L = L - dL + dD \tag{1.3.41}$$

$$dL = dD \tag{1.3.42}$$

Material balance of the component A yields

$$Lx = (L- dL)(x - dx) + y\, dD \tag{1.3.43}$$

$$Lx = Lx - Ldx - xdL + dLdx + ydD$$

dLdx being very small can be neglected.

$$0 = -Ldx - xdL + ydD \tag{1.3.44}$$

Put the value of dD as dL from Equation (1.3.44)

$$0 = -Ldx - xdL + ydD \tag{1.3.45}$$

$$Ldx = (y - x)\, dL \tag{1.3.46}$$

$$\frac{dL}{L} = \frac{dx}{y-x} \tag{1.3.47}$$

Integrating the above equation between two extreme conditions:

$L = F$, $x = x_F$ and $L = W$, $x = x_w$

$$\int_{w}^{F} \frac{dL}{L} = \int_{x_w}^{x_F} \frac{dx}{y-x} \tag{1.3.48}$$

$$In \frac{F}{W} = \int_{x_w}^{x_F} \frac{dx}{y-x} \tag{1.3.49}$$

y is in equilibrium with x *i.e.* y is the vapour phase composition in equilibrium with the liquid phase composition (x)).

Equation (1.3.49) is known as the Rayleigh equation. It is used to determine F, W, $x_F$ or $x_W$ when three of these are known. The RHS of Equation (1.3.49), *i.e.*, the integral on the RHS side is evaluated graphically by plotting 1/(y - x) as an ordinate and x as an abscissa, *i.e.*, plotting 1/(y - x) v/s x and determining the area under the curve between $x - x_F$ and $x = x_w$. The required data for the above procedure are taken from the vapour- liquid equilibrium relationship.

If $y_{D', avg.}$ is the composition of the composited distillate, then, Material balance of the component A is:

$$x_F\,F = y_{D',avg.}\,D + x_W\,W \tag{1.3.50}$$

Though the simple or differential distillation as a method of separation is not effective, many such units are used, especially where (i) the components to be separated have widely different boiling points and (ii) methods giving sharp separations are not necessary.

**1.3.6.2.** ***Constant Relative Volatility:*** During the course of distillation, although the temperature in the distillation still/kettle increases, the relative volatility does not change much and so some average volatility α can be used, over the concentration range involved, which makes the direct integration of RHS of Equation (1.3.50) possible and graphical integration can thus be avoided. The relationship between the composition of vapour (y) and that of liquid (x), in equilibrium, in terms of relative volatility is

$$y = \alpha x / 1 + (\alpha - 1)\,x \tag{1.3.51}$$

So y from the above equation can be substituted in Equation (1.3.50).

$$In\ \frac{F}{W} = \int_{x_w}^{x_F} \frac{dx}{y - x} \tag{1.3.52}$$

$$In\ \frac{F}{W} = \int_{x_w}^{x_F} \frac{dx}{\dfrac{\alpha x}{1+(\alpha-1)x} - x} \tag{1.3.53}$$

$$In\ \frac{F}{W} = \int_{x_w}^{x_F} \frac{[1+(\alpha-1)x]dx}{\alpha x - x[1+(\alpha-1)x]} \tag{1.3.54}$$

Consider the term $\dfrac{1+(\alpha-1)x}{\alpha x - x[1+(\alpha-1)x]}$ of the RHS of Equation 1.3.54

$$\frac{1+(\alpha-1)x}{\alpha x - x[1+(\alpha-1)x]} = \frac{1+\alpha\,x - x}{\alpha x - x - \alpha x^2 + x^2}$$

Putting the above value of RHS back into Equation (1.3.54), we get

$$In\ \frac{F}{W} = \frac{1}{\alpha - 1}\left[\int_{x_w}^{x_F} \frac{1}{x}dx + \alpha \int_{x_w}^{x_F} \frac{dx}{1-x}\right]$$

Integrating, we get

$$In\ \frac{F}{W} = \frac{1}{\alpha - 1}\left[In\frac{x_F}{x_w} - \alpha\ \ In\ \frac{1-x_F}{1-x_w}\right] \tag{1.3.55}$$

Equation (1.3.55) relates F, W, $x_F$ and $x_w$ through a.

Equation (1.3.55) can be arranged in another useful form as:

Add and subtract in $[(1 - x_w)/(1 - x_F)]$ in bracket on RHS of Equation (1.3.55).

$$In\frac{F}{W} = \frac{1}{\alpha - 1}\left[In\frac{x_F}{x_w} - In\frac{1-x_w}{1-x_F} - In\frac{1-x_w}{1-x_F} - \alpha\, In\frac{1-x_F}{1-x_w}\right.$$

Combining and rearranging the terms in bracket,

$$\ln\frac{F}{W} = \frac{1}{\alpha - 1}\ln\frac{x_F(1 - x_W)}{x_W(1 - x_F)} + \ln\frac{1 - x_W}{1 - x_F} \qquad (1.3.56)$$

This can be rearranged to another from as:

$$In\frac{F}{W} - In\frac{1-x_w}{1-x_F} = \frac{1}{\alpha - 1}\; In\frac{x_F(1-x_w)}{x_w(1-x_F)}$$

$$\ln\frac{Fx_F}{Wx_W} = \alpha\ln\frac{F(1 - x_F)}{W(1 - x_W)} \qquad (1.3.57)$$

Equation (1.3.57) relates the number of moles of A remaining in the residue, W $x_w$ to that of B remaining, W (1 – $x_w$), F $x_F$ is the number of moles of A initially, let it be $No_A$; F (1 – $x_F$); is the number of moles of B initially, let it be $No_B$, W $x_w$ is the number of moles of A initially, let it be $N_A$, (after distillation is over), W (1 – $x_w$) is the number of moles of B initially, let it be $N_B$.

$$\ln\frac{N_{0_A}}{N_A} = \alpha\ln\frac{N_{0_B}}{N_B}$$

$$\frac{N_{0_A}}{N_A} = \left(\frac{N_{0_B}}{N_B}\right)^{\infty}$$

$$\left(\frac{N_B}{N_{0_B}}\right)^{\infty} = \frac{N_A}{N_{0A}}$$

$$\frac{N_B}{N_{0_B}} = \left(\frac{N_A}{N_{0_A}}\right)^{1/\infty} \qquad (1.3.58)$$

1.3.6.3. *Differential Condensation*

In this case, a vapour feed is slowly condensed and the condensate is drawn in equilibrium with the vapour residue. Let F be the moles of vapour of composition $y_F$ and D be the moles of vaporous residue of composition $y_D$ then the treatment similar to that described above yields.

$$\ln\frac{F}{W} = \int_{y_F}^{y_D}\frac{dy}{y - x^*} \qquad (1.3.59)$$

### 1.3.6.4. *Flash of Equilibrium Distillation*

Flash distillation is normally carried out in a continuous manner. In this method, a liquid mixture is partially vaporised, the vapour and liquid are allowed to attain equilibrium and finally withdrawn separately. Feed is heated in a tubular heat exchanger. The hot liquid mixture is then fed to a separator via a pressure reducing valve whereby the pressure is reduced and a vapour is formed at the expense of liquid adiabatically. The liquid is withdrawn from the bottom of the separator and the equilibrium vapour leaves the separator from the top which is then liquified in a condenser. Flash distillation is commonly used in petroleum industry, handling multicomponent system in the pipe stills.

Consider one mole of a liquid mixture, having $x_F$ mole fraction of the more volatile component, is fed to a flash distillation unit. Let f be the fraction of the feed that is vaporised and is of composition y. Then (1 - f) will be the molse of the residual liquid obtained. Let x be the mole fraction of the residual liquid obtained. Let x be the mole fraction of the more volatile component in the liquid.

Then, a material balance of the more volatile component gives.

$$x_F = f.y + (1 - f)x$$
$$x_y = -(1 - f)x + x_F$$

$$y = -\frac{(1-f)}{f}x + \frac{x_F}{f} \tag{1.3.60}$$

Equation (1.3.60) is the material balance/operating line for flash distillation with a slope equal to - (1 - f)/f and an intercept equal to $x_F/f$.

The point of intersection of the operating line and the diagonal (x = y) is,

$$y = -\frac{(1-f)x}{f} + \frac{x_F}{f}$$
$$x = y$$
$$x = -\frac{(1-f)x}{f} + \frac{x_F}{f}$$
$$xf = -x + xf + x_F$$
$$x = x_F$$

and $$y = x_F$$

for f = 1, feed totally vaporised (feed 100 mole per cent vaporised)

$$Slope = -\frac{(1-f)}{f} = 0$$

and hence the operating line is parallel to x- axis through the point ($x_F$, $x_F$) on the diagonal. For f = 0, no feed is vaporised. Slope = ∞ (Infinite) and hence the operating line will be parallel to y - axis through the point ($x_F$, $x_F$) on the diagonal.

Overall material balance over Liquid flow to plate m + 1 = Vapour flow from plate m + 1 + Bottom product withdrawn

$$L_m = V_{m+1} + W \tag{1.3.61}$$

Similarly, material balance of more volatile component (A):

$$L_m x_m = V_{m+1} y_{m+1} + W x_w \tag{1.3.62}$$

Rearranging the Equation (1.3.62),

$$y_{m+1} = \frac{L_m}{V_{m+1}} x_m - \frac{W x_W}{V_{m+1}} \tag{1.3.63}$$

We know from the Equation (1.3.55) , that $V_{m+1} = L_m - W$

$$y_{m+1} = \frac{L_m}{L_m - W} x_m - \frac{W x_w}{L_m - W} \tag{1.3.64}$$

The Equation (1.3.64) is known as the material balance line of operating line of the stripping section.

1.3.6.5. *Analysis of Fractionating Columns*

The theoretical/ideal stages/plates required for a given degree of separation can be obtained by: 1. McCabe - Thiele method. 2. Lewis - Sorel method. 3. Ponchon - Savarit method.

1.3.6.5.1. *Mccabe - Thiele Method*

It is a graphical procedure of obtaining theoretical plates. Assumptions underlying this method are: Constant molal heat of vaporisation, No heat losses, No heat of mixing.

Consider any plate n in the rectifying section. A heat balance across plate n is $L_{n-1} H_{L_{n-1}} + V_{n+1} H_{v_{n+1}} = L_n H_{L_n} + V_n H_{v_n}$ + heat losses + heat of mixing

Heat losses are negligible for good insulation. Heat of mixing is zero for an ideal system. $H_L$ and $H_V$ - enthalpies of liquid and vapour per mol.

If the molar heat of vaporisation is constant (independent of the composition) then one mol of vapour $V_{n+1}$ on condensing releases sufficient heat to liberate one mol of vapour $V_n$. It follows that $V_n = V_{n+1}$. There fore, the molar flow of vapour up the column is constant. As the temperature change from one plate to the next is small, $H_{LN}$ can be taken equal to $H_{Ln-1}$

$$L_{n-1} H_{L_{n-1}} + V_{n+1} H_{v_{n+1}} = L_{n+1} H_{L_{n+1}} + L_n H_{L_{n-1}}$$

$$L_{n-1} = L_n$$

So the moles of liquid reflux in this section are also constant. Therefore, $L_n$ and $V_{n+1}$ are constant in the rectifying section and $V_{m+1}$ and $L_m$ are constant in the stripping sectoin.

*i.e.* molar flow rates of vapour and liquid are constant in each section of the column.

$$L_n = L_{n-1} = L,\ L_m = L_{m+1} = L'$$

$$V_n = V_{n+1} = V,\ L_{m+1} = V_m = V'$$

(The above assumptions lead to a concept of constant molal vapour flow and constant molal liquid flow in any section of the column. In other words, for one mole of vapour condensed, one mole of liquid is vaporised. Thus $V_n = V_{n+1} L_n = L_{n+1}$.... etc).

The subscripts n, n + 1, m, m + 1 ... etc. may be dropped and the operating lines than plotted on the equilibrium diagram are straight. The operating line of the rectifying section becomes:

$$y = \frac{L}{L+D}x + \frac{Dx_D}{L+D} \tag{1.3.65}$$

As $R = \frac{L}{D}$, the above equation becomes:

$$y = \frac{R}{R+1}x + \frac{x_D}{R+1} \tag{1.3.66}$$

The point of intersection of the operating line for the rectifying section and the 45° diagonal (x = y) obtained as,

$$y = \frac{R}{R+1}x + \frac{x_D}{R+1}$$

Putting $y = x$

$$x = \frac{R}{R+1}x + \frac{x_D}{R+1}$$

$$(R + 1)\,x = R\,x + x_D$$

$$x = x_D$$

and $$y = x_D$$

Thus, the operating line of the rectifying section is to be drawn on the equilibrium diagram through the point $(x_D, x_D)$ on the 45° diagonal and with a slope equal to $R/(R + 1)$ or an intercept on y- axis equal to $x_D/(R + 1)$.

The equation of the operating line of the stripping section becomes:

$$y = \frac{L'}{L'-W}x - \frac{Wx_w}{L'-W} \tag{1.3.67}$$

The point of intersection of the operating line of the stripping section and the 45° diagonal is obtained as floolws:

$$y = \frac{L'}{L'-W}x - \frac{Wx_w}{L'-W}$$

Diagonal: $y = x$

$$x = \frac{L'}{L'-W}x - \frac{Wx_w}{L'-W}$$

$L' x - W. x = L' w - W x_w$

$x = x_w$

and $y = x_w$

Thus, the operating line of the stripping section is to be drawn through the point $(x_w, x_w)$ on the diagonal and with slope equal to L'/L' - W.

1.3.6.5.1. *Step-wise Procedure for Obtaining Theoretical PLATES*

1. By material balances, evaluate the terms D, W, L, etc.
2. Draw the equilibrium curve and the diagonal with the help of x - y data given. If the relative volatility is given, generate the x - y data.
3. Draw the operating line of the rectifying section through the point P $(x_D, x_D)$ on the diagonal and with an intercept equal to $x_D/(R + 1)$ or $Dx_D/(L + D)$ or a slope equal to $R/(R + 1)$.
4. Draw the operating line of the stripping section through the point M $(x_w, x_w)$ on the diagonal and with a slope equal to L'/L' - W.
5. Starting from the point $(x_D, x_D)$ on the diagonal, draw a horizontal line to meet the equilibrium curve at point 'a'. Drop a vertical from point 'a' to meet the operating line at point 1.
6. Proceed in this way that is constructing the triangles (representing the stages) between the equilibrium curve and the operating line of rectifying section, till we are above the point of intersection of two operating lines. Once we cross this point of intersection, construct the triangles between the equilibrium curve and the operating line of the stripping section, *i.e.*, dropping verticals on the operating line of the stripping section.
7. Proceed in the same manner till we reach exactly or cover the point M $(x_w, x_w)$.
8. Count the triangles constructed between $x_D$ and $x_w$. Each triangle on the x - y diagram represents a theoretical plate.
9. If the number of triangles constructed are 'n', then 'n' represents the theoretical number of plates including reboiler (since the vapour and liquid leaving the reboiler are in equilibrium, the reboiler is equivalent to one theoretical plate) and n - 1 represents the number of theoretical plates in the column to achieve a desired degree of separation.

The graphical step -by-step construction utilising alternately the operating lines and the equilibrium curve to find the number of ideal plates/stages for a given separation is known as the McCabe-Thiele method.

**1.3.6.5.2. *Limitations of Mccabe-Thiele Method*:** (1) It is not used when the relative volatility is less than 1.3 or greater than 5. (2) It is not used when more than 25 theoretical stages/plates are required, and (3) It is not used when the operating reflux ratio is less than 1.1 times the minimus reflux ratio.

1.3.6.5.2. *Lewis - Sorel Method of Determination of Number of Theoretical Stages / Plates*

The equation of the operating line of the rectifying section is:

$$y_{n+1} = \frac{L_n}{L_n + D} x_n + \frac{D x_n}{L_n + D} \tag{1.3.68}$$

The equation of the operating line of the stripping section is:

$$y_{m+1} = \frac{L_m}{L_m - W} x_m - \frac{W x_w}{L_m - W} \tag{1.3.69}$$

**1.3.6.5.2.1. *Steps for Using Lewis – Sorel Method***

1. From the data in a any problem, evaluate the terms - D, W, L etc.
2. From the x - y data provided (or can be generated knowing the relative volatility), draw an equilibrium diagram.
2. From the x - y data provided (or can be generated knowing the relative volatility) draw an equilibrium diagram.
3. Put the values of $L_n$ (L), $x_D$, D in equation (1.3.68), so that it will be a relationship between $y_{n+1}$ and $x_n$.
4. Similarly, put the values $L_m$ (L') W, $x_w$ in equation (1.3.69), so that it will be a relationship between $y_{m+1}$ and $x_m$ $L_m$ is to be evaluated by taking into consideration the condition of feed. For example, if it is a liquid at its bubble point, $L_m = L_n + qF$, $q = 1\backslash L_m = L_n + F$, $L' = L + F$.
5. The composition of distillate ($x_D$) given in the problem statement represents the composition of vapour ($y_n$) as it is obtained by condensing the vapour leaving the top of column.
6. From the vapour phase composition- $y_n = x_D$, find the liquid phase composition $x_{n+1}$ from the equilibrium diagram and put this value of $x_{n+1}$ in the equation of the operating line of the rectifying section to get $y_{n+2.}$
7. Find $x_{n+2}$ value from the equilibrium diagram corresponding to $y_{n+2}$ .
8. Calculate $y_{n+3}$ and proceed with the same equation till we reach the point when the liquid phase composition equal to or less than $x_F$.
9. Then make use of the material balance equation or operating line of the stripping section and proceed in the same way as described above till we get the liquid phase composition equal to or below $x_w$. Suppose we end with $y_{n+10}$, [m = n, for convenience] it means that there will be 10 plates in the column.

**1.3.6.5.2.2. *Feed Plate and Feed Line*:** The plate on which feed is introduced is called the feed plate. The feed to the column may be introduced as: (a) cold liquid, (b) liquid at its bubble point (saturated- liquid) (c) partially vaporised (partly vapour and partly liquid), (d) saturated vapour at its dew point and (e) superheated vapour. The condition of feed introduced on the feed plate alters the phase flow rates. Introduction on the feed as a cold liquid and liquid at its bubble point increases the liquid flow rate in the stripping section. Feed-partially vaporised increases the vapour flow rate in the rectifying section and also the liquid flow rate in the stripping section. The feed as a saturated vapour and a superheated vapour increases vapour flow rate in the rectifying section.

The phase flow rates are shown based on the fact that the column liquid is always at its bubble point and the vapour is always at its dew point. To calculate the change in phase flow rates by the introduction of feed, a factor 'q' is introduced (The 'q' is a measure of the thermal condition of the feed and is defined as the number of moles of saturated liquid resulting in the stripping section for each mole of feed introduced. Thus for a feed F, we get:

$$L' = L + qF \quad (1.3.70)$$

and $$V = V' + (1 - q) F \quad (1.3.71)$$

$$q = \text{fraction of feed that is liquid.}$$

The values of 'q' for the various thermal condition of feed are: (a) *Cold Liquid*: Q>1, (b) *Saturated liquid*: q = 1, (c) *Feed partially vapour (partially flashed, i.e., vapour- liquid mixture)*: 0 < q < 1, (d) *Saturated vapour feed*: q = 0, and (e) *superheated vapour feed*: q < 0. The value of q for any particular feed condition can the calculated from:

$$q = \frac{\text{Energy to convert 1 mole of feed to saturated vapour}}{\text{Molal latent heat of vaporisation}}$$

The liuqid flow in the stripping section is:

$$L' = L + qF \quad (1.3.72)$$

$$L' L = qF \quad (1.3.73)$$

Similarly, the vapour flow in the rectifying section is:

$$V = V' + (1 - q) F \quad (1.3.74)$$

$$V - V' = (1 - q) F \quad (1.3.75)$$

$$V = L + D \text{ - (overall material balance upper section)} \quad (1.3.76)$$

$$V_y = Lx + Dx_D \text{ - (material balance of A)} \quad (1.3.77)$$

$$V' = L' - W \text{ - (overall material balance lower section)} \quad (1.3.78)$$

and $$V'_y = L'x - Wx_w \text{ - (material balance of A)} \quad (1.3.79)$$

Subtracting Equation (1.3.79) from Equation (1.3.71),

$$y (V - V') = x (L - L') + D\, x_D + Wx_W. \quad (1.3.80)$$

But we know that:

$$x_F F = D\, x_D + W\, x_W \quad (1.3.81)$$

The Equation (1.74) becomes:

$$y (V - V') = x (L - L') + x_F F' \quad (1.3.82)$$

Putting the values of V - V' and L - L' from Equations (1.3.71) and (1.3.70) into Equation (1.76)

$$y\,(1 - q)\,F \quad = \quad x\,(-\,qF) + x_F\,F \tag{1.3.83}$$

$$y = \frac{-q}{1-q}\,x + \frac{x_F}{1-q} \tag{1.3.84}$$

Equation (1.3.82) is known as the Feed line or q-line.

The point of intersection of the feed line and the diagonal (y = x) is:

$$y \;=\; x \;=\; \frac{-q}{1-q}\,x + \frac{x_F}{1-q} \tag{1.3.85}$$

$$x = x_F$$

and $\quad y = x_F$

The feed line thus passes through the point N ($x_F$, $x_F$) on the 45° diagonal and has slope equal to -1/1-q.

For q = 1 , slope = ∞ (infinite) and hence the feed line is parallel to y-axis through the points N($x_F$, $x_F$) on the diagonal on the equilibrium diagram.

For q = 0, slope = 0 and hence the feed line is parallel to the x-axis through the points N ($x_F$, $x_F$) on the diagonal. Typical feed line or q-line constructions for the various values of 'q' are shown. Plotting the feed line and operating lines to determine the number of stages: (1) the feed line on the equilibrium diagram, and (2) the operating line of the rectifying section as usual, it will cut the feed line at certain point then (3) the operating line of the stripping section can be easily plotted. This line is to be plotted through the point M ($x_w$, $x_w$) on the diagonal and it, will pass through the point of intersection of the feed line and the operating line of the rectifying section. The point of intersection of lines-1, 2 and 3 falls in a triangle which represents a feed plate. The triangle representing the feed plate has one corner on the operating line of the rectifying section and one on the operating line of the stripping section. Feed should be introduced at any point between A and B but at the same time we have to keep in mind the maximum enrichment and minimum number of plates. Introduction of the feed near A or B unnecessarily results into a large number of plates in the rectifying or stripping section. The optimum location will be the one at which the liquid phase composition value crosses the feed composition value $x_F$.

**1.3.6.5.2.3.** ***Partial Condenser:*** The total condenser is probably most commonly used on fractionating columns but in cases where the overhead product is desired as a vapour or where condensation of the entire vapour would be difficult, a partial condenser is used. When the component to be removed (requiring large pressure at reasonable condenser temperatures) as a distillate has very low boiling point, then the complete condensation is difficult. In such cases, some part of the overhead vapour is condensed to provide a required reflux and the remainder of the overhead vapour is removed as a distillate. With a partial condenser or dephlegmator, the distillate is removed in the vapour state. When sufficient time of contact between liquid reflux and vapour product is provided then these two streams leave the condenser in equilibrium with each other and thus the partial condenser acts as one theoretical stage for the separation, *i.e.* the partial condenser is equal to an additional stage (for enrichment) in the distillation unit. This is true normally, and so the partial condenser is represented by the triangle on the xy diagram.

Let $x_C$ be the composition of the liquid reflux and $y_D$ be the composition of the distillate withdrawn (vapour product). Vapour leaving the partial condenser has a composition $y_D$ which is in equilibrium with $x_C$, the composition of liquid reflux. The operating line for the rectifying section with a total condenser is:

$$y_{n+1} = (R/R + 1)\, x_n + x_D/R + 1$$

The operating line for the rectifying section with a partial condenser is:

$$y_{n+1} = (R/R + 1)\, x_n + y_D/R + 1$$

only difference is $x_D$ is replaced by $y_D$.

This operating line passes through the point $(y_D, y_D)$ on the diagonal where $y_D$ is the composition of the distillate withdrawn (in the vapour state).

In case of partial condenser, we have to adopt the same procedure of construction of stages as that described previously for the total condeser [only difference is $x_n$ is to be replaced by $y_D$- (composition of vapour product)] and count the number of stages between $y_{-D}$ and $x_w$. The total number of theoretical plates counted starting from $y_D$ to $x_W$ will include a reboiler as well as a partial condenser (each of them acting as one theoretical stage).

1.3.6.5.3. *Reflux Ratio*

**1.3.6.5.3.1. *Infinite Reflux Ratio or Total Reflux Ratio*:** The total reflux operation is of interest to know the minimum number of plates required (it sets a lower limit on the number of plates required) and is not a practical method of operation as at total reflux the product rate is zero. A column operating under total reflux has enough material is charged to the distillation assembly and the column is operated under total reflux. During the operation of the column, the vapour issuing from the top is condensed and the entire condensed stream is fed back to the column as reflux. Also all the liquid going to the reboiler is vaporised and is fed to the column. Thus, under the condition of total reflux F = 0, D = 0 and W = 0, therefore $V_{n+1}$ $L_n$ throughout the column. The slope (L/V) of both the operating lines becomes unity and hence both the operating lines coincide with the diagonal and a minimum number of stages (theoretical plates) are then required for a desired degree of separation $(x_D - x_W)$. The total reflux operation corresponds to a maximum reboiler heat supply and condenser cooling capacity for the separation.

**1.3.6.5.3.2. *Minimum Reflux Ratio*:** At total reflux, the operating lines of both sections coincide with the diagonal and to effect a desired separation, the number of stages (theoretical plates) required are minimum. As the reflux ratio is reduced, the operating lines move towards the equilibrium curve along the feed line and the number of stages increase. Ultimately, further reduction in the reflux ratio results into a condition wherein the top operating line touches the equilibrium curve. The point of intersection of the operating lines lies on the equilibrium curve and at this point step becomes very close together, so that a zone of nearly constant composition is formed near the feed plate and an infinite number of plates are then required. This represents a condition of the minimum reflux. Thus, the minimum reflux ratio is that reflux ratio at which an infinite number of plates are required for a desired separation. At minimum reflux ratio, the requirement of heat supply to the reboiler and that of coolant supply to the condenser are minimum.

*Calculation of minimum Reflux Ratio, ($R_M$):*

For liquid at its bubble point, the feed line is vertical (parallel to y-axis) and the point C lies on the equilibrium curve and has the co-ordinates ($x_F$, $y_F$). Draw the equilibrium curve and the diagonal and draw the feed line, a vertical through point P ($x_F$, $x_F$) of the diagonal. The point of intersection of the feed line with the equilibrium curve is also a point of intersection of the operating lines at minimum reflux. Slope of the operating line is given as:

$$\frac{R_m}{R_m+1} = \frac{x_D - y_F}{x_D - x_F} \tag{1.3.86}$$

$$R_m (x_D - x_F) = (R_m + 1)(x_D - y_F)$$

$$R_m = \frac{x_D - y_F}{y_F - x_F} \tag{1.3.87}$$

Similarly for feed as a saturated vapour, the feed line is horizontal.

Slope of the line is given as:

$$\frac{R_m}{R_m+1} = \frac{x_D - y'}{x_D - x'} \tag{1.3.88}$$

$$R_m = \frac{x_D - y'}{y' - x'}$$

$$= \frac{x_D - x_F}{x_F - x'} \; (y' = x_F) \tag{1.3.89}$$

**1.3.6.5.3.3. *Optimum Reflux Ratio*:** Any reflux ratio between infinite reflux ratio requiring a minimum number of plates and minimum reflux ratio requiring an infinite number of plates is a workable system which requires finite stages for a desired degree of separation. At minimum reflux ratio, as infinite number of plates is required, the fixed cost is also infinite while the cost of heat supply to the reboiler and the condenser coolant is minimum. As the reflux ration is increased, the number of plates decreases and the fixed cost decreases at first, passes through a minimum and then increases as with higher reflux ratio the diameter of the column and sizes of reboiler and condenser increase. The operating cost increases continuously as the reflux ratio is increased as it is directly proportional to (R + 1). At total reflux, though the number of plates is minimum, the cost of heat supply to reboiler and condenser coolant is maximum and also large capacity reboiler and condenser are needed. The total cost which is the sum of the fixed cost and the operating cost also decreases to a minimum and then increases with reflux ratio. The optimum reflux ratio occurs at a point where the sum of the fixed cost and the operating cost is minimum. The optimum reflux ratio is defined as that reflux ratio at which the sum of the fixed cost and operating cost is minimum. As a rough approximation, the optimum reflux ratio is usually 1.1 to 1.5 times the minimum reflux ratio.

**1.3.6.5.4. *Fenske-Underwood Equations*:** Minimum Reflux Ratio ($R_m$): The minimum reflux ratio, requiring infinite number of plates, can be obtained analytically when the relative volatility is taken as constant over the concentration range concerned ($x_D - x_w$). So let us obtain the required relationship with which we can calculate $R_m$. For the normal equilibrium curve which is concave downward throughout its length, the operating line of the rectifying section, at minimum reflux, passes through the point of intersection of the q-line and the equilibrium curve. The equation of the operating line for the rectifying section dropping the subscripts is:

$$y = \frac{R}{R+1}x + \frac{x_D}{R+1} \qquad (1.3.90)$$

The q-line equation is

$$y = \frac{-q}{1-q}x + \frac{x_F}{1-q}$$

$$y = \frac{-q}{1-q}x - \frac{x_F}{q-1} \qquad (1.3.91)$$

Putting value of y from Equation (1.3.90) into Equation (1.3.91),

$$\frac{q}{1-q}x - \frac{x_F}{q-1} = \frac{R}{R+1}x + \frac{x_D}{R+1}$$

$$\frac{q}{1-q}x - \frac{R}{R+1}x = \frac{x_D}{R+1} + \frac{x_F}{q-1}$$

$$x = \frac{x_D(q-1) + x_F(R+1)}{(R+q)}$$

From Equation (1.3.91), we have

$$\left[y + \frac{x_F}{q-1}\right]\frac{(q-1)}{q} = x \qquad (1.3.92)$$

Putting the value of x from Equation (1.3.90) into Equation (1.3.92)

$$y = \frac{R}{R+1}\left[y + \frac{x_F}{q-1}\right]\frac{(q-1)}{q} + \frac{x_D}{(R+1)}$$

$$q\ (R+1)y = \left[Ry + \frac{R\,x_F}{q-1}\right]\ [q-1]\ = q\,x_D$$

$$y = \frac{Rx_F + q\ x_D}{(R+q)}$$

$$y = \frac{Rx_F + q\ x_D}{R+q}$$

So the coordinates of the point of intersection of the operating lines with the q-line are:

$$x\ = \frac{x_D(q-1) + x_F(R+1)}{(R+q)} \text{ and } y = \frac{Rx_F + qx_D}{(R+q)}$$

So when this point of intersection of the operating line (of the rectification section) with the q-line lies on the equilibrium curve then R will be $R_m$ and the coordinates represent the equilibrium values. The coordinates of the point of intersection (or feed line and operating line of rectification section) on the equilibrium curve (replacing R by $R_m$): (x, y) are given as

$$x\ = \frac{x_D(q-1) + x_F(R+1)}{(R+q)} \text{ and } y = \frac{Rx_F + qx_D}{(R+q)}$$

As x and y represent the equilibrium values, they can be substituted in the definition of α.

We have,

$$\alpha\ = \frac{y(1-x)}{x(1-y)}$$

$$\alpha\ = \frac{\frac{1}{x} - 1}{\frac{1}{y} - 1} \qquad (1.3.93)$$

Put the values of $x_a$ and $y_a$ in Equation (1.3.93) as $x_a$(=x) and $y_a$ (= y) are the coordinates of the point of intersection of operating line and q-line lying on the equilibrium curve.

$$\alpha = \frac{\dfrac{(R_m+q)}{x_D(q-1)+x_F(R_m+1)} - 1}{\dfrac{(R_m+q)}{(R_m x_F + q x_D)} - 1}$$

Rearranging, we get.

$$\frac{\alpha[x_D(q-1)+x_F(R_m+1)]}{(R_m+q) - [x_D(q-1)+x_F(R_m+1)]} = \frac{R_m x_F + q x_D}{R_m(1-x_F)+q(1-x_D)} \qquad (1.3.94)$$

We have arranged the RHS in desired form as given in Equation (1.3.94). Now rearrange the denominator of the LHS of Equation (1.3.94) in a suitable form and then put in Equation (1.3.94).

Denominator of LHS of Equation (1.3.94) = $(R_m + q) - [x_D(q - 1) + x_F (R_m + 1)]$

$$(R_m + q) - [x_D(q-1) + x_F (R_m + 1)] = R_m + q - x_D q + x_D - x_F R_m + x_F$$

Add 1 and subtract 1.

$$= R_m + q - x_D - q + x_D - x_F + 1 - 1$$

Collecting the terms, we get

$$= R_m (1 - x_F) + (1 - x_F) + 1 (1 - x_D) - 1 (1 - x_D)$$

$$= (R_m + 1) + (1 - x_F) + (q - 1) (1 - x_D)$$

Put the value of denominator of the LHS in Equation (1.3.94) to get a desired relationship.

$$\frac{R_m x_F + q x_D}{R_m(1-x_F)+q(1-x_D)} = \frac{\alpha[x_D(q-1)+x_F(R_m+1)]}{(R_m+1)(1-x_F)+(q-1)(1-x_D)} \qquad (1.3.95)$$

Equation (1.3.95) is the desired relationship for $R_m$ and can be conveniently solved for any value of q. With the help of the above equation, we can calculate $R_m$ (minimum reflux) for a given $x_F$, $x_D$, $\alpha$ and feed condition (q).

Case I: $R_M$ for q = 0, *i.e.*, Feed is a saturated vapour at its dew point.

$$\frac{R_m x_F + q x_D}{R_m (1-x_F)+q(1-x_D)} = \frac{\alpha[x_D(q-1)+x_F(R_m+1)]}{(R_m+1)(1-x_F)+(q-1)(1-x_D)}$$

For saturated vapour feed, put q = 0 in the above equation and then the above relationship reduces to:

$$\frac{R_m x_F}{R_m (1-x_F)} = \frac{\alpha[-x_D+x_F(R_m+1)]}{(R_m+1)(1-x_F) - (1-x_D)}$$

$R_m/R_m$ on LHS of above equation will cancel and rearranging it, we get,

$$x_F (R_m+ 1) (1 - x_F) - x_F - (1 - x_D) = \alpha [- x_D + x_F (R_m + 1)] (1 - x_F)$$

Divide the LHS and RHS by $x_F(1 - x_F)$.

$$(R_m + 1) - \frac{(1 - x_D)}{(1 - x_F)} = \frac{\alpha}{x_F}[-x_D + x_F(R_m + 1)]$$

$$(R_m + 1) - \frac{(1 - x_D)}{(1 - x_F)} = \frac{-\alpha x_D}{x_F} + \alpha(R_m + 1)$$

Rearranging, we get

$$\frac{\alpha x_D}{x_F} - \frac{(1 - x_D)}{(1 - x_F)} = \alpha(R_m + 1) - (R_m + 1)$$

$$R_m + 1 = \frac{1}{\alpha - 1}\left[\frac{\alpha x_D}{x_F} - \frac{(1 - x_D)}{(1 - x_F)}\right]$$

$$R_m = \frac{1}{\alpha - 1}\left[\frac{\alpha x_D}{x_F} - \frac{(1 - x_D)}{(1 - x_F)}\right] - 1 \qquad (1.3.96)$$

Equation (1.3.96) is the desired relationship for $R_m$ when the feed is a saturated vapour.

Case II: For q = 1: Feed is a liquid at its boiling point, *i.e.*, feed is a saturated liquid.

$$\frac{R_m x_F + q\,x_D}{R_m(1 - x_F) + q(1 - x_D)} = \frac{\alpha[x_D(q - 1) + x_F(R_m + 1)]}{(R_m + 1)(1 - x_F) + (q - 1)(1 - x_D)}$$

For saturated liquid feed, *i.e.* liquid at its bubble point, put q = 1 and then the above relationship reduces to

$$\frac{R_m x_F + x_D}{R_m(1 - x_F) + (1 - x_D)} = \frac{\alpha x_F(R_m + 1)}{(R_m + 1) + (1 - x_F)}$$

$$R_m\, x_F(R_m + 1)(1 - x_F) + x_D(R_m + 1)(1 - x_F) = \alpha\, x_F(R_m + 1)[R_m(1 - x_F) + \alpha\, x_F(R_m + 1)(1 - x_D)$$

Divide each term of the RHS and LHS of the above Equation by $(R_m + 1)(1 - x_F)$.

$$R_m\, x_F + x_D = \alpha x_F\, R_m + \frac{\alpha x_F(1 - x_D)}{1 - x_F}$$

$$\alpha x_F\, R_m - R_m\, x_F = x_D - \frac{\alpha x_F(1 - x_D)}{1 - x_F}$$

Divide each term of the LHS and RHS by $x_F$.

$$\alpha R_m - R_m = \frac{x_D}{x_F} - \frac{-x_D}{(1 - x_F)}$$

$$R_m = \frac{1}{(\alpha - 1)}\left[\frac{x_D}{x_F} - \frac{\alpha(1 - x_D)}{(1 - x_F)}\right] \qquad (1.3.97)$$

Equation (1.3.97) is the desired relationship for $R_m$ for a saturated liquid feed (*i.e.*, q = 1).

***Alternative Approach:*** we will obtain the same relationships with the help of slope of the operating line for a minimum reflux and using the relationship: $y = \alpha x/ [1 + (\alpha -1) x]$ for Case I: Feed is a liquid at its bubble point (saturated liquid feed) and Case II: Feed is a saturated vapour at its dew point (Feed Condition of q = 0). We will get following Equations:

$$R_m = \frac{1}{(\alpha - 1)} \left[\frac{x_p}{x_f} - \frac{\alpha(1 - x_p)}{(1 - x_f)}\right] \tag{1.3.98}$$

$$R_m = \frac{1}{(\alpha - 1)} \left[\frac{\alpha x_p}{x_f} - \frac{(1 - x_p)}{(1 - x_f)}\right] - 1 \tag{1.3.99}$$

**1.3.6.5.4.1.** ***The Number of Plates at Total Reflux by Fenske's Method:*** In cases where the relative volatility is taken as constant over the range of temperature encountered in a column, Fenske derived an analytical expression for calculating the number of plates required at total reflux for achieving a desired degree of separation. For a binary system comprising of A and B with A as the more volatile component, the relative volatility is given by,

$$\alpha = \frac{y_A / x_A}{y_B / x_B}$$

but $x_B = 1 - x_A$ and $y_B = 1 - y_A$

$$\alpha = \frac{y_A / (1 - y_A)}{x_B / (1 - x_A)}$$

$$\frac{y_A}{1 - y_A} = \alpha \frac{x_A}{1 - x_A}$$

$$\frac{y}{1 - y} = \alpha \frac{x}{1 - x} \tag{1.3.100}$$

x and y are the equilibrium compositions of liquid and vapour phases in terms of molefraction of more volatile component (A) respectively. Equation (1.3.100) for plate n + 1 can be written as

$$\frac{y_{n+1}}{1 - y_{n+1}} = \alpha_{n+1} \frac{x_{n+1}}{1 - x_{n+1}} \tag{1.3.101}$$

We have, the operating line for rectifying section:

$$y_{n+1} = \frac{L_n}{V_{n+1}} x_n + \frac{x_D D}{V_{n+1}}, \; V_{n+1} = L_n + D$$

At total reflux, D = 0 and L = V, so we get,

$$y_{n+1} = x_n$$

So replacing $y_{n+1}$ by $x_n$, Equation (1.3.101) becomes

$$\frac{x_n}{1-x_n} = \alpha_{n+1}\frac{x_{n+1}}{1-x_{n+1}} \tag{1.3.102}$$

At the top of the column if a total condenser is used, the composition of the vapour leaving the top plate and entering a condenser and that of the distillate removed from it will be the same.

$y_1 = x_D$, y is the composition of the vapour leaving the plate - 1 (*i.e.* top plate)

Equation (1.3.101) becomes

$$\frac{x_D}{1-x_D} = \alpha_1\frac{x_1}{1-x_2} \tag{1.3.103}$$

For plate - 2, Equation (1.3.302) becomes

$$\frac{x_1}{1-x_1} = \alpha_2\frac{x_2}{1-x_2} \tag{1.3.104}$$

For plate - n, Equation (1.3.307) becomes

$$\frac{x_{n-1}}{1-x_n} = \alpha_n\frac{x_n}{1-x_n} \tag{1.3.105}$$

Combining Equation (1.3.308) and all Equations upto plate - n, we get

$$\frac{x_D}{1-x_D} = \alpha_1\frac{x_1}{1-x_1} = \alpha_1\alpha_2\left(\frac{x_2}{1-x_2}\right)$$

$$\frac{x_D}{1-x_D} = \alpha_1\alpha_2\alpha_3....\alpha_n\left(\frac{x_n}{1-x_n}\right)$$

$$\frac{x_D}{1-x_D} = \alpha_1\alpha_2\alpha_3....\alpha_{n-1}\alpha_n\left(\frac{x_n}{1-x_n}\right)$$

If some average relative volatility $\alpha_{avg}$. is used throughout the column, then

$$\alpha_{avg} = \alpha_1 = \alpha_2 = ... = \alpha_n$$

$$\frac{x_D}{1-x_D} = (\alpha_{avg.})_n\frac{x_n}{1-x_n} \tag{1.3.106}$$

To reach the bottom product W of composition $x_w$, $n_{min}$ (*i.e.* $n_m$) plates and a reboiler are required. The reboiler itself represents an equilibrium stage (reboiler acts as one equilibrium stage). Equation (1.3.106) then gives

$$\frac{x_D}{1-x_D} = (\alpha_{avg.})^{nm+1}\, n\frac{x_w}{1-x_w} \tag{1.3.107}$$

$n_m$ is the total minimum number of theoretical plates required.

$$\ln [x_D/(1-x_D)] = \ln (\alpha_{avg.})^{nm+1} x_w/(1 - x_w)]$$

$$\ln [x_D/(1 - x_D)] = (n_m + 1) \ln (\alpha_{avg}.) + \ln [x_w/(1 - x_w)]$$

$$(n_m + 1) \ln (\alpha_{avg.}) = \ln [- x_D /(1 - x_D)] - \ln [- x_w/(1 - x_w)]$$

$$= \ln [[- x_D / (1- x_D)] / [- x_w/(1 - x_w)]]$$

$$(n_m + 1) In (\alpha_{avg.}) = In \left[\frac{x_D(1-x_w)}{x_w(1-x_D)}\right]$$

$$n_m + 1 = \frac{In\left[\frac{x_D(1-x_w)}{x_w(1-x_D)}\right]}{In (\alpha_{avg.})} \quad (1.3.108)$$

$$n_m = \frac{In[x_D(1-x_w)/x_w(1-x_D)]}{In(\alpha_{avg.})} - 1$$

Equation (1.3.109) is known as the Fenske's equation. This Equation gives the total minimum number of theoretical plates required to produce products of compositions $x_D$ and $x_w$. $n_m$ - total minimum number of theoretical plates required at total reflux thus can be calculated from the knowledge of terminal compositions $x_D$ and $x_w$. For a moderate variation of α from the top to the bottom of the column, α geometric mean of α at the top and the bottom products *i.e.* $\alpha_{1-}$ $\alpha\text{-}_w$ is used in place of $\alpha_{avg}$.

**1.3.6.5.4.2. *Side Streams*:** Side streams from the fractionating column are generally removed in case of multi-component systems to obtain the products of intermediate composition (*i.e.* petroleum products). These are taken out from intermediate locations along the column height.

To get the idea regarding a construction of theoretical stages in such situations, let us consider the fractionation of a binary system with one side stream in the rectifying section.

Let S be the molar flow rate of the side stream removed and $x_s$ be the composition of the side stream with respect to the more volatile composition (A). Column has three sections/or column is divided into three sections:

(i) Rectifying section above side stream

(ii) Rectifying section below side stream

(iii) Stripping section.

(1) Assuming constant molar overflow, the operating line for the rectifying section above the side stream is given by:

$$y_{n+1} = \frac{L_n}{V_{n+1}} x_n + \frac{D\, x_n}{V_{n+1}}$$

$$y_{n+1} = \frac{L}{V} x_n + \frac{D\, x_D}{V}$$

$$V = L + D$$

V is the vapour flow up the column and L is the liquid flow down the column in this portion of the rectifying section.

(2) Let $L_s$ be the liquid flow down the column below the side stream in the rectifying section. Let $V_s$ be the vapour flow below the side stream in the rectifying section.

The material balance for any plate in the rectifying section between the feed plate and the plate from which the side stream is taken out (*i.e.* the rectifying section above the feed plate and the below side stream) is:

$$V_s = Lg + S + D \tag{1.3.109}$$

For the component A, it becomes :

$$V_s \cdot y_{n+1} = L_s\, x_n + S\, x_s + D\, x_D$$

$$y_{n+1} = \frac{L_s}{V_s} x_n + \frac{Sx_S + Dx_D}{V_s} \tag{1.3.110}$$

This is the equation of the operating line applicable for a portion of the rectifying section between feed and side stream. Let us obtain the point of intersection of the above given operating line and the diagonal y = x.

Subscripts n, n + 1, the operating line equation is

$$y = \frac{L_s}{V_s} x + \frac{Sx_S + Dx_D}{V_s}$$

$$y = x$$

$$x = \frac{L_s}{V_s} x + \frac{Sx_S + Dx_D}{V_s}$$

$$(V_s - L_s) = Sx_S + Dx_D$$

$$x = \frac{Sx_S + Dx_D}{V_s - L_s}$$

$$V_s = L_s + S + D$$

$$\bar{V}_s - L_s = \bar{S} + D$$

$$x = y = \frac{Sx_S + Dx_D}{S + D}$$

This is the required point of intersection of the operating line with the diagonal. Let us find the point of intersection of two operating lines of the rectifying section.

$$y = \frac{L}{V} x + \frac{Dx_D}{V}$$

$$y = \frac{L_s}{V_s} x + \frac{Sx_S + Dx_D}{V_s}$$

$$\frac{L}{V}x + \frac{Dx_D}{V} = \frac{L_s}{V_s}x + \frac{Sx_S + Dx_D}{V_s}$$

$$V_s = V$$

$$L\,x + D\,x_D = L_{s-}\,x + S\,x_s + D\,x_D$$

$$(L - L_s\,x = S\,x_s$$

But $$L_S = L - S$$

$$(L - L + S)\,x = S\,x_s$$

$$x = x_s$$

So operating lines intersect at $x = x_s$.

(3) The operating line for the stripping section is given in the usual manner,

$$y_{m+1} = \frac{L_m}{V_{m+1}}x_m - \frac{W\,x_w}{V_{m+1}}$$

$$y_{m+1} = \frac{L'}{V'}x_m - \frac{W\,x_w}{V'} \qquad (1.3.111)$$

L′ is the liquid flow down the column in the stripping section and V is the vapour flow up the column in the stripping section.

The side stream is normally removed as liquid.

$L_n = L_s + S$

$L_s = Ln - S$

(i) $L_s = L - S$, where L is liquid reflux to the column, L = RD.

(ii) $V_s = V$, where V = L + D

(iii) $L' = L_s + qF$

(if feed is a liquid at its bubble point, $L' = L_s + F$)

$V_s = V = V' + (1 - q)F$

$V' = V - (1 - q)F$

**1.3.6.5.4.3. *Use of Open Steam*:** In the fractionation of an aqueous solution wherein the non-aqueous component is removed as the distillate and water is removed as the bottom product, the heat requirement of the fractionation may be provided by introducing steam directly at the bottom of the fractionating column, *i.e.*, underneath the bottom plate (open steam). This procedure thus eliminates the need of a reboiler (heating surface) normally used. But with this arrangement, for a given reflux ratio and distillate composition, more plates will be required in the column. Even though the plates required are more, the cost of plates is less than the cost of the reboiler and its maintenance. For equimolal vaporisation and overflow: $L_m = L_{m-1} - L'$ and $V_{m+1} - V_m = V'$ (liquid flow (rate) falling from each plate and vapour flow (rate) rising from each plate are constant).

If S is the molar flow rate of steam to the fractionator (assume saturated at a column pressure) then the overall material balance is

$$F + S = D + W$$

The overall material balance of the component A is

$$_F\ x_F + S\ y_s = D\ x_D + W\ x_w$$

As the steam is free of A, $y_s = 0$

$$Fx_F = Dx_D + Wxw \tag{1.3.112}$$

The rectifying section of the column is unaffected by the use of open steam, so the operating line for this section is located in the usual manner. The material balance of the component A below plate m in the stripping section is

$$L'\ x_m + Sy_s - V'\ y_{m+1} + Wx_w$$

as $y_s = 0$ - for steam containing no A

$$L'\ x_m = V'\ y_{m+1} + Wx_w$$

We have:

$$L' = W$$

$$L'\ x_m = V'\ y_{m+1} + L'\text{-}\ x_w$$

$$L'\ (x_m - x_w) = V'\ y_{m+1}$$

$$\frac{L'}{V'} = \frac{L'}{S} = \frac{y_{m+1}}{x_m - x_w},\ as\ V' = S \tag{1.3.113}$$

So the operating line of the stripping section passes through the point ($x = x_w$, $y = 0$). As this point is on x-axis, a graphical construction must be continued upto the x-axis of the x-y diagram. The slope of the operating line is L'/V' (= L'/S). If steam entering the column is superheated, it will vapourise liquid on the plate at the bottom to an extent necessary to bring it to saturation. If lm is the molar latent heat at a column pressure, then

$$V' = S'\left[1 + \frac{H_{VS'} - H_{VS}}{\lambda m}\right] \tag{1.3.114}$$

Where $H_{vs'}$ is the enthalpy of superheated steam and Hvs is the enthalpy of saturated steam. S' is the molar flow rate of superheated steam and L' = V' - S' + W.

### 1.3.6.5.4. *Ponchon - Savarit Method*

When molar latent heat is not constant and there is a substantial heat of mixing, then for establishing the relationship between number of plates, product composition and liquid/vapour ratio, we have to adopt the method developed by Ponchon and Savarit. Determination of the number of plates by this method is applicable for both constant and varying molar overflow situations. The Ponchon-Savarit method is based on the use of an enthalpy-composition diagram (refer Robert E. Treybal 1981. Mass transfer operations, McGraw Hill) and it therefore requires detailed enthalpy data. The enthalpy-composition diagram consists of the dew-point curve (H v/s y), the bubble-point curve (H v/s x) and the tie lines joining these curves. A continuous distillation unit to which F moles of feed of composition $X_F$ is fed continuously and from which D moles of distillate of composition $X_D$ and W moles of bottom product $X_V$. Overall material balance: The plates are numbered top-down, subscripts n and m indicating the rectifying section and the stripping section respectively. Qc is the heat removed in a condenser and $Q_B$ is the heat added to a reboiler.

The overall material balance over the column is:

$$Fx_F = Dx_D + Wx_w \quad (1.3.115)$$

The overall energy balance over the column is :

$$FH_P = DH_D + Qc + WH_w - Q_B \quad (1.3.116)$$

(Assuming negligible heat losses) So heat to be added to the reboiler is:

$$Q_B = DH_D + Q_C + WH_W - FH_F \quad (1.3.117)$$

Equation (1.3.117) gives the reboiler heat *i.e.* heat to be added at the reboiler.

The material balance over the condenser is

$$V_1 = Lo + D$$

$$V_1 = (R + 1)\, D \quad (1.3.118)$$

For component A:

$$V_1y_1 = D\, x_D + L_0\, x$$

The enthalpy balance around the condenser is:

$$V_i\, H_v = Qc + LoH_L + DH_D$$

but

$$H_L = H_D$$

$$V_i\, H_v = Qc + LoH_D + DH_D$$

$$Q_c = (R + 1)\, DH_V - (L_0 + D)H_D,$$

$$Q_C = (R + 1)\, DH_V - (L_0 + D)H_D, \text{ as } L_0/D = R$$

$$Qc = D\, [(R + 1)H_v - RH_D - H_D] \quad (1.3.119)$$

Equation (1.3.119) provides the heat load of the condenser.

(i) *Rectifying/enriching section*: For the complete section above plate n, where n is any plate in this section. The overall material balance is:

$$V_{n+1} = L_n + D$$

and the material balance of the component A is:

$$V_{n+1}\, y_{n+1} = L_n x_n + Dx_D$$

$$V_{n+1}\, y_{n+1} - L_n x_n = Dx_D \quad (1.3.120)$$

The LHS of Equation (1.3.120) is the difference in the rate of flow of component A, up-down, *i.e.*, net flow of A upward. For a given distillation, as the RHS of Equation (1.3.120) is constant, it follows that the net rate of flow of component A upward is constant irrespective of plate number in this section and is always equal to A which is permanently withdrawn from the top.

$$D = V_{n+1} - L_n$$

$$V_{n+i}\, y_{n+i} = L_n X_n + (V_{n+1} - L_n)\, x_D$$

$$\frac{L_n}{V_{n+1}} = \frac{x_D - y_{n+1}}{x_D - x_n}$$

Similarly, the enthalpy balance with negligible heat loss is

$$V_{n+i}\, H_{yn+1} = L_n H_n + Qc + DH_D$$

Let $Q'$ heat removed in the condenser and heat associated with the distillate removed per mole of distillate.

$$Q' = \frac{Q_C + D\,H_D}{D} = \frac{Q_C}{D} + H_D$$

$$V_{n+1} H_{Vn+1} = L_n H_{Ln} = DQ' \tag{1.3.121}$$

The LHS of Equation (1.3.121) is the difference in rate of flow of heat, up-down, *i.e.*, the net rate of heat flow upward. For given conditions, the RHS of Equation (1.3.126) is constant, so the net rate of heat flow upward is constant irrespective of plate number in this section. The net rate of heat flow upward is equal to the heat taken out permanently with the distillate and at the condenser.

$$V_{n+1} H_{Vn+1} = L_n H_{Ln} + (V_{n+1} - L_n)\, Q'$$

$$V_{n+1} Q' - H_{Vn+1}) = L_n (Q' - H_{Ln})$$

$$\frac{L_n}{V_{n+1}} = \frac{Q' - H_{vn+1}}{Q' - H_{Ln}} \tag{1.3.122}$$

So combining Equations (1.3.126) and (1.3.127), we get

$$\frac{L_n}{V_{n+1}} = \frac{x_D - y_{n+1}}{x_D - x_n} = \frac{Q' - H_{vn+1}}{Q' - H_{Ln}} \tag{1.3.123}$$

So Equation (1.3.123) on the enthalpy-composition diagram, *i.e.*, on the Hxy diagram (refer Robert E. Treybal 1981. Mass transfer operations, McGraw Hill) is a straight line passing through points $V_{n+1}(H_{vn+i}\, y_{n+1)}$ $L_n$ $(H_{Ln}, x_n)$ and $\Delta_D(Q, x_D)$.

$\Delta_D(Q', x_D)$ is called as a difference point as its coordinates represent the difference in rates of flow:

$Q'$ = net heat out/net moles out

$x_D$ = net moles A out/net moles out

$\Delta_D$ is a fictitious stream with amount equal to the difference between vapour and liquid streams between two plates, of composition $x_D$ and of enthalpy $Q'$.

$V_{n+1} - Ln = \Delta_D$ (1.3.124)

On the xy diagram, Equation (1.3.123) is the equation of a straight line of slope $L_n/V_{n+1}$ and passing through points $(y_{n+1}, x_n)$ and $(x_D, x_D)$. In case of a total condenser, the distillate D and reflux $L_0$ are having the same composition and enthalpy content, *i.e.* on the Hxy diagram they are having the same coordinates $(x_D, H_D)$. The point D indicates that both the distillate and reflux are below the bubble point (*i.e.* they are subcooled liquids). When both D and $L_0$ are at the bubble point (*i.e.* if only the latent heat is removed in the condenser) then the point D will be on the saturated liquid curve ($H_L$ v/s x). Locate point AD on the Hxy diagram with its coordinates $(x_D, Q')$- Extend a line through points D, $V_1$ upward and on this line mark $A_D$ at which $H = Q'$. The points $A_D$, D and $V_x$ are on the same line as they are having the same composition $(x_D)$. When the saturated vapour $V_1$ issuing from plate-1 to the condenser is condensed completely then it has the same composition as that of D and $L_0$. The saturated vapour $V_1$ is shown on the saturated vapour curve (*i.e.* $H_v$ v/s y) with its coordinates ($x_D$, $H_{V1}$ - enthalpy content of vapour). The liquid $L_1$ leaving plate-1, which is in equilibrium with Vi, is located on the bubble point curve/saturated liquid curve at the end of a tie line-1. For drawing a tie line; from Vi, drop a vertical on the xy diagram to cut the diagonal at point A then proceed horizontally towards the equilibrium curve to cut it at B and then draw a vertical through B to cut $H_L$ v/s x curve at $L_1$. Join $V_1$ $L_1$ - $V_1L_1$ is the required tie line. As Equation (1.3.123) is applicable to any plate in the rectifying section, $V_2$ can be located on the saturated vapour curve by drawing a line from $L_1$ to $D_D$. The three quantities $V_{n+1}$, $L_n$ and $A_D$ will always lie on the straight line through $A_D$. The tie line through $V_2$ locates point $L_2$ on the saturated liquid curve. The point P is produced and all such points produce the operating curve that passes through $y = x = x_D$. The difference point D is used in this way for further construction till the feed plate is reached.

We have, $$V_{n+1} = L_n + D$$

and $$\frac{L_n}{V_{n+1}} = \frac{Q'-H_{vn+1}}{Q'-H_{Ln}} = \frac{x_D - y_{n+1}}{x_D - x_n}$$

$$1+\frac{D}{L_n} = \frac{Q'-H_{Ln}}{Q'-H_{vn+1}}$$

$$\frac{D}{L_n} = \frac{Q'-H_{Ln}}{Q'-H_{vn+1}} - 1$$

$$\frac{D}{L_n} = \frac{H_{vn+1}-H_{Ln}}{Q'-H_{vn+1}} -$$

$$\frac{L_n}{D} = \frac{Q'-H_{Ln}}{H_{vn+1}-H_{Ln}} \qquad (1.3.125)$$

When the above equation is applied at the top plate, we get the external reflux ratio.

$$R = \frac{L_0}{D} = \frac{Q' - H_{v1}}{H_{v1} - H_{Lo}} = \frac{line\,length\ \Delta_D V_1}{line\,length\ V_1 L_0}$$

$$= \frac{line\ length\ \Delta_D V_1}{line\,length\ V_1 D} \tag{1.3.126}$$

For a given reflux ratio, the line-length ratio given by Equation (1.3.126) can be used to locate the difference point $A_D$ and the y-coordinate of $A_D$ can then be used to obtain the heat load of the condenser.

### 1.3.6.5.4.1. *Partial Condenser*

When a partial condenser is used, a distillate is withdrawn in the vapour state (as a saturated vapour) and the condensed liquid provides the reflux. Only some part of the overhead vapour leaving the top plate is condensed to provide the reflux and the remainder of it is removed as the distillate. In the case of a partial condenser, $\Delta_D$ is plotted at an abscissa $y_{D,}$ where $y_D$ is the composition of the vapour product/distillate withdrawn. Normally, the vapour leaving the condenser as the product is in equilibrium with the liquid reflux, the liquid reflux $L_0$ will be at the end of the tie line C. $V_1$ is located by the construction line $L_0\ \Delta_D$. As each of reboiler and partial condenser acts as one theoretical stage, the number of theoretical plates required in the column will be (n + 2) - 2 (*i.e.* the number of tie lines (between $y_D$ and $x_w$) – 2).

The reflux ratio (external reflux ratio)

$$= R = L_o / D = \frac{line\ length\ \Delta_D V_1}{line\ length\ \ V_1 L_o}$$

*The Stripping Section*: For the complete section below plate m, where m is any plate in the stripping section, the overall material balance is

$$L'_m = L'_{m+1} + W \tag{1.3.127}$$

Material balance of component A is:

$$L'_m\ x_m = V'_{m+1}\ y_{m+1} + Wx_w \tag{1.3.128}$$

The LHS of Equation (1.3.128) is the difference in rate of flow of component A, down-up *i.e*, the net rate of flow of component A downward. As the RHS of Equation (1.3.128) is constant for a given distillation condition, the difference in rate of flow is independent of plate number so it is the same for any plate in this section and is equal to the rate of removal of A as the bottom product.

An enthalpy balance over the same portion of the column is

$$L'_m H_{Lm} + Q_B = V_{m+i}\ H_{Vm+1} + W\ H_w \tag{1.3.129}$$

Let Q" be the net flow of heat out outward at the bottom per mole of residue/bottom product.

$$Q'' = \frac{WH_w - Q_B}{W} = H_w - Q_B / W \tag{1.3.130}$$

With Equations (1.3.129) and (1.3.130), we get

$$L'_m H_{Lm} - V'_{m+i} H_{vm+i} = WQ" \quad (1.3.131)$$

The LHS of Equation (1.3.131) is the difference in rate of heat flow, down-up. This Equation is applicable for any plate in the column and the LHS is equal to the net rate of heat flow out at the bottom. Let us eliminate W from Equations (1.3.127) and (1.3.128):

$$L'_m = V'_{m+1} + W$$

$$L'_m x_m = V'_{m+1} y_{m+1} + W x_w$$

$$L'_m x_m = V'_{m+1} + y_{m+1} + (L'_m - V'_{m+1}) x_w$$

$$L'_m x_m = V'_{m+1} y_{m+1} + L'_m - x_w - V'_{m+1} x_w$$

$$\frac{L'_m}{V'_{m+1}} = \frac{y_{m+1} - x_w}{x_w - x_w} \quad (1.3.132)$$

We have,

$$L'_m = V'_{m+1} + W$$

$$L'_m H_{Lm} - V'_{m+i} H_{Vm+i} = WQ''$$

Eliminating W between these two equations gives

$$L'_m H_{Lm} - V'_m H_{Vm+1} = (L'_m - V'_{m+1)} Q''$$

$$L'_m H_{Lm} - V'_m H_{Vm+1} = (L'_m Q'' - V'_{m+1)} Q''$$

$$\frac{L'_m}{V'_{m+1}} = \frac{H_{vm+1} - Q''}{H_{Lm} - Q''} \quad (1.3.133)$$

Combining Equations (1.3.129) and (1.3.130) gives

$$\frac{L'_m}{V'_{m+1}} = \frac{H_{vm+1} - Q''}{H_{Lm} - Q''} = \frac{y_{m+1} - x_w}{x_m - x_w} \quad (1.3.134)$$

Equation (1.3.134) is a straight line on the Hxy diagram passing through points $V_{m+i}$ $(H_{V\,m+i}, y_{m+1})$ Lm, $(H_{Lm}, x_m)$ and $\Delta_w$ $(Q'', x_w)$.

$\Delta_w$ is a difference point with coordinates $(Q'', x_w)$.

$Q''$ = net heat out/net moles out, at the bottom

xy = moles A out/net moles out, at the bottom

$\Delta_w$ is a fictitious stream which is equal to the net flow outward (W) of properties $(Q'', x_w)$ (*i.e.* of composition $x_w$ and of enthalpy $Q''$).

$$L'_m - V'_{m+1} = \Delta w$$

Equation (1.3.134) is a straight line passing through points $(y_{m+i}, x_m)$ and $(x_w, x_w)$ with a slope of $L'_m/V'_{m+1}$. Equation (1.3.134) is applicable to all plates in the stripping section. On the Hxy diagram (refer Robert E. Treybal, 1981. Mass transfer operations, McGraw Hill) locate point W on the saturated liquid curve at ($x_w$, H,w) and point $\Delta_w$ at ($x_w$,- Q″). W and Δw are located on the same line as they have the same composition $x_w$. All the operating lines represented by Equation (1.3.134) pass through $D_w$ (the difference point). The line from $V'_{m+1}$ (representing the vapour leaving the reboiler and entering the plate at the bottom of the column) to $\Delta_w$ intersects the saturated liquid curve at point $L'_{m'}$, the liquid leaving the bottom plate. The vapour $v'_M$ leaving the bottom plate is in equilibrium with the liquid $L'_m$ and is at the end of the tie line $n_M$. The procedure of construction of equilibrium stages similar to that described for the rectifying section is applied here. The number of ideal plates is determined by alternating construction lines to $\Delta_w$ and tie lines, each tie line represents an equilibrium stage.

The complete firactionator:

Overall material balance over the fractionater is

$$F = D + W \tag{1.3.135}$$

Overall material balance of the component A is

$$Fx_F = Dx_D + Wx_w \tag{1.3.136}$$

Overall enthalpy balance for negligible heat losses over the fractionator is

$$Q_B = D\,H_D + W\,H_w + Q_C - F\,H_F$$

We have,

$$Q' = (Q_C/D) + H_D$$

$$Q'' = H_W - (Q_B/W)$$

$$Q_B = (H_W - Q'')\,W$$

and

$$Q_c = (Q' - H_D)D$$

$$(H_W - Q'')\,W = DH_D + WH_W + (Q' - H_D)\,D - FH_F$$

$$W\,H_w - Q''\,W = D\,H_D + W\,H_w + Q'\,D - H_D\,D - F\,H_F$$

$$FH_F = DQ' + WQ'' \tag{1.3.137}$$

Eliminating F between Equations (1.3.135) and (1.3.136) gives

$$x_F\,(D + W) = D\,x_D + W\,x_w$$

$$x_F\,D + x_F\,W = D\,x_D + W\,x_w$$

$$W\,(x_F - x_w) = D\,(x_D - x_F)$$

$$D/W = \frac{x_F - x_w}{x_D - x_F} \tag{1.3.138}$$

Eliminating F between Equations (1.3.135) and (1.3.136) gives

$$(D + W)\,H_F = DQ' + WQ''$$

$$DH_F + WH_F = DQ^1 + WQ''$$

$$W(H_F - Q'') = D(Q' - H_F)$$

$$\frac{D}{W} = \frac{H_F - Q''}{Q' - H_F} \tag{1.3.139}$$

Combining Equations (1.3.138) and (1.3.139), we get

$$D/W = \frac{x_F - x_w}{x_D - x_F} = \frac{H_F - Q''}{Q' - H_F} \tag{1.3.140}$$

On the Hxy diagram (refer robert E. Treybal, 1981. Mass transfer operations, McGraw Hill). Equation (1.3.140) is a straight line passing through point $\Delta_D$ ($Q'$ $x_D$), F ($H_F$, $x_F$), and $\Delta_w$ ($Q''$, $x_w$). The location of point F ($x_F$, $H_F$) indicates that the feed is a liquid below its bubble point (cold feed). In the case of a saturated liquid, the feed point F will lie on the bubble-point curve/saturated liquid curve and in the case of a saturated vapour feed (feed-saturated vapour), F will lie on the saturated vapour curve/dew-point curve. In any case, note that $\Delta_D$ ($x_D$, $Q'$), $\Delta_w$ ($x_w$, $Q''$) and F ($x_F$, $H_F$) must lie on a single straight line.

**1.3.6.5.4.2. *Minimum Reflux Ratio*:** At minimum reflux ratio, requiring an infinite number of plates to achieve the desired degree of separation, both the heat requirement of the reboiler and the cooling water requirement of the condenser are minimum. For the lowest possible heat load of the reboiler and the condenser cooling load, on the Hxy diagram (refer robert E. Treybal, 1981. Mass transfer operations, McGraw Hill). The tie line passes through F ($x_F$, $H_F$). The tie line passing through F intersects a vertical line through $x = x_D$ at $\Delta_{Dm}$, which gives the value of ordinate $Q'_m$. Once $Q'_m$ is determined, the minimum reflux ratio, $R_m$, is determined, the minimum reflux ratio, $R_m$, is calculated with the help of the following equation:

$$R_m = \frac{\text{line length } \Delta_D V_1}{\text{line length } V_1 D} = \frac{Q'_m - H_{v1}}{H_{v1} - H_D} \tag{1.3.141}$$

Locating tie line through F: Through F (on the saturated liquid curve for feed at its bubble point) drop a vertical line of the xy diagram which will cut the equilibrium curve at A. Through A, move horizontally to the diagonal y = x at point B ($y_F$) and then through B on the diagonal, draw a vertical to cu the saturated vapour curve at $V_F$. Join F - $V_F$ and extend it to meet the vertical line through

$x = x_D$ at $\Delta_{Dm}$ and read the ordinate Q: of $D_D$

Take any value of x on $H_L$ v/s x, go to the equilibrium curve to get y, proceed vertically on $H_v$ v/s y and see that F, x and y lying on the same straight line or not. Repeat the procedure till they lie on the same line. Here F, x' on $H_L$ v/s x and $y^1$ on Hy v/s y lie on the same straight line (refer robert E. Treybal, 1981. Mass transfer operations, McGraw Hill).

***Total Reflux Ratio*:** At total reflux ratio, requiring minimum number of plates to achieve the desired degree of separation, the head load of the reboiler and the cooling load of the condenser are infinite. The total reflux operation is of interest only because that it sets a lower limit on the number of plates required and is not a practical method of operation as at total reflux the product rate is zero. At total reflux, $\Delta_D$ and $\Delta_w$ points are at infinity (∞) and consequently the lines joining points $L_n$ ($x_n$, $H_{Ln}$) and $D_D$ as well as points ($x_m$, $H_{Lm}$) and $\Delta_w$ are all vertical. In the fractionation of a water solution to obtain the more volatile non-aqueous component as the distillate and where water is removed as the bottom product, the heat required can be provided by introduction of steam directly to the bottom of the column, thus eliminating the need of a reboiler-normally used.

The material balance over the entire column is:

$$F + S = D + W \tag{1.3.142}$$

The material balance of the component A is

$$x_F F + y_s S = x_D D + x_w W \tag{1.3.143}$$

The steam is free of A, so

$$y_s = o$$

$$x_F F = x_D D + x_w W$$

The energy/enthalpy balance over the entire column is

$$FH_F + SH_S = DH_D + WH_W + Q_C$$

$$FH_F + SH_S = D(H_D + Qc/D) + W H_w \tag{1.3.144}$$

$H_s$ is the enthalpy of steam used and S is the molar rate of steam introduced. On the Hxy diagram, the point $\Delta_D (x_D, Q')$ is located in the usual manner as the rectifying section of the column is unaffected by the use of open steam and all equations previously derived for this section are all applicable here.

Here $\Delta_w$ has its usual meaning which is nothing but the net flow outward.

$$\Delta_W = L'_M - V'_{M+1} = W - S \tag{1.3.145}$$

and its coordinates are:

$$x\Delta_w = \text{net moles A out/net moles out} = W\, xw/(W - S)$$

$$Q'' \frac{\textit{net heat out}}{\textit{net moles out}} = \frac{W\,Hw - S\,Hs}{W - S}$$

where $H_s$ is the enthalpy of the steam used.

We have:

$$L'_m - V'_{m+1} = D_w$$

$$L'_m x_m - V'_{m+1} y_{m+1} = D_w \,.\, xw$$

$$L'_m H_{LM} - V'_{m+1} H_{v\,m+1} = D_w \,.\, Q''$$

$$\frac{L'_m}{V'_{m+1}} = \frac{y_{m+1} - x_{\Delta w}}{x_m - x_{\Delta w}} = \frac{H_{m+1} - Q''}{H_{Lm} - Q''} \tag{1.3.146}$$

m may be any plate in the stripping section.

A graphical construction for the number of theoretical plates required will be the same as that previously described for the section above the feed plate (rectifying section). From the section below the feed plate (stripping section), the only difference is that a point $A_w$ with coordinates $x\Delta_w$, $Q''$ is used for determining point ($y_{m+i}$, $Hy_{m+i}$) corresponding to the point ($x_m$, $H_{Lm}$).

**1.3.6.5.4.2. *Determination of Amount of Steam Used*:** The amount of steam used can be determined by means of Equations (1.3.142), (1.3.143) and (1.3.144).

Let $M = D + S$

$$Mx_M = Dx_D + Sy_s \tag{1.3.147}$$

$$MH_M = D\,(H_D + (Qc/D) + W\,HW \tag{1.3.148}$$

So the point M ($x_M$, HW) must lie on a straight line joining the points W ($x_w$, $H_w$) and $\Delta D$(XD, Q′). From Equations (1.3.142), (1.3.143) and (1.3.144), it is clear that the point M must also lie on the straight line through points ($y_s$, $H_s$) and F ($x_F$, $H_F$). These points are known. Hxy diagram (refer robert E. Treybal, 1981. Mass transfer operations, McGraw Hill) the steam used is shown slightly superheated (point S). S will be at R for the saturated steam at a given pressure. The intersection of the two lines S - F and W ~ $A_D$ determines the point M ($x_M$, $H_M$). Furthermore,

$$\frac{S}{F} = \frac{H_M - H_F}{H_s - H_m} = \frac{x_M - x_F}{y_s - x_m} \tag{1.3.149}$$

All quantities in Equation (1.3.149) can be known except S, so that S can be determined. The vaporisation efficiency is defined as:

$$\eta_v = p_A / P_A^0$$

The vaporisation efficiency is rarely unity. Therefore, the steam requirement for atmospheric steam distillation should be computed from the following equation:

$$\frac{W_B}{W_A} = \left(\frac{101.325 - \eta_v p_A}{\eta_v p_A}\right)\frac{M_B}{M_A} \tag{1.3.150}$$

## 1.3.7. *Types of Distillation Column and its Characteristics*

***Distillation Towers/Columns*:** Distillation (fractionation) is the most widely used mass transfer operation in the chemical industry. In case of this operation, the gas-liquid contacting devices for mass transfer are: plate columns and packed columns.

A column is also known as a tower. It is essentially a tall vertical cylindrical shell, with number of nozzles. The column consists of a series of plates or trays or variety of packings (usually random packings).

Based upon the contacting devices used for mass transfer to take place, the towers/columns are divided into two types namely, plate or tray columns and packed columns.

### 1.3.7.1. *Plate Columns*

The process of distillation with rectification makes use of either a plate (tray) column or packed column, for intimate phase contacting, the former being more common. In the plate columns, the process of mass transfer is stage wise and in packed towers/columns, the process of mass transfer is continuous. Each plate in a vertical column represents a stage as on each plate, phases (gas and liquid) are brought into intimate contact, mass is transferred from one phase to another (to and from both the phases) and finally phases are separated and leave the plate/tray in equilibrium (ideal case). A plate column is a vertical column divided into sections by a number of horizontal plates. In this column the liquid enters the plate from the top, flows over the plate and then from plate to plate via downcomers. A certain height of a liquid pool is maintained on each plate with the help of outlet weir. Gas/vapour enters in the column from bottom inlet and flows upward through the opening in the plate, it bubbles through the pool of liquid on the plate, disengages from liquid and passes to the plate above. In this way, vapour and liquid are brought into intimate contact on the every plate.

*Various plates used for contacting of phases are*: (1) Bubble cap plate/tray (2) Sieve plate and (3) Valve plate.

In plate columns, the minimum plate spacing is 300 mm. For columns of diameter 1.5 m or more, spacing of 450 or 600 mm is used for easier tray access. In plate columns, plates are provided with downcomer and weirs. The function of the downcomer is to provide a passage for the downward flow of the liquid from a plate above to the plate below. The downcomer area for the each plate is limited to 10% of the total area of the plate. Pipe downcomer, segmental downcomer and chord type downcomer are various types of downcomers that can be used. A depth of liquid on the plate required for gas contacting is maintained by an overflow (outlet) weir. An inlet weir helps to distribute liquid as it enters the plate from the downcomer and prevents impingement of liquid on a contacting device.

**1.3.7.1.1.** ***Bubble Cap Plate (Tray)*:** In case of bubble cap tray, gas/vapour flows through a riser, reverses flow under cap, moves downward through the annular space between the riser and the cap, and finally it bubbles into a pool of liquid through a series of slots (openings) provided along the periphery of the cap at the bottom. The slots may be circular, rectangular, or trapezoidal in shape. Bubble cap disperses the gas phase into the liquid as the fine bubbles, it prevents the liquid drainage through the gas passage at low gas rates. It directs the gas flow first horizontally and then vertically upward through the pool of liquid in the form of bubbles. The bubble cap columns can be operated with high and constant stage efficiencies over a wide range of the vapour and liquid flow rates. Bubble cap plates are now rarely used for new installations because of (i) their relatively high cost and (ii) high pressure drop.

**1.3.7.1.2.** ***Sieve Plate*:** Very common plates in use in recent years are the perforated plates because of their (i) simplicity and (ii) low cost. A sieve plate is the perforated plate employing the perforations for dispersing gas into liquid on the plate. It is a metal sheet having hundreds of round perforations in it of the size ranging between 3 to 12 mm (5 mm is a common size. The total area of holes ranges from 5 to 15% of the plate area. In the sieve plate column, liquid flows across the plate and gas passes vertically upward through the holes, in the form of bubbles through the liquid on the plate. Gas flow through the perforations prevents the down-flow of liquid through perforation in the plate but at low gas flow rates weeping (draining of the liquid through perforations in plate) is severe which in turn reduces stage efficiencies, and this is a limitation of the sieve plate. Gas-liquid contact is poor with the sieve plate as compared to that with bubble cap plate because in the sieve plate gas is directed vertically and in the bubble cap plate gas is first directed horizontally and then moves vertically upward. The pressure drop with sieve plate is low as compared to it with bubble cap columns. This is the most common type of plate (or tray) contacting device. It is usually operated over a wide range of flow rates between weeping and flooding points.

**1.3.7.1.3. *Valve Plate*:** It is the recent development in the perforated plates. A valve plate is nothing but a perforated metal sheet wherein the perforations are covered with liftable caps (valves). The valves are metal discs and are held in the plate by means of legs which restrict the upward motion of caps. The valve provides variable orifices for vapour flow *i.e.* opening in the plate will be small at the low vapour rate and large at high vapour/gas rate. Thus, weeping will be greatly reduced with this plate at low gas rates. Valve plates can be operated over a wide range of phase flow rates. Gas-liquid contact with the valve plate is much better than it with sieve plate as liftable caps direct the gas horizontally into the liquid on the plate while in sieve plate gas passes vertically upward through the liquid. The cost of the valve plate is in between that of bubble cap and sieve plate, approximately twenty per cent higher than that of the sieve plate.

### *Liquid flow patterns over plate/tray:*

**1.3.7.2. *Packed Columns*:** For distillation operation, packed columns are especially useful whenever we have to carry out operation at low pressure (vacuum distillation) and whenever we are dealing with heat sensitive materials. Packed columns are cheaper than plates for column diameter less than 600 mm. A packed column consists of a cylindrical shell filled with some sort of packings (rings or saddles) which offer a large interfacial area for mass transfer. A common arrangement of the packed column for distillation operation, wherein the vapour flows upward and the refluxed liquid flows downward giving a continuous contact. The ability of a given packing to effect the desired mass transfer between gas and liquid phases (*i.e.* to effect the desired degree of separation) is usually expressed as the height equivalent to one theoretical plate (HETP). In plate columns wherein a process of enrichment is stage wise, the vapour leaving the plate is richer with respect to more volatile component than the vapour entering the plate by one equilibrium stage. In packed columns, the same enrichment of the vapour will occur in a certain height of packing which is termed as the height equivalent to one theoretical plate or height equivalent to a theoretical plate. Thus, in packed columns, one equilibrium step is represented by a certain height of a packed bed and the required height of packing for the desired degree of separation is given by: HETP x number of ideal plates required. HETP is defined as the height of packing (packed bed) giving the enrichment of vapour same as that can be obtained by one theoretical plate (in stage-wise operation). HETP can be estimated with the help of the following empirical equation such as

$$HETP = k_1 - G_t^{k2} \bullet D_t^{k3} \bullet Z^{1/3} \bullet \alpha \bullet \mu_T/\rho_L \quad (1.3.151)$$

where $k_1$, $k_2$ and $k_3$ are empirical constants for packing and are function of type and size of packing, G - Superfacial gas mass velocity, $D_t$ - Tower diameter, Z - Height of packing

$\alpha$ - Relative volatility

$\mu_L$ and $\rho_L$ are viscosity and density of liquid respectively

HETP is an experimental quantity, a characteristic of each packing and depends on the type and size of the packing, flow rates of each fluid and concentration.

### 1.3.8. LIQUID FLOW PATTERN

The type of liquid flow pattern over a plate depends upon liquid to gas flow ratio. Various flow patterns are:

**1.3.8.1. *Cross flow*:** It is most frequently used. Provides a good length of the liquid path. Here the liquid flows across a plate and then falls to the plate below through a passage at the other side of the plate.

**1.3.8.2. *Reverse flow*:** Provides a very long liquid path especially suitable for low liquid to gas ratios.

**1.3.8.3. *Split flow*:** Especially suitable for handling high liquid to gas ratios. Here the liquid flowing across the plate is splitted into at least two paths.

**1.3.9. *Downcomer and weir*:** A downcomer is the passage for liquid flow from a plate above to the plate below which may be in the form of circular pipes or segments of the tower/column set aside (for liquid flow) by vertical plates (partitions). The discharge end of the downcomer is generally projected far enough into the liquid on the plate so as to avoid vapour (gas) from rising up the downcomer to short-circuit the plate above. For this purpose, inlet weir may be used. A exit weir or outlet weir is used for maintaining a certain depth of liquid (liquid pool) on each plate needed for the gas contacting. Straight rectangular weirs with length equal to 60 to 80% of the tower diameter are frequently used.

**1.3.10. *Entrainment*:** As gas/vapour bubbles through the pool of liquid on a plate, a large amount of turbulence is set up and the liquid particles can be entrained by the rising gas. Carry over of liquid particles by the rising gas from a plate below to the plate above is known as en trainment and is much more pronounced at high gas flow rates. Entrainment reduces the concentration change per plate and hence, decreases the plate efficiency. Tray spacing vapour flow rates, depth of liquid on the plate are the factors on which the amount of entrainment depends.

**1.3.11. *Method of Transfer Units:*** Fractionation by means of a packed column involves continuous countercurrent contact of the vapour and liquid whereas fractionation by means of a plate column involves stagewise contact of the vapour and liquid. So in the packed column there is a continuous change in the composition of the components and in the plate column there is a discontinuous change in the composition of components (affected at each plate). The theoretical unit of separation in a packed column is the transfer unit and not the theoretical plate (as it in a plate column). The efficiency of a packing consequently can be measured as the height of packing to give a separation of one transfer unit (height of transfer unit). The height of a packed column can be based on the number of transfer units and the height of transfer unit. The height of a packed column can easily be obtained by multiplying the number of transfer units by the height of transfer unit.

$$Z = \text{Number of transfer units x height of transfer unit}$$

The number of transfer units is a measure of the ease or difficulty of separation. More difficult the separation, the more such units would be required. Hence, if very high purity products are desired or for a high degree of separation, a large number of transfer units will be required. The height of transfer unit is a measure of the separation effectiveness of a particular packing for chemicals being processed. If the rate of mass transfer is high and the surface area for mass transfer is large then the height of transfer unit is small. For determination of the height of transfer unit, we require a knowledge of mass transfer coefficients in units consistent with those in which the composition of phases is expressed.

Consider a packed column in which the composition of the more volatile component increases from $y_b$ to $y_t$. Let A be the empty column cross-sectional area, Z be the height of packed section of a column, 'a' be the active interfacial surface per unit volume of packing (interfacial area per unit volume of packing). Consider a differential element of height dZ, through which two phases pass countercurrently, wherein a differential change in vapour composition, dy, takes place, At the same time, a differential change in liquid composition is dx. For steady-state operation, the material balance of the more volatile component A over the differential height dZ assuming constant molar overflow *i.e.* neglecting change in molar flow rate in any particular section is

$$V\,dy = L\,dx \qquad (1.3.152)$$

Where V is the vapour flow rate in the enriching section of the column in moles/time and L is the molar flow rate of the liquid in mole/time. V is constant in the entire enriching section. It will be also constant in the stripping section but of a different magnitude depending upon the feed conditions.

Each term of Equation (1.3.152) is the indication of moles of A diffusing *i.e.* the rate of addition or depletion of A to the vapour phase or from the liquid phase and is a statement of the rate of transfer of A across the transfer area in dZ. This must be equal to the total driving force divided by the total resistance.

So the rate of change of the component A within the phase must be equal to the rate of transfer to the phase.

Thus, for V-phase (vapour phase):

$$V\,dy = k_G\,(y_i - y)\,dS = K_G\,(y^* - y)\,dS \tag{1.3.153}$$

where dS is the interfacial transfer area associated with the differential column height dZ.

dV = differential volume of packing in height dZ = AdZ

dS = differential interfacial area for mass transfer = aAdZ in height dZ

$$V\,dy = k_G\,(y_i - y)\,aA\,dZ = K_G\,(y^* - y)\,aA\,dZ$$

$$Vdy = k_G\,a\,(y_i - y)\,AdZ = K_G\,a\,(y^* - y)\,A\,dZ \tag{1.3.154}$$

where (yi - y) is the driving force over the gas film. (y*- y) is the overall driving force for the gas phase/vapour phase - (overall driving force in terms of vapour composition). $K_G$ is the overall gas phase mass transfer coefficient in mol/(time)- (unit driving force). $K_Ga$ is the volumetric overall mass transfer coefficient in mol/ (volume) (time) (unit driving force).

In distillation operation, vapour and liquid compositions are expressed in terms of molefraction, so the units of $K_Ga$ will be mol/(volume) (time) (molefraction). Use of $K_Ga$ requires the gas phase composition (y*) in equilibrium with the bulk liquid composition and use of $K_Ga$ requires the gas phase composition ($y_i$) at the interface. It is possible only to measure the bulk composition and determination of interface composition is ordinarily impossible.

$$V\,dy = K_G\,a\,(y^* - y)\,A\,dZ \tag{1.3.155}$$

$$(V/A)\,dy = K_G\,a\,(y^* - y)\,dZ$$

$$dZ = \frac{V/A}{K_G\,a}\left[\frac{dy}{y^* - y}\right]$$

Integrating between limits,

$$\int_0^z dZ = \int_{yb}^{yt} \frac{V/A}{K_G\,a}\left(\frac{dy}{y^* - y}\right)$$

The molar flow rates (V and L) are assumed to be constant in the enriching section. The overall mass transfer coefficient is a function of these phase flow rates. If the physical properties do not vary appreciable, $K_G$ is also constant.

$$\int_0^z dZ = \frac{V/A}{K_G\,a}\int_{yb}^{yt}\left(\frac{dy}{y^* - y}\right)$$

$$Z = \frac{V/A}{K_G\,a}\int_{yb}^{yt}\frac{dy}{y^* - y} \tag{1.3.156}$$

In Equation (1.3.156), the integral term represents the integrated ratio of the change in vapour composition to the available driving force (in terms of vapour composition) tending to bring this about. This term is defined by Chilton and Colburn as the number of transfer units (Nt). In our present case, Nt is $Nt_{QG}$.

$$Nt_{OG} = \int_{yb}^{yt} \frac{dy}{y^* - y} \tag{1.3.157}$$

$Nt_{OG}$ is the number of overall gas transfer units (*i.e.* number of transfer units based on gas phase). The quantity outside the integral of Equation (1.3.156) is called the height of a transfer unit (Ht). In present case, Ht is $Ht_{OG}$.

$$Ht_{OG} = \frac{V / A}{K_G a} \tag{1.3.158}$$

Thus, Equation (1.165) can be written as

$$Z = Ht_{OG} . N_{toG} \tag{1.3.159}$$

A similar series of equations can be written in terms of composition of the more volatile component in liquid phase. The column height can be evaluated using changes in composition of A in liquid phase and corresponding rate.

$$Vdy = K_G a(y^*-y)A\,dZ = L\,dx = K_L a\,(x - x^*)A\,dZ \tag{1.3.160}$$

$$Ldx = K_L a\,(x - x^*)\,A\,dZ \tag{1.3.161}$$

Where x - x* is the overall driving force in terms of liquid phase composition.

$K_L$ a is the overall volumetric mass transfer coefficient for liquid phase in mol /(time) • (volume) • (unit driving force), KL is the overall liquid phase mass transfer coefficient in mol/ (time) • (area) • (unit driving force).

So for liquid phase, we can write,

$$Nt_{OG} = \int_{x_b}^{x_t} \frac{dx}{x - x^*} \tag{1.3.162}$$

$N_{tOL}$ is the number of overall liquid phase transfer units (number of transfer units based on liquid phase).

$$H_{tOL} = \frac{L / A}{K_L\, a} \tag{1.3.163}$$

$H_{tOL}$ is the height of an overall liquid transfer unit.

$H_{toG}$ and $H_{toL}$ are having dimension of length while $N_{toL}$ and $N_{tOG}$ are dimensionless.

**1.3.12. *Plate Efficiency*:** By using either the McCabe-Thiele method or the Ponchon-Savarit method, we obtain the number of ideal/theoretical/equilibrium stages required for achieving the desired degree of separation. In case of an ideal stage, the vapour leaving a plate is in equilibrium with the liquid which leaves it. In actual practice, complete equilibrium is never reached on any plate. This is due to short contact time and incomplete mixing on any plate. The vapour leaving an actual plate is usually weaker in the more volatile component than the vapour in equilibrium with the liquid leaving it. Thus, for achieving the desired degree of separation in actual practice, we require more number of plates in a column than the ideal plates, To apply either of the methods to an actual case, it is necessary to convert the number of ideal plates to the actual plates. The factor that is used for converting the ideal plates into the actual plates is the plate efficiency. The reboiler delivers vapour directly from the boiling liquid and is thus considered as a perfect plate.

Types of plate efficiency:

(i) ***Overall efficiency/overall plate efficiency*** - pertains to the entire column.

(ii) ***Murphree efficiency*** - pertains to any single plate.

(iii) ***Local/point efficiency*** - pertains to a specific location on a given plate.

**1.3.12.1.** ***Overall efficiency*** is the ratio of the number of ideal plates (theoretical plates) required to effect a given degree separation to the number of actual plates required to effect the same separation.

$$\eta_M \frac{\text{theoretical / ideal plates required}}{\text{actual plates required}} \times 100$$

For example, if we need six theoretical plates to effect the desired separation and the overall efficiency is 60% then the actual number of plates required are 6/0.60 = 10. The overall efficiency may vary from 30 to 100 percent.

### 1.3.12.2. *Murphree efficiency*

It is the efficiency of any single plate based on the vapour phase composition. For the n[th] plate in the column, the Murphree efficiency is given by

$$\eta_M = \frac{y_n - y_{n+1}}{y^*_n - y_{n+1}} \tag{1.3.164}$$

where, $y_n$ is the average composition of the vapour leaving plate.

$y_{n+i}$ is the average composition of the vapour entering plate

$y_n{}^*$ is the composition of vapour in equilibrium with liquid leaving plate.

It applies to any individual plate in a column and is defined as the actual change in the average vapour composition achieved by a given plate divided by the change in the average vapour composition if the vapour leaving the plate were in equilibrium with the liquid leaving the plate. This efficiency is less convenient to use as it involves plate to plate compositions rather than the terminal compositions.

**1.3.12.3.** ***Point efficiency*:** It pertains to a particular place on the plate surface (applies to single location on a given plate). It is defined by

$$\eta_M = \frac{y_n, local - y_{n+1}\ local}{y^*_{local} - y_{n+1},\ local} \tag{1.3.165}$$

where $y_{n,}$ local is the composition of vapour leaving specific location on plate n, $y_{n+i}$, local is the composition of vapour entering plate n at the same location, $y^*_{local}$ is the composition of vapour in equilibrium with the liquid at the same location (*i.e.* in equilibrium with $x_{local}$). To obtain satisfactory plate efficiencies, it is highly necessary that the plates must operate properly. Adequate and intimate contact between vapour and liquid is essential. Each of the unfavourable phenomena such as entrainment, excessive foaming, weeping, short-circuiting and flooding lower the plate efficiency.

## 1.4. LIQUID-LIQUID EXTRACTION

### 1.4.1. *Introduction*

The separation of the components of a liquid mixture by treatment with a solvent in which one or more of the desired components is preferentially soluble is known as liquid–liquid extraction. There are three stages involved in extraction;

(a) Bringing the feed mixture and the solvent into intimate contact,
(b) Separation of the resulting two phases, and
(c) Removal and recovery of the solvent from each phase.

There are different arrangements of solvent systems such as

(a) *Three component systems*: The two to be separated and the solvent.
(b) *Four component systems*: The two to be separated and solvent employing two components.
(c) *Four component systems*: In this the mixture to be separated is distributed between two immiscible solvents.

The above three arrangements are subjected to two types of physical arrangement of the apparatus employed.

- *Differential contacting equipments*: In these types of equipments, contacting and separation are combined into a single piece of equipment such as a column, which is then operated continuously.
- *Stagewise equipments*: Mixing of the two liquid phases by agitation, followed by settling in a separate vessels by gravity (*e.g.* a mixer settler).

### 1.4.2. *Distillation vs Extraction*

Distillation is the most widely employed separation techniques in industries for the separation of various components. However, in many situations employing distillation is not justified owing to cost and economic concerns. In those cases it is preferable to employ other separation techniques and the most opted in extraction. However, it is always helpful to keep in mind this parallel process of distillation whenever extraction is studied. These two separation techniques are analogues. In distillation the phases are separated by addition of heat whereas in solvent extraction a selective solvent is employed for the purpose. In distillation the vapor created by heating is condensed to get the separated phase whereas in extraction the two phases are settled so that the both the phases disengages and form two separated layers. There are many cases where extraction is preferential to distillation. These are:

(a) Where distillation would require excessive amounts of heat, such as, for example, when the relative volatility is near unity.
(b) When the formation of azeotropes limits the degree of separation obtainable in distillation. Further for systems with low relative volatility like acetic acid water system, extraction is more preferable.
(c) In dilute solutions of the desired component, particularly where the desired component is less volatile, with distillation large amount of heat will be required to evaporate the water. On the other hand extraction will selectively remove the desired component leaving aside water.
(d) When heating must be avoided.
(e) Extraction can be successfully used as an alternative to distillation under vacuum at very low temperature to avoid thermal decomposition.
(f) Distillation usually separates substances which are chemically very similar as the vapor is created from liquid by addition of heat. When the components to be separated are quite different in nature, extraction is advantageously used. *e.g.* separation of aromatic and paraffinic hydrocarbons of nearly the same molecular weight are impossible to separate by distillation because their vapor pressure are nearly the same but can be successfully be separated by extraction by a number of solvents.

Over chemical methods extraction proves its superiority in view of that the chemical methods consume lot of chemicals and produce by products which lead to disposal problems. On the other hand extraction involves no chemical consumption or by product production and is less costly.

### 1.4.3. *Applications of Extraction*

Extraction has wide number of applications such as:

1. Separation of aromatics from kerosene-based fuel oils to improve their burning qualities
2. Separation of aromatics from paraffin and naphthenic compounds to improve the temperature-viscosity characteristics of lubricating oils.
3. To obtain relatively pure compounds such as benzene, toluene, and xylene from catalytically produced reformats in the oil industry.
4. In the production of anhydrous acetic acid.
5. In the extraction of phenol from coal tar liquors
6. In the metallurgical (purification of uranium fuel and for the recovery of spent fuel elements in the nuclear power industry) and biotechnology industries.

### 1.4.5. *Liquid-liquid Extraction Equipments*

Fig. 1.4.1 Shows a extractor using packed bed. The heavier phase is entered from the top and the lighter phase from the bottom. The interface is maintained at above the heavier phase inlet and adjusted by interface maintainer. Continuous stirring has controlled by the motor. D.C. motor is of variable speed. Feed tanks are provided for reactant storage and the reactants are compressed by the air. Safety valve is also for release the excess air. The sample taken from the top and bottom are analyzed.

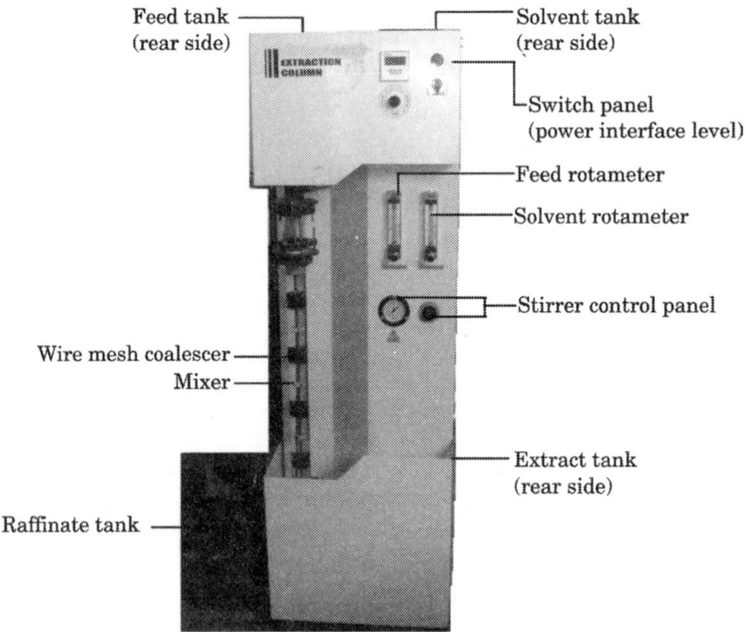


**Fig. 1.4.1:** Laboratory scale extraction column (Suresh, unpublished figure)

In liquid where density differences are likely to be one-tenth as large or less, good dispersion of systems of high interfacial tension is impossible in simple towers and mass transfer rates are poor. For such systems dispersion is best brought about by mechanical agitation of the liquids, where upon good mass transfer rates are developed. One of the problem using the perforated plate and indeed packed column is that re-dispersion of the liquids after each stage is very poor. To overcome this Scheibel developed a unit. In the York-Scheibel's unit extractor the regions surrounding the agitators are packed with wire mesh to encourage coalescence and separation of the phases. Most of the extraction takes place in the mixing sections, but some also occurs in the calming sections so that efficiency of each mixer-settler units is very high. In case, light liquid forms the continuous phase and heavy liquid the dispersed phase, the liquid-liquid interface should be accordingly adjusted. Equations for $N_{OR}$ and $N_{OE}$. Consider a differential height dZ of packing.

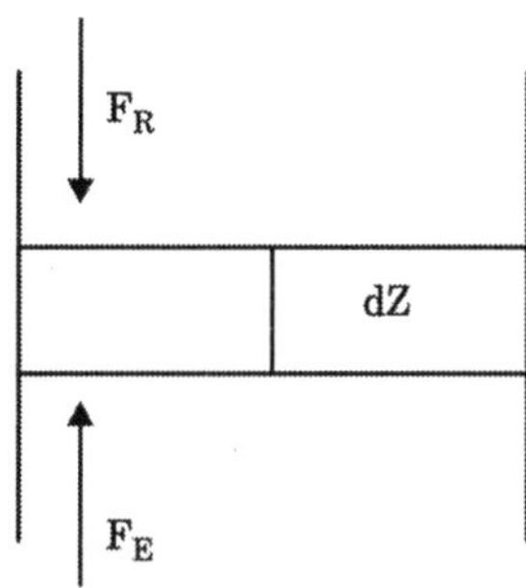


Let S = Column cross sectional area

The amount of solute transferred is given by:

$$F_R\,Sdx = K^a_{OR}\,(x - x^*)\,SdZ \tag{1.4.1}$$

$$F_R\,Sdy = K^a_{OE}\,(y - y^*)\,SdZ \tag{1.4.2}$$

Where $K^a_{OR}$ and $K^a_{OE}$ are the volumetric mass transfer coefficients of the raffinate and extract phase respectively and x* and y* are the equilibrium values of x and y respectively. The number of transfer units of the extract $N_{OE}$ and raffinate $N_{OR}$ can be computed from:

$$N_{OR} = \int_{XR}^{XF} \frac{dx}{x - x^*} \tag{1.4.3}$$

$$N_{OE} = \int_{yO}^{yE} \frac{dy}{y - y^*} \tag{1.4.4}$$

The efficiency of the extraction can be enhanced by the combination of mixing and setting of the phases in a stage wise fashion. This is done by contacting the liquid in a mixer. The heavy phase is entered from the top and light phase from the bottom (countercurrent). Sometimes this can be accomplished in a sieve tray tower similar to that used in distillation (Fig. 1.4.2), where the density difference cause the separation of the phases, so that we obtain the heavy liquid below and light liquid floating upward. An agitator can result maximum mixing of the phases resulting dispersion of one phase as fine droplets into another phase in a continuous manner. The mixtures obtained in this state are called emulsion. The emulsion from the mixer is brought to the settler where the droplets begin to coalesce after the solute has redistributed across the liquid-liquid interface. Lighter phase obtained from top of the settler is called Extract Phase and the heavier phase obtained at the bottom is called Raffinate Phase. This design constitutes the Mixer-Settler extraction (Fig. 1.4.3). This design can be extended in a continuous stage wise manner *i.e.* the mixer-settler tanks are joined continuously as a countercurrent cascade of stages as in Fig. 1.4.4.

## 1.4.6. *Extraction Process Description*

The solvent and solution (feed) are mixed together and then allowed to separate into the two phases—the extract E containing the required solute in the added solvent and the raffinate R, the weaker solution with some associated solvent.

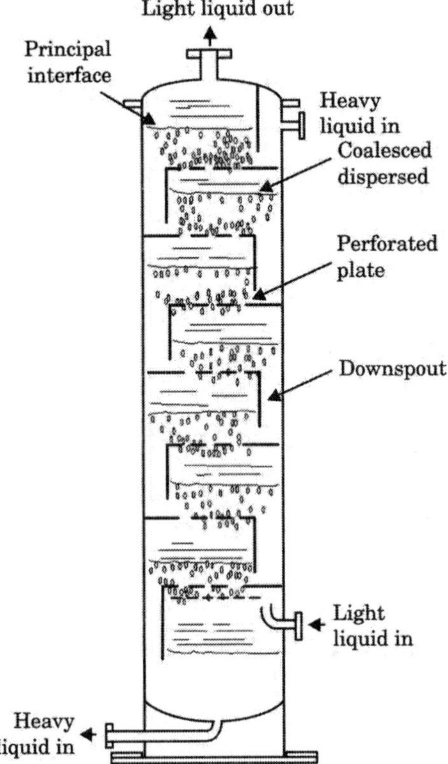


**Fig. 1.4.2:** Schematic diagram of equipment with sieve tray for liquid-liquid extraction

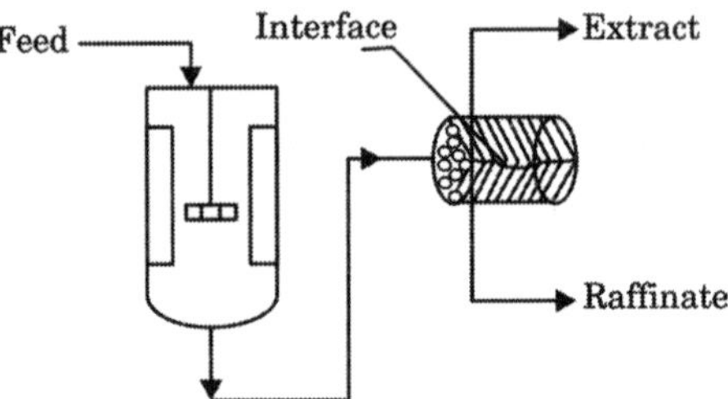


**Fig. 1.4.3:** Single stage mixer-settler

### 1.4.6.1. *Mechanism of Transfer of Solute from One Phase to Other*

The transfer of solute occurs both via molecular and eddy diffusion. Most of the theories developed in literature are for gas liquid systems, where solute has been considered to be the dissolved gas and solvent as the liquid phase. For the absorption of solute without chemical reaction, the rate of absoption can be given as:

$$\text{Rate} \ \alpha \ (A^* - A_o) \tag{1.4.5}$$

or

$$\text{Rate} = k_L a(A^* - A_o) \tag{1.4.6}$$

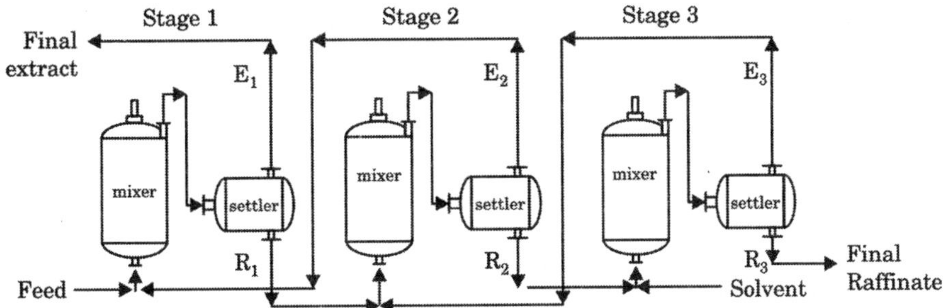


**Fig. 1.4.4:** Three stage counter-current mixer-settler extraction cascade

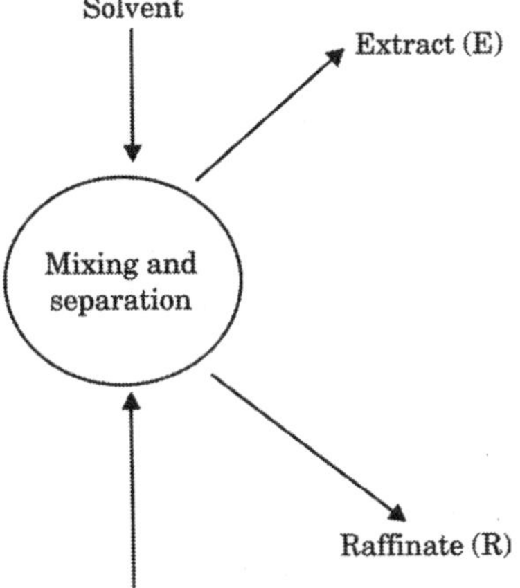


Where A* is the concentration of solute corresponding to equilibrium with the liquid phase at the interface, a is the interfacial area between the aqueous and organic phases and $A_o$ is the average concentration of solute in the liquid phase. There are different models that predict the value of proportionality constant and predict the rate of transfer of solute from bulk of the gas phase to the liquid phase

1.4.6.1.1. *Film Theory*

There is a stagnant film of thickness 'δ' at the surface of liquid adjacent to the interface, while the rest of liquid is kept uniform in concentration by agitation, the concentration in the film falls from A* at the interface to Ao at the inner edge. There is no convection in the film and dissolved gas crosses by molecular diffusion only.

$$\text{Rate} = \frac{D_A}{\delta}(A^* - A_o) \tag{1.4.7}$$

or

$$\text{Rate} = k_L(A^* - A_o) \tag{1.4.8}$$

Where $k_L = D_A/\delta$; δ is function of hydrodynamic properties of system such as geometry, liquid agitation and physical properties etc. This model takes a great credit since it incorporates an essential feature of real systems that gas gets into the liquid by molecular diffusion, but it suffer from few shortcomings. This is not a very realistic model and the uniform thickness of film taken as assumption never really exists.

1.4.6.1.2. *Still Surface Models*

In any transport process there is a region close to surface where transport is by molecular diffusion alone. At increasing distance from the surface, transport by turbulent movement becomes relatively more important and eventually predominates.

(a) *King still surface model*: This is two parameter model and according to it transport is the combined result of molecular diffusivity and eddy diffusivity which is proportional to some power of distance from surface. In these circumstances $k_L$ α {D}; where x varies form 0 to 1. Due to two parameters in this model, these are difficult to obtain than the film theory which has only one parameter.

(b) *Surface rejuvenation model (Andrew and Dankwerts)*: This is again a two parameter model and according to this model

(i) Diffusion takes place for a period at diminishing rate into stagnant liquid.

(ii) At convective disturbance then replaces the liquid upto a certain depth below the surface by liquid of the bulk concentration.

(iii) Quiescent diffusion recommences with a steeper concentration gradient and hence a more rapid rate of absorption than before the disturbance

(iv) The above continues until the next disturbance

1.4.6.1.3. *Surface Renewal Model*

According to this model there is replacement of liquid at the surface by fresh liquid of bulk composition by turbulent motion from interior. While the element is at the surface and is exposed to the gas, it absorbs gas as though it was quiescent and infinitely deep and rate of absorption is function of time of exposure of element. Initially rate is rapid and infinite and decreases with time.

(A) *Higbie model*: This model assumes that every element at the surface is exposed to the gas for the same length of time δ, before being replaced by liquid of bulk composition. During this time it absorbs the same amount (Q) of gas per unit area as though it was stagnant and infinitely deep. The average rate of absorption is given by Q/δ. δ is determined by hydrodynamics properties. The relation between δ and $k_L$ is derived as follows:

Amount of A absorbed by unit are of surface is given as

$$Q = 2(A^* - A_o)\sqrt{\frac{D_A \theta}{\pi}} \tag{1.4.9}$$

Rate of absorption =

$$R = \frac{Q}{\theta} = 2(A^* - A_o)\sqrt{\frac{D_A}{\pi\theta}} \tag{1.4.10}$$

Hence

$$k_L = \frac{R}{A^* - A_o} = 2\sqrt{\frac{D_A}{\pi\theta}} \quad (1.4.11)$$

The disadvantage of this model is that it is unrealistic in specifying the same time of exposure for all the elements of the surface.

(B) *Danckwerts model*: This is the modification of the Higbie model. According to this model, the chance of element of surface being replaced with fresh liquid is independent of length of time for which it has been exposed. There is a stationary distribution of ages in which the fraction of the surface which at any instant has been exposed to the surface of times between $\theta$ and $\theta$+d$\theta$ and is $s^*e^{-s}\theta\, d\theta$, where s is the fraction of area of the surface which is replaced with fresh liquid in unit time. Instantaneous rate of absorption per unit area is

$$R = (A^* - A^o)\sqrt{\frac{D_A}{\pi\theta}} \quad (1.4.12)$$

And the average rate of absorption into the surface is

$$\overline{R} = s\int_0^\infty Re^{-s\theta}\, d\theta \quad (1.4.13)$$

$$= s(A^* - A^o)\sqrt{\frac{D_A}{\pi}}\int_0^\infty \frac{e^{-s\theta}}{\sqrt{\theta}}d\theta$$

$$= (A^* - A^o)\sqrt{D_A s}$$

$$= K_L(A^* - A^o) \quad (1.4.14)$$

The advantage of this model is that only one parameter is there *i.e.* s.

1.4.6.2. *Discussion Among the Various Models*

All the models are equally good. Film model have advantage that only ordinary differential equation needs to be solved. However, numerical computational techniques have facilitated the use of other methods also. Higbie's model results are obtained from the results of transient absorption by putting

$$\overline{R} = Q(\theta)/\theta \quad (1.4.15)$$

Where $\theta = 4D_A/\delta k_L^2$ or by directly using the value of E (E is average rate of absorption into the agitated liquid in the presence of reaction to the average rate of absorption without reaction).

### 1.4.7. *Equilibrium Data*

The addition of a new solvent to a binary mixture of a solute in a solvent may lead to the formation of several types of mixture:

(a) A homogeneous solution may be formed and the selected solvent is then unsuitable.

(b) The solvent may be completely immiscible with the initial solvent.

(c) The solvent may be partially miscible with the original solvent resulting in the formation of one pair of partially miscible liquids.

(d) The new solvent may lead to the formation of two or three partially miscible liquids.

(b) And (c) are the most promising to be separated by LLE.

With type (b), the equilibrium relation is conveniently shown by a plot of the concentration of solute in one phase against the concentration in the second phase. In (c) and (d) equilibrium data is usually represented by triangular diagrams.

1.4.7.1. *Equilateral Triangular Coordinates*

An equilateral triangle is used for representing compositions, use being made of the fact that the sum of the perpendiculars from any point within the triangle to the three sides equals the altitude. The length of the altitude is then allowed to represent 100 per cent composition and the length of the perpendiculars from any point the percentages of the three components. Refer to Fig. 1.4.5, the apexes of the triangle represent the pure components A, B, and C, respectively. These symbols represent the weight of these components such as pounds of A or kg of A etc.

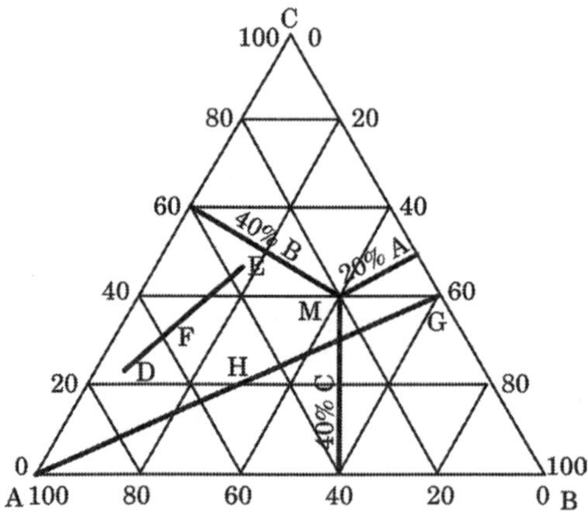


**Fig. 1.4.5:** Triangular diagram

Any point on the side of the triangle represents a binary mixture of the two components marked at the ends of the side: thus, point M is a mixture of 20 per cent A, 40 per cent 5, and 40 per cent C. Points outside the triangle represent imaginary mixtures and have constructional significance only.

Several other characteristics of such plots are significant. A mixture at D, when added to one at E, will form a third mixture F on the straight line DFE.

Furthermore, the relative weights of D and E mixed will cause the point F to be so located that

$$\frac{\text{Weight of E}}{\text{Weight of D}} = \frac{\text{Length of DF}}{\text{Length of FE}} \tag{1.4.16}$$

Similarly, if E is removed from F, the point D representing the residue is on the straight line EF extended through F, and the above relationship of weights and line-segments again applies. Formation of One Pair of Partially Miscible Liquids.

In a system of this sort, the liquid pairs A-C and B-C are miscible in all proportions at the prevailing temperature; A and B are partially miscible, and points D and E represent the saturated solutions in the binary system. A typical example is available in the system benzene (A)-water (B)-ethanol (C). All mixtures of the components represented by points in the area outside the curve DNPLE are homogeneous single liquid- phase solutions, while mixtures within the area bounded by the curve and the line DE form two insoluble liquid layers. The curve DNPLE represents the saturated solutions and is called the solubility or bimodal curve. It is ordinarily concave throughout, as shown in Fig. 1.4.6 but several cases exhibiting a change in curvature have been recorded, such as the system water-ethylene glycol-amyl alcohol. A mixture of over-all composition M will form the two immiscible liquid solutions of compositions L and N, respectively, and the point M is therefore on the straight line LN, which is a tie line. All mixtures represented on the line LN form conjugate layers of the same composition, while the relative weights of the two layers can be calculated analytically from the compositions or graphically from the tie-line segments as indicated above. The area of heterogeneity is to be imagined as filled with an infinite number of tie lines, only a few of which are shown in Fig. 1.4.6. These are not parallel and ordinarily change their slope slowly in one direction with changing concentration. In a few systems there is a reversal of the slope, as in the case of water-ethyl acetate-ethanol at 0°C and ethanol ethyl ether-water. In the case of the system of Fig. 1.4.6 it is clear that component C, when added to

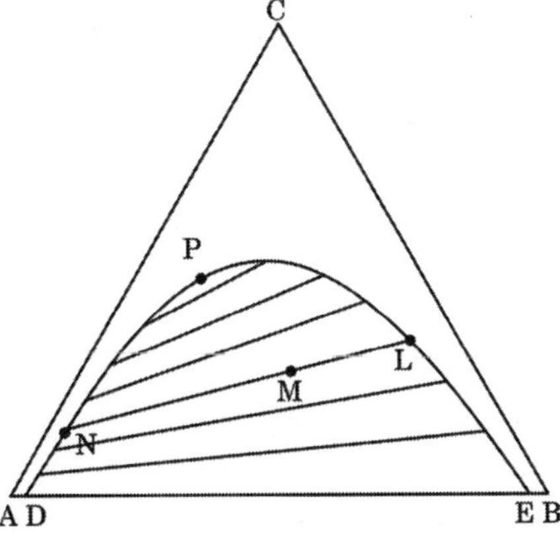


**Fig. 1.4.6:** Triangular diagram for partially miscible liquids

a heterogeneous liquid mixture of A and B, distributes itself unequally between the two conjugate layers, with a greater concentration in the B-rich solutions. As more C is added to such a mixture, the mutual solubility of A and B increases. At point P, the plait point, the two branches of the solubility curve merge, not ordinarily at the maximum value of C on the curve. The tie lines shrink in length at higher concentrations of C until at the plait point they vanish. Since at the plait point two liquid layers of identical composition and density form, the point is a true critical condition. F, E and R in the triangular diagram will designate feed, extract and raffinate solutions, respectively with respect to weight and compositions. X will designate weight fractions with appropriate subscripts such as $X_{AB}$ represents weight fraction of A in a B rich solution and $X_{CE}$, the weight fraction of C in the extract etc. Mole and mole fraction may also be substituted for weight and weights fraction if desired.

1.4.7.2. *Simple Distribution Diagrams*

In type (b) because of complete immiscibility of the organic phase (solvent) and aqueous phase (solute in the other phase), there shall be over crowding of the points near the apexes of the triangular diagram that may cause difficulties in graphical computations. In these cases, concentration of solute in organic phase is plotted against the acid in the aqueous phase. These diagrams are called as distribution diagrams.

## 1.4.8. *Selection of Solvent*

There are certain desirable characteristics of solvent that could be employed for the extraction process. Usually no single solvent possesses all these properties but it is desirable that they should at least posses more of these.

1. *Selectivity*: This refers to the ability of solvent to extract one component of a solution in preference to another. Consider the following triangular diagram (Fig. 1.4.7) where feed M (mixture of A and C) is contacted with solvent B and results after shaking and settling, two separated mixtures R and T. E and D represents the solvent free solution corresponding to T and R respectively. On the other hand in the same

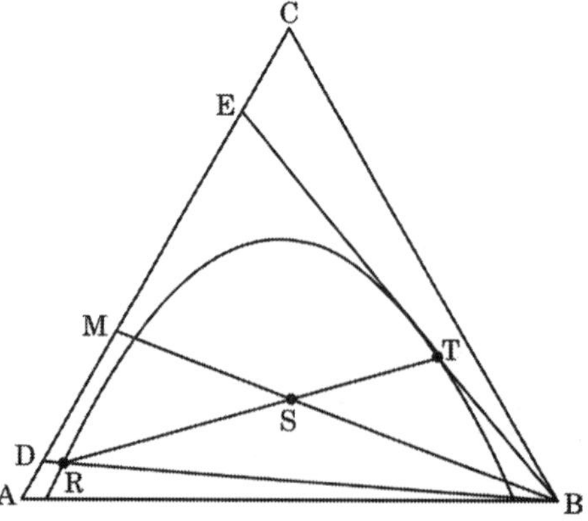


**Fig. 1.4.7:** Selectivity considerations of B as solvent

figure if A is considered as the solvent to separate feed M which is the mixture of C and B, the difference which is easily seen is that the E and D are not as far placed as in the previous figure. The distance between E and R indicates the effectiveness of the separation, and thus B is more selective in separating A and C the A for separating B and C.

If the distribution diagrams have been used for the selectivity study of the solvent, the method of selection is equivalent of that in distillation, where vapor liquid curve is drawn and the system having higher relative volatility has a higher distance between it and 45 degree line. The same procedure when adopted to liquid liquid extraction, by plotting weight fraction of solute in extract against that in raffinate and more the curve is farther from the 45 degree line more selective is the solvent. The selectivity of B for C, in general is defined as

$$\beta = \frac{X_{CB} \times X_{AA}}{X_{CA} \times X_{AB}} \tag{1.4.17}$$

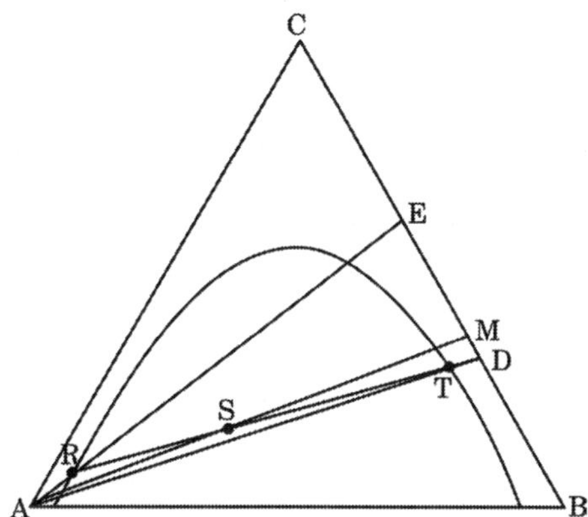


**Fig. 1.4.8:** Selectivity considerations of A as solvent

2. *Recoverability*: The product free of solvent is the desire of carrying out solvent extraction and not the mixture of solvent and product. Further the solvent must be reused to decrease the cost of plant. To separate the solvent from solute the technique used are either distillation or another solvent extraction as the recovery unit. To employ the distillation it is preferable to use solvent having low volatility in comparison to solute so that in distillation solute can be easily removed. The inverse of that could also be employed but would require large heat particularly in situations where the solute is in dilute form. Other method such as solvent extraction or crystallization for recovery of solute from solvent could also be attractive in certain situations.
3. *Density*: The different in density of the two solutions is very essential in view of better disengaging after they are shaken and settled and also the terms of increase in capacity of the contacting equipments.
4. *Interfacial tension*: Interfacial tension should be high enough so that the two phases could disengage on settling but should not be too high so

that there is difficulty of the dispersion of one into another. Low values of interfacial in undesirable in view of formation of stable emulsion.

5. *Reactivity*: Presence of reaction in the system could lead to stable product formation that hinders the removal of solvent and hence should be avoided. However there are certain reactions which fall in the range of Vander wall forces and hence could be easily reversed and hence could lead to the separation of the solvent for reuse. These shall be discussed in the section of reactive extraction in Chapter 3 in detail.
6. *Immiscibility of the solvent with the solution*: In the triangular diagram, the immisibility range is achieved in the outside of the dome of the solubility curve. So for the systems where the dome is high there is the possibility of using the solvent to get the solvent free product as far as possible from the raffinate product. Thus good degree of separation is achieved if the solvent is immiscible with the solution.
7. *Corrosiveness, inflammability and toxicity*: Solvent should not be corrosive to the material of the system, should be inflammable in view of safety and hazard basis and should be non poisonous in view of handling for the flora and fauna.
8. *Viscosity*: The solvent should have low viscosity. Thus could lead to low pumping costs, higher heat transfer rates, better handling and higher extraction since it could better solvate the desired component.
9. *Vapor pressure and freezing point*: Should be low so that they can be stored and used for extraction at atmospheric pressure and outside temperature.
10. *Cost*: Usually cost factor should not come if there is complete recycle of the solvent however, if there is some loss in any section of solvent extraction plant this factor must be included. Too expensive chemical for extraction would lead to increase in the cost of operation of the plant.

### 1.4.9. *Effect of temperature and pressure*

The influence of pressure on the extraction process is usually very small, except at very high pressures, and hence can be ignored. For most of systems the mutual solubility of solution and solvent increases with increase in temperature and above some temperature (critical solution temperature) they dissolve completely. The increase in temperature decreases the area under the solubility curve and also affects the slopes of tie lines. It is recommended to carry out any extraction process below the critical solution temperature.

### 1.4.10. *Method of stage calculations*

(i) *For single contact cocurrent*

The stage considered here is the theoretical stage which is the one in which the contact between the phases is sufficiently intimate and maintained for a sufficient period of time and a distribution equilibrium is established. Three types of design diagrams can be used in the stage calculations. Here only triangular diagrams are considered which are the basically employed in the stage calculations. A typical triangular diagram is shown in Fig. 1.4.9. Following are the sign conventions for the triangular diagram

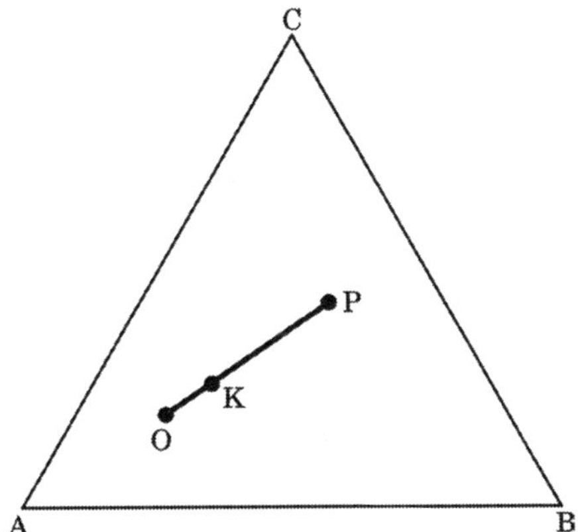


**Fig. 1.4.9:** Triangular diagram

- Symbol A, B and C represents pure components
- Point on the side represents a binary mixture
- Point in the diagram represents a ternary mixture
- F, E and R, represents feed, extract and raffinate
- X=weight fraction compositions
- $X_{AB}$= weight fraction A in a B rich solution etc.
- Length of a line on a phase diagram between two points and P will be represented by OP, A mixture at O, when added to one at P, results in a new mixture at K on the straight line OP, such that

$$\frac{O}{P} = \frac{\overline{KP}}{\overline{OK}} \tag{1.4.18}$$

Fig. 1.4.10. Shows the typical stage. Steps in the stage calculation are as follows

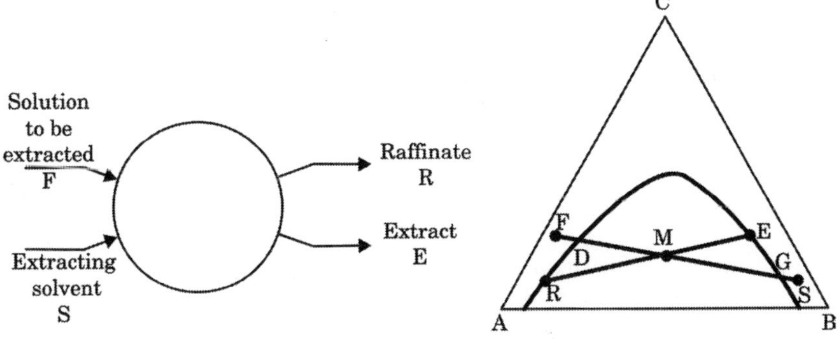


**Fig. 1.4.10:** Single contact co-current triangular diagram and flowsheet

Step 1: Material balance for the operation

$$F + S = R + E = M \quad (1.4.19)$$

Step 2: Location of M on FS:

(a) Graphically

$$\frac{F}{S} = \frac{\overline{MS}}{\overline{FM}} \quad (1.4.20)$$

(b) By material balance for C and B

$$FX_{CF} + SX_{CS} = MX_{CM} \quad (1.4.21)$$

$$X_{CM} = \frac{FX_{CF} + SX_{CS}}{M} = \frac{FX_{CF} + SX_{CS}}{F + S} \quad (1.4.22)$$

$$X_{BM} = \frac{FX_{BF} + SX_{BS}}{M} = \frac{FX_{BF} + SX_{BS}}{F + S} \quad (1.4.23)$$

Step 3: Location of E and R:

By trial and error by tie line interpolation

Step 4: Calculating the weights of E and R:

(a) Graphically

$$\frac{R}{E} = \frac{\overline{EM}}{\overline{RM}} \quad (1.4.24)$$

(b) By material balance on C

$$EX_{CE} + RX_{CR} = MX_{CM} \quad (1.4.25)$$

$$E = \frac{M(X_{CM} - X_{CR})}{X_{CE} - X_{CR}} \quad (1.4.26)$$

$$R + E = M$$

Cases:

(a) Determine the quantity of solvent needed for a given feed in order to produce a predetermined M (By rearranging the equation in step 2)

$$S = \frac{F(X_{BM} - X_{BF})}{X_{BS} - X_{BM}} \quad (1.4.27)$$

(b) *Calculating of minimum amount of solvent*: Point D results in a maximum of raffinate of composition D and no extract and hence here minimum amount of solvent is obtained

By locating M at point D

$$S_{min} = \frac{F(X_{BD} - X_{BF})}{X_{BS} - X_{BD}} \tag{1.4.28}$$

(c) *Calculating of maximum amount of solvent*: Point G results in a minimum of raffinate of composition G and resulting in a maximum of extract of composition G and no raffinate

By locating M at point G

$$S_{max} = \frac{F(X_{BG} - X_{BF})}{X_{BS} - X_{BG}} \tag{1.4.29}$$

(ii) *For multiple contact co-current*

Fig. 1.4.11. Shows the stage calculations for multiple contact cocurrent. The procedure for stage calculation remains same as explained for single stage in the earlier article but the only difference is that the raffinate from stage 1 becomes the feed for the stage 2. Hence, for stage 2 calculations, R shall be joined to S and location of M will be on this line. For other stages too the same procedure shall be followed.

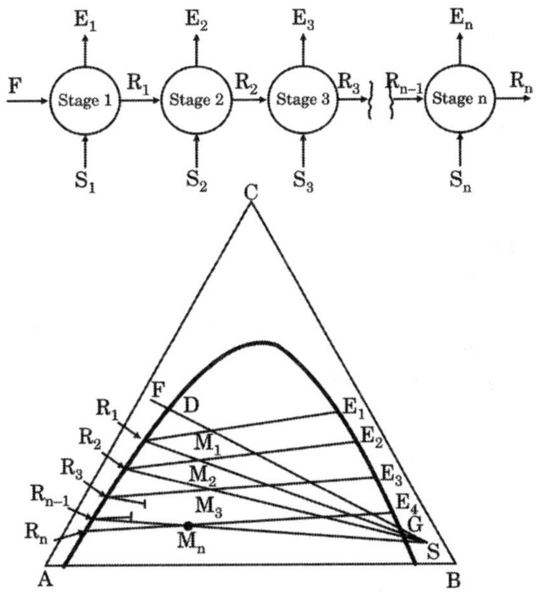


**Fig. 1.4.11:** Multiple contact co-current triangular diagram and flowsheet

(iii) *Co-current multiple contact with Immiscible solvents*

In case the solvent is completely immiscible, plotting on triangular diagrams results in overcrowding of the points near the faces of triangles and hence use of triangular diagrams are not feasible. In such cases following procedure is adopted. The coordinate systems which shall be used in such cases are:

$$y = \frac{X_{CB}}{X_{BB}}; \text{ weight of c/ weight of B in extract} \quad (1.4.30)$$

$$x = \frac{X_{CA}}{X_{AA}}, \text{ weight of c/ weight of A in extract} \quad (1.4.31)$$

The solubility curve shall be converted in this format and plotted as shown in Fig. 1.4.12.

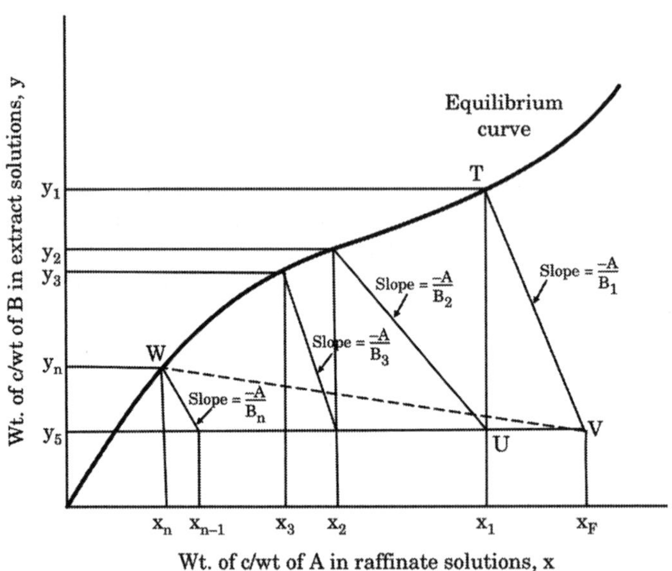


**Fig. 1.4.12:** Stage calculations for immiscible solvents

Since no amount of solution (A) goes into solvent side (B) and no amount of solvents enters solution side and only solute is partitioned in between the two phases, A Kg and B Kg shall remain in all stages. For any stage (m) making out C balance

$$Ax_{m-1} + B_m y_s = B_m y_m + Ax_m \quad (1.4.32)$$

or

$$\frac{y_m - ys}{x_m - x_{m-1}} = -\frac{A}{B_m} \quad (1.4.33)$$

where $y_m$ and $x_m$ are the concentration of extract and raffinate from the m[th] stage. This is the equation of straight line on the coordinated of Fig. 1.4.12 of slope ($-A/B_m$) passing through points ($x_m$, $y_m$) and ($x_{m-1}$, $y_m$). ($x_m$, $y_m$) lies on the solubility curve as the stage is ideal stage. Thus for first stage from point ($x_F$,$y_s$) a straight line of slope ($-A/B_1$) shall be drawn which intersects the solubility curve at ($x_1$, $y_1$). The second stage can be obtained from point U and following the same procedure until the final raffinate concentration is reached.

(iv) *Continuous Countercurrent multiple contact*

Material balance on the m[th] stage can be written as

$$R_{m-1} + E_{m+1} = R_m + E_m \quad (1.4.34)$$

$$R_{m-1} - E_m = R_m - E_{m+1} = O \quad (1.4.35)$$

Material balance on the 1[st] stage

$$F + E_{m+1} = R_m + E_1 \quad (1.4.36)$$

$$F - E_1 = R_m - E_{m+1} = O \quad (1.4.37)$$

*Drawing the equilibrium lines*

Any extract $E_{m+1}$ can be located from any raffinate $R_m$ by extending the line $OR_m$ to the B-rich solubility curve. As with all ideal stages, extract $E_m$ and raffinate $R_m$ will be in equilibrium and on opposite ends of a tie line. Consequently, $R_1$ may be located at the opposite end of a tie line through $E_1$, $E_2$ by line $OR_1$ extended, $R_2$ by a tie line through $B_2$, etc. The cascade of stages for multiple contact counter current operation is shown in Fig. 1.4.13. Steps in the stage calculation are as follows (Fig. 1.4.14).

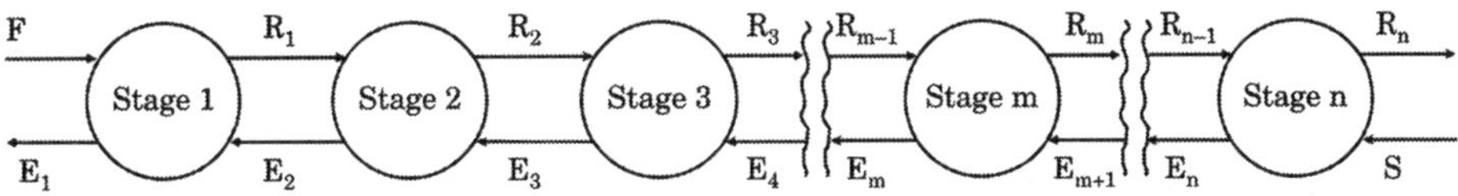


**Fig. 1.4.13:** Continuous counter current multiple contact flowsheet

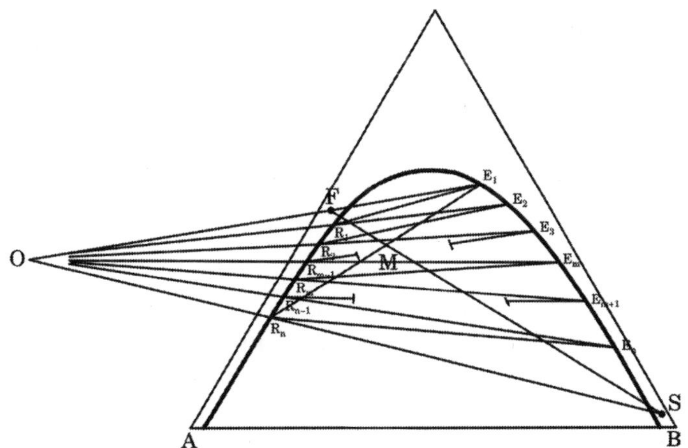


**Fig. 1.4.14:** Triangular diagram for continuous counter current multiple contact

Step 1: Material balance for the operation

$$F + S = R_n + E_1 = M \quad (1.4.38)$$

$$F - E_1 = R_n - S = O \quad (1.4.39)$$

Thus F, $R_n$ and S are located on the triangular diagram. To find the position of $E_1$, location of M is essential.

Step 2: Location of M on FS: It can be obtained graphically or by material balance for B and C as described in earlier sections.

Step 3: The values of $X_{CM}$, $X_{CE1}$ and $X_{CRN}$ can be read from the graph

Step 4: $E_1$ and $R_n$ can be obtained by material balance on C as

$$E_1 = \frac{M(X_{CM} - X_{CR_n})}{X_{CE_1} - X_{CR_n}} \quad (1.4.40)$$

Step 5: Calculations for other concentrations at the respective E's and R's can be read from triangular diagram

Step 6: The corresponding E's and R's can be obtained as

$$E_{m+1} = R_m + E_m - R_{m-1} \quad (1.4.41)$$

$$R_{m-1} X_{ARm-1} + E_{m+1} X_{AEm+1} = R_m X_{ARm} + E_m X_{Aem} \quad (1.4.42)$$

Combining the above two equations

$$R_m = \frac{R_{m-1}(X_{AR_{m-1}} - X_{AE_{m+1}}) + E_m(X_{AE_{m+1}} - X_{AE_m})}{X_{AR_{m1}} - X_{AE_{m+1}}} \quad (1.4.43)$$

Thus, $R_1$ by above equation by letting m -1 = F and m = 1; $E_2$ by step 1 with m = 1; $R_2$ by the above equation with m = 2; $E_3$ by step 1 with m = 2, etc.

## 1.4.11. SUPERCRITICAL FLUID EXTRACTION

### 1.4.11.1. *Introduction*

The use of supercritical fluids (SCF) for extraction purposes was introduced in the late nineteenth century. A supercritical fluid is a substance at temperatures and pressures beyond its critical point at which the liquid phase of the substance will not exist. At these temperatures and pressures, the supercritical fluid has properties between gas- and liquid-phase characteristics. These properties make supercritical fluids efficient extraction solvents with high mass transfer characteristics. SFE is used on a large scale for the extraction of some food grade and essential oils and pharmaceutical products from plants. It is relatively rapid because of the low viscosities and high diffusivities associated with supercritical fluids. The extraction can be selective to some extent by controlling the density of the medium and the extracted material is easily recovered by simply depressurizing, allowing the supercritical fluid to return to gas phase and evaporate leaving no or little solvent residues. Carbon dioxide is the most common supercritical solvent. Since the end of the 1970s, supercritical fluids has been used to isolate natural products, but for a long time the applications concentrated on only few products. Now the development of processes and equipment is beginning to pay off and industries are getting more and more interested in supercritical techniques. In SCF the physicochemical properties of a given fluid, such as density, diffusivity, dielectric constant and viscosity can be easily controlled by changing the pressure or the temperature without ever crossing phase boundaries. The dissolving power of a supercritical fluid depends on its density, which unlike that of liquid solvents is highly adjustable by changing pressure or temperature. Especially in the vicinity of the critical point a small isothermal raise in pressure increases the fluid density remarkably while the effect diminishes with increasing distance from the critical values. Thus, the solvent strength of a fluid can be controlled more easily than that of a liquid solvent. Recently, the supercritical fluid has approved higher diffusion coefficient and lower viscosity and surface tension than a liquid solvent, which leads to a more favorable mass transfer. The demand for new analytical techniques in food and environmental sciences is strongly related to the higher demand of information on processing, quality control, adulteration, contamination, environmental and food regulations and the need for faster, more powerful, cleaner and cheaper analytical procedures are required by chemists, regulatory agencies and quality control laboratories to meet these demands can be achieved by SFE. Thus, this review paper has been justified.

## 1.4.11.2. *Properties and Fundamentals of Supercritical Fluids*

### 1.4.11.2.1. *Solvent Strength*

The density of a supercritical fluid is extremely sensitive to minor changes in temperature and pressure near the critical point. The densities of the fluids are closer to that of organic liquids but the solubility of solids can be 3-10 orders of magnitude higher. The solvent strength of a fluid can be expressed by the solubility parameter, which is the square root of the cohesive energy density and is defined rigorously from first principles. A plot of the solubility parameter for carbon dioxide versus pressure would resemble that a plot of density versus pressure. This confirms that the solvation strength of a supercritical fluid is directly related to the fluid density. Thus the solubility of a solid can be manipulated by making slight changes in temperatures and pressures. Another attractive feature of supercritical fluids is that the properties lie between that of gases and liquids. A supercritical fluid has densities similar to that of liquids, while the viscosities and diffusivities are closer to that of gases. Thus, a supercritical fluid can diffuse faster in a solid matrix than a liquid, yet possess a solvent strength to extract the solute from the solid matrix.

### 1.4.11.2.2. *Phase behavior*

An understanding of the phase behavior is important since the phase behavior observed in supercritical fluids considerably differ from the behavior observed in liquids. One such behavior is the retrograde region. For an isobaric system, an increase in the temperature of a solution increases the solubility of the solute over certain ranges of pressure (consistent with the typical liquid systems) but decreases the solute solubility in other pressure ranges. This anomalous behavior wherein the solubility of the solute decreases with a temperature increase is called the retrograde behaviour (Fig. 1.4.15). Thus, the following generalizations may be made regarding the solute solubility in supercritical fluids. Solute solubility approach and may exceed that of liquid solvents. Solubility generally increase with increase in pressure. An increase in the temperature of the supercritical fluid may increase, decrease or have no effect on the solubility of the solute depending upon the pressure. $CO_2$ is not a very good solvent for high molecular weight and polar compounds. To increase the solubility of such compounds in supercritical carbon dioxide, small amounts (ranging from 0 to 20 mol%) of polar or non-polar co-solvents called modifiers may be added. The co solvent interacts strongly with the solute and significantly increases the solubility.

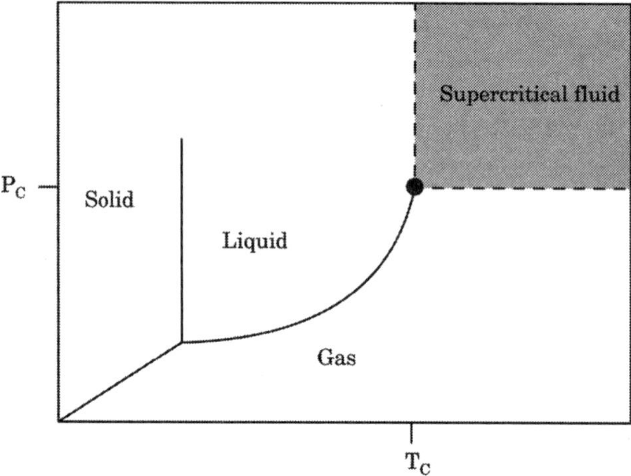


**Fig. 1.4.15:** Phase diagram for a single substance. Pc, critical pressure; Tc, critical temperature

1.4.11.2.3. *Co-solvents or modifiers*

A more common practice in SFE is to change the polarity of the supercritical fluid and increase their solvating power towards the analyte of interest by employing polar modifiers (co-solvents). For example, the addition of relatively small percentages (1-10%) of methanol to carbon dioxide expands its extraction range to include more polar analyses. The modifiers can also reduce the analyzed matrix interactions improving their quantitative extraction.

There are two main procedures to study with co-solvents or modifiers in SFE are; the first one, and the most common, accounts for a mixing of the modifier with the $CO_2$ flow while the second mixes the modifier with raw material in the extraction cell. This procedure is always associated to a static extraction step in which the modifier, in intimate contact with the sample matrix, is able to substitute the analyte molecules bound in active centers of the matrix and release them into the supercritical fluid phase.

1.4.11.2.4. *Extraction with supercritical fluids*

Supercritical extraction has been applied to a large number of solid matrices. The desired product can be either the extract or the extracted solid itself. The advantage of using supercritical fluids in extraction is the ease of separation of the extracted solute from the supercritical fluid solvent by simple expansion. In addition, supercritical fluids have liquid like densities but superior mass transfer characteristics compared to liquid solvents due to their high diffusion and very low surface tension that enables easy penetration into the porous structure of the solid matrix to release the solute. Extraction of soluble species (solutes) from solid matrices takes place through four different mechanisms. If there are no interactions between the solute and the solid phase, the process is simple dissolution of the solute in a suitable solvent that does not dissolve the solid matrix. If there are interactions between the solid and the solute, then the extraction process is termed as desorption and the adsorption isotherm of the solute on the solid in presence of the solvent determines the equilibrium.

A third mechanism is swelling of the solid phase by the solvent accompanied by extraction of the entrapped solute through the first two mechanisms, such as extraction of pigments or residual solvents from polymeric matrices. The fourth mechanism is reactive extraction where the insoluble solute reacts with the solvent and the reaction products are soluble hence extractable, such as extraction of lignin from cellulose. Extraction is always followed by another separation process where the extracted solute is separated from the solvent. Another important aspect in supercritical extraction relates to solvent/solute interactions. Normally the interactions between the solid and the solute determine the ease of extraction, *i.e.*, the strength of the adsorption isotherm is determined by interactions between the adsorbent and the adsorbate. However, when supercritical fluids are used, interactions between the solvent and the solute affect the adsorption characteristics due to large negative partial molar volumes and partial molar enthalpies in supercritical fluids. The thermodynamic parameters that govern the extraction are found to be temperature, pressure, the adsorption equilibrium constant and the solubility of the organic in supercritical fluid. Similar to the retrograde behavior of solubility in supercritical fluids, the adsorption equilibrium constants can either decrease or increase for an increase in temperature at isobaric conditions. This is primarily due to the large negative partial molar properties of the supercritical fluids. In addition to the above factors, the rate parameters like the external mass transfer resistances, the axial dispersion in the fluid phase, and the effective diffusion of the organics in the pores also play a crucial role in the desorption process. A thorough understanding of these governing parameters is important in the modeling of supercritical fluid extraction process and in the design, development and future scale-up of the process.

#### 1.4.11.2.5. *SFE Process*

Raw material is charged in the extraction tank which is equipped with temperature controllers and pressure valves at both ends to keep desired extraction conditions. The extraction tank is pressurized with the fluid by means of pumps, which are also needed for the circulation of the fluid in the system. From the tank the fluid and the solubilized components are transferred to the separator where the solvation power of the fluid is decreased by increasing the temperature, or more likely, decreasing the pressure of the system. The product is then collected via a valve located in the lower part of the separator(s).

### 1.4.11.3. *Advantages and drawbacks of supercritical $CO_2$ (SC-$CO_2$)*

There are a large number of compounds that can be used as a fluid in supercritical techniques, but by far the most widely used is carbon dioxide. From the viewpoint of pharmaceutical, nutraceutical and food applications it is a good solvent, because it is non-toxic, non-flammable, inexpensive, easy to remove from the product and its critical temperature and pressure are relatively low (Tc=31.1°C, Pc=72 bar) make it important for food and natural products sample preparation, is the ability of SFE using $CO_2$ to be operated at low temperatures using a non-oxidant medium, which allows the extraction of thermally labile or easily oxidized compounds. It is environmental friendly and generally recognized as safe by FDA and EFSA. These properties make it suitable for extracting, for example, thermally labile and non-polar bioactive compounds but, because of its non-polar nature, it cannot be used for dissolving polar molecules. The solubility of polar compounds and the selectivity of the process can be increased by adding small quantities of other solvents, such as ethanol, in the fluid that named as co-solvent or modifier. On one hand, it decreases the processing times, increases yields and makes it possible to use milder processing conditions, but on the other, it complicates system thermodynamics and increases capital costs.

The use of high purity SFE-grade $CO_2$ is not required but impurity and moisture in industrial grade $CO_2$ can accumulate and may interfere with further analytical operations (gas or liquid chromatography). Thus, an on-line fluid cleanup system may be used to remove trace contaminants. An important drawback of SC-$CO_2$ and most of the other supercritical fluids is that predominantly, a non-polar extraction fluid, such as $CO_2$, is used. Therefore, a logical trend to widen the application range of this technique is the study of new methods to decrease analyte polarity to make them more soluble in non-polar supercritical fluids. In this sense, chemical *in situ* derivatization has been applied to improve the selectivity of the extraction towards a specific group of compounds. So the solvent power of SC-$CO_2$ can be summarized by a few rules:

- It dissolves non-polar or slightly polar compounds.
- The solvent power for low molecular weight compounds is high and decreases with increasing molecular weight.
- SC-$CO_2$ has high affinity with oxygenated organic compounds of medium molecular weight.
- Free fatty acids and their glycerides exhibit low solubilities.
- Pigments are even less soluble.
- Water has a low solubility (<0.5% w/w) at temperatures below 100°C.
- Proteins, polysaccharides, sugars and mineral salts are insoluble and;
- SC-$CO_2$ is capable of separating compounds that are less volatile, have a higher molecular weight and/or are more polar as pressure increases.

Special applications of supercritical fluids to food processing: As mentioned before carbon dioxide is the most common supercritical fluid in the food industry.

Due to the non-toxicity and low critical temperature, it can be used to extract thermally labile food components and the product is not contaminated with residual solvent. Further, the extract's color, composition, odor, texture are controllable and extraction by supercritical fluid carbon dioxide retains the aroma of the product.

Supercritical fluid extraction provides a distinct advantage not only in the replacement but also extracts oils that are lower in iron and free fatty acid. Some application of SFE in food is mentioned below:

1.4.11.3.1. *Removal of fat from foods*

Edible oils and their components has been the target of supercritical fluid processing since the early 70s. Although triacylglycerides are only fairly soluble in SC-$CO_2$, the advantages of organic solvent-free processing have stimulated research and development in various areas.

1.4.11.3.2. *Enrichment of vitamin E from natural sources*

SFE offers several advantages for the enrichment of tocochromanols over conventional techniques such as vacuum distillation, in particular a lower operating temperature. As starting material one can use various edible oils or their distillates. These contains several tocotrienols and tocopherols at a total concentration of approximately 500 ppm.

1.4.11.3.4. *Removal of alcohol from wine and beer, and related applications*

De-alcoholized wine or beer is achieved by removing ethanol from water. Distillation is well known for this purpose with the disadvantage that aroma compounds will also be removed. New techniques like membrane separation (pervaporation) emerge, and in between these is SFE with $CO_2$.

1.4.11.3.5. *Encapsulation of liquids for engineering solid products*

A liquid product can be entrapped by adsorption onto solid particles (liquid at the outside of solid particles), by agglomeration (liquid in the free volumes between the solid particles), or by impregnation (liquid within the pore system of the solid particles). Microspheres or larger capsules can be formed, totally encapsulating the liquid. The solid material provides a coating for the liquid inside. Such particulate products can be achieved by means of supercritical fluid processing. An example is the so called concentrated powder form-process, wherein $CO_2$ is mixed (dissolved) in the liquid feed by static mixing. The $CO_2$-liquid feed mixture is then sprayed into a spray chamber at ambient conditions together with the substrate material. The $CO_2$ is suddenly released from the liquid, and the liquid forms small droplets. During the spraying process, solid substrate and liquid droplets are intensively mixed and combined to a solid particulate product of the type described above.

1.4.11.3.6. *Application of SFE in food safety*

At present, food safety includes many different issues such as detection of frauds, adulterations and contaminations. Among these topics, detection of food pollutants is important not only for consumers but also for administrations, control laboratories, and regulatory agencies. In order to protect consumers' health, regulations establish strict limits to the presence of pollutants in foods that must be carefully observed and determined. Generally, the analysis of food pollutants is linked to long extraction and cleanup procedures commonly based on the use of, *e.g.*, soxhlet and/or saponification.

1.4.11.3.7. *Supercritical Particle Formation*

In the 1970s and 1980s, supercritical processes were mainly focused on extraction and separation but, within the last 10 years, interest in supercritical fluid processing has also arisen in other areas. Here some of the new applications of SFE are mentioned.

1.4.11.4. *Nanoparticle formation using supercritical fluids*

SFE were used in a number of different circumstances for the preparation of micro-nanodispersed organic systems. Of industrial use are their special solvent properties and the technical feasibility of large temperature gradients.

1.4.11.4.1. *Supercritical drying*

Supercritical drying goes beyond the critical point of the working fluid in order to avoid the direct liquid-gas transition seen in ordinary drying. It is a process to remove liquid in a precisely controlled way, similar to freeze drying. It is commonly used in the production of aerogel and in the preparation of biological specimens for scanning electron microscopy. As a substance crosses the boundary from liquid to gas (phase diagram), the substance volatilizes and so the volume of the liquid decreases. As this happens, surface tension at the solid-liquid interface pulls against any structures that the liquid is attached to. $CO_2$ is often used in the development of SFE instead of the organic solvents normally employed in conventional extraction methodologies. The main advantages of using carbon dioxide fluid are: a reduced potential for the oxidation of extracted solutes, higher selectivity, increased sample throughput, shorter extraction time and a low critical temperature. The latter is beneficial in extracting thermally labile compounds, such as natural vegetable products. Supercritical carbon dioxide also has chemical inertness, suitable solvent strength, permits the separation of compounds of widely different polarities and molecular masses, has low cost and, what is more, can be removed from the extracted products without leaving any chemical residue. In addition, carbon dioxide is both non-toxic and non-explosive and its use can reduce the consumption of organic solvents; this is especially useful for the production of natural products used in foods and pharmaceuticals. There are several excellent articles on the use of supercritical carbon dioxide methodologies in various analytical areas, including the extraction of vegetable oil and of fat-soluble vitamins. Knowledge of the solubility of vegetable oil and its fat-associated products in supercritical $CO_2$ along with pressure and temperature are important for the successful application of this technology. The extraction conditions for specific solutes or classes of solutes from the vegetable matrix can also be optimised by changing the flow rate of the supercritical fluid or its pressure or temperature. Recently, an increasing interest in both the detection and the search for new oil and vitamin extraction techniques has been reported, and supercritical $CO_2$ extraction methods have been successfully applied in the extraction of oils from orange peel, hazelnut, olive, blackcurrant and vineyard grape. Despite the large number of matrices processed, only some models of supercritical $CO_2$ have been published. Nevertheless, to optimize the extraction conditions, a relationship between the composition of vegetable oils and their solubility in supercritical $CO_2$ must be taken in account. Almond is one of the major nut tree crops of the Mediterranean region. Almond kernel, analysed by using a conventional solvent method for the isolation of oil and fat-soluble vitamins, is shown to consist mainly of (g/kg): lipids (450–600), proteins (200), and carbohydrates (200), with other elements, such as tocopherols, present in much smaller quantities (tocopherols 0.4–0.8).10–12 The major lipid species were found to be (g/kg total fatty acids): oleic acid, linoleic acid (100–170), and palmitic acid (55–70). SFE has matured considerably as a processing technique over the last two decades due to fundamental thermodynamic, phase equilibria, and mass transfer studies, which have provided a basis for design of industrial scale extraction processes.

For the extraction of oils from seeds and other natural products, SFE provides the processor with the ability to alter the solvent power of the fluid, and hence its selectivity, by varying the fluids density as a function of extraction pressure and temperature. This factor, coupled with the superior mass transport properties exhibited by solutes dissolved in supercritical media, assures rapid removal of solutes from natural product matrices. SFE uses clean, safe, nonflammable, noncorrosive, nontoxic and environmentally friendly and nonpolluting solvents that do not leave behind any harmful residues. Its near-ambient critical temperature (31.1°C), good solvation power, low viscosity and high diffusivity make it ideally suitable for thermolabile natural products especially in food and pharmaceutical applications. SFE is suitable for extraction of the non-polar compounds having a relative molecular mass lower than 500 Da. There are few reports on SC-$CO_2$ extraction of microbial metabolites directly from the fermented biomass such as microbial lipids from the alga *Scenedesmus obliquus*. Freeze-drying prior to SFE has been shown to be advantageous for extraction of lipids from several microbial biomasses. Although most of biochemical and chemical productions which have produced natural and synthetic methods sensitive to heat, SFE is an alternative method to be produced that materials. This method is more suitable than crystallization, adsorption, chromatography, and evaporation methods for environment conditions. Fine chemicals in plant, can be found in cell, or out of the cell wall as synthesized. The extraction process from porous solid materials by supercritical fluids usually includes dissolution of solid or liquid component into pore liquid, diffusion in porous materials, and external mass-transfer around the solid particle.

Extraction system involves a bulk solid phase and a fluid phase. The fluid phase comprises the supercritical solvent and the dissolved extract. In many cases, the fluid can be assumed identical with the supercritical solvent. However, this has to be examined for each case. The solid phase remains within the extraction vessel (semi-continuous operation). Mass transport occurs between the two phases. Substances are transported from the solid to the fluid phase. Following parameters, such as dispersion of extract concentration along extractor and depend time, concentration of the extractable substance in the bulk phase, and radial and axial dispersion, pretreatment of the solid and operation parameters, have interest for the extraction processes. The SFE process is based on the contact between a solid raw material and a pressurized solvent, which removes the compounds of interest from the solid phase. After this removal, the extract is separated from the solvent through pressure reduction at low temperature. The optimization of operational conditions used in SFE is of great importance to make this technique economically viable, since the investment to build a SFE unit is high when compared to conventional methods. The mathematical modeling of experimental data of SFE has the objective to determine parameters for process design, such as equipment dimensions, solvent flow rate and particle size, in order to make the estimation of the viability of SFE processes in industrial scale, through the simulation of overall extraction curves (OECs). Many models for SFE OECs have been proposed, being them empirical or based on the mass balance inside the extraction bed. In this case, each author gives his own interpretation of the mass transport phenomena that happen during the process, in order to solve the differential mass balance equations, obtaining the OECs.

Recently, several studies investigating the supercritical fluid extraction of natural substance from solid matrices (seeds, flowers, leaves, etc.) have been noted. In most of these studies carbon dioxide is the solvent used because of its relatively low critical temperature (304.1K), non-toxicity, non-flammability, good solvent power, ease of removal and low cost. In addition, carbon dioxide has a low surface tension and viscosity and high diffusivity. Replacing the organic solvent with supercritical carbon dioxide provides a safety concern associated with solvent extraction, but would result in an increase in capital and operating costs. Clove (*Eugenia caryophillus*) is a native plant from Indonesia, which was spread to other tropical regions of the planet. Clove oil is rich in eugenol, which is used in the manufacturing of pharmaceutical products and by dentists. In food industry, clove oil is used as aroma, and may be also applied as anti-microbial agent in meat products and cheese. In SFE with $CO_2$ as solvent, clove has been used as a model for kinetic and thermodynamic studies of SFE systems, due to its richness in volatile oil (about 15%), which is composed by basically four substances. Eugenol is the main compound in clove oil, representing over 50% of the total extract composition. Other compounds found are α- caryophyllene, α-humulene and eugenyl acetate. Volatile oils extracted from other vegetable raw materials may have hundreds of different compounds, so clove oil, with its quite simple compositions, can be treated in SFE modeling as a single pseudo-compound. The methods employed for the extractions of the pigments studied, using conventional organic solvents, are not only limited by the ruling legislation but also require various stages of purification; their utilisation is diminishing progressively for environmental, health and safety reasons. Generally, the authors reported in the literature the use of processes of $CO_2$ supercritical extraction to obtain natural essential oils as an alternative to processes using high temperatures or liquid solvents Extraction with carbon dioxide in supercritical conditions constitutes a technology of low environmental impact; carbon dioxide possesses many undisputed advantages as a solvent, among which the more significant are its low toxicity, low cost and facility for subsequent separation from the product extracted. Moreover, an extra dimension of quality is added to the products obtained by this technique since they are not subjected to any kind of excessive heating, which usually has negative effects on thermolabile compounds. The choice of ethanol as co-solvent is based on the effective co-solvent in the supercritical extraction of hydroxycarotenoids originating from various matrices. Its presence, in trace quantities, in the final extract does not compromise the use of the carotenoids obtained in nutritional or pharmaceutical applications.

The research work already carried out has demonstrated that extraction with supercritical carbon dioxide of pigments, such as carotenoids, from carrots, cabbage and microalgae enables good extraction yields to be obtained. However, previous studies dedicated to the extraction of chlorophylls are scarce. Some of the most significant results obtained from research conducted on processes for the extraction of colorants with antioxidant properties, in particular, anthocyans, from red grape marc, and carotenoids and chlorophylls, from *microalgae* of

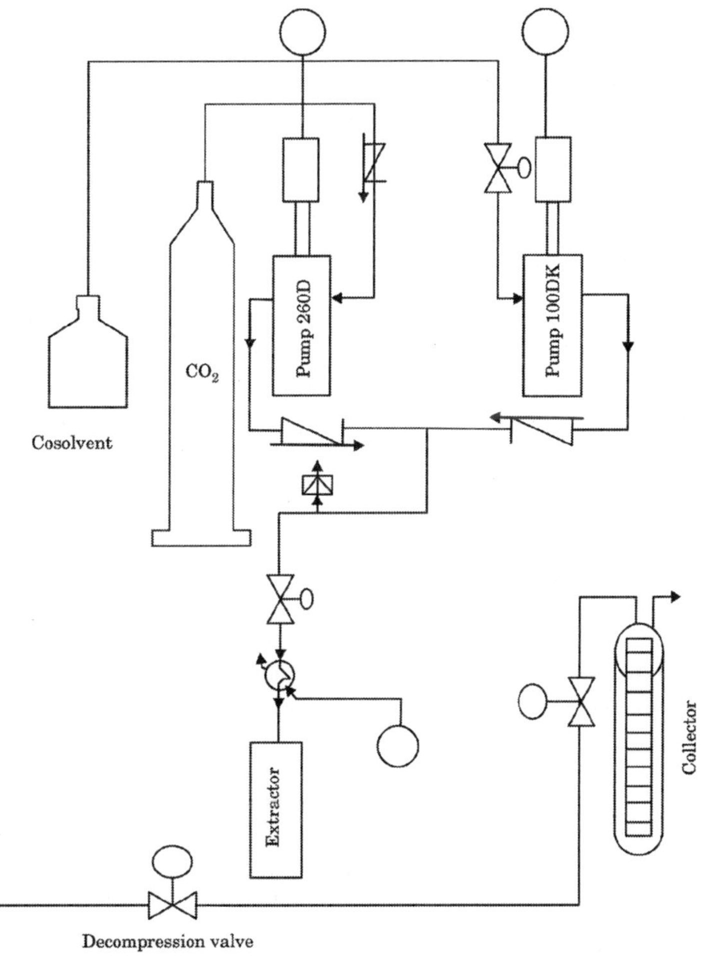


**Fig. 1.4.16:** Flow diagram of the supercritical carbon dioxide and co-solvent.

marine origin *Nannochloropsis gaditana*, *Synechococcus* sp. and *Dunaliella salina*, using supercritical carbon dioxide, either alone or mixed with co-solvents (water, methanol, ethanol). This plant comprises an extractor, a controller, a restrictor and two syringe pumps (shown in Fig.1.4.16) to supply the carbon dioxide and the ethanol, respectively. The flow rate is regulated by a micrometric valve, thermostatically controlled to 50°C, to maintain the flow rate constant at 4.5 mmol/min set in the experiments with microalgae, and at 22.5 mmol/min for the extraction of anthocyans. The extraction times employed are 3 hours for the microalgae and 2 hours for the anthocyans. The samples extracted are collected in glass tubes that contain ethanol, and are stored at 4°C in darkness, until their subsequent analysis. With the object of having a reference value with which to compare the experimental results obtained in the supercritical extraction process, experiments were conducted utilising as solvent methanol and N, N-dimethyl formamide at atmospheric pressure. The total concentrations of anthocyans, carotenoids and chlorophylls have been determined by measuring the absorbance of the various samples in a Spectrophotometer.

The engineering design of processes of supercritical fluid extraction requires knowledge of the thermodynamic constraints (solubility) as well as the mass transfer parameters. Because the flow pattern in a fixed bed is complex, it has been found useful to develop semi-empirical correlations to describe the mass transfer between the solute dissolved in the solid phase and that in the solvent phase. The SFE modeling was developed by means of different approaches: empirical models, models based on a heat transfer analogy, and models based on differential mass balance. Empirical kinetics models have been proposed by several author.

The comparison of the other methods shows that supercritical extraction is an efficient method by which natural oil can be obtained. Some of examples of functional activity and extraction conditions shown in Table 1.4. Supercritical fluid extraction was performed in a 150 $cm^3$ laboratory extractor, at 280 ± 2 bar, 40 ± 0.1°C and a $CO_2$ flow rate, 0.5 standard $dm^3$/min; and the extraction (processing) time, 3 h. With the investigated grade of milling of grape seeds (dmean = 0.35 mm), 61% extracted oil obtained. Hexane was shown as a more selective solvent, while ethanol as a more safety solvent. Five cultivars of hop were extracted by the method of supercritical fluid extraction using carbon dioxide (SFE–$CO_2$) as extractant. For the qualitative and quantitative analysis of the obtained hop extracts, the GC-MS method was used. Two of four themost common compounds of hop aroma ($\alpha$-humulene and $\alpha$-caryophyllene) were detected in samples of series A. In addition, isomerized $\alpha$-acids and a high content of $\alpha$-acids were detected. The $\alpha$-acids content in the samples of series B was the highest in the extract of the Magnum cultivar (it is a bitter variety of hop). The low contents of acids in the other entire hop samples resulted in extracts with low $\alpha$-acids content, *i.e.*, that contents were under the prescribed $\alpha$-acids content. Hazelnuts (*Corylus avellana L.*) and walnuts (*Juglans regia L.*) are important vegetable products from trees found in European countries.

Paprika oleoresin was fractionated by extraction with supercritical carbon dioxide (SCF-$CO_2$). Higher extraction volumes, increasing extraction pressures, and similarly, the use of cosolvents such as 1% ethanol or acetone resulted in

**Table 1.4:** Some of examples of functional activity and extraction conditions of supercritical carbon dioxide process

| *Compounds* | *Functional activity* | *Extraction conditions* |
|---|---|---|
| Paenol | Anti-inflammatory, anti-diabetic, cardiovascular protective Antioxidant, antimicrobial | $CO_2$ + methanol, 150 bar, 55°C, 20 min (static) + 90 min (dynamic) |
| Volatile oil | Antioxidant | $CO_2$ + ethyl acetate (10%), 250 bar, 45°C |
| Coumarin | Anti-fungicidal, anti-tumor | $CO_2$, 248 bar, 55°C (optimized through neural network model) |
| Volatile, fatty acids, tocopherols | Antioxidant | $CO_2$, 300 bar, 35°C |
| Essential oil | Antimicrobial | $CO_2$, 400 bar, 40°C, 35 min |
| Rosavin | Antioxidant, anti-stress, among others | $CO_2$ + water (10%), 200 bar, 80°C, 3 h |
| Phytosterols | Anti-inflammatory, anti-cancer | $CO_2$ + ethanol, 250 bar, 45°C, 1 h (static) + 1 h (dynamic) |
| Valerenic acid | Tranquilizing | $CO_2$ + ethanol, 360 bar, 37°C, 20 min |
| Glycosides | Anti-inflammatory, diuretic, among others | $CO_2$, 211 bar, 8°C, 60 min |
| Phenolics | Antioxidant | $CO_2$, 400 bar, 60°C, 20 min |
| Seed oil | Antioxidant | $CO_2$, 200 bar, 50°C, 2.5 h (dynamic) |
| Germ oil | Antioxidant | $CO_2$, 250 bar, 50°C, 2 h (dynamic) |
| Carotenoids | Antioxidant | $CO_2$, 250 bar, 60°C, 2 h (dynamic) |
| Xanthones | Antioxidant | $CO_2$ + ethanol (4%), 200 bar, 40°C |
| Xanthones | Natural herbicide | $CO_2$, 500 bar, 50°C, 15 min |
| Sabinene | Antifungal, antibacterial | $CO_2$ + methanol (1.5%, v/v), 100 bar, 55°C, 30 min (dynamic) |
| Flavonoids | Antioxidant | $CO_2$ + ethanol (3%, v/v), 200 bar, 40°C |
| α-Bisabolol | Anti-inflammatory | $CO_2$, 150 bar, 40°C (dynamic) |
| Cytotoxic compounds | Anti-cancer drugs | $CO_2$, 350 bar, 35°C |
| Antilehimaniosis drugs | Anti-cancer drugs | $CO_2$ + ethanol, 250 bar, 45°C |
| Antiviral compounds | Anti-tomatitis virus | $CO_2$, 200-500 bar, 60-100°C |
| Coagulation related compounds | Antithromboticand antiaterogenic | $CO_2$, 450 bar, 60°C |
| Triterpenoids | Anti fatigue | $CO_2$ + co-solvents, 250-300 bar, 50-65°C |

higher pigment yields. Within the 2000-7000 psi range, total oleoresin yield always approached 100%. Pigments isolated at lower pressures consisted almost exclusively of α-carotene, while pigments obtained at higher pressures contained a greater proportion of red carotenoids (capsorubin, capsanthin, zeaxanthin, α-cryptoxanthin) and small amounts of α-carotene.

## 1.5. LEACHING

### 1.5.1. *Introduction:*

Leaching is defined as extraction of a soluble constituent from a solid by means of a solvent. There are two purposes of leaching

1. Either for the production of a concentrated solution of a valuable solid material,
2. Or in order to remove an insoluble solid, such as a pigment, from a soluble material with which it is contaminated.

Leaching involves three steps

- Change of phase of the solute as it dissolves in the solvent,
- Its diffusion through the solvent in the pores of the solid to the outside of the particle, and
- The transfer of the solute from the solution in contact with the particles to the main bulk of the solution.

Usually the first two steps are fast and does not affect the overall rate.

## 1.5.2. FACTORS INFLUENCING THE RATE OF EXTRACTION

### 1.5.2.1. *Particle size*

The smaller the size, the greater is the interfacial area between the solid and liquid, and therefore the higher is the rate of transfer of material and the smaller is the distance the solute must diffuse within the solid as already indicated. On the other hand, the surface may not be so effectively used with a very fine material if circulation of the liquid is impeded, and separation of the particles from the liquid and drainage of the solid residue are made more difficult.

### 1.5.2.2. *Solvent*

Solvent should be selective and have sufficiently low viscosity so that it could circulate freely and involve actively in leaching out solute from solid

### 1.5.2.3. *Temperature*

In most cases, the solubility of the material which is being extracted will increase with temperature to give a higher rate of extraction. So it is desirable to operate the leaching equipment at higher temperature. However, the upper limit of temperature is determined by secondary considerations, such as, for example, the necessity to avoid enzyme action during the extraction of sugar.

### 1.5.2.4. *Agitation of the fluid*

To increases the eddy diffusion and therefore the transfer of material from the surface of the particles to the bulk of the solution, agitation is desirable. Further, agitation of suspensions of fine particles prevents sedimentation and more effective use is made of the interfacial surface.

## 1.5.3. MASS TRANSFER IN LEACHING OPERATIONS

Rate of transfer from the particles to the bulk of the liquid is obtained by using film theory:

$$\frac{dM}{dt} = \frac{k'A(c_s - c)}{b} \tag{1.5.1}$$

A is the area of the solid–liquid interface,

b is the effective thickness of the liquid film surrounding the particles,

c is the concentration of the solute in the bulk of the solution at time *t*

$c_s$ is the concentration of the saturated solution in contact with the particles,

M is the mass of solute transferred in time *t*, and

k is the diffusion coefficient.

For a batch process in which V, the total volume of solution, is assumed to remain constant,

$$dM = Vdc \tag{1.5.2}$$

$$\frac{dc}{dt} = \frac{k'A(c_s - c)}{bV} \tag{1.5.3}$$

Time *t* taken for the concentration of the solution to rise from its initial value $c_0$ to a value c is found by integration, on the assumption that both *b* and *A* remain constant

$$\int_{c_o}^{c} \frac{dc}{c_s - c} = \int \frac{k'A}{bV} dt$$

$$\ln \frac{c_s - c_o}{c_s - c} = \frac{k'A}{bV} t \tag{1.5.4}$$

If pure solvent is used initially, $c_0 = 0$,

$$1 - \frac{c}{c_s} = e^{(-\frac{k'A}{bV})t} \tag{1.5.5}$$

Thus the solution approaches a saturated condition exponentially

## 1.5.4. METHOD OF CALCULATION

### 1.5.4.1. *Equilibrium stage:*

Solid is leached with more than enough solvent and contacted for sufficient time that all the solute dissolves and there is no preferential adsorption of either solvent or solute by the solid

- All the solute will be dissolved
- Mixture is then a slurry of insoluble solid immersed in a solution of solute in the solvent

Insoluble phase is then separated physically by settling, filtration or drainage.

In practice stage is not 100% efficient due to

- Solute may be incompletely dissolved because of inadequate contact time
- Liquid-solid mechanical operation is not 100% practical

### 1.5.4.2. *Practical equilibrium*

Components

| | |
|---|---|
| Pure solvent: | A |
| Insoluble carrier solid | B |
| Solute | C |

Triangular diagram cannot be used due to frequent crowding near the corner, so Janecke diagram is used. The coordinates used in Janecke diagram are as follows

- Concentration of insoluble solid B in any mixture: N mass B/mass (A+C)
- Solute C compositions expressed in weight fraction on B free basis

x= wt. fraction C in effluent solution from stage (B free basis)
y= wt. fraction C in solid/slurry from stage (B free basis)

### 1.5.4.3. *Coordinate system*

Consider a mixture at $M_1$ (conc. of clear solution =x and insoluble solid=$N_{M1}$) (Figs. 1.5.1 and 1.5.2). If it is allowed to settle it will form a clean liquid drawn at $R_1$ having the composition of clean liquid x and solid B at $E_1$ in which the solution will also have the same composition as the clear liquid. $E_1R_1$ is a vertical line and is a tie line.

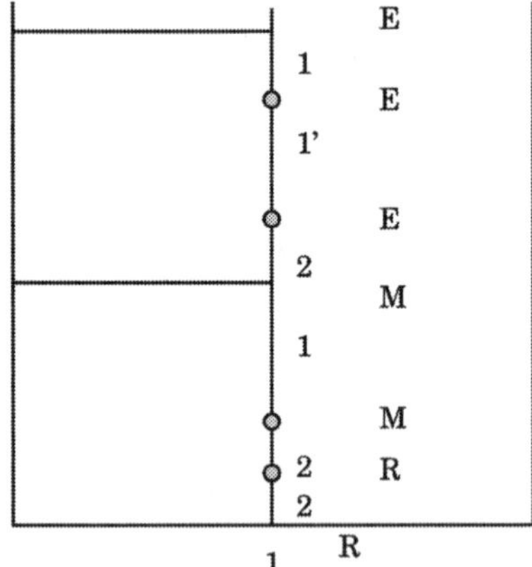


**Fig. 1.5.1:** Tie line tracing of leaching operation

#### *Single stage leaching*

Overall solution balance

- $F + R_o = E_1 + R_1 = M_1$ (1.5.6)

Location of M

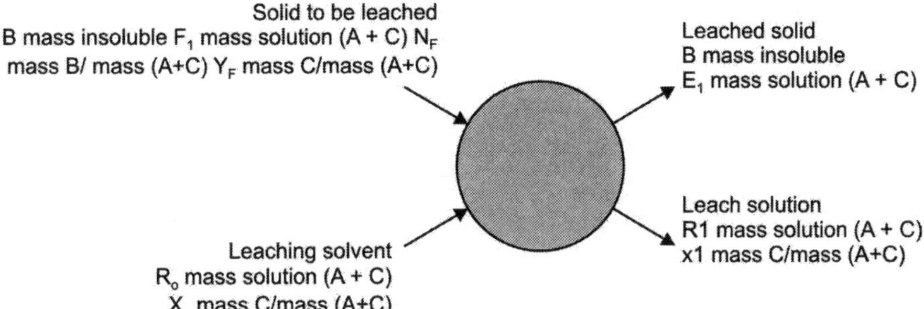


**Fig. 1.5.2:** Flowsheet for single stage leaching operation

- $N_{M1} = B/(F + R_o)$ (1.5.7)
- $y_{M1} = (y_F F + R_o x_o)/(F + R_o)$ (1.5.8)

After M is located, tie line is drawn which passes through $M_1$ using equilibrium data.

Thus N and x or y are located for $E_1$ and $R_1$. The weights of $E_1$ and $R_1$ are obtained from

- $B = N_F F = E_1 N_1$ (1.5.9)

Thus $E_1$ is determined and substituting in overall material balance would yield $R_1$

### *Multistage cross current leaching*

The stage calculation shall remain same as for single stage but the only difference in that the solid with solute leaving the stage 1 becomes the feed for the stage 2. Hence, for stage 2 calculations, appropriate terms shall have to be used in the equations given in method above.

### *Multistage counter current leaching*

The stage calculations are shown in Fig. 1.5.3.

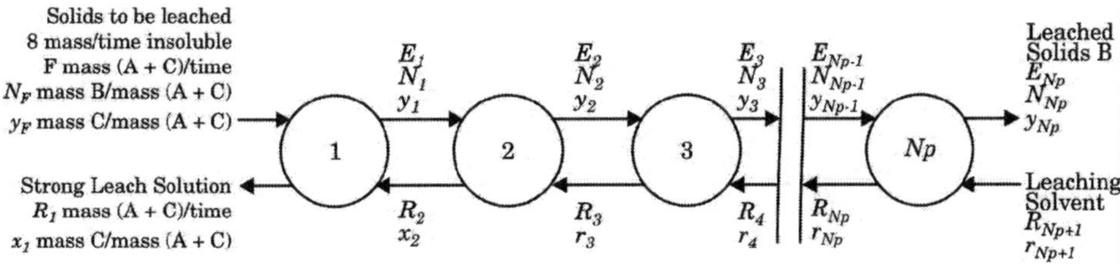


**Fig. 1.5.3:** Flowsheet for multistage leaching operation

Step 1: Overall plant balance

$$F + R_{Np+1} = R_1 + E_{Np} = M \quad (1.5.10)$$

Solution (A + C) balance

$$Fy_F + R_{Np+1}x_{Np+1} = R_1x_1 + E_{Np}y_{Np} = My_M \quad (1.5.11)$$

Step 2: Location of M

B balance: $MN_M$=B

$$(F + R_{Np+1})N_M = B \quad (1.5.12)$$

$$N_M = B / (F+R_{Np+1}) \quad (1.5.13)$$

C Balance

$$My_M = Fy_F + R_{Np+1}x_{Np+1} \quad (1.5.14)$$

$$y_M = (Fy_F + R_{Np+1}x_{Np+1})/(F + R_{Np+1}) \quad (1.5.15)$$

Step 3: Locate $E_{Np}$ by extending line $R_1M$ by intersection with N vs y curve

Step 4: Join $FR_1$ and $E_{Np}R_{Np+1}$ to meet when extended at ΔR

$$F\text{-}R_1 = E_{Np} - R_{Np+1} = \Delta R \quad (1.5.16)$$

Step 5: Locate the tie line that passes through $R_1$ and have another point at $E_1$ on N vs y curve

Step 6: Join $E_1$ ΔR that will give point $R_2$ on N vs x curve

Step 7: Similarly the tie through $R_2$ gives $E_2$ and so on

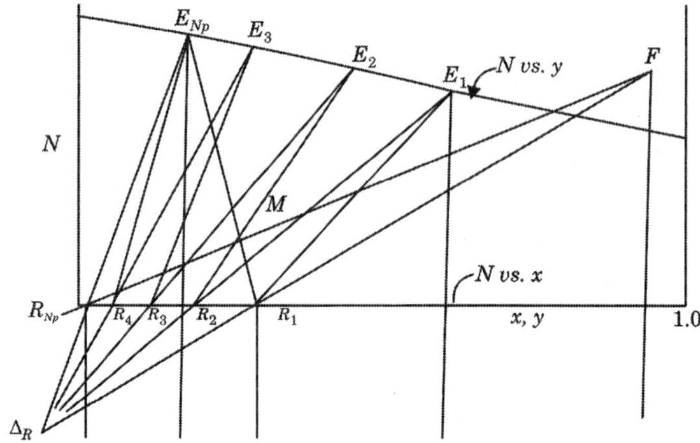


**Fig. 1.5.4:** Stage calculation for multistage leaching operation

## 1.6. EVAPORATION

### 1.6.1. *Introduction*

Evaporation is defined as vaporizing a portion of solvent to produce a concentrated solution of thick liquor. Usually heating in evaporators are done by steam (at low pressure 3 atm abs) condensing on the metal tubes. In majority of cases, evaporation refers to the removal of water from aqueous solution. Typical example includes concentration of aqueous solutions of sugar, NaCl, milk etc. In some cases evaporation is a pre-step to crystallization. The primary purpose of evaporation is to concentrate the solution so that upon cooling, crystal can be formed.

### 1.6.2. *Heat transfer by vaporization without mechanical agitation*

When steam is passed through a tube dipped inside water, initially due to high temperature inside the pipe, the layer near the pipe outside surface gets heated. At the start the heat shall be transferred due to convection only. With passage of time the bubbles starts to form but will not rise to the surface as long as the buoyancy effect is small. With the due passage of time due to buoyancy the bubbles loose the holding and rises to the surface. This shall happen only if the liquid in the bubble is hotter than the saturation temperature in the incipient bubbles *i.e.* heat flows from the liquid to the bubbles. The numbers of bubbles formed are function of texture of the tube surface and hence higher the roughness of surface more is the bubbles formed.

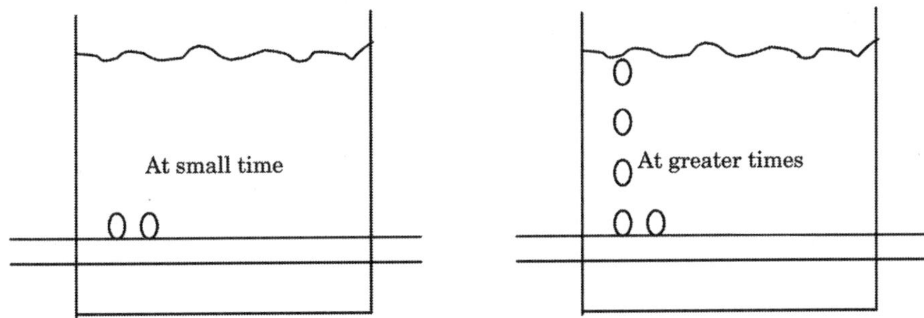


**Fig. 1.6.1:** Bubble formation in the tank containing tube in which steam is passing

Heat can be transferred by free convection or by convection produced by rising stream of bubbles. Bubble formation and growth depends on the surface tension. It is important that the measurement of surface tension is determined against what *i.e.* against fluid ($ST_{fluid\text{-}fluid}$) or metal ($ST_{fluid\text{-}metal}$) since these two quantities are very different. Here as can be seen that liquid is in contact with the pipe, so surface tension is measured against metal ($ST_{liq\text{-}metal}$). If $ST_{liq\text{-}metal}$ is low *i.e.* liquid tends to wet the surface bubble are readily occluded by the liquid and hence rises. In $ST_{liq\text{-}metal}$ is intermediate, a balance exists between bubbles and tubes and hence large bubble needs to develop to get free from the surface. If $ST_{liq\text{-}metal}$ is high the bubble will not leave the surface.

### 1.6.3. *Typical boiling curve*

The typical boiling curve is as shown in Fig. 1.6.2. As can be seen, initially, higher is the difference between tube wall and vaporization temperature, higher is the heat flux. But at higher temperature difference, as against condensation or convection where a continuous increase in flux is obtained, in case of evaporation/boiling because of blanketing or vapor binding (gas film is formed that provides resistance to heat flow) smaller vaporization coefficients results and hence flux decreases.

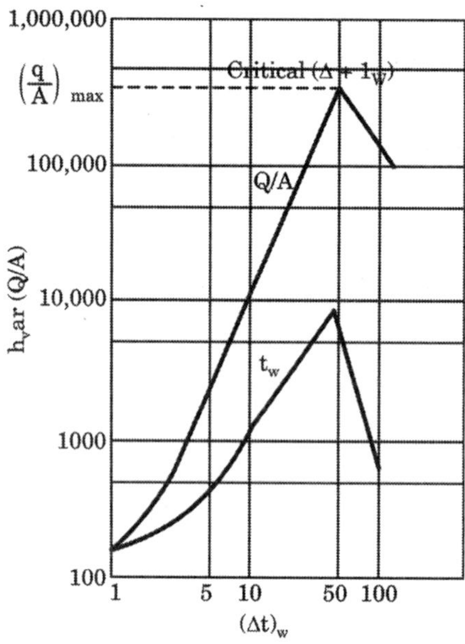


**Fig. 1.6.2:** Typical boiling curve

## 1.6.4. *Factors affecting rate of mass transfer by vaporization from pools*

1. Nature of surface and distribution of bubbles
2. Properties of liquid (surface tension, coefficient of expansion, viscosity etc.)
3. Influence of temperature distribution upon evolution and vigor of the bubbles.

## 1.6.5. *Classification of vaporizing equipments*

The categorizations of vaporizing equipment are shown in Fig. 1.6.3.

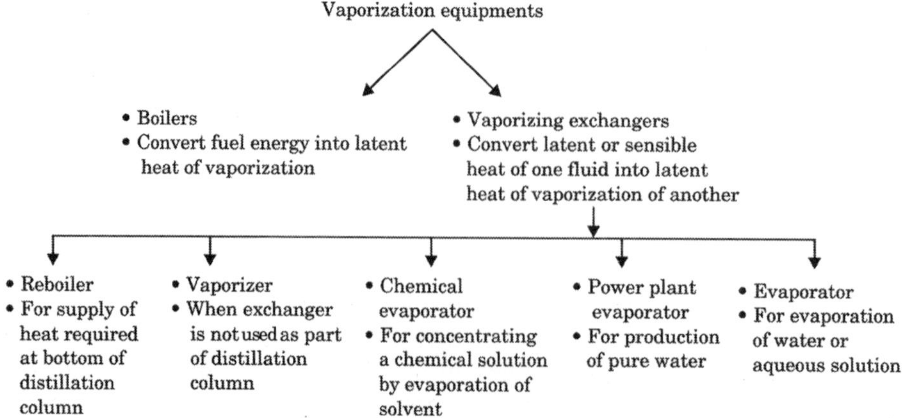


**Fig. 1.6.3:** Classification of vaporizing equipments

### 1.6.6. *Chemical evaporators*

Important terms related to evaporator design.

**Difference between chemical evaporator and power plant evaporator**

| *Power plant evaporator* | *Chemical evaporator* |
|---|---|
| Separate pure water from raw impurities called as blow down | Evaporate dilute solutions *e.g.* caustic soda, table salt, sugar etc. to get a concentrated solution |
| Unevaporated portion of feed is residue | Unevaporated portion of feed is product |
| Have blow down | No blow down and liquid is fed into multiple forward or backward feed evaporators, thus greater heat recover is obtained |

### *Boiling point rise (BPR)*

Though chemical evaporators are capable of high heat efficiency, under certain conditions of high utilization of temperature potential, they become uncapable leading to the requirement of greater surfaces.

### *Reason*

Concentrated aqueous solutions undergoes boiling point rise above the saturation temperature (T*, at which water should boil) corresponding to pure water at the same pressure. *E.g.*

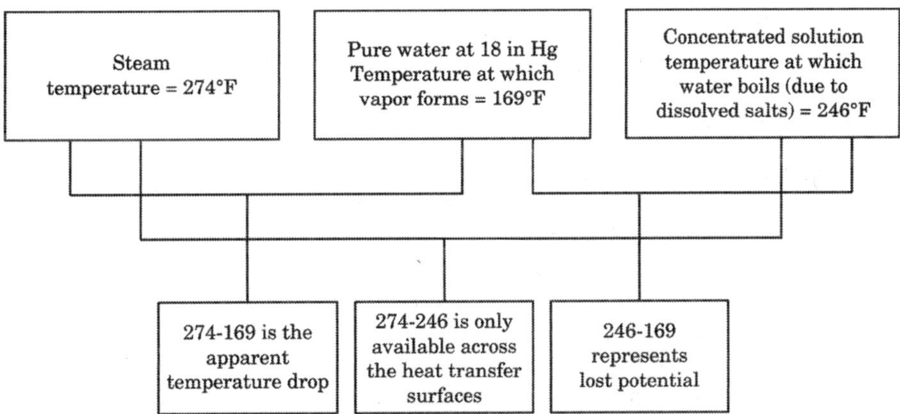


Pure water at 18 in Hg boils at 169°F but due to presence of dissolved salts it boils at 246°F. If the temperature of steam used for vaporization is 244 K, instead of true apparent temperature drop, which is 274-169 = 105°F, only 274-246 = 28°F shall be available for the process. Thus 246-169 represents the lost potential. Hence, to vaporize a portion of feed, the work which should be done in one evaporator for the same, it is required to have supply either more heat or use more number of evaporators. Further, the overall heat transfer coefficient shall be based on this actual temperature difference rather than the apparent temperature difference. Thus BPR represents a major constraint in the design of evaporators.

If BPR is above 5°F, latent heat of vaporization of steam cannot be obtained from steam tables. They are evaluated as follows

## 1. *Duhring rule*

$$\frac{\lambda_w}{\lambda_s} = \frac{\Delta t_w'}{\Delta t_s'} \left(\frac{t_w'}{t_s'}\right)^2_{abs} \tag{1.6.1}$$

$\lambda;_w$ = latent heat of 1 lb of pure water at temperature = $t_w$' and pressure = $p_s$

$\lambda_s$ = latent heat of pure water from solution at temperature = t and pressure = $p_s$

$\Delta t_w', \Delta t_s'$ = rate of change of the two boiling point curves over the same pressure range

## 2. *Othmer method (based on Clausius Clapeyron equation)*

$$\frac{\lambda_s}{\lambda_w} = \frac{d\log p_s}{d\log p_w} \tag{1.6.2}$$

$d \log p_s, d \log p_w$ are the respective absolute vapor pressure of the solution and pure water.

*Note*: BPR can be computed only for dilute solutions which are relative ideal. For real solutions data are obtained experimentally.

## *Fluid properties*

Fluid properties are important in the design of evaporators. Residues in higher concentrations have high viscosities. Some the important physical properties of evaporating liquid is as follows, *Concentration*: Boiling is the combination of vaporization and free convection and overall heat transfer coefficient is the function of concentration and temperature at which evaporation occurs. With evaporation, the concentration of the liquor increases and hence viscosity. Higher viscosities results in negligible Grashof's number (Gr $\propto 1/\mu^2$), hence for evaporators operating under natural convection, the problem of liquid circulation persists. Thus, in those cases it is better to employ forced circulation. Further, in some cases crystal formation can occur and hence it may lead to clogging of tubes. In addition, BPR is much higher in the case of concentrated solutions.

*Foaming and scaling*: If the residue are highly foaming and scaling care must be taken. Foams can lead to entrainment of liquid with the vapor leaving. This may lead to the lost of the material out of evaporator with the vapors. If the solution deposits scale, the useful heat transfer area shall decrease and many a time subsequent shutdowns are required owing to low overall coefficient. Additional costs are involved for cleaning scales that leads to increase in the plant operating costs.

***Temperature sensitivity*:** For the concentration of heat sensitive material, care must be taken to reduce the residence time of the material.

### *Hydrostatic head*

Boiling point is function of pressure at which liquid vaporizes at the interface. If there is great layer of liquid above the tube bundles it will exert hydrostatic pressure upon liquid in contact with the tube surface. This added pressure raises the boiling temperature at heat transfer surface above that necessary to produce vaporization corresponding the pressure at liquid vapor interface. Hydrostatic head reduces the useful temperature difference effective upon the heat transfer surface. Thus size of heat transfer surface must be designed accordingly. Hydrostatic head may be determined as

$$\Delta t_h = 0.03 \frac{T_R v}{\lambda_s} \Delta p \quad (1.6.3)$$

$\Delta t_h$ = hydrostatic elevation of the boiling point, °F

$\Delta p$ = hydrostatic head, ft

$v$ = specific volume of water vapor at $T_R$, $ft^3/lb$

$T_R$ = solution boiling temperature, °R

$\lambda_s$ = latent heat of vaporization corresponding to the saturation pressure Btu/lb

Influence of hydrostatic pressure will be greater as vaccum upon system is increased since velocity varies considerably with pressure whereas $\lambda_s$ is very little.

## 1.6.7. *Method of operation*

(i) ***Single effect evaporator*:** Fig. 1.6.4 shows the simplified diagram of single effect evaporator

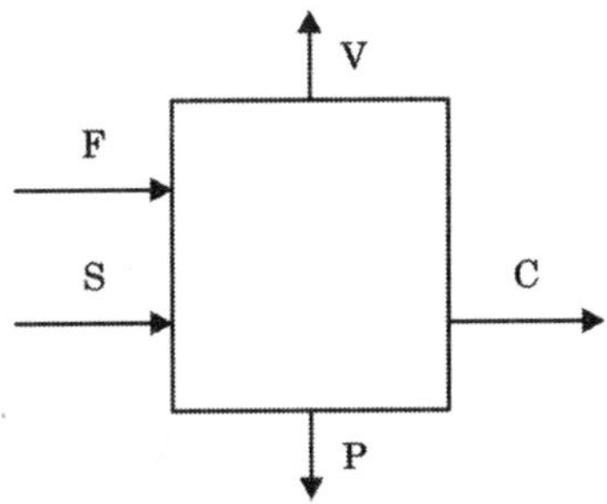


**Fig. 1.6.4:** Flowsheet of single effect evaporator

Feed and steam enters the evaporator. Condensate steam leaves as drip. Because of a complete mixed system, both the streams leaving and inside of the evaporator are at same temperature and composition. Feed is generally at boiling point and for 1 kg of steam condensing, 1 kg of vapor evaporated (for dilute solution only). Single effect evaporators are generally used when the required capacity (number of kg of water vaporized per hour) is relatively small and the cost of steam is relatively cheap compared to evaporator cost.

**ii) Multiple effect evaporator:**

***Forward feed***

In this arrangement multiple evaporators are joined is series to form triple effect system. Vapor from the first effect serves as the heating medium for the second effect and so on. The noncondensable from the last effect are removed using vacuum pump. Feed (dilute) enters the first effect where the pressure in the vapor space is highest and moved to the other stages where the pressure is lower than the first stage. In the last stage the vapor space pressure is minimum. In each effect, the temperature drop across the heating surface corresponding to the pressure drop in that effect. Pipes were used to connect the various evaporators. Dilute feed gets concentrated as it moves through the multiple effect evaporators. Thick product is removed from the last effect. Pump is required to move the feed from effects to effects; however, pressure gradient can also be used to move the feed through the effects.

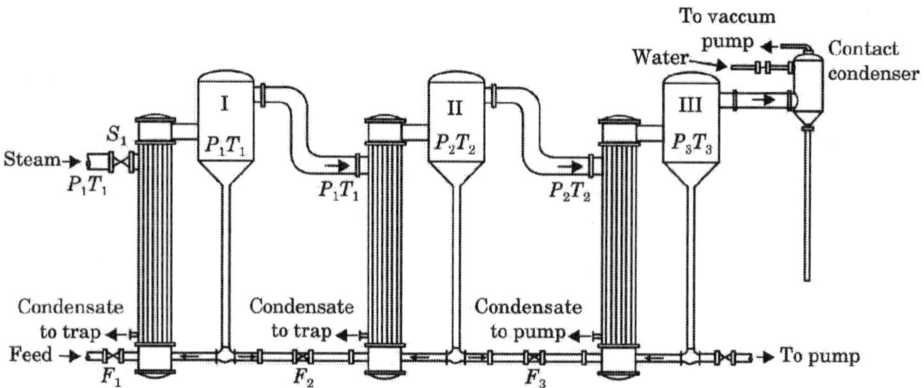


**Fig. 1.6.5:** Triple effect evaporator

The heat transmitted by the heating surface per hour is given as:

$$q_1 = U_1 A_1 \Delta T \tag{1.6.4}$$

If the feed is at boiling point all the heat shall be transferred to convert the liquid in the feed to vapor. The temperature of feed leaving the first effect is very near to the temperature of vapor leaving it. Under steady state operation all the heat that has been spent in creating vapor in the first effect shall be given up when the same vapor condenses in the second effect and so on. Thus

$$q_2 = U_2 A_2 \Delta T \tag{1.6.5}$$

and

$$U_1 A_1 \Delta T = U_2 A_2 \Delta T \tag{1.6.6}$$

The same reasoning can be expanded to the other effects. In practice the area for heat transfer in all the effects are same and hence

$$U_1 \Delta T = U_2 \Delta T = U_2 \Delta T = \frac{q}{A} \tag{1.6.7}$$

### *Backward feed*

In this the feed enters the last effect and coldest and concentrated product leaves the first effect. Pumping is required in each stage exit to pump the concentrated mass to the next stage and so on. Backward feed often gives a higher capacity than the forward feed when the thick liquor is viscous, but may give lower economy (number of kg vaporized per kg of steam fed to the unit) than forward feed.

## 1.6.8. *Method of calculation*

Multiple effect evaporators: In industries it is preferable to use multiple effect evaporators against single effect since a single effect evaporator is wasteful of energy because the latent heat of the vapor leaving is not used but is discarded. However, much of this latent heat can be recovered and reused by employing multiple effects. The following give the calculations for heat transfer area for double effect evaporator (Fig. 1.6.6).

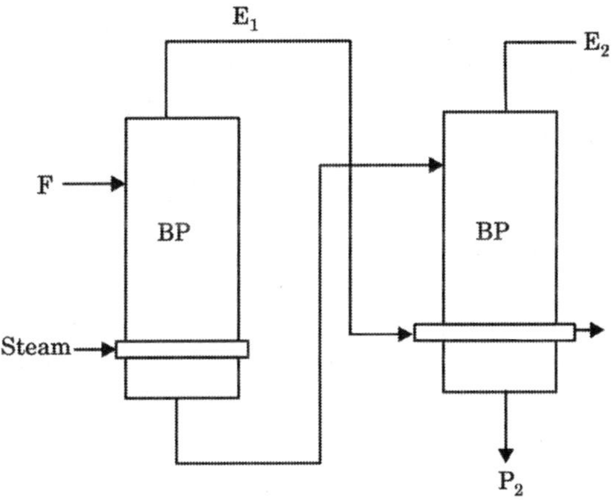


**Fig. 1.6.6:** Flowsheet of double effect evaporator

Overall component balance

$$F = E_1 + E_2 + P_2 \tag{1.6.8}$$

First effect

$$F_1 = E_1 + P_1 \tag{1.6.9}$$

Second effect

$$P_1 = P_2 + E_2 \tag{1.6.10}$$

Component continuity equation

$$Fx_F = P_2 x_{P2} \tag{1.6.11}$$

Energy balance

$$Fh_f + S\,\lambda_s = E_1 H_1 + P_1 H_1 \tag{1.6.12}$$

$$P_1 h_1 + E_1 \lambda_{E1} = E_2 H_2 + P_2 H_2 \tag{1.6.13}$$

ΔT ovarall determination

$$U_1 \Delta T_1 = U_2 \Delta T_2 \tag{1.6.14}$$

Finding the area

$$A_1 = \frac{S\lambda_s}{U_1 \Delta T_1} \tag{1.6.15}$$

$$A_2 = \frac{E_1 \lambda_1}{U_2 \Delta T_2} \tag{1.6.16}$$

## 1.7 CRYSTALLIZATION

### 1.7.1. *Introduction*

Crystallization is a separation and purification technique employed to produce a wide variety of materials. Crystallization may be defined as a phase change in which a crystalline product is obtained from a solution. A solution is a mixture of two or more species that form a homogenous single phase. Solutions are normally thought of in terms of liquids, however, solutions may include solids suspension. Typically, the term solution has come to mean a liquid solution consisting a solvent, which is a liquid, and a solute, which is a solid, at the conditions of interest. The solution to be ready for crystallization must be supersaturated. A solution in which the solute concentration exceeds the equilibrium (saturated) solute concentration at a given temperature is known as a supersaturated solution. There are four main methods to generate supersaturation that are the following:

- Temperature change (mainly cooling),
- Evaporation of solvent,
- Chemical reaction, and
- Changing the solvent composition (*e.g.* salting out).

Crystallization of important in many regards

- Variety of materials are marked in crystalline form
- Crystal formed from a impure solution is itself pure
- Easy packing and storing

Crystal's formation occurs from solution named as magma, which is a two phase mixture of mother liquor and crystals of all sizes. The most important parameter in crystallization is the size. If the crystal is to be further processed, reasonable size and size uniformity is important for filtering, washing, reacting with other chemicals, transporting, storing etc. Further size is important in term of its customer acceptance. The size of crystal is given in terms of characteristic length which is the product of sphericity and diameter of the particle. Sphericity is given as

$$\Phi_s = \frac{6V_p}{s_P D_p} \qquad (1.7.1)$$

Crystal has constituents particles arranged in three dimensional array of space lattic. They appear as polyhedrons with sharp corners and flat sides or faces. Size of face and edges of various crystals formed from same material may be widely different but the angle made by corresponding faces of same material are equal and is the characteristics of that material. Based on the angle the crystals are classified as cubic, hexagonal, tetragonal, orthorhombic or monoclinic etc. Usually the crystal are 100% pure but still some impurity can remain. The reason of this is the occlusion of mother liquor within the solid mass when mother liquor of low purity dry on the surface and results in the contamination.

## 1.7.2. *Equilibria*

Equilibria relationship of bulk crystals is give in terms of solubility curve. There are three types of solubility curves for materials (Fig. 1.7.1).

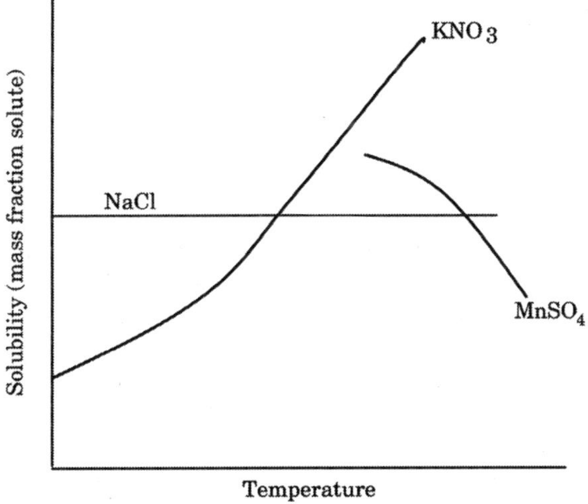


**Fig. 1.7.1:** Equilibria curve

Phase equilibria of substance which crystallized with water of hydration is different because several hydrated are formed. Fig. (1.7.2) show the equilibria curve for magnesium sulphate showing its various hydrates as function of temperature and composition of the mixture.

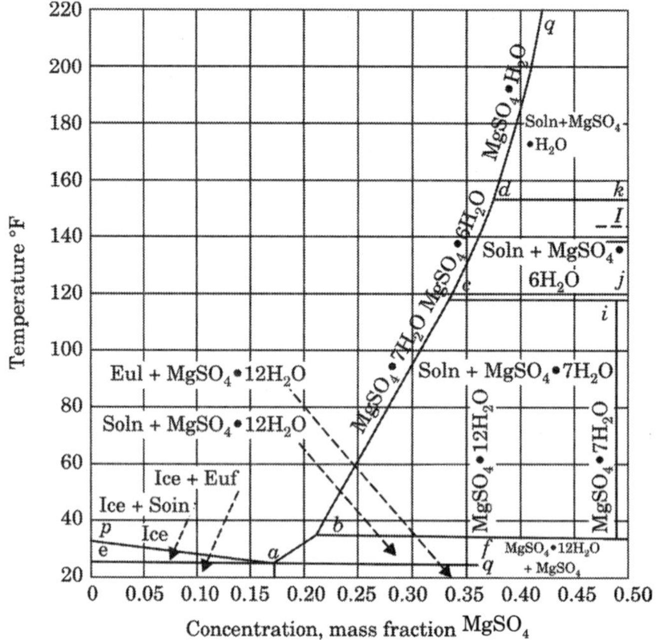


**Fig. 1.7.2:** Equilibrium curve for Magnesium sulphate hydrates

pae- represents mixture of ice and saturated solution. Any solution containing less than 16.5% $MgSO_4$ precipitates ice when temperature reaches pa line. eagfhij represents complete solidification of liquid solution to form various solid phases. Any solution more concentrated than 16.5% $MgSO_4$ precipitates on cooling, a solid when temperature reaches this line. abcdq - represents solubility curve

## 1.7.3. *Yield*

Crystal and mother liquor are in long contact to reach equilibrium and mother liquor is saturated at the final temperature of process. Yield is calculated from the concentration of original solution and its solubility at final temperature. When crystal growth is slow (this happens when solution is viscous or where crystal collects in bottom so that little crystal surface is exposed to supersaturation), mother liquor retains appreciable supersaturation and actual yield is less than calculated from solubility curve. If the crystals are anhydrous, yield calculation is simple as solid phase contains no solvent, however, when crop contains water of crystallization, it is important to take account of accompanying water as this is not available for retaining solute in solution. Solubility data however ignore water of crystallization.

## 1.7.4. *Supersaturation*

*There are two steps in crystal formation*: i) Nucleation *i.e.* the birth of new particle and ii) growth to macroscopic size. These two determines crystal size distribution and are the driving potential for both the rates in supersaturation

How to generate supersaturation?

(i) If solubility increases strongly with temperature *e.g.* inorganic salts and organic substances, supersaturation can be attained by cooling the saturated solution.

(ii) If solubility is independent of temperature *e.g.* common salt, supersaturation is attained by evaporating a portion of solvent.

(iii) If neither cooling nor evaporation is desirable, addition of third component is adopted to create crystallization. There are two situation under this.

a. *Salting*: If third component act physically by forming with the original solvent a mixed solvent in which the solubility of the solvent is sharply reduced.

b. *Precipitation*: If the third component react with original solute and form a insoluble substance.

### *Units of supersaturation*

Supersaturation is expressed as the concentration difference between that of supersaturated solution in which crystal is growing and that of a solution in equilibrium with crystal. In terms of mole fraction it is represented as

$$\Delta y = y - y_s \tag{1.7.2}$$

where $\Delta y$ is supersaturation (mole fraction of solute), y is mole fraction of solute in solution and $y_s$ is molar concentration of solute in saturated solution. In terms of moles of solute in unit volume of solution, it can be given as

$$\Delta c = c - c_s \tag{1.7.3}$$

where $\Delta c$ is molar supersaturation, c is molar concentration of solute in solution and $c_s$ is molar concentration of solute in saturated solution. $\Delta c$ can be expressed in terms of $\Delta y$ as

$$\Delta c = \rho_M y - \rho_s y_s \tag{1.7.4}$$

$\rho_m$ are $\rho_s$ molar densities of solution and supersaturated solution respectively. Two other term are important, concentration ratio and fractional supersaturation and can be defined as

$$\alpha = \frac{c}{c_s} = 1 + \frac{\Delta c}{c_s} = \frac{y}{y_s} = 1 + \frac{\Delta y}{y_s} = 1 + s \tag{1.7.5}$$

Percentage of supersaturation is expressed as 100 * s. In practice this is usually less than 2%.

### 1.7.5. *Nucleation*

Rate of nucleation is defined as the number of new particles of crystal formed per unit time per unit volume of magma. There are three kinds of nucleations

1.7.5.1. *Spurious nucleation*

**a) *Due to macroscopic attrition*:** Impact of propeller breaks soft or weak crystals and hence give rise to new crystals. The disadvantage of this process is that it degrade the quality of product, however the important feature of it is that it is the only source of new crystal formation that is independent of supersaturation.

**b) *Initial breeding*:** In this method seed crystals originating from previous crystallization are added to the crystallization system. These seeds carry on their surface many small crystals (formed during drying or storage after manufacture). These wash off and subsequently grow in supersaturated solution.

**c) *Needle breeding*:** This is a growth related spurious nucleation and it occur at large supersaturation or accompanies poor magma circulation. It is characterized by abnormal needle like or whisker like growths from the end of crystals. These spikes are imperfect crystals bound to parent crystal by weak forces which break off to give crystals of poor quality.

**d) *Veiled growth*:** This is another growth related imperfection unrelated to nucleation and occurs at moderate supersaturation. The reason for this being the occlusion of mother liquor onto the faces of the crystal giving a milky surface and an impure product. The reason for this is too rapid crystal growth, which traps mother liquor into the crystal face.

All the above type of imperfections can be removed by growing crystals at low supersaturation or by using well designed and operated pumps and agitators.

1.7.5.2. *Primary nucleation*

Nucleation represents the birth of very small bodies of new phases withing a supersaturated homogeneous existing phase. It is a consequence of rapid local fluctuation on a molecular scale in a homogeneous phase that is in a state of metastable equilibrium. There are two type of primary nucleation

a) Homogeneous nucleation: Formation of new particles within a phase uninfluenced in any way by solids of any sort, including the wall of contained or every very minute particle of foreign substances is called as homogeneous nucleation. However such type of nucleation never happens. The fundamental of such type of crystal birth is represented in Fig. 1.7.3. Small crystals (less than 1 μm size) posses a significant amount of energy per unit mass and hence their solubility is large than big crystals and they are in equilibrium (unstable) with supersaturated solution. If the solution of both small and large crystals are present, the smaller crystal will dissolve and the larger one will grow until the smaller one disappears. This is called as Ostwald ripening. Relationship between solubility of substance and particle size is given by relation name as Kelvin equation defined as

$$\ln \alpha = \frac{4V_m \sigma}{\upsilon RTL} \qquad (1.7.6)$$

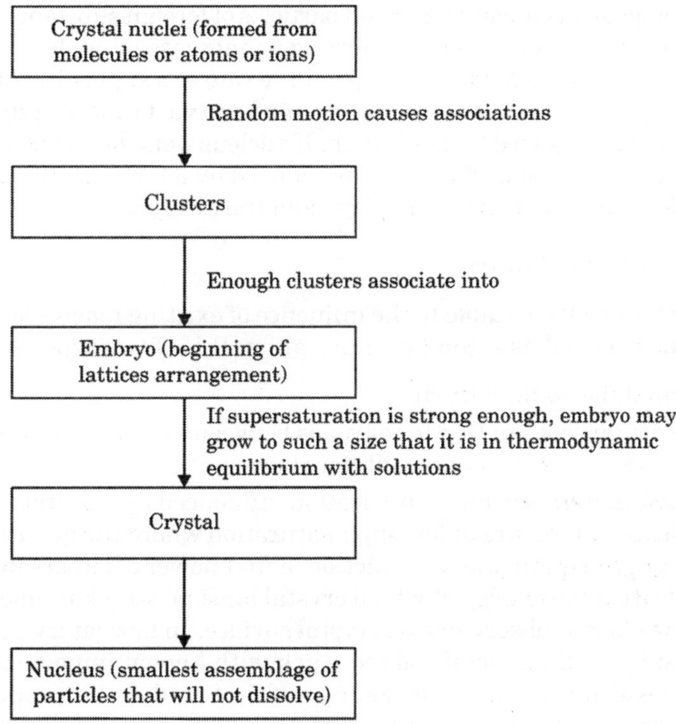


**Fig. 1.7.3:** Fundamental of crystal growth by primary nucleation

where α represents the ratio of concentration of supersaturated and saturated solutions (α = 1 + s), σ is average surface tension, ν is number of ion per molecule of solute, L is crystal size, $V_m$ is molar volume of crystals.

Rate of nucleation is given as

$$B_o = c \exp\left[\frac{-16\,\pi\,\sigma^3\,V_m^2 Na}{3\nu^2 (RT)^3 (\ln \alpha)^2}\right] \tag{1.7.7}$$

$B_o$ is nucleation rate (Number/cm$^3$.s), R is gas constant (8.314 * $10^7$ erg/g.mol.K), Na is Avogadro Number (6.022*$10^{23}$ mlecules/g.mol), c is frequency factor and is statistical measure of rate of formation of embryos that reaches the critical size. It is proportional to concentration of individual particles and rate of collision of these particles with an embryo of the critical size required to form stable nucleus. It's exact values is unknown. From anology its values is approx. $10^{25}$ nuclei.cm$^2$.s. σ numerical valuc is uncertain to predict, still they may be estimated from solid state theory. For ordinary salt σ is approximately 80-100 erg/cm$^2$.

Using the values stated above, Eq. (1.7.7) is used to predict the value of 's' for $B_o$ = 1 (Nucleation rate of 1 nucleus per sec per cm$^3$) and the value of 's' calculated comes out to be very large, which is impossible for materials of usual solubility's. This proves that homogeneous nucleation never occurs.

b) Heterogeneous nucleation: Solid particles of foreign substances influence the nucleation process by catalyzing an increase in nucleation rate at a given supersaturation or giving a finite rate at a supersaturation where homogeneous nucleation would occur after a vast time. Catalyst reduces the energy required for nucleation. If nucleus wets the surface of catalyst, the work of nucleus formation is reduced by a factor that is function of angle of wetting between nucleus and the catalyst.

1.7.5.3. *Secondary nucleation*

Nucleus formation attributable to the influence of existing macroscopic crystals in the magma is called as secondary nucleation. It is of two types

a. Attributable to fluid shear
b. Contact nucleation: Due to collisions between existing crystals with each other and walls/impellers/agitators blades.

***Contact nucleation*:** Secondary nucleation influenced by agitation is called as contact nucleation. It occurs at low supersaturation where the growth of crystal is optimum for good quality and is function of first power of supersaturation. In contact nucleation the energy at which crystal must be struck is amazingly low and no visible effect is observable on crystal surface. In laboratory experiments it is initiated by making small rods to stuck with known impact energy. The number of resulting nuclei is measure and is found to be proportional to supersaturation and contact energy. Contact nucleation is actually combination of breakage of microscopic dendritic growths on the surface of the growing crystals and an interference by the contacting object (this deflects or dislodges particles ranging in size from embryo to small crystals) with clusters of solid particles moving to become organized into crystals.

1.7.6. *Crystal growth*

The crystal growth and the steps (Diffusion + reaction) are explained in Fig. 1.7.4.

Driving force for mass transfer = $y - y'$

Driving force for reaction = $y' - y_s$

Equation for mass transfer

$$\text{Molar flux } N_A = \frac{\dot{m}}{S_p} = k_s(y' - y_s) \tag{1.7.8}$$

Where $S_p$ is surface area of crystal, $k_y$ is mass transfer coefficient and $\dot{m}$ is rate of mass transfer

Equation for surface reaction

$$\text{Molar flux } N_A = \frac{\dot{m}}{S_p} = K(y - y_s) \tag{1.7.9}$$

The resistance of two steps may be added to give overall coefficient as

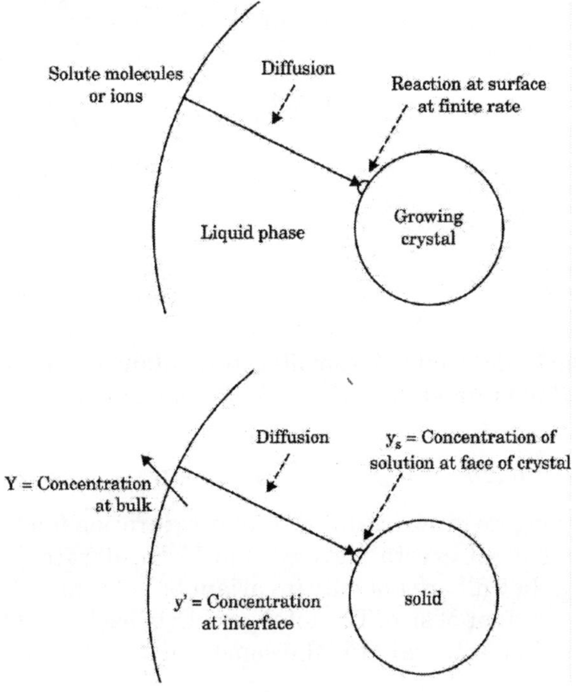


**Fig. 1.7.4:** Steps in crystal growth

$$\frac{\dot{m}}{S_p} = K\,(y - y_s) \tag{1.7.10}$$

Eliminating y' from Eq. 1.7.8 and 1.7.9 and followed by substitution in Eq.1.7.10

$$K = \frac{1}{\dfrac{1}{k_y} + \dfrac{1}{k_s}} \tag{1.7.11}$$

### *Growth rate*

Volume of crystal is proportional to cube of characteristic length (L), hence

$$v_P = a\,L^3 \tag{1.7.12}$$

hence mass (m) is defined as

$$m = v_P * \rho_M = a\,L^3 \rho_M \tag{1.7.13}$$

where $\tilde{n}_M$ is molar density. Differentiating Eq. 1.7.13,

$$\dot{m} = \frac{dm}{dt} = 3\,a\,L^2 \rho_M\,\frac{dL}{dt} \tag{1.7.14}$$

dL/dt is defined as growth rate (G). Now,

$$S_P = \frac{6\,v_P}{L} = 6\,a\,L^2 \tag{1.7.15}$$

Using Eq. 1.7.15 and Eq. 1.7.14 in Eq. 1.7.11,

$$K = \frac{3\,a\,L^2\,\rho_M\,G}{6\,a\,L^2\,(y\text{-}y_s)} \tag{1.7.16}$$

$$G = \frac{2\,K\,(y\text{-}y_s)}{\rho_M} \tag{1.7.17}$$

$k_y$ and ks can be determined from literature. Many literatures are available for, however still the numerical value of kind that can be used in the design is scarce.

*ΔL law of crystal growth*

If all crystals in magma grow in uniform supersaturation field and at the same temperature and if all crystals grow from birth at rate governed by the supersaturation, they all are not only invariant but also have the same growth rate that is independent of size. The total growth of each crystal in the magma during the same time interval $\Delta t$ is the same and

$$\Delta L = G\Delta t \tag{1.7.18}$$

This law is highly idealized model and is certainly not realistic in all situations. Further the constant growth rate is expected for crystals between 50-500 μm size.

1.7.7. *Material balance on crystallizer*

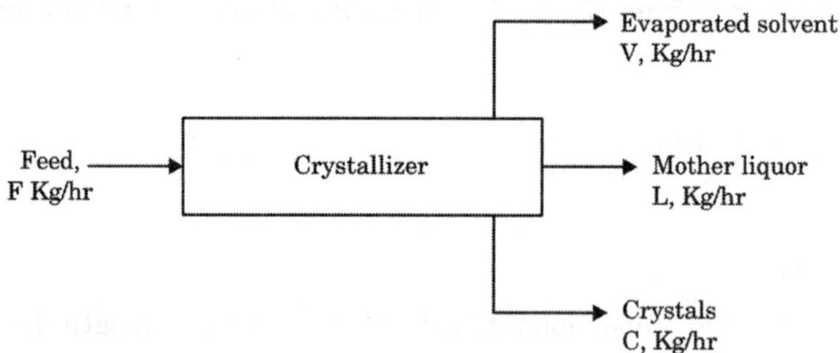


F, V, L, C = mass flow rates of feed solution, evaporated solvent, mother liquor and crystals obtained

$X_F'$ = weight fraction of anhydrous solute in feed solution

$X'$ = solubility of solute at final temperature expressed as weight ratio of anhydrous salt to solvent.

$M_1$ = molecular weight of anhydrous solute salt

$M_2$ = molecular weight of hydrated salt

L' = Kg/hr of solvent in mother liquor

Overall material balance

$$F = C + L + V \tag{1.7.19}$$

### *Material balance for solvent*

Solvent in feed = solvent evaporated + solvent in mother liquor + solvent in hydrated crystals

$$F\,(1 - X_F') = V + L' + \frac{M_2 - M_1}{M_2} C \tag{1.7.20}$$

Rearranging

$$L' = F\,(1 - X_F') - V - \frac{M_2 - M_1}{M_2} C \tag{1.7.21}$$

### *Material balance of solute*

Solute in feed = solute in product crystal + solute in mother liquor

$$F\,X_F' = \frac{M_1}{M_2} C + [F\,(1 - X_F') - V - \frac{M_2 - M_1}{M_2} C\,]X' \tag{1.7.22}$$

$M_1/M_2$ = ratio of molar masses of anahydrous salt to hydrate. If the crystals are anhydrous then,

$$F\,X_F' = C + [\,F\,(1 - X_F') - V)X' \tag{1.7.23}$$

If no evaporation takes place V = 0

$$\%\text{ yield crystals} = \frac{\text{Amount of solute cyrstallized} * 100}{\text{Amount of solute in feed solute}} \tag{1.7.24}$$

## 1.8 DRYING OF SOLIDS

### 1.8.1. *Introduction*

The removal of moisture from a solid, a solution or a gas to make it dry is often necessary in various industrial operations, particularly in chemical industrial process. Thus drying may be defined as an operation in which the liquid, generally water present in wet solid is removed by vaporization to get a moisture free solid. Drying of solid is not a complete removal of moisture. Drying is a liquid-solid separation process. Drying is separation different from other liquid- liquid separation process. There are many separation process which is not called drying. The removal of water from a purified soap lye to make glycerine is called evaporation because the final product is not a solid. Production of condensed milk is evaporation but production of milk powder is drying. Other example of drying is the dehydration of a gas by absorption or adsorption of the moisture. Natural gas is died by absorption of moisture in triethyle glycol in a tray tower. With a crystalline product, it is essential that the crystals are not damaged during the drying process, and, in the case of pharmaceutical products, care must be taken to avoid contamination. Shrinkage, as with paper, cracking, as with wood, or loss of flavour, as with fruit, must also be prevented. With the exception of the partial drying of a material by squeezing in a press or the removal of water by adsorption, almost all drying processes involve the removal of water by vaporisation, which requires the addition of heat.

### 1.8.1.1. *Mechanism of drying*

The main function that governs the drying process is the heat and mass transfer involved. Consider a drop of water suspended in air at the tip of thin wire. If the vapour pressure of the water at the temperature is higher than the partial pressure of water vapour in the ambient air, evaporation of water will occur. The latent heat required for evaporation will be supplied by partly by the surrounding air (by convective heat transfer); partly by water drop due to this temperature of water falls. As the temperature falls the drop vapour pressure also decrease and cause of reduction of driving force of mass transfer and increase in heat transfer driving force from the ambient air to water drop so the contribution of heat transfer from the ambient air towards the water drop increases, if the sufficient time is allowed, the drop will attain a steady state temperature. At steady state the heat flux from the ambient is equal to the latent heat of required for the evaporation. The temperatures remain constant known as wet bulb temperature.

### 1.8.1.2. *Important Definition or Terms:*

***Equilibrium moisture*:** Air entering has some relative humidity if we have to dry the material; we cannot do it below this relative humidity as drying is an equilibrium process. The portion of water in the wet solid that cannot be removed by inlet air because of humidity of the latter is called as equilibrium moisture.

***Free moisture*:** Difference between the total moisture content of the solid and equilibrium water content

$$X = X_T - X^*$$

***Unbound moisture*:** In the equilibrium curve, when the curve intersect the 100% humidity line. This point is the maximum moisture that the substance can carry if the water content is above that or equal to at 100% humidity; it will exert a vapour pressure equal to that exerted by liquid water at same temperature. This water is unbound moisture.

***Bound moisture*:** Water corresponding to concentration lower than that indicated by the intersection of the curve with the 100% humidity line is called as bound moisture. This water exerts vapour pressure less than the liquid water at the same temperature.

### 1.8.1.3. *Mechanism of moisture movement during drying*

When surface evaporation occurs, there must be a movement from the depths of the solid to the surface. The nature of the movement influences the drying

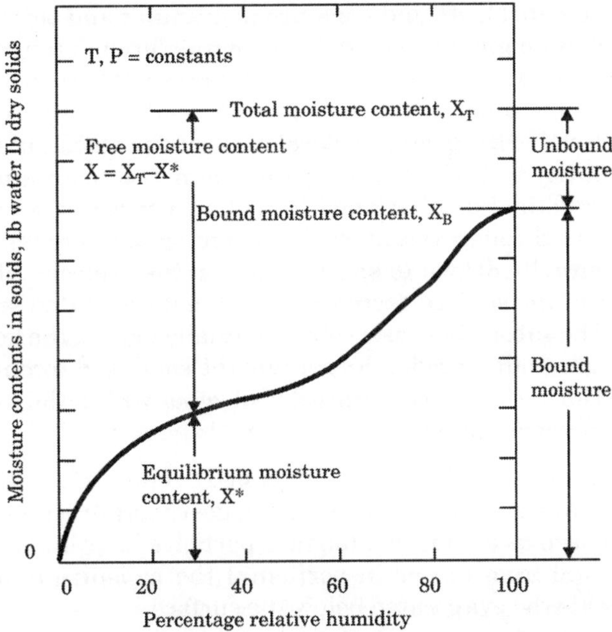


**Fig. 1.8.1:** Isotherm for equilibrium moisture content

during the falling rate periods. In order to appreciate the diverse nature of the falling rate period portion of the drying curve which has been observed, let us review some of the theories advanced to explain moisture movement and their relation to the falling rate curves.

***Liquid diffusion:*** Diffusion of liquid moisture may result because of concentration gradient existing between the depth of solid, where the concentration is high and the surface where is low. These gradients are set up during drying from the surface. This method of moisture transport is probably limited to cases where single phase solid solution are formed with the moisture, as in case of glue, gelatine, and the like, and to certain cases where bound moisture is dried, as in the last portion of water from clays, flour, textile, paper and wood. The general mechanism of this process is

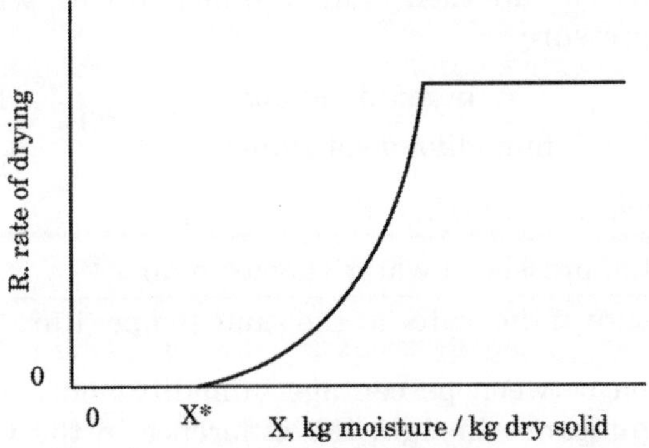


**Fig. 1.8.2:** Drying rate curve

***Capillary movement*:** Unbound moisture in granular and porous solids such as clays, sand, paint pigments, and the like, moves through the capillaries and interstices of the solids by a mechanism involving surface tension, the way oil moves through a lamp wick. The capillaries extend from small reservoirs of moisture in the solid to the drying surface. As drying proceeds, at first moisture moves by capillarity to the surface rapidly enough to maintain a uniformly wetted surface and the rate of drying is constant. The water is replaced by air entering the solid through relatively few openings and cracks. The surface moisture is eventually drawn to spaces between the granules of the surface, the wetted area at the surface decreases, and the unsaturated-surface drying period follows. The subsurface reservoirs eventually dry up, the liquid in a zone or plane which gradually recedes deeper into the solid, and second falling-rate period results. During this period, diffusion of vapor within the solid will occur from the place of vaporization to the surface.

***Vapor diffusion*:** Especially if heat is supplied to one surface of a solid while drying proceeds from another, the moisture may evaporate beneath the surface and diffuse outward as a vapour. Moisture particles in granular solids which have been isolated from the main portion of the moisture flowing through capillaries may also be evaporated below the surface.

***Pressure*:** Shrinkage of the outside layers of a solid on drying may squeeze moisture to the surface. Usually we can only speculate about which mechanism is appropriate to a particular solid and must rely on more or less empirical treatment of the experimental rates of drying.

## 1.8.2. *Pyschroemetry*

### 1.8.2.1. *General pyschrometric properties*

Humidity H, mass of water per unit mass of dry air.

$$\text{H} = \frac{P_w}{(P - P_w)} = \frac{18P_w}{29(P - P_w)} \quad \frac{\text{moles of water vapour}}{\text{moles of dry air}} \tag{1.8.1}$$

Where $P_w$ is the partial pressure of water vapor and P is the total pressure. Humidity of saturated air $H_0$. This is the humidity of air when it is saturated with water vapour. The air then is in equilibrium with water at the given temperature and pressure.

$$\text{Percentage humidity} = \frac{\text{humidity of air}}{\text{humidity of saturated air}} \times 100 = \frac{\text{H}}{\text{H}_0} \times 100 \tag{1.8.2}$$

Percentage relative humidity, R =

$$\frac{\text{partial pressue of water vapour in air}}{\text{vapour pressure of the water at the same temperature}} \times 100 \tag{1.8.3}$$

The distinction between percentage humidity and percentage relative humidity is of significance though; the difference in the values of the two quantities does not usually exceed 7 to 8 per cent.

Humid volume is the volume of unit mass of dry air and its associated vapour. Then, under ideal conditions, at atmospheric pressure:

$$\text{Humid volume} = \frac{22.4}{29}\left(\frac{T}{273}\right) + \frac{22.4H}{18}\left(\frac{T}{273}\right) m^3 / kg \tag{1.8.4}$$

Where T in degrees K,

$$\text{Or:} = \frac{359}{29}\left(\frac{T}{492}\right) + \frac{359H}{29}\left(\frac{T}{492}\right) ft^3 / lb \tag{1.8.5}$$

Where T in degrees Rankine

Saturated volume is the volume of unit mass of dry air, together with the water vapour required to saturate it. Humid heat is the heat required to raise unit mass of dry air and associated vapour through 1 degree K at constant pressure or 1.00 + 1.88H kJ/kg K. Dew point is the temperature at which condensation will first occur when air is cooled.

***Wet bulb temperature*:** If a stream of air is passed rapidly over a water surface, vaporisation occurs, provided the temperature of the water is above the dew point of the air. The temperature of the water falls and heat flows from the air to the water. If the surface is sufficiently small for the condition of the air to change inappreciably and if the velocity is in excess of about 5 m/s, the water reaches the wet bulb temperature $\theta_w$ at equilibrium. The rate of heat transfer from gas to liquid is given by:

$$Q = hA(\theta - \theta_w) \tag{1.8.6}$$

The mass transfer of vaporization is given by:

$$G_\upsilon = \frac{h_D A M_w}{RT}(P_{wo} - P_w)$$

$$= \frac{h_D A M_A}{RT}[(P - P_w)_{mean}(H_w - H)]$$

$$= h_D A \rho_A (H_w - H) \tag{1.8.7}$$

The rate of heat transfer required to effect vaporisation at this rate is given by:

$$G_\upsilon = h_D A \rho_A (H_w - H)\lambda \tag{1.8.8}$$

At equilibrium, the rates of heat transfer given by equations 1.8.6 and 1.8.8 must be equal, and hence:

$$(H - H_w) = -\frac{h}{h_D \rho_A \lambda}(\theta - \theta_w) \tag{1.8.9}$$

In this way, it is seen that the wet bulb temperature θw depends only on the temperature and humidity of the drying air. In these equations:

h is the heat transfer coefficient, $h_D$ is the mass transfer coefficient, A is the surface area, θ is the temperature of the air stream, $\theta_w$ is the wet bulb temperature, $P_{w0}$ is the vapour pressure of water at temperature θw, $M_A$ is the molecular weight of air, $M_w$ is the molecular weight of water, *R* is the universal gas constant, T is the absolute temperature, H is the humidity of the gas stream, $H_w$ is the humidity of saturated air at temperature $\theta_w$, $\rho_A$ is the density of air at its mean partial pressure, and λ is the latent heat of vaporisation of unit mass of water.

### 1.8.2.2. *Adiabatic saturation temperature*

Water is continuously sprayed in the humidification chamber through the spray pipe. The water at the bottom of the chamber is recirculated by a pump. An air stream continuously enters into the chamber and is brought in intimate contact with the water spray. The air stream attains thermal equilibrium with the water at temperature Ts and also gets saturated with water vapour at that temperature before it leaves the chamber. A small quantity of water at the temperature Ts may be fed continuously in order to compensate for the vaporization loss of water. The walls of the chamber are well-insulated and the system operates adiabatically. The temperature Ts attained by the air (which is the same as the temperature of water) is called the 'adiabatic saturation temperature'. (Also denoted as $T_{as}$). Suppose Y' is the humidity of the inlet and Tg is its temperature; b is the heat of vaporization of water at the adiabatic saturation temperature,Ts. If Ts is also the reference temperature (chosen for convenience) and Y' is the humidity of the saturated exit air,

Enthalpy of the inlet air = cH (Tg-Ts) + bY', in kJ per kg dry air (1.8.10)

Enthalpy of exit air = cH (Ts-Ts) + bYs'= bYs' (1.8.11)

Because the water temperature does not change, its enthalpy remains constant. The enthalpy of air also remains constant as there is no heat exchange with the surroundings. At steady state, we have

cH(Tg - Ts) + bYs'=bYs', this implies (Tg -Ts)=(Ys' -Ys)b/cH (1.8.12)

### 1.8.2.3. *Equilibrium moisture content of solid*

Drying equilibrium is the relation between the moisture content of a solid and the humidity of the ambient drying medium at equilibrium. Drying by vaporization occurs till the moisture content of the solid is more than the equilibrium value under the prevailing state (temperature and moisture content) of the ambient medium. The moisture present in the solid exerts a pressure which is equal to or less than the vapour pressure of water at a particular temperature. This depends upon the nature of the solid, its moisture content and the temperature. For a non-hygroscopic material the equilibrium moisture content is zero at all temperature and humidity condition of the ambient medium. Equilibrium moisture content in solid can be determined experimentally by keeping the material suspended in a gas of a constant humidity and temperature for a sufficiently long time. Weights of the solid at equilibrium as well as in bone dry (a solid totally free from moisture) conditions are to be noted. Moisture content is generally expressed as the mass of moisture per unit mass of bone dry solid. Equilibrium moisture content of few substances is plotted in figure against the relative humidity of the ambient medium to get moisture:-

From the graph (Fig. 1.8.3) we can see that the point B is having moisture content 0.3 and relative humidity 1. On the right of point B, the curve is horizontal. For moisture content of the solid equal to or more than 0.3, the equilibrium vapour pressure exerted by the solid is equal to the vapour pressure.

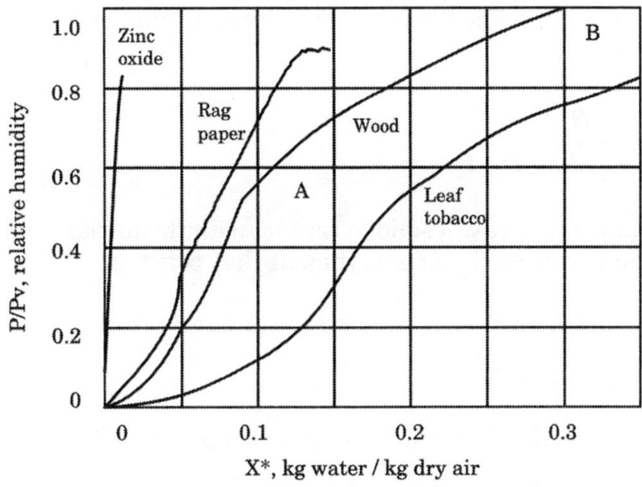


**Fig. 1.8.3:** Equilibrium moisture content of some solids at 25°C

1.8.2.4. *Moisture-Evaporation Temperature*

The combination of the concept of the adiabatic- saturation temperature and the wet bulb temperature can be used to track the gas temperature and the moisture- evaporation (solid temperature), $T_v$, with respect to location or time when removing surface moisture from a wet solid. If the moisture is water, $T_v$ will be constant and equal to the constant $T_w$ of the gas. If the moisture is an organic compound, $T_v$ still equal to $T_w$, but the temperature will not be constant and will decrease as the gas temperature decreases.

Let $T_0$ and $H_0$ be the hypothetical entering conditions of a gas being used to dry a wet solid in an adiabatic, direct- heat dryer. At any point in the dryer, the condition of the gas is given by:

$$T_o - T_g = \Delta H_s^{vap} \frac{\left(H_g - H_o\right)}{\left(C_s\right)_o} \quad (1.8.13)$$

Assuming quasi- steady state transport condition at any point, the gas – liquid interface conditions of $T_v$ and $H_v$ are related by the following form:

$$T_v - T_g = k_y M_B \Delta H_s^{vap} \frac{(H_g - H_v)}{h} \tag{1.8.14}$$

For a continuous dryer, the above equation holds regardless of whether the flow of the gas and wet solid is countercurrent or concurrent. The equation —(1.8.13) and —(1.8.14) can be combined to give:

$$\frac{T_v - T_g}{T_o - T_g} = \left(\frac{k_y M_B (C_s)_o}{h}\right)\left(\frac{H_g - H_v}{H_g - H_o}\right) \tag{1.8.15}$$

The coefficient of the right side of the equation (1.8.15) is the inverse of the psychrometic ratio. For the air – water system, that ratio can be assumed to be 1.0 giving

$$\frac{T_v - T_g}{T_o - T_g} = \left(\frac{H_g - H_v}{H_g - H_o}\right) \tag{1.8.16}$$

Now assume that the wet solid in contact with the initial or entering gas is at wet bulb temperature, $T_w$, of gas, then at that point

$$\frac{T_w - T_g}{T_o - T_g} = \left(\frac{H_g - H_w}{H_g - H_o}\right) \tag{1.8.17}$$

But since $T_o$ is fixed, $T_v$ must remain at the value $T_w$ regardless of the value of $T_g$.

### 1.8.3 *Drying Periods*

In drying, it is necessary to remove free moisture from the surface and also moisture from the interior of the material. If the change in moisture content for a material is determined as a function of time, a smooth curve is obtained from which the rate of drying at any given moisture content may be evaluated. The form of the drying rate curve varies with the structure and type of material.

#### 1.8.3.1 *Constant rate period*

During the constant rate period, it is assumed that drying takes place from a saturated surface of the material by diffusion of the water vapour through a stationary air film into the air stream. In order to calculate the rate of drying under these conditions, the relationships obtained for diffusion of a vapour from a liquid surface into a gas may be used. The simplest equation of this type is:

$$W = k_G A (P_s - P_w) \tag{1.8.18}$$

Where $k_G$ is the mass transfer coefficient. Since the rate of transfer depends on the velocity u of the air stream, raised to a power of about 0.8, then the mass rate of evaporation is:

$$W = k_G A\left(P_s - P_w\right)u^{0.8} \tag{1.8.19}$$

where: A is the surface area,

$P_s$ is the vapour pressure of the water, and

$P_w$ is the partial pressure of water vapour in the air stream

This type of equation, used for the rate of vaporisation into an air stream, simply states that the rate of transfer is equal to the transfer coefficient multiplied by the driving force. It may be noted, however, that $(P_s - P_w)$ is not only a driving force, but it is also related to the capacity of the air stream to absorb moisture. These equations suggest that the rate of drying is independent of the geometrical shape of the surface. However, the ratio of the length to the width of the surface is of some importance, and the evaporation rate is given more accurately as:

(a) For values of u = 1–3 m/s:

$$W = 5.53 \times 10\text{–}9L^{0.77}B(P_s - P_w)(1 + 61u^{0.85})\ \text{kg/s} \tag{1.8.20}$$

(b) For values of u < 1 m/s:

$$W = 3.72 \times 10\text{–}9L^{0.73}B^{0.8}(P_s - P_w)(1 + 61u^{0.85})\ \text{kg/s} \tag{1.8.21}$$

where: $P_s$, the saturation pressure at the temperature of the surface (N/m$^2$), $P_w$, the vapour pressure in the air stream (N/m$^2$), and L and B are the length and width of the surface, respectively (m).

For most design purposes, it may be assumed that the rate of drying is proportional to the transfer coefficient multiplied by $(P_s - P_w)$. If the temperature of the surface is greater than that of the air stream, then $P_w$ may easily reach a value corresponding to saturation of the air. Under these conditions, the capacity of the air to take up moisture is zero, while the force causing evaporation is $(P_s - P_w)$. As a result, a mist will form and water may be re-deposited on the surface. In all drying equipment, care must therefore be taken to ensure that the air or gas used does not become saturated with moisture at any stage. The rate of drying in the constant rate period is given by:

$$W = \frac{dw}{dt} = \frac{hA\Delta T}{\lambda} = k_G A\left(P_s - P_w\right) \tag{1.8.22}$$

Where: W is the rate of loss of water, h is the heat transfer coefficient from air to the wet surface, T is the temperature difference between the air and the surface, $\lambda$ is the latent heat of vaporisation per unit mass, $k_G$ is the mass transfer coefficient for diffusion from the wet surface through the gas film, A is the area of interface for heat and mass transfer, and $(P_s - P_w)$ is the difference between the vapour pressure of water at the surface and the partial pressure in the air.

It is more convenient to express the mass transfer coefficient in terms of a humidity difference, so that $k_G A\,(P_s - P_w) = kA\,(H_s - H)$. The rate of drying is thus determined by the values of h, $\Delta T$ and A, and is not influenced by the conditions inside the solid.

### 1.8.3.2 *Falling rate period*

Below the critical moisture content drying curve obey falling rate period. The shape varies from material to material. Shape also depends on thickness of material and on other external variables. In some cases there is distinct break in the curve during falling rate period signifying a change in drying mechanism. The break occurs at second critical point. When the average moisture content of the solid has reached a value $X_c$, the critical moisture content, the surface film of moisture has been so reduced by evaporation that further drying causes dry spots appear upon the face, these spots occur increasingly larger portion of the exposed face as drying proceeds. Since, however, the rate N is computed by means of constant gross surface A, the value of N must fall even though the rate per unit or wet surface remain constant. This give rise to the first part of unsaturated surface drying (c to d) (Fig. 1.8.4). On further drying, the rate at which moisture can move through the solid, as a concentration gradient exist is the controlling step. in any event, the rate of drying falls even more rapidly than before, as from D to E. At point E, the moisture content of the solid has fallen to the equilibrium value X* for the prevailing air humidity and drying stops.

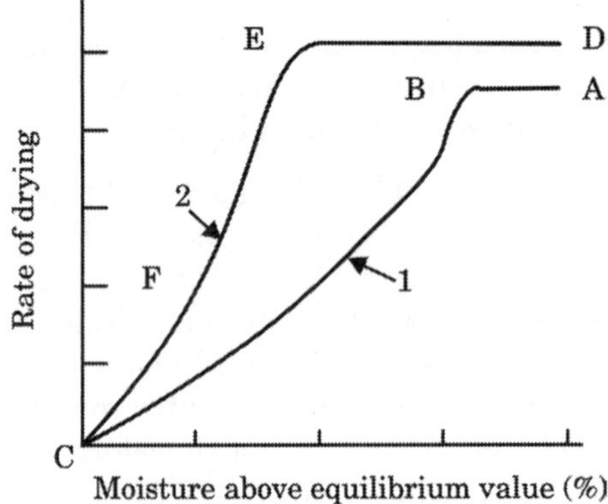


**Fig. 1.8.4:** Rate of drying of a granular material

***First falling-rate period:*** The points B and E in Fig. 1.8.5 represent conditions where the surface is no longer capable of supplying sufficient free moisture to saturate the air in contact with it. Under these conditions, the rate of drying depends very much on the mechanism by which the moisture from inside the material is transferred to the surface. In general, the curves in Fig. 1.8.5 will apply, although for a type 1 solid, a simplified expression for the rate of drying in this period may be obtained.

***Second falling-rate period:*** At the conclusion of the first falling rate period it may be assumed that the surface is dry and that the plane of separation has moved into the solid. In this case, evaporation takes place from within the solid and the vapour reaches the surface by molecular diffusion through the material. The forces controlling the vapour diffusion determine the final rate of drying, and these are largely independent of the conditions outside the material.

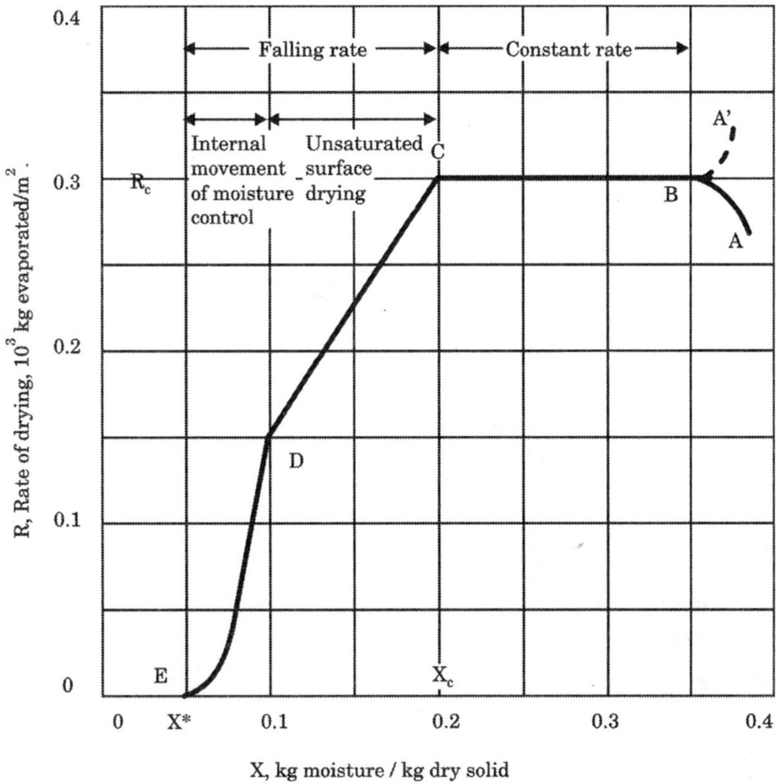


**Fig. 1.8.5:** Typical rate of drying curve, under constant condition

## 1.8.4 *Drying Rate curve*

Drying rate is constant as long as delivery of water from interior to surface is sufficient. At critical point, surface layer of water begin to recede into solid. This starts with larger pores. Though rate of evaporation per unit wetted area remains unchanged, rate based on total area (wet + dry) is less than in the constant rate period. The total drying curve divided into two parts.

### 1.8.4.1 *Rate of drying*

Rate of drying depends on the same factors that are active during constant rate period, since mechanism of evaporation is unchanged. Vaporization zone is at or near surface. Water in pores is continuous phase and air is dispersed phase. Rate of drying curve is linear

Fig. 1.8.5 given below explains the rate of drying and the amount of moisture content left with the passage of time:

- With the passage of time $X_T$ falls as follow (line A)
- Feed is heated to vaporization temperature and then the graph is linear, then curve toward horizontal and then levels off

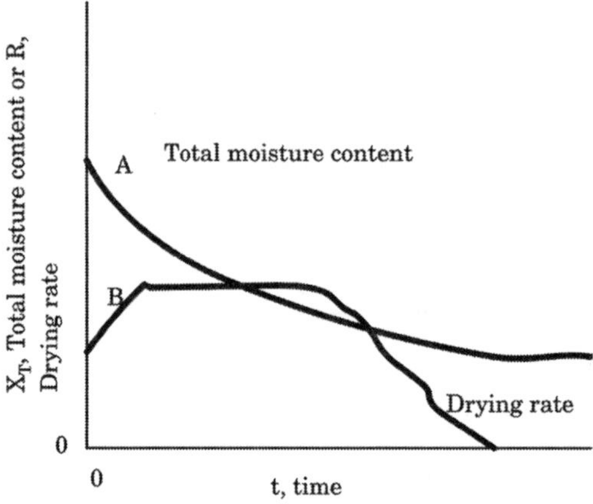


**Fig. 1.8.6:** Rate of drying curve

- Drying rate is shown in B and is horizontal for much length indication the constant rate of drying

Rate of drying curve conforms to diffusion model and is concave upward. Vaporization rate is practically independent of velocity of air.

***Constant rate of drying (A-B):*** Drying rate is plotted against the free moisture content for the porous and non- porous materials. For non – porous material (Fig. 1.8.7a) the rate of drying increases sharply for some time and then becomes constant when the moisture content drops to minimum in the material. For the porous materials (Fig. 1.8.7b) the rate of drying first increases sharply then increases linearly for some time and at the end becomes constant when the moisture content drops to minimum.

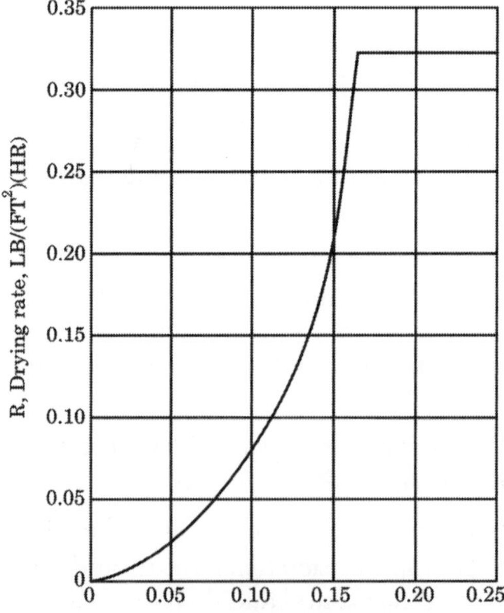


**Fig. 1.8.7:** (a)Non porous solid

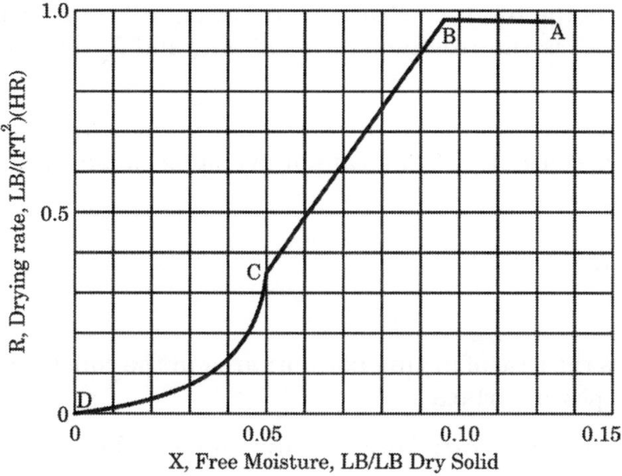


**Fig. 1.8.7:** (b)Porous solid

***From the graph it is clear that:*** Drying rate is independent of moisture content, Reason: solid is so wet that a continuous film of water exists over the entire drying surface and this water acts as if the solid were not there, For non porous solid- this is mainly superficial water, For porous solid: this is supplied constantly from the interior of the solid, Transport mechanism: evaporation and hence correlations for evaporation are applicable,

1.8.4.2 *Critical moisture content*

Point terminating the constant rate period is called as critical point. Solid is not so wet that a continuous film of water exists over the entire drying surface. For non porous solid- superficial water is evaporated completely at critical point. For porous solid: this is no constant supply from the interior of the solid. rate of moisture flow to the surface no longer equals the rate of evaporation

1.8.4.3 *Calculation of Drying time*

Drying time fixes the size of the equipment needed for a given capacity. Drying time is found from rate of drying curve which is obtained experimentally. If a material is dried by passing hot air over a surface which is initially wet, the rate of drying curve in its simplest form is represented by BCD, shown in Fig. 1.8.8.

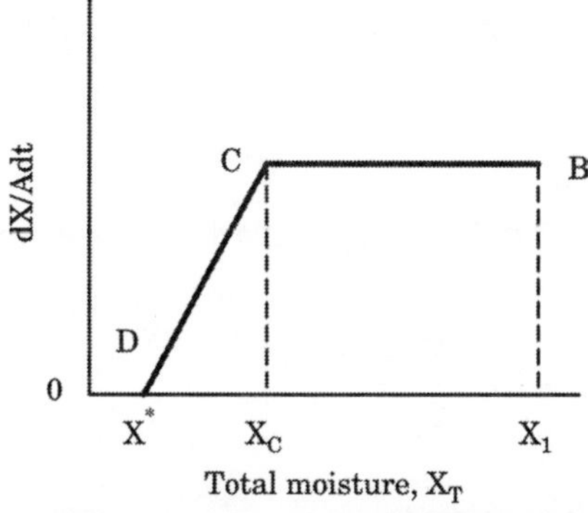


**Fig. 1.8.8:** Drying curve use in the estimation of time of drying

Where: $X_T$ is the total moisture, X* is the equilibrium moisture content (point D), $X_T$ – X* is the free moisture content, and $X_c$ the critical moisture content (point C).

During the period of drying from the initial moisture content $X_1$ to the critical moisture content, X*, the rate of drying is constant, and the time of drying $t_c$ is given by:

$$t_c = \frac{X_1 - X_c}{R_c A} \tag{1.8.23}$$

Where $R_c$ is the rate of drying per unit area in the constant rate period, A is the area of exposed surface

***Falling-rate period:*** During this period the rate of drying is, approximately, directly proportional to the free moisture content ($X_T$ – X*), or:

$$-\left(\frac{1}{A}\right)\frac{dX}{dt} = m\left(X - X^*\right) \tag{1.8.24}$$

$$-\frac{1}{mA}\int_{X^*}^{X_T}\frac{dX}{\left(X - X^*\right)} = \int_0^{t_f} dt \tag{1.8.25}$$

$$\frac{1}{mA}\ln\left[\frac{X_c - X^*}{X_T - X^*}\right] = t_f \tag{1.8.26}$$

$$\text{Or } t_f = \frac{1}{mA}\ln\left[\frac{X_c - X^*}{X_T - X^*}\right] \tag{1.8.27}$$

## *Total time of drying*

The total time t of drying from $X_1$ to $X_T$ is given by t = ($t_c$ + $t_f$). The rate of drying $R_c$ over the constant rate period is equal to the initial rate of drying in the falling rate period, so that $R_c = m\,(X_c - X^*)$

$$\text{Then, } t_c = \frac{X_1 - X_c}{mA\left(X_T - X^*\right)}$$

$$t = \frac{X_1 - X_c}{mA\left(X_T - X^*\right)} + \frac{1}{mA}\ln\left(\frac{X_c - X^*}{X_T - X^*}\right)$$

$$= \frac{1}{mA}\left[\left(\frac{X_1 - X_c}{X_T - X^*}\right) + \ln\left(\frac{X_c - X^*}{X_T - X^*}\right)\right] \tag{1.8.28}$$

### 1.8.5 *Classification and selection of dryers*

Because of the very wide range of dryer designs available, classification is a virtually impossible task (Fig. 1.8.9). However, the means by which material is transferred through the dryer is a basis of this classification, with a view to presenting a guide to the selection of dryers. Probably the most thorough classification of dryer types has been presented a decimalised system based on the following factors: (a) Temperature and pressure in the dryer, (b) The method of heating, (c) The means by which moist material is transported through the dryer, (d) Any mechanical aids aimed at improving drying, (e) The method by which the air is circulated, (f) The way in which the moist material is supported, (g) The heating medium, and (h) The nature of the wet feed and the way it is introduced into the dryer. In general, continuous operation has the important advantage of ease of integration into the rest of the process coupled with a lower unit cost of drying. As the rate of throughput of material becomes smaller, however, the capital cost becomes the major component in the total running costs and the relative cheapness of batch plant becomes more attractive. In addition, a batch dryer is much more versatile and it can often be used for different materials. The humidity may be controlled during the drying operation, and this is especially important in cases where the humidity has to be maintained at different levels for varying periods of time.

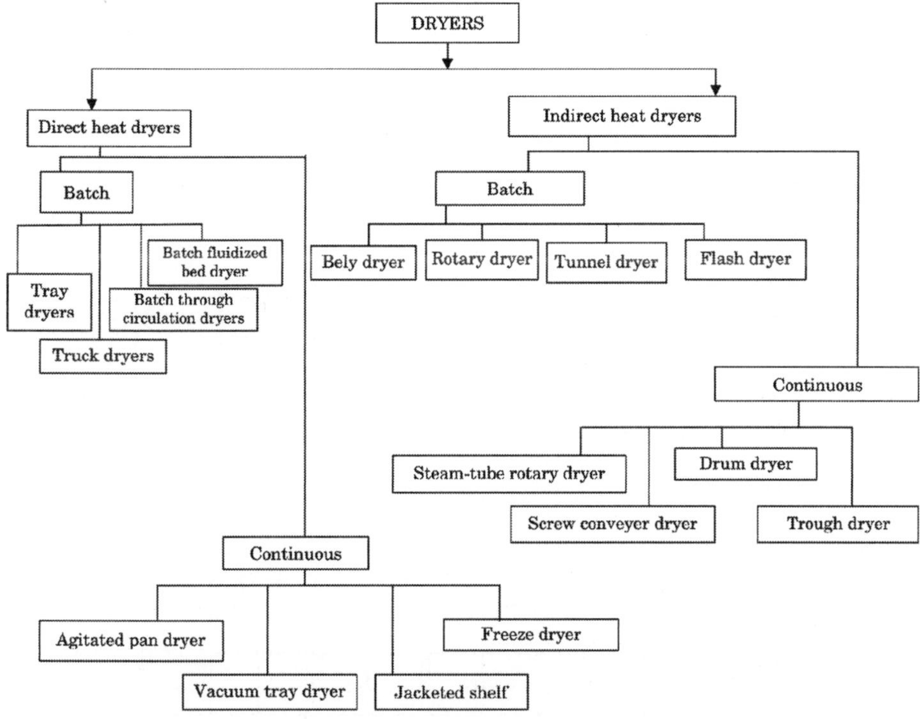


**Fig. 1.8.9:** Classification of dryers

**1.8.5.1.** ***Direct heating dryers*:** Direct heating, in which the material is heated primarily by convection from hot gases has several advantages. Firstly, directly heated dryers are, in general, less costly, mainly because of the absence of tubes or jackets within which the heating medium must be contained. Secondly, it is possible to control the temperature of the gas within very fine limits, and indeed it is relatively simple to ensure that the material is not heated beyond a specified temperature. This is especially important with heat-sensitive materials. Against this, the overall thermal efficiency of directly heated dryers is generally low due to the loss of energy in the exhaust gas and, where an expensive solvent is evaporated from the solid, the operation is often difficult and costly. Losses also occur in the case of fluffy and powdery materials, and further problems are encountered where either the product or the solvent reacts with oxygen in the air.

### 1.8.5.2 *Direct heat Batch dryers*

***Tray dryer*:** Rectangular chamber containing two trucks that support racks that load material to dry. ***Typical specification*:** Rack contain shallow trays: 750 mm square and 50-150 mm deep; Air velocity: 2-5 m/s by fan and motor; Baffles distributes air uniformly over the stack of trays; Make up air added through exhaust duct. Disadvantages: Used when production rate is small; High labor cost for loading and unloading; Expensive to operate; Drying by circulation of air is slow and drying cycles is 4-48 h per batch. Tray dryers are after used for drying of valuable products such as dyes and pharmaceuticals.

### 1.8.5.3 *Direct heat Continuous dryers*

***Tunnel Dryers*:** In tunnel dryers, a series of trays or trolleys is moved slowly through a long tunnel, which may or may not be heated, and drying takes place in a current of warm air (Fig. 1.8.10). Tunnel dryers are used for drying paraffin wax, gelatine, soap, pottery ware, and wherever the throughput is so large that

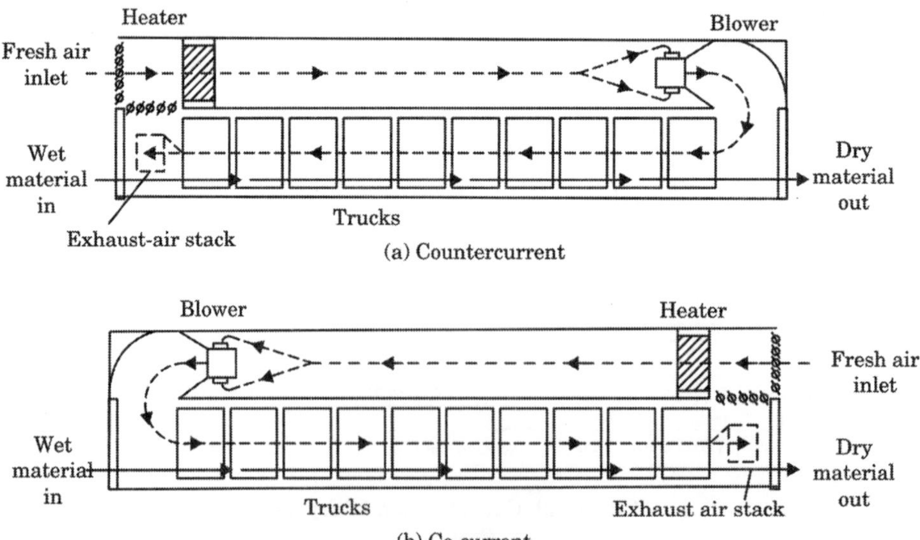


**Fig. 1.8.10:** Tunnel dryer

individual cabinet dryers would involve too much handling. Alternatively, material is placed on a belt conveyor passing through the tunnel, an arrangement which is well suited to vacuum operation.

***Tower dryers:*** Series of circular trays mounted one above another on central rotating shaft. Solid feed dropped on the topmost tray is exposed to stream of hot air that passes across the tray. Solid is scraped off and dropped to tray below. Typical specification: Air velocity: 0.6-2.4 m/s, Bottom two trays act as cooling section for dry solids. Turbodryer functions partially by cross circulation drying as in tray dryer and partly by showering the particles through the hot gas as they tumble from one tray to another.

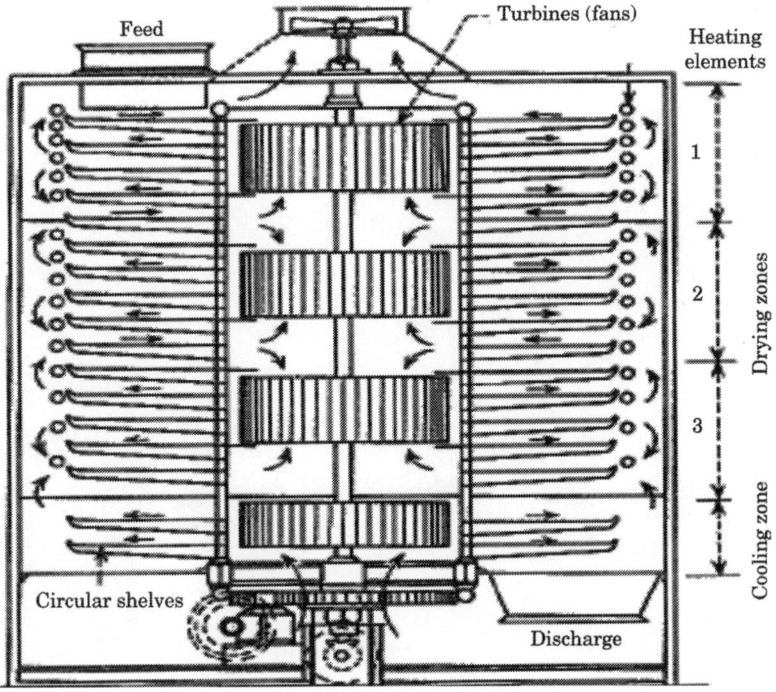


**Fig. 1.8.11:** Tower Dryer

***Screen conveyer dryer:*** Layer 25-150 mm thick is dried slowly by carrying on a traveling metal screen through a long drying chamber (Fig. 1.8.12). Chamber contains series of separate sections each with its own fan and heater. At inlet air is passed upward through screen and at discharge to prevent loss of dusty dried material air is passed downward. Typical specifications: 2 m long and 4-50 m wide, drying time 5-120 min, minimum screen size = 30 mesh. Handle variety of solid continuously; Reasonable cost; Steam consumption low. Advantages: When the drying conditions must be appreciably changed as the moisture content of solid is reduced.

***Direct Heat Rotary dryer:*** It consists of revolving cylindrical shell, which is horizontal or slightly inclined (Fig. 1.8.13). Wet feed enters one end of cylinder

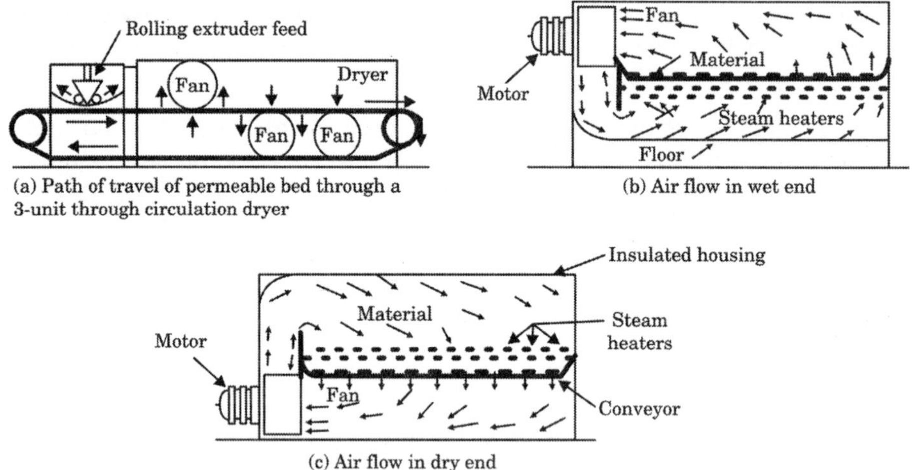


(a) Path of travel of permeable bed through a 3-unit through circulation dryer

(b) Air flow in wet end

(c) Air flow in dry end

**Fig. 1.8.12:** Screen conveyer dryer

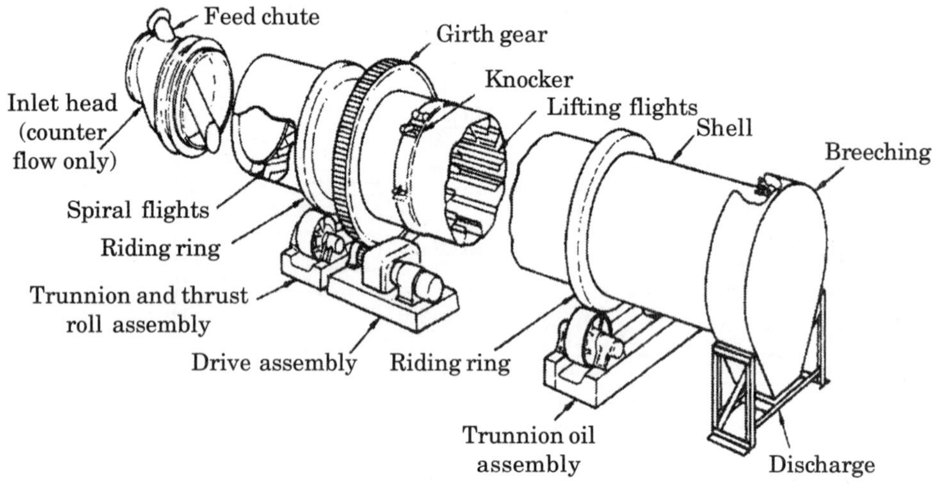


**Fig. 1.8.13:** Direct heat rotary dryer

and dry material discharges from other. As the shell rotates, internal flights lift the solid and shower them down. While in contact with the shell it gets heated by external jacket, during showering, it gets heated by direct contact with gas. Typical specification: Allowable mass velocity = 2000-25000 kg/m$^2$h; Inlet gas temp = 120-175°C for steam heated air and 550-800°C for flue gas; Diameter = 1-3 m; Peripheral speed of the shell = 20-25 m/min. For the continuous drying of materials on a large scale, 0.3 kg/s (1 tonne/h) or greater, a rotary dryer, which consists of a relatively long cylindrical shell mounted on rollers and driven at a low speed, up to 0.4 Hz is suitable. The shell is supported at a small angle to the horizontal so that material fed in at the higher end will travel through the dryer under gravity, and hot gases or air used as the drying medium are fed in either at the upper end of the dryer to give co-current flow or at the discharge end of the machine to give countercurrent flow. The shell of a rotary dryer is usually constructed by welding rolled plate, thick enough for the transmission of the torque required to cause rotation, and to support its own weight and the weight of material in the dryer. With counter-current operation, since the gases are often exhausted by a fan, there is a slight vacuum in the dryer, and dust-laden gases are in this way prevented from escaping. With co-current flow, the rate of passage of the material through the dryer tends to be greater since the gas is travelling in the same direction. Contact between the wet material and the inlet gases gives rise to rapid surface drying, and this is an advantage if the material tends to stick to the walls. This rapid surface drying is also helpful with materials containing water of crystallisation. The dried product leaves at a lower temperature than with counter-current systems, and this may also be an advantage. The rapid lowering of the gas temperature as a result of immediate contact with the wet material also enables heat sensitive materials to be handled rather more satisfactorily.

***Screw conveyor dryers:*** It consists of horizontal screw conveyor enclosed in a cylindrical jacketed shell. Solid is fed in one end is conveyed slowly through the heated zone and discharges from the other end. *Normal specification*: Shell 75-600 mm dia, 6 m long, rate of rotation = 2-30 r/min, heat transfer coefficient = 15-60W/ $m^2$ °C. *Applications*: For solids that are too fine and too sticky for rotary dryer.

***Spray dryers:*** Water may be evaporated from a solution or a suspension of solid particles by spraying the mixture into a vessel through which a current of hot gases is passed. In this way, a large interfacial area is produced and consequently a high rate of evaporation is obtained. Drop temperatures remain below the wet bulb temperature of the drying gas until drying is almost complete, and the process thus affords a convenient means of drying substances which may deteriorate if their temperatures rise too high, such as milk, coffee, and plasma. Furthermore, because of the fine state of subdivision of the liquid, the dried material is obtained in a finely divided state. In spray drying, it is necessary to atomise and distribute, under controlled conditions, a wide variety of liquids, the properties of which range from those of solutions, emulsions, and dispersions, to slurries and even gels. Most of the atomisers commonly employed are designed for simple a liquid, which is a mobile Newtonian liquid. When atomisers are employed for slurries, pastes, and liquids having anomalous properties, there is a great deterioration in performance and, in many cases, atomisers may be rapidly eroded and damaged so as to become useless. There is therefore much to be gained by considering various types and designs of atomiser so that a suitable selection can be made for the given duty. The performance of a spray dryer or reaction system is critically dependent on the drop size produced by the atomiser and the manner in which the gaseous medium mixes with the drops. In this context an atomiser is defined as a device which causes liquid to be disintegrated into drops lying within a specified size range, and which controls their spatial distribution. For solution and slurries slurry or liquid solution dispersed into a stream of hot gas in the form of mist of fine droplets moisture rapidly vaporized from droplets leaving residual particles of dry solid droplets formed by pressure nozzles. To prevent droplets or wet particles of solid from striking solid surface, diameter is made large (2.5-9 m). Normal specification:

Spray disk = 300mm, rotates at 5000-10000 r/min; Gas leaving is passed to cyclone for removal of entrained solid. Cylindrical chamber is having conical bottom to collect the solids.

**1.8.5.4 *Indirectly Heated Dryer*:** Hot gases pass through the innermost cylinder, and then return through the annular space between the outer cylinders. This form of dryer can be arranged to give direct contact with the wet material during the return passage of the gases. Flights on the outer surface of the inner cylinder, and the inner surface of the outer cylinder, assist in moving the material along the dryer. This form of unit gives a better heat recovery than the single flow direct dryer, though it is more expensive. In a simpler arrangement, a single shell is mounted inside a brickwork chamber, through which the hot gases are introduced. The steam-tube dryer, shown in Fig. 1.8.14, incorporates a series of steam tubes, fitted along the shell in concentric circles and rotating with the shell. These tubes may be fitted with fins to increase the heat transfer surface although material may then stick to the tubes. The solids pass along the inclined shell, and leave through suitable ports at the other end. A small current of air is passed through the dryer to carry away the moisture, and the air leaves almost saturated. In this arrangement, the wet material comes in contact with very humid air, and surface drying is therefore minimised. This type of unit has a high thermal efficiency, and can be made from corrosion resisting materials without difficulty.

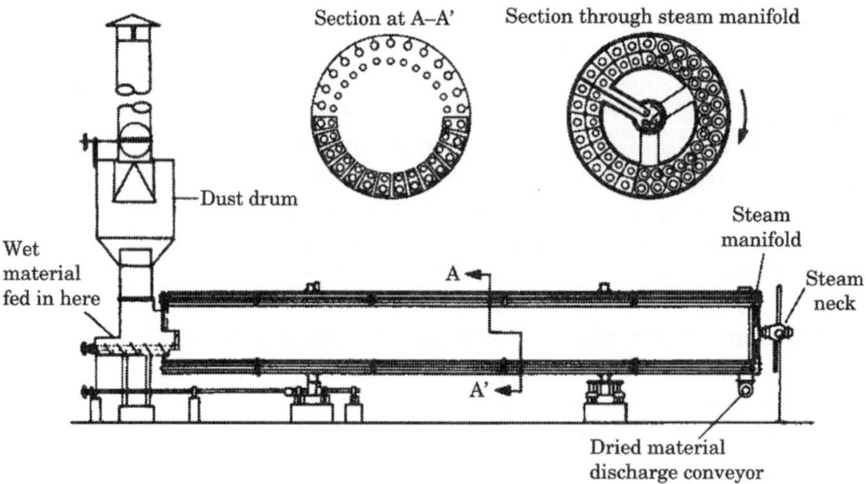


**Fig. 1.8.14:** Steam tube rotary dryer

***Drum dryers*:** Heated metal rolls on the outside of which a thin layer of liquid is evaporated to dryness constitute the drum dryer (Fig. 1.8.15). Dried solid is scraped off the rolls as they slowly revolve. It is effective with dilute solutions, concentrated solution of highly soluble material and moderately heavy slurries and not suitable for solutions of salts with limited solubility or for slurries of abrasive solids that settles out and create excessive pressure between drums. *Normal specification*: 0.6–3 m diameter, 0.6-4m long, revolving at 1-10 r/min, time for solid contact = 6–15 s, HT coefficient= 1200-2000 W/m$^{2\circ}$C, drying capacity = 5-50 kg per square meter of drying surface per hour.

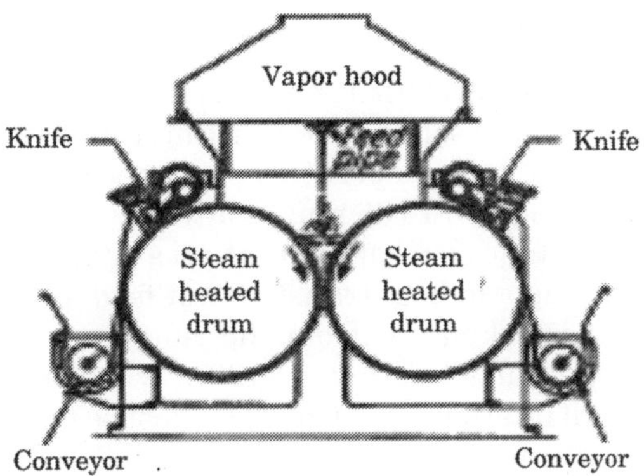


**Fig. 1.8.15:** Drum dryer

1.8.5.5 *Air Suspended Dryers*

***Industrial spray dryers*:** Spray dryers are used in a variety of applications where a fairly high grade product is to be made in granular form. In the drying chamber the gas and liquid streams are brought into contact, and the efficiency of mixing depends upon the flow patterns induced in the chamber. Rotating disc atomisers are most commonly used. Counter-current dryers give the highest thermal efficiencies although product temperatures are higher in these units. This limits their use to materials which are not affected by overheating. Co-current dryers suffer from relatively low efficiencies, although they have the advantage of low product temperatures unless back-mixing occurs. In the case of materials which are extremely sensitive to heat, great care has to be taken in the design of the chamber to avoid overheating. Combustion gases are frequently used directly although, in some cases, such as the preparation of food products, indirectly heated air is used. Maximum temperatures are then normally limited to lower values than those with direct heating.

***Pneumatic dryers*:** In pneumatic dryers, the material to be dried is kept in a state of fine division, so that the surface per unit volume is high and high rates of heat transfer are obtained. Solids are introduced into the dryer by some form of mechanical feeder, such as a rotating star wheel, or by an extrusion machine arranged with a high-speed guillotine to give short lengths of material, such as 5–10 mm. Hot gases from a furnace, or more frequently from an oil burner, are passed into the bottom of the dryer, and these pick up the particles and carry them up the column. The stream of particles leaves the dryer through a cyclone separator and the hot gases pass out of the system. In some instances, final collection of the fine particles is by way of a series of bag filters. The time of contact of particles with the gases is small, typically 5 s, even with a lengthy duct—and the particle temperature does not approach the temperature of the hot gas stream. In some cases the material is recycled, especially where bound moisture is involved. Evaporative capacities, which are high, are greatly affected by the solids–air ratio. Wet pulverized solid is transported for few seconds in hot gas stream. There is rapid drying with drying time 3-4 sec. Temperature of gas is too high 650°C at inlet. However too small contact time rarely raised temperature of solid above 50°C. Application: To sensitive materials that in other dryers would have to be dried indirectly by a much cooler heating medium.

***Fluidised bed dryers:*** The principal characteristics of such beds include cross flow of solid and drying gas, a solids residence time controllable from seconds to hours, and suitability for any temperature. It is necessary that the solids be free flowing, of a size range 0.1 to 36 mm. Multistage cross-flow operation is a possibility, as is a two-stage counter-current arrangement as in (Fig. 1.8.16). Advantage: Rapid mixing and heat transfer. Wet feed is admitted to the top of the bed; dry product is taken out from the side, near the bottom. Typical specification: Average particle stay: 30-120 s when only surface is to be evaporated; 15-30 min if internal diffusion. Care must be taken for small particles as problem of carry away is severe. The principles of fluidisation are applied in this type of dryer, shown typically in Fig. 1.8.17. Heated air, or hot gas from a burner, is passed by way of a plenum chamber and a diffuser plate, fitted with suitable nozzles to prevent any back-flow of solids, into the fluidised bed of material, from which it passes to a dust separator. Wet material is fed continuously into the bed through a rotary valve, and this mixes immediately with the dry charge. Dry material overflows through a downcomer to an integral

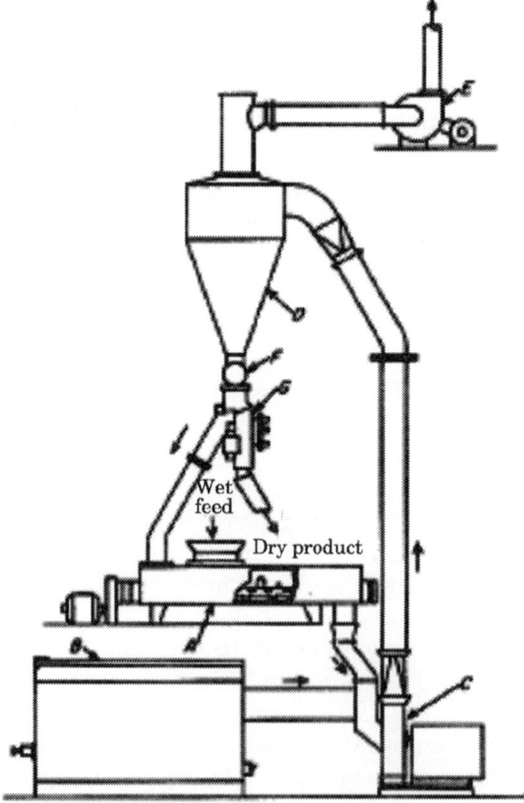


**Fig. 1.8.16:** Pneumatic-conveyor (Flash) dryer

after-cooler. An alternative design of this type of dryer is one in which a thin bed is used.

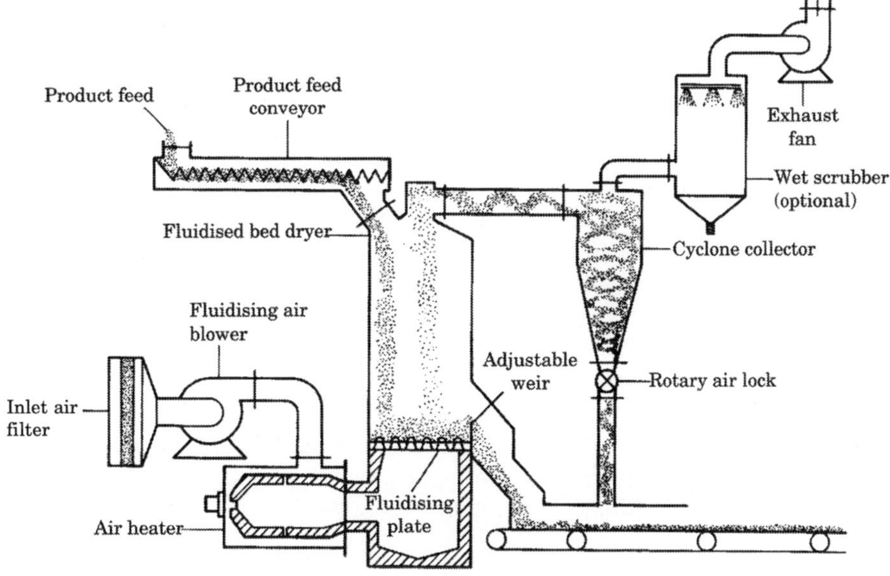


**Fig. 1.8.17:** Fluidized bed dryer

## 1.9. MEMBRANE PROCESSES

### 1.9.1. *Introduction*

Systematic studies of the membrane phenomena can be traced to the seventeenth century philosopher scientist (see Table 1.9.1). Abbe Nolet coined the word 'osmosis' to describe permeation of water through a diaphragm 1748. Through the nineteenth and the early twentieth centuries, membranes had no industrial or commercial uses, but were used as laboratories to develop physical/chemical theories. For *e.g.*, the measurements of solution osmotic pressure made with

**Table 1.9.1:** Some of the historical development of the membrane technology before the Golden Age of membrane technology

| ***Year*** | ***Important development*** |
|---|---|
| 1748 | Abbe Nollet – water diffuses from dilute to concentrated solution |
| 1846 | The first synthetic (or semi synthetic) polymer studied by Schoenbein & produced commercially in 1869. |
| 1855 | Fick employed cellulose nitrate membrane in his classic study *Ueber Diffusion.* |
| 1866 | Fick, Traube, artificial membranes (nitrocellulose) |
| 1907 | Bechhold, pore size control, "ultrafiltration" |
| 1927 | Sartorius Company, membranes available commercially |
| 1945 | German scientists, methods for bacterial culturing |
| 1957 | USPH, officially accepts membrane procedure |
| 1958 | Sourirajan, first success in desalinating water |

membranes by Tiaube and Tfeffer were used by van't Hoff in 1887 to develop his limit law, which explains the behavior of ideal dilute solutions; this work led directly to the van't Hoff equation. At about the same time, the concept of perfectly selective semi permeable was used by Maxwell and others in developing the kinetic theory of gases. In 1907, Bechhold Devised, developed a technique to prepare nitrocellulose membranes of graded pore size which he determined by bubble test. Other early workers, particularly Elford, Zsigmondy and Bachmann and Ferry improved on Bechhold technique and by early 1930's micro pores collodion membranes were commercially available. During the next years, this early micro filtration membranes technology was expanded to other polymers, notably cellulose acetate. Membrane found their first significant application in the testing of drinking water at the end of World War II. Drinking water supplies serving large communities in Germany and else where in Europe had broken down, and filters to test for water safety were needed urgently. The research effort to develop these filters, sponsored by the US army, was later exploited by the Millipore Corporation, the first and still the largest US micro filtration membrane producer. By 1960, the elements of modern membrane science had been developed, but membranes were used only a few laboratory and small, specialized industrial applications.

In general, membrane acts as a barrier between two bulk phases for separation. Transport through the membrane takes place when a driving force is applied to the components in the feed. In most of the membrane process, the driving force is the pressure difference or a concentration difference across a membrane. It controls the relative rate of transport of various species and thus, as with all separation, gives one product depleted in certain components and a second product concentrated in these components. The performance of a membrane is defined in two simple factors, flux and retention or selectivity. In an ideal situation, we aim to use a membrane with high selectivity or retention with a high flux or permeability is required but this is not possible. Hence we maximize one of the components by compromising on the other. Of the many types of membrane separation processes and membrane-based equilibrium separation processes available for separation, membrane reactors have been studied using the following: reverse osmosis (RO), nanofiltration (NF), ultrafiltration (UF), microfiltration (MF), electrodialysis (ED), liquid membranes (LM), pervaporation (PV), gas permeation, vapor permeation, molecular sieving, Knudsen diffusion (and molecular diffusion), gas membrane, membrane solvent extraction, and membrane gas absorption/stripping. Whilst effective product separation is crucial to economic operation in the process industries, certain types of materials are inherently difficult and expensive to separate (Table 1.9.2). Important examples include: (a) finely dispersed solids, especially those which are compressible, and which have a density close to that of the liquid phase, have high viscosity, or are gelatinous. (b) Low molecular weight, non-volatile organics or pharmaceuticals and dissolved salts. (c) Biological materials which are very sensitive to their physical and chemical environment. The processing of these categories of materials has become increasingly important in recent years, especially with the growth of the newer biotechnological industries and with the increasingly sophisticated nature of processing in the food industries. Agricultural drainage water reclamation would reduce the quantity of imported water and minimize drainage water volume. However, high levels of calcium and sulfate ions in the agricultural drainage water could potentially form scale on the membrane surface as their concentrations increase due to water recovery and concentration polarization effects. In recent literatures, the conditions for gypsum scale formation on nanofiltration (NF) membrane surfaces were systematically studied for the reclamation of agricultural drainage water. The work involved glassware experiments to observe basic characteristics of gypsum crystallization, and experimental studies of gypsum scale formation on a high performance NF membrane with actual samples and model solutions of the agricultural drainage water. Scaling occurred at low-water recoveries, and was mostly due to calcium sulfate (gypsum) formation at the membrane surface.

**1.9.2. *Classification of Membrane*:** Membrane is defined essentially as a barrier, which separates two phases and restricts transport of various chemicals in a selective manner. A membrane can be homogenous or heterogeneous, symmetric or asymmetric in structure, solid or liquid; can carry a positive or negative charge or be neutral or bipolar. Transport through a membrane can be affected by convection or by diffusion of individual molecules, induced by an electric field or concentration, pressure or temperature gradient. The membrane is a main part in all the membrane processes and interphase between the two bulk phases. A membrane can be thick or thin, or may be any one or combination of the following: nonporous solid, microporous or macroporous solid with a fluid in the pores, a liquid phase with or without second phase or a gel. The membrane thickness of hollow fibers is of the order of the dimensions of one of the bulk phase in contrast to the other types of membranes those are thin than the bulk phases. The membrane thickness may vary from as small as 10 microns to few hundred micrometers. The principal types of membrane are shown schematically in Fig. 1.9.1 and can be classified as below: • Isotropic membranes–

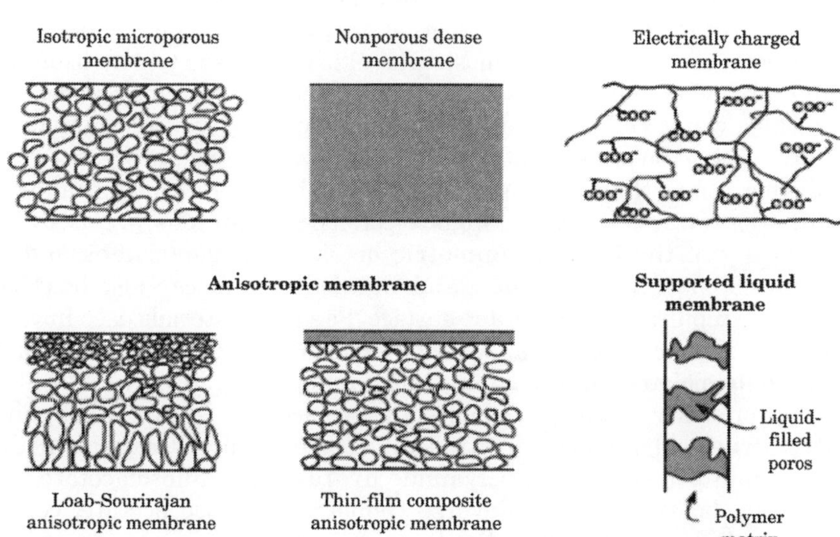


**Fig. 1.9.1:** Schematic diagrams of principal types of membranes

Microporous membranes– Nonporous, dense membranes – Electrically charged membranes; • Anisotropic membranes; • Ceramic, metal and liquid membranes.

Membrane reactors have been investigated since the 1970s. The early investigations employed primarily polymeric membranes and enzymatic reactions. Later investigations show an abundance of petrochemically relevant systems and inorganic membranes. Whole cell fermentation-based chemical and biochemical productions as well as degradation of pollutants biologically or otherwise have also been studied in membrane reactors. 1. Microporous Membrane: Behaves almost like a fiber and separates by a sizing mechanism determined by the pore diameter and particle size *e.g.* ceramics, graphite, metal oxide, and polymer. 2. Homogeneous Membrane: This is a dense film through which a mixture of molecule is transported by pressure, concentration or potential gradient. Using these membranes chemical species of similar size and diffusivity can be separated efficiently when there concentration difference is significant. 3. Asymmetric membranes: This comprise of a very thin layer on a highly porous thick substructure. The thin skin acts as a selective membrane. 4. Electrically charged membrane: These are necessary ion exchange membranes consisting of highly swollen gels carrying fixed positive or negative charges; they are mainly used in electrodialysis. 5. Microfiltration: It is essentially sterile filtration with pores so small that micro organism cannot pass through the diameter of smallest bacteria. Pseudomonas diameter is 0.3 μm. 6. Ultrafiltration: Capable of retaining species in the range of 300-50000 Daltons of molecular weight with pore size ranging from – 10-1000 $A^0$. These are mostly described by nominal molecular weight cut off 100–100000 MWCO which means the membranes have > 90% rejection. 7. Nanofiltration: It is a looser RO. It is capable of concentrating sugar, divalent salts, proteins, bacteria, dyes and other constituents that have weight greater then 1000 Daltons. Membranes used for nanofiltration are of cellulose acetate and aromatic polyamide type characteristics a rejection of salt from 95%for divalent salts to 90% for monovalent salts and approximately 300 mol weight cut-off for organics. Advantage of nano filtration over reverse osmosis is it typically operates at higher recoveries. 8. Reverse Osmosis: Reverse Osmosis is widely used in the desalination and demineralization of saline water. The feed may be sea water and brackish water and the product drinking water or tap water may be used and pretreated in the production of deionized water. Hydraulic pressure on the feed side maintains the necessary driving force, hydrophilic membranes with high affinities towards water, but high rejections towards ions, are used. These are generally of the types asymmetric or skinned membranes and TFC membranes.95-99% inorganic salts and charged organic can also be rejected due to charge repulsion at membrane surface. Polymer materials used in various pressure driven membranes: Micro filtration: Regenerated cellulose, polycarbonate, polytetrafluroethylene, polyvinyl difluoride, polyvinyl chloride and polypropylene. Ultra filtration: cellulose acetate, aromatic polyamide, polyacronitrile, polyamide, polyvinyl difluoride. Nano filtration: modified polyamide and polysulphones, polyamide hydrazide, cellulose acetate blend, polyvinyl alcohol derivatives. Reverse Osmosis: cellulose acetate, cellulose triacetate, cellulose acetate blend, polyamide, polyamide hydrazide, polyurea composite.

### 1.9.3. *Driving force and Principle of membrane process*

The membrane phase, which is set between two bulk phases, has ability to control mass transfer between the two bulk phases in a membrane process. The movement of any species across the membrane is a result of one or more driving forces which can be gradients in concentration, pressure, temperature or electrical potential. The flux-force relationship is described by a linear phenomenological equation between flux ($J$) and the driving force:

$$J = -A\frac{dX}{dx} \tag{1.9.1}$$

Where $A$ is called phenomenological coefficient and (d$X$/d$x$) is the driving force, expressed as the gradient of $X$ (concentration, pressure, temperature) along a coordinate $x$ perpendicular to the transport barrier.

The basic principle of membrane separation is mainly the same although the membrane can have different structure (porous/non-porous) and may vary in material (organic/inorganic). The separation is achieved via a membrane. Transport through the membrane takes place when a driving force is applied to the compounds in the feed. The driving force is mostly a pressure difference or a concentration (or activity) difference across the membrane. In Fig. 1.9.2. is shown schematic drawing of membrane separation process with different driving forces.

**1.9.3.1. The performance and efficiency of the membrane process** is explained by selectivity and flux.

*Flux or permeation rate* is the volumetric flow rate of fluid passing through the membrane per unit area of membrane per unit time.

*Selectivity* is a measure of relative permeation rates of different components through the membrane.

*Retention* is the fraction of solute in the feed retained by the membrane.

The selectivity of a membrane is given as a retention factor *(R)*, usually for dilute aqueous solutions, or by a separation factor ($\alpha$), mainly for gas and organic mixtures. The retention factor is expressed by equation:

$$R = 1 - \frac{C_{permeate}}{C_{feed}} \tag{1.9.2}$$

where: $C$ is the solute concentration in each phase noted, while the separation factor is given by the respective concentrations in the feed ($xA$ , $xB$) and permeate ($y A$ , $yB$) as:

$$\alpha_{A/B} = \frac{x_B \cdot y_A}{x_A \cdot y_B} \tag{1.9.3}$$

It should be mentioned that the amount of driving force depends on the gradient in potential through the membrane. The main potential differences in membrane processes are the chemical potential difference ($\Delta\mu$), which includes

**Table 1.9.2.2:** Some properties of membrane process

| *Parameter* | *Microfiltration* | *Ultrafiltration* | *Nanofiltration* | *Reverse Osmosis* |
|---|---|---|---|---|
| Pore size | 0.01–1.0 μm | 0.001–0.01 μm | 0.0001–0.001 μm | <0.0001 μm |
| Molecular cutoff | >100,000 | 1,000–300,000 | 300–1,000 | 100–300 |
| Operating pressure | <30 psi | 20–100 psi | 50–300 psi | 225–1,00 psi |
| Membrane materials | Ceramics, polypropylene, polysulfone, polyvinylidenedifluoride | Ceramics, polysulfone, polyvinylidenedifluoride, cellulose acetate, thin film composite | Cellulose acetate, thin film composite | Cellulose acetate, thin film composite, polysulfonated polysulfone |
| Membrane configuration | Tubular, hollow fibre | Tubular, hollow fibre, spiral wound, plate and frame | Tubular, spiral wound, plate and frame | Tubular, spiral wound, plate and frame |
| Types of materials removed | Clay, bacteria, viruses, suspended solids | Proteins, starch, viruses, colloid silica, organics, dyes, fats, paint solids | Starch, sugar, pesticides, herbicides, divalent anions, organics, BOD, COD, detergents | Metal cations, acids, sugars, aqueous salts, amino acids, monovalent salts, BOD, COD |
| Suspended solids removal | Yes | Yes | Yes | Yes |
| Dissolved organic removal | Yes | Yes | Yes | Yes |
| Dissolved inorganic removal | None | Yes | Yes | Yes |
| Microorganism removal | Protozoan cysts, algae, bacteria | Protozoan cysts, algae, bacteria, virus | All | All |
| Osmotic pressure effects | None | Slight | Moderate | High |
| Concentration capabilities | High | High | Moderate | Moderate |
| Permeate purity | High | High | Moderate-High | High |
| Energy usage | Low | Low | Low-Moderate | Moderate |
| Membrane stability | High | High | Moderate | Moderate |

pressure differences across the membrane, concentration gradients between two phases or temperature differences, and the electrical potential.

## 1.9.4. FACTORS AFFECTING MEMBRANES

### 1.9.4.1. *Recovery/Conversion:*

It is used to define the percentage of the feed water that is converted to permeate and is calculated in the following way:

$$Y = Q_P/Q_F \times 100 \tag{1.9.4}$$

Where, Y =% Recovery or Conversion, $Q_p$ = Product stream flow rate, $Q_F$ = Feed stream flow rate.

**1.9.4.2. *Temperature*:** Temperature changes affect osmotic pressure and water flux. Water flux is directly proportional to temperature. a rule of thumb is that membrane capacity increases about 3% per degree Celsius increase in water temperature

**1.9.4.3. *Pressure*:** Increase in pressure results in increase water flow per unit of membrane area. Although transport of salt across the membrane is not affected by pressure, increase water flow with pressure dilutes the salt passing through the membrane, which results in lower permeate salt concentration.

**1.9.4.4. *Concentration Polarisation*:** This is the result from the build of boundary layer of more highly concentrated solute on the membrane surface than in the bulk liquid. This occurs because water permeation of membrane surface leaves the more concentrated solute layer which must diffuse back in the bulk liquid owing to the higher flux rates. Proper membrane device design to maintain high fluid flow parallel to membrane surface and promote mixing at the surface is important to minimize the boundary layer thickness.

This increases the osmotic pressure at the surface resulting in reduction in water flux and increase in water transport across the membrane. If the concentration of the sparingly soluble salt in the boundary layer exceeds their solubility limits, precipitation or scaling will occur on the membrane surface. At such higher concentrations, colloidal materials become less stable and may agglomerate and foul the membrane surface.

### 1.9.4.5. *Membrane fouling*

***Operation and Maintenance:*** To maintain membrane performance and extend membrane life, pretreatment chemicals may be necessary, depending on the characteristics of the stream. In addition, chemicals may be required to achieve clean water specifications. Filtering wastewater may be necessary to remove suspended solids before wastewater is fed to the RO modules. Membrane performance can be enhanced by control of pH, removal of certain dissolved species and colloidal materials such as clays and oils, and dissolved or suspended organic. In any RO system, depending on the capacity and size of modules, a number of parallel modules may be needed. Membrane fouling can result from the formation of a fouling layer on the membrane surface, or from internal changes of the membrane material. Both forms of fouling can cause membrane permeability to decline. Scaling is a form of fouling that occurs when dissolved species are concentrated in excess of their solubility limit. Chemical agents can be added to slow the formation of precipitates. Acidification is used to prevent the formation of carbonates of low solubility, such as magnesium carbonate. An ion exchanger is sometimes used to trade cations of low solubility salts for cations that are more soluble, for example, sodium sulfate may be traded for calcium sulfate. Prevention of biological growth is necessary to prevent damage to the membrane. Biological growth can be inhibited with chlorination, but some RO membranes are chlorine sensitive, so water must be dechlorinated before entering the RO module. Other disinfectants are ozone, formaldehyde, ultraviolet light, copper sulfate, and sodium bisulfate. Fouling is the major cause of reduction in flux during several membrane separations in particular microfiltration and ultrafiltration where feed pretreatment is not used. This is essentially the deposition of retained particles on to the membrane surface or pores .The fouling behavior is specific to the system of interest and consequently varies according to the contaminants to the foulant, whether organic macromolecules, biological inorganic salts, particulates.

***Fouling control and correction:*** Fouling cannot be avoided but there are several methods to control the extent of fouling or correct the fouling by cleaning. The methods can be summarized as, pretreatment, adjustment of membrane properties, membrane cleaning and modification of operating conditions. Compared to conventional absorption or stripping processes such as bubble columns and packed beds, the membrane processes provide many advantages including larger interfacial area per unit volume, independent control of gas and liquid flowrates without any flooding, loading, foaming, and known gas–liquid interfacial area, etc. Recognition of these advantages is reflected by a number of investigations on the use of membrane modules for gas absorption and stripping reported in the literature. Furthermore, due to the tangential nature of the flow in membrane contactors, the gas stripping process does not require operation at a high-pressure drop. This also contributes to lower capital cost and ease of operation. Hollow fiber membrane contactors have been used to remove volatile compounds (VCs) such as $NH_3$, $H_2S$, $SO_2$, HCN, etc., from water and wastewaters. The hollow fibers used are commonly microporous and hydrophobic. In this process, the feed and stripping solution are flowing on either side of a membrane. Since the membrane is hydrophobic, it prevents aqueous solution, which has higher surface tension, to penetrate the gas-filled pores. The VCs will volatilize from the feed, diffuse through the gas-filled membrane pores, and react with the stripping solution. Compared with conventional techniques for removing of VCs, hydrophobic hollow fiber membrane contactor offers a number of advantages. Large surface area per volume which provides fast removal of the volatile contaminants; membrane pores filled with gas rarely dramatically impede transport of VCs; no secondary pollutants are produced and the VCs can be recovered and reused if it is necessary. Membranes used for removal of dissolved volatile species from water usually are hydrophobic in nature so as to provide less mass transfer resistances. Polypropylene (PP) and polytetrafluoroethylene membranes are the ideal candidates because of their good hydrophobicity. PP has become a new membrane material in recent years due to its both good hydrophobicity and feasibility to form asymmetric membranes via phase inversion methods. Polypropylene gives high chemical stability and high temperature resistance and low extractable levels.

### 1.9.5. *Dialysis and electrodialysis*

In dialysis, solutes diffuse from the aqueous donor side of a porous membrane to the aqueous receiving side as a result of a concentration gradient. Separation between the solutes is obtained as a result of differences in diffusion rates arising from differences in molecular size. Dialysis is therefore most effective in removing large molecules like proteins from small ones. It is a technique that has few applications for environmental analysis except in biological samples since in the former, both the analytes and the sample matrix compounds are small molecules. The analytes are also continually diluted in the acceptor unless a precolumn is incorporated. Electrodialysis technique tries to redress the drawbacks of dialysis in environmental analysis, *viz.* its nonselectivity and the dilution of the analytes in the receiving phase. It is a porous membrane technique whose separation power relies on differences in molecular size of the compounds and also on their charges. The electrical potential applied across the separation membrane makes sure that properly charged analytes in the feed solution are drawn through the membrane to the receiving side. With a flowing feed solution and stagnant receiving phase, additional selectivity and enrichment of charged analytes can be obtained. It has its own limitations such as limited enrichment factors, pH change in the feed during extraction, and thermal degradation of the membrane at high potential. The cited problems make electrodialysis less attractive for analytical applications.

The above techniques have some disadvantage or the other and hence the use of *membranes* is introduced which has the following advantages

(1) Appreciable energy saving
(2) Clean technology with operation ease
(3) Ambient operating temperature
(4) Simple phase operation
(5) Produces high quality product
(6) Flexibility in designing system

***Dialysis:*** Dialysis is the process of separating elements in a solution by diffusion across a semipermeable membrane (diffusive solute transport) down a concentration gradient. This is the principal process for removing the end-products of nitrogen metabolism (urea, creatinine, uric acid), and for repletion of the bicarbonate deficit of the metabolic acidosis associated with renal failure in humans. The preponderance of diffusion as the result of gradient is shown by the displacement of the *arrow*. Dialysis is driven by a differential concentration gradient, discriminating biomolecules across a 'semi-permeable' membrane. The membrane can be considered to display pores that retain molecules above a given size, and is permeable to small molecules. Membrane should be chosen with a Molecular Weight Cut-Off (MWCO) between the MW of molecules to recover in the sample compartment, and the MW of the molecules to remove (usually salts, by-products of reaction). Dialysis is performed by conventional tools as tubings, or flat sheets using a suitable support, or with specialized devices dedicated to applications for small samples, multiple samples, in vivo applications, equilibrium studies. In hemodialysis, blood from the patient is circulated through a synthetic extracorporeal membrane and returned to the patient. The opposite side of that membrane is washed with an electrolyte solution (dialysate) containing the normal constituents of plasma water. The apparatus contains a blood pump to circulate the blood through the system, proportioning pumps that mix a concentrated salt solution with water purified by reverse osmosis and/or deionization to produce the dialysate, a means of removing excess fluid from the blood (mismatching dialysate inflow and outflow to the dialysate compartment), and a series of pressure, conductivity, and air embolus monitors to protect the patient. Dialysate is warmed to body temperature by a heater.

Most membranes are derived from cellulose. (The earliest clinically useful hemodialyzers were made from cellophane sausage casing.) Other names of these materials include cupraphane, hemophan, cellulose acetate. They are usually sterilized by ethylene oxide or gamma irradiation by the manufacturer. They are relatively porous to fluid and solute but do not allow large molecules (albumin, vitamin B12) to pass freely. There is some suggestion that cupraphane membranes sterilized by ethylene oxide have a high incidence of biosensitization, meaning that the patient may have a form of allergic reaction to the membrane. Polysulfone, polyacrylonitrile, and polymethylmethacrylate membranes are more biocompatible and more porous (high flux membranes). They are most often formed into hollow fibers. Blood travels down the center of these fibers, and dialysate circulates around the outside of the fibers but inside a plastic casing. Water for dialysis must meet critical chemical and bacteriologic standards.

### 1.9.6. *Microfiltration*

**1.9.6.1. *Definition*:** Microfiltration is the membrane process which most closely resembles conventional coarse filtration. The pores sizes of microfiltration membranes range from 10 to 0.05 μm, making the process suitable for retaining suspensions and emulsions.

**1.9.6.2. *Principle and process*:** In filtration the flow of fluid through the filter medium is proportional to the applied pressure difference (ΔP) across the membrane. The rate of flow is dictated by the resistance of the filter to the flow of fluid and any resistance associated with the trapped particulate. The filter medium characteristic is defined in terms of its resistance or permeability, k. The flow or the flux, J, through the filter is given by

$$J = k\,\Delta P/\mu \tag{1.9.5}$$

The flow rate or flux is thus also dependent on the viscosity, m of the fluid being filtered. Microfiltration is widely applied in a dead-end mode of operation. In this the feed flow is perpendicular to the membrane surface and the retained (filtered) particles accumulate on the surface forming layer of retained solid or filter cake. With increase in time the cake thickness increases and permeation rate decreases as resistance of cake increases. Eventually the membrane filter reaches an impractical, low filtration rate and is either cleaned or replaced. The effect of a build up of solid cake on the membrane surface can be reduced by the use of cross flow or tangential flow. In such separation feed stream flows along, tangential to a membrane surface. As a result of the application of an appropriate driving force, *i.e.* differential pressure, a permeating species passes through the membrane. This permeate is then collected as a second product phase. As a result of this permeation, the feed is gradually reduced in concentration of the permeating species along the membrane. The retained solution above the membrane may also be product stream which can go to further processing. Cross flow velocities of several m/sec can be used in practice to minimize the impact of the particulate material. Even so there is general a decline in flux rate because the microfiltration membrane suffer from the problem of fouling, *i.e.* the accumulation of material at the surface of the membrane. Procedures are usually required to clean and sterilize the membrane, which must be able to withstand the associated mechanical, chemical and thermal stresses. Thus as well as organic polymers, inorganic materials such as ceramics, carbon, metals are used as membranes although these are generally thicker than organics, although usually asymmetric in structure

**1.9.6.3. *Membranes for microfiltration*:** Microfiltration membranes may be prepared from a large number of different materials based on either organic materials (polymers) or inorganic (ceramics, metals, glasses).

**1.9.6.3.1. *Hydrophobic polymeric membranes*:** (1) Polytetrafluoroethylene (Teflon), (2) poly (vinylidene fluoride), (3) polypropylene, (4) polyethylene.

**1.9.6.3.2. *Hydrophilic polymeric membranes*:** (1) Cellulose esters, (2) polycarbonate, (3) polysulfone/polyethersulfone, (4) polyimide/polyetherimide, (5) (aliphatic) polyamide, (6) polyetheretheketone; *Ceramic membranes*: (1) alumina, (2) zirconia, (3) titania, (4) silicium carbide. Other materials such as glass, carbon and various metals (stainless steel, palladium, and tungsten, silver) have also been used for preparing microfiltration membranes. Microfiltration membranes have pores in the range of 0.1-2μm.

**1.9.6.4. *Industrial Applications:*** Microfiltration is used in a wide variety of industrial applications were particles of size greater than 0.1μm, have to be retained from a liquid. The most important applications are still based on dead end filtrations using cartridges and single membranes applied in all kind of analytical laboratories. For large scale applications dead end filtration will slowly be replaced by cross flow filtration. One of the main industrial applications is the sterilization and clarification of all kinds of beverages and pharmaceuticals in food and pharmaceuticals industries. This can be done at any temperature, even at low temperatures. Microfiltration is also used for removing particles during the processing of ultra pure water in the semiconductor industry. New fields of applications are bio-technology and bio-medical technology. In bio-technology, microfiltration is suitable in harvesting and as a part of membrane bioreactor (involving a combination of biological conversion and separation). In biomedical field, plasmapheresis which involves separation of plasma with its value products from blood cells appears to have an enormous potential. A number of applications are summarized below: Cold sterilization of beverages and pharmaceuticals, Cell harvesting, Clarification of fruit juice, wine and beer, Ultra pure water in the semiconductor industry, Metal recovery as colloidal oxides or hydroxides, Waste water treatment, Continuous fermentation, Separation of oil water emulsions, Dehydration of lattices.

### 1.9.7. *Ultrafiltration*

**1.9.7.1.** ***Introduction:*** Ultrafiltration (UF) is the process of separating extremely small particles and dissolved molecules from fluids. The primary basis for separation is molecular size, although in all filtration applications, the permeability of a filter medium can be affected by the chemical, molecular or electrostatic properties of the sample. Ultrafiltration can only separate molecules which differ by at least an order of magnitude in size. Molecules of similar size can not be separated by ultrafiltration. Materials ranging in size from 1K to 1000K molecular weight (MW) are retained by certain ultrafiltration membranes, while salts and water will pass through. Colloidal and particulate matter can also be retained. Ultrafiltration membranes can be used both to purify material passing through the filter and also to collect material retained by the filter. Materials significantly smaller than the pore size rating pass through the filter and can be depyrogenated, clarified and separated from high molecular weight contaminants. Materials larger than the pore size rating are retained by the filter and can be concentrated or separated from low molecular weight contaminants. Ultrafiltration is typically used to separate proteins from buffer components for buffer exchange, desalting, or concentration. Ultrafilters are also ideal for removal or exchange of sugars, non-aqueous solvents, the separation of free from protein-bound ligands, the removal of materials of low molecular weight, or the rapid change of ionic and/or pH environment. Depending on the protein to be retained, the most frequently used membranes have a nominal molecular weight limit of 3 kDa to 100 kDa. Ultrafiltration is far gentler to solutes than processes such as precipitation. UF is more efficient because it can simultaneously concentrate and desalt solutes. It does not require a phase change, which often denatures labile species, and UF can be performed either at room temperature or in a cold room. Ultrafiltration membranes have pore size from 0.1 to 0.01 microns. They can retain solute upto 1000 daltons. Pressure is the driving force for the flow of mass through the membrane and flow is proportional to the pressure difference across the membrane and inversely proportional to the viscosity. The primary basis for separation is molecular size although secondary factors such as molecule shape and charges can apply a role. Colloidal and particulate can be retained. Bacteria and viruses can also be filtered.Ultrafiltration membranes can be used to purify, and collect, both the fluid material passing through the filter and material retained by filter. Materials smaller than the pore size rating pass through the filter and thus can be depyrogenated, clarified, and separated from high molecular weight contaminants. Materials larger than the pore size rating are retained by the filter and can be concentrated or separated from low molecular weight contaminants. Ultrafiltration membranes are often operated in tangential flow mode-feed material sweeps tangentially across the upstream surface of the membrane as filtration occurs, thereby maximizing flux rates and filter rates and filter life. The membranes are advantageous in terms of long life as they can be easily regenerated with strong cleaning agents. These membranes separate metals like nickel, copper, chromium, cadmium, arsenic etc by the process called as complexation ultrafiltration. In this process we use an complexing or binding agent to separate the required metal.

**1.9.7.2. *Definition*:** Ultrafiltration is a membrane process capable of retaining solutes as small as 1000 daltons, while passing solvent and smaller solutes. By convention, UF is distinguished from RO in that UF does not retain species for which bulk solution osmotic pressure is significant and distinguished from microfiltration in that UF does not exhibit some retention for soluble macromolecules regardless of pore size.

UF usually is applied to aqueous streams which may contain soluble macromolecules, colloids, salts, sugars, and so on. It is used to concentrate or fractionate, often simultaneously.

$$\frac{R = 1 - \text{permeate concentration}}{\text{Feed concentration}}$$

R = retention or rejection

**1.9.7.3. *Principle and process*:** Pressure is the driving force for the flow of mass through the membrane and, in the absence of macro solutes; flow is proportional to the pressure difference across the membrane and inversely proportional to the viscosity. The primary basis for separation is molecular size although secondary factors such as molecule shape and charges can apply a role. Colloidal and particulate can be retained. Ultrafiltration membranes can be used to purify, and collect, both the fluid material passing through the filter and material retained by filter. Materials smaller than the pore size rating pass through the filter and thus can be depyrogenated, clarified, and separated from high molecular weight contaminants. Materials larger than the pore size rating are retained by the filter and can be concentrated or separated from low molecular weight contaminants. Ultrafiltration membranes are often operated in tangential flow mode-feed material sweeps tangentially across the upstream surface of the membrane as filtration occurs, thereby maximizing flux rates and filter rates and filter life. The membranes are advantageous in terms of long life as they can be easily regenerated with strong cleaning agents.

**1.9.7.4. *Membranes for Ultrafiltration:*** Ultrafiltration (UF) is a low pressure membrane technology (5 to 25 psi driving force) that enables cost effective compliance with the Safe Drinking Water Act. It produces more consistent and higher quality water than is obtainable with conventional media filters. Because of the extremely small pore size, permeate quality is typically superior to that produced by microfiltration. It is this small pore size that makes UF a positive barrier to viruses, bacteria, *Giardia* cysts and parasites like *Cryptosporidium.* A barrier means that the membrane structure is such that a properly operated system prevents contaminant breakthrough to the product water side, regardless of how long the system is in service, or what feed pressure is used. Since packaged UF systems operate automatically, they can run unattended for long periods of time with little risk of degradation of the water that is sent to distribution. Ultrafiltration membranes are commonly asymmetric polymeric membranes. Most of Ultrafiltration membranes used commercially today days are made from polymeric materials by a phase inversion process. Membrane manufacturers seek a variety of pore sizes and the inherent flux characteristics to the customer's requirements. Resistance to temperature, pH, solvents and aggressive chemicals has been the major goal. Because of its high residual stress and specific surface, a membrane is rarely as chemically resistant as its parent polymer, and some caution is appropriate. Some of the membrane materials are listed below: Polysulfone/polyethersulfone/sulfonated polysulfone, Poly(vinylidene fluoride), Polyacrylonitrite and related co-polymers, Cellulosics(cellulose acetate), Polyimide/poly(etheramide), Apliphaticpolyamides, Polyetherthеketone. In addition to such polymeric materials, inorganic (ceramic) materials have been used for ultrafiltration membranes, especially alumina and zirconia. Since the lower limit of preparing membranes is about 0.1μm in pore diameter, this technique cannot be used to prepare ultrafiltration membrane. Such sintered porous structures can be used as sub layers for composite ultrafiltration membranes, a technique frequently employed in preparation of ceramic ultrafiltration membrane. On the other hand, ultrafiltration membranes themselves are often used as sub layers in composite membranes for reverse osmosis, nanofiltration, gas separation and pervaporation. Ultrafiltration is often applied for concentration of macro molecular solutions where large molecules are retained by the membrane while small molecules and the solvent permeate freely. An important point which must be considered is that as in microfiltration the process performance is not equal intrinsic membrane properties in actual separations. The reason for this is again the occurrence of concentration polarization and fouling. The macromolecular solute retained by the membrane accumulates at the surface of the membrane resulting in the concentration build up. At steady state, flow the solute to the membrane is equal to the diffusional backflow from membrane to the bulk. Further pressure increase will not result in an increase in flux because the resistance of the boundary layer has increased so that a limiting flux value ($J_\infty$) is attained. As in microfiltration, these boundary layer phenomena mainly determine the process performance. Thus intrinsic properties are not all that important in membrane development, but rather its chemical and thermal resistance and ability to reduce fouling tendency. The number of membrane application increase as they become more resistant to higher temperatures (>100° C) to a wide range of pH (1 to 14) and to organic solvents. Furthermore, as in microfiltration, module and system design are very important for reducing fouling as much as possible at a minimal cost.

### 1.9.8. *Reverse Osmosis*

**1.9.8.1. *Introduction:*** Reverse osmosis is a membrane process capable of separating a solvent from a solution by forcing the solvent through a semi permeable membrane by applying a pressure greater then the osmotic pressure of solute. It is also called as Hyperfiltration. It acts as a barrier to all dissolved salts and inorganic molecules as well as organic molecules with a molecular weight greater than 300 Daltons. Membrane pore size is of the order of 50 to 150 µm. It works on ambient temperatures. Operating pressures are 14 bar for brackish water and 50-55 bar sea water. Reverse osmosis is the separation process that uses pressure to force a solvent through a membrane that retains the solute on one side and allows the pure solvent to pass through other side. More formally it is the process of forcing a solvent from a region of high solute concentration through a membrane to a region of low solute concentration by applying a pressure in excess of osmotic pressure. This is reverse of the normal osmosis process which is the natural movement of solvent from an area of low solute concentration, through a membrane, to an area of high solute concentration when no external pressure is applied. The membrane here is semi permeable, meaning it allows the passage of solvent but not of solute. The membranes used for reverse osmosis have a dense barrier layer in the polymer matrix where most separation occurs.

**1.9.8.2.** ***Defination*****:** The concepts of 'osmosis' and 'reverse osmosis' have been known for many years. Osmosis is a natural phenomenon in which a solvent (usually water) passes through a semipermeable barrier from the side with lower solute concentration to the higher solute concentration side. Water flow continues until chemical potential equilibrium of the solvent is established. At equilibrium, the pressure difference between the two sides of the membrane is equal to the osmotic pressure of the solution. To reverse the flow of water (solvent), a pressure difference greater than the osmotic pressure difference is applied as a result, separation of water from the solution occurs as pure water flows from the high concentration side to the low concentration side. This phenomenon is termed reverse osmosis. It is also called as Hyperfiltration. It acts as a barrier to all dissolved salts and inorganic molecules as well as organic molecules with a molecular weight greater than 300 Daltons. Membrane pore size is of the order of 50 to150 µm. It works on ambient temperatures. Operating pressures are 14bar for brackish water and 50-55bar sea water. In fact, studies on osmosis were carried out as early as 1748 by the French scientist Nollet, and many researchers investigated these phenomena over the next two centuries. However, the use of reverse osmosis (RO) as a feasible separation process is a relatively young technology. In fact, only in the late 1950's did the work of Reid show that cellulose acetate RO membranes were capable of separating salt from water, even though the water fluxes obtained were too small to be practical. Then, in the early 1960's, Loeb and Sourirajan developed a method for making asymmetric cellulose acetate membranes with relatively high water fluxes and separations, thus making RO separations both possible and practical. Since then, the development of new-generation membranes such as the thin-film, composite membrane that can tolerate wide pH ranges, higher temperatures, and harsh chemical environments and that have highly improved water flux and solute separation characteristics has resulted in many RO applications. In addition to the traditional seawater and brackish water desalination processes,

RO membranes have found uses in wastewater treatment, production of ultrapure water, water softening, and food processing as well as many others. An estimate indicated that sales of RO membrane products had grown to $118 million yearly in 1990, with great potential for continued growth. The driving force for the development and use of RO membranes is the advantages that these have over traditional separation processes such as distillation, extraction, ion exchange, and adsorption. Reverse osmosis is a pressure-driven process so no energy-intensive phase changes or potentially expensive solvents or adsorbents are needed for RO separations. Reverse osmosis is a process that is inherently simple to design and operate compared to many traditional separation processes. Also, simultaneous separation and concentration of both inorganic and organic compounds is possible with the RO process. In addition, with nanofiltration membranes selective solute separations based on charge and molecular weight/size differences are possible. Finally, reverse osmosis technology can also be combined with ultrafiltration, pervaporation, distillation, and other separation techniques to produce hybrid processes that result in highly efficient and selective separations.

**1.9.8.3. *Reverse Osmosis Process Description*:** Reverse osmosis (RO) is a physical separation process in which properly pretreated source water is delivered at moderate pressures against a semipermeable membrane. The membrane rejects most solute ions and molecules, while allowing water of very low mineral content to pass through. This process also works as an absolute barrier for cysts and viruses. The process produces a concentrated reject stream in addition to the clean permeate product. Reverse osmosis systems have been successfully applied to saline groundwaters, brackish waters, and seawater, as well as for removal of inorganic contaminants such as radionuclides, nitrates, arsenic, and other contaminants such as pesticides. In an RO system (shown in Fig. 1.9.2), a higher concentration solution on one side of a semipermeable plastic membrane is subjected to pressure, causing freshwater to diffuse through the membrane leaving behind a more concentrated solution containing a majority of the dissolved minerals and other contaminants. The major energy requirement for reverse osmosis is to pressurize the source, or "feed" water. Depending on the characteristics of the feedwater, different types of membranes may be used. Because the feedwater has to pass through very narrow passages in the membrane module, larger suspended solids must be removed during the initial treatment phase (pretreatment). A reverse osmosis membrane acts as the semipermeable barrier to flow in the RO process, allowing selective passage of a particular species (solvent, usually water) while partially or completely retaining other species (solutes). Chemical potential gradients across the membrane provide the driving forces for solute and solvent transport across the membrane: $-\Delta \mu_s$, the solute chemical potential gradient, is usually expressed in terms of concentration; and $-\Delta \mu_w$, the water (solvent) chemical potential gradient, is usually expressed in terms of pressure difference across the membrane. Although the phenomenon of osmosis has been observed and studied for over 200 years, its application to the desalination of water has been studied seriously only since about 1950 when the U.S. Department of the interior began sponsoring research in the area. In the 20 years since formulation of the first practical, synthetic asymmetric semipermeable membrane, reverse osmosis technology (RO) has evolved as a highly competitive process involving multi-million dollar institution around the world. Whereas RO was originally conceived of as a method of producing potable water from the sea or from brackish sources, the technology is now being applied on a large scale in municipal and industrial wastewater treatment and reuse and has found additional applications, on a smaller scale, in a wide variety of industrial processes.

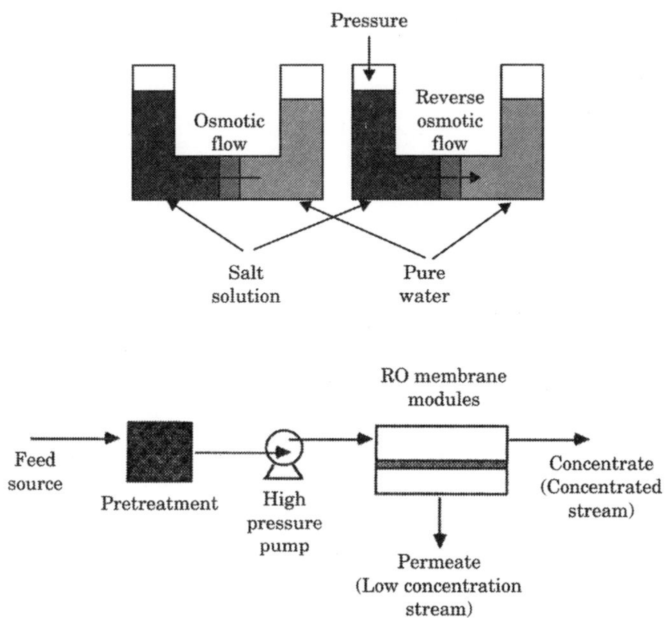


**(b) Reverse osmosis membrane process**

**Fig. 1.9.2:** Schematic of (a) Reverse osmosis and (b) Reverse osmosis membrane process

### 1.9.8.4. *Equations related of Reversis Osmosis*

*Transport Equations*: Transport through semi permeable membrane at constant temperature is described by following equation

$$\Delta Q_{w} = K_{w}(\Delta p - \Delta\pi)^{*} A/\tau \qquad (1.9.6)$$

Where, $Q_{w}$ = water flow rate through the membrane [m³/sec], $K_w$ = membrane permeability coefficient for water [L²], ΔP = hydraulic pressure differential across membrane [N/m²], Δπ = osmotic pressure differential across membrane [N/m²], A = membrane surface area [ m²], τ = membrane thickness [μm]

***Osmotic Pressure*:** Osmotic pressure is a colligative property. It depends on solute concentration, solute temperature and types of ions present. For dilute solutions osmotic pressure is approximated using the van't Hoff equation:

$$\pi = V_i C_i RT \qquad (1.9.7)$$

Where, π = osmotic pressure [N/m²], $C_i$ = molar concentration of the solute [mol/l], $V_i$ = no. of ions formed if the solute dissociates, R = gas constant [J/mol

K], T = absolute temperature [K], For practical purposes this equation is useful for brackish water.

***Salt transport:*** Salt transport across a membrane is proportional to the concentration difference

$$Q_s = K_s(\Delta C)\, A / \tau \qquad (1.9.8)$$

Where, $Q_S$ = salt flow through membrane [m³/sec], Ks = membrane permeability coefficient for salt [L²], ΔC = salt concentration differential across membrane [mol/litre].

The transport of salt across a membrane is commonly expressed as salt passage or salt rejection. Salt passage is the percentage of the salt in the feed that passes through the membrane in to permeate, and is calculated as follows:

$$SP = C_P/C_F \times 100 \qquad (1.9.9)$$

Where, SP = % Salt passage, $C_P$ = Salt concentration in product stream [mol/litre]

$$C_p = Q_s/Q_w \qquad (1.9.10)$$

$C_f$ = Salt concentration in feed stream [ mol/litre]

The percentage of salt rejection is 100 percent minus the percentage of salt passage.

$$SR = 100\text{-}SP \qquad (1.9.11)$$

SR = Salt rejection

**1.9.8.5. *Reverse Osmosis Process Terminology*:** The RO process is relatively simple in design. It consists of a feed water source, feed pretreatment, high pressure pump, RO membrane modules, and, in some cases, posttreatment step. The feed; the product stream called the permeate; and the concentrated feed stream, called the concentrate or retentate. The water flow through the membrane is reported in terms of water flux $J_w$, where

$$J_w = \frac{\text{volumetric or mass permeation rate}}{\text{membrane area}} \qquad (1.9.12)$$

Solute passage is defined in terms of solute flux, $J_{s:}$

When Selecting RO/NF Systems, the following should be considered: (1). *Membrane Selection*: Two types of membranes are most commonly used. These are cellulose acetate based and polyamide composites. Membrane configurations typically include spiral wound and hollow fiber. Operational conditions and useful life vary depending on the type of membrane selected, quality of feedwater, and process operating parameters. Most current manufacturers only have the spiral wound option. (2). *Useful Life of the Membrane*: Membrane replacement and power consumption represent major components in the overall water production costs. The relative contributions depend primarily on feedwater salinity. In well designed and operated RO systems, membranes have lasted 5 to over 10 years in suitable applications. (3). *Pretreatment Requirements*: Acceptable feedwater characteristics are dependent on the type of membrane chosen and operational parameters of the system. Without suitable pretreatment or acceptable feedwater quality, the membrane may become fouled or scaled, and consequently its useful life is shortened. Pretreatment is usually needed for turbidity reduction, iron or manganese removal, stabilization of the water to prevent scale formation, microbial control, chlorine removal (for certain membrane types), and pH adjustment. As a minimum pretreatment, cartridge filters are used for protection of the membranes against particulate matter. (4). *Treatment Efficiency*: Reverse osmosis is highly efficient in removing metallic salts and ions from raw water. Efficiencies, however, do vary depending on the ion being removed and the membrane utilized.

For most commonly found ions, removal efficiencies will range from 85% to over 99%. Organics removal is dependent on the molecular weight, shape and charge of the organic molecule, and the characteristics of the membrane utilized. Organic removal efficiencies may range from as high as 99% to less than 50%, depending on the membrane type and treatment objective. (5). *Bypass Water*: Reverse osmosis permeate will be virtually demineralized. If the raw water does not contain unacceptable contaminants, the design may provide for a portion of the raw water to bypass the unit and blend with RO permeate in order to maintain a stable water within the distribution system and improve process economics. Bypass/Blend will reduce equipment size and power requirements. (6). *Post-Treatment*: Post-treatment typically includes degasification for carbon dioxide (if excessive) and hydrogen sulphide removal (if present), pH and hardness adjustment for corrosion control, and disinfection as a secondary pathogen control and for distribution system protection. (7). *Desalting By-Product*: By-product water or the "concentrate" may range from 10% to 60% of the raw water pumped to the reverse osmosis unit. For most brackish waters and ionic contaminant removal applications, the by-product is in the 10-25% range while for seawater it could be as high as 60%. The by-product volume should be evaluated in terms of availability of source water and cost of disposal. Acceptable methods of byproduct disposal typically include discharge to municipal sewer system or to waste treatment facilities, discharge to sea, or by deep well injection, depending on the by-product concentration and availability of the discharge option being considered. 8. *Pilot Plant Study*: Prior to initiating the design of a reverse osmosis treatment facility, the state reviewing agency should be contacted to determine if a pilot plant study will be required. In many cases, a pilot plant study will be necessary to determine the best membrane to use, type of pretreatment as well as post- treatment, bypass ratio, amount of reject water, system recovery, process efficiency, and other design and operational parameters.

### 1.9.9. *Nanofiltration*

**1.9.9.1. *Introduction:*** Nanofiltration is essentially a form of RO, *i.e.* it is a pressure driven process applied in the area between the separation capabilities of RO membranes and UF membranes. NF is similar to RO only the network structure is more open. It is used in the separation of ions from solutes such as small molecules of sugar. Retention for monovalent salts such as sodium salts and chloride salts is very low but the retention of bivalent salts such as calcium salts and carbonate ions remains very high. In addition the retention is high as well for micropolluants or microsolutes such as herbicides, insecticides, pesticides and for other low molecular components such as dyes. Nanofiltration membranes have pore size 0.001 μm to 0.1 μm. Nanofiltration is a pressure driven separation process. The filtration process takes place on a selective separation layer formed by an organic semi permeable membrane. The driving force of the separation process is the pressure difference between the feed (retentate) and the filtrate (permeate) side at the separation layer of the membrane. Nanofiltration (NF) is one of the four membrane technologies, which utilize pressure to effect separation of contaminants from water streams. The other three are microfiltration, ultrafiltration and reverse osmosis (RO). All of these technologies utilize semi-permeable membranes that have the ability to hold back (reject) dissolved and/or suspended solids from a water stream containing these contaminants. Each of the four technologies has its own characteristics regarding the kinds of contaminants it will remove (see Table 1.9.2.3). For example, NF rejects ions (dissolved ionic contaminants), but not to the same degree as RO. The ability to reject ions isn't a simple filtration mechanism, but a more complex process that's not fully understood.

**1.9.9.2.** ***Definition:*** Nanofiltration is a pressure driven separation process. The filtration process takes place on a selective separation layer formed by an organic semi permeable membrane. The driving force of the separation process is the pressure difference between the feed (retentate) and the filtrate (permeate) side at the separation layer of the membrane. Pores have not been observed in NF membranes under any microscope, however, water can still pass through the membrane and multivalent salts and low molecular weight organics are rejected. It is difficult to predict the performance of NF membranes, especially if more than three solutes are present in the solution, since membrane rejection is influenced by the size, structure and charge of the components in solution (Fig. 1.9.3). As a result, piloting is highly recommended for NF applications,

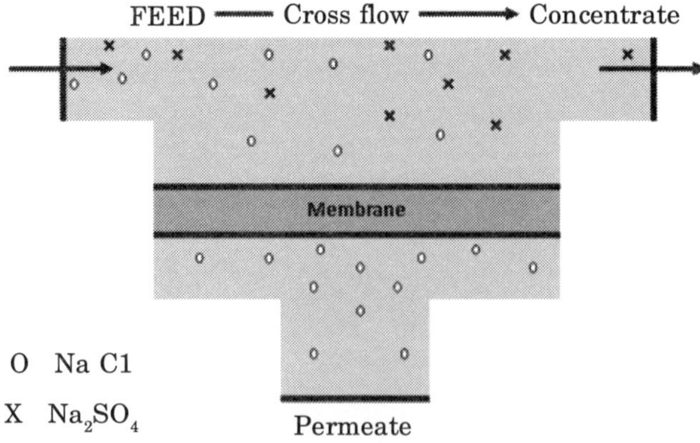


**Fig. 1.9.3:** Nanofiltration process

even if a detailed feed water analysis is available. It is clear that the application of both, RO and NF processes is different; when a high retention is required for NaCl with high feed concentration RO is preferred. Since water permeability is much higher in NF the capital cost for a certain application will be lower. *Nanofiltration* filters out multivalent solutes having a range of 0.001 μm to 0.1 μm.

### 1.9.9.3. *Mechanism*

Recognize that a salt is a compound of two or more ions with an electronic charge. Valence is the number of charges on the ions that form the specific salt, which is not always sodium chloride (NaCl); sodium and chloride are monovalent ions because they have only one charge, whereas ions such as calcium and sulfate are multivalent because they have more than one charge. A defining characteristic of NF membranes is that they reject multiva-lent ions to a significantly greater degree than monovalent ions. The specific rejection of ions varies from one membrane manufacturer to another, but a multivalent ion rejection of 95 percent with a monovalent ion rejection of only 20 percent isn't unusual for NF membranes. Most of these membranes available today are in spiral wound construction only, although it's expected that capillary fiber nanofilters will soon be on the market.

### 1.9.10. *Pervaporation, Vapour Permeation and Gas Permeation*

Although the first pervaporation plants were installed for the dehydration of bio-ethanol, already in 1985 a first plant for dehydration of ethylacetate started its operation in the chemical industry. The first plants were still isolated, that is working from a storage feed tank to a product storage tank. With relative small capacities of a few tons of solvent per day they could easily by-passed if any problems would occur.With increasing experience and confidence in the new technology, quite soon solvent dehydration by means of pervaporation and vapour permeation became an essential step in a production. Today the technology is regarded as a reliable process and, in numerous applications, is even an integrated part of a production process.Organic solvents are used for a variety of purposes in the chemical industry *e.g.* for synthesis of pharmaceuticals, to precipitate materials from aqueous solutions, for cleaning purposes and for drying of final products. Spent solvents nearly always contain some water. Dehydration is an essential step in their recovery but difficult since most of the more common solvents form azeotropeswith water. Final water removal by distillation is then impossible or complicated. Conventional entrainer distillation is not a real option for pharmaceutical or fine che mical production. The addition and afterwards removal of the entrainer is difficult and the residual concentration will have to be monitored continuously. Furthermore, entrainer distillation systems require a certain minimum capacity to be economical. Quite often this capacity is above the amount of solvent which will have to be treated at a single location. The only solution is then in many cases to ship out the spent solvent and buy fresh one, with all the related problems of logistics and storage.

**Table. 1.9.2.3:** Some application of membrane processes

| *Applications* | *Permeate* | *Concentration (Retentate)* | *Benefits of Membrane* |
|---|---|---|---|
| | | ***Microfitration*** | |
| | Low turbidity, clear juice | Juice, suspended solids, colloidal haze particles | Removes suspended solids and turbidity while allowing the passage of color, sugar and taste. |
| Wine filtration | Low turbidity, clear, flavorful wine | Wine, suspended solids, colloidal haze particles | Removes suspended solids and turbidity while allowing the passage of color, alcohol and taste. |
| Industrial Waste Water Treatment | Water & dissolved solids | Water, suspended solids, insoluble metal hydroxide solids | Removes suspended solids and insoluble metal hydroxide solids from wastewater to allow the permeate to be discharged to a local POTW. |
| Fermentation broth clarification | Water & dissolved solids | Water & suspended solids | Removes suspended solids from the fermentation broth leaving clarified liquid. |
| | | ***Ultrafiltration*** | |
| Potable Water | Potable water free of suspended solids, turbidity, bacteria and large viruses | Water, particulate solids | Certified for 4-log removal of bacteria, *Giardia, Cryptosporidium* and viruses from drinking water. Turbidity < 0.1 NTU |
| Oily Wastewater | Oil free (< 50 mg/L) water | Water, suspended solids, insoluble metal hydroxides, etc. | Consistent permeate water quality even when feed composition subject to upsets. Highly concentrated oil emulsion for waste disposal. |
| Electrocoat Paint | Water, dissolved salts and solvents | Water, paint resins and pigments | Allows the recovery of valuable paint resins and pigments |
| Milk | Lactose and salts solution | Protein concentrate | Protein concentration at lower energy cost than evaporation. |
| Whey | Lactose and salts solution | Whey Protein Concentrate (WPC) | Protein concentration at lower energy cost than evaporation. |
| Apple juice clarification | Low turbidity, clear juice | Juice, suspended solids, pulp, colloidal haze particles | Removes suspended solids and turbidity while allowing the passage of juice, sugar and taste. |
| Reuse of municipal wastewater | Water free of suspended solids, bacteria & virusesTurbidity < 0.1 NTUSDI$_{15}$ <3 | Water, particulate solids | Polishes biologically treated municipal effluent for discharge, reuse, or feed to reverse osmosis |

**Table. 1.9.2.3:** (*Contd...*)

**Table. 1.9.2.3:** *(Contd...)*

| *Applications* | *Permeate* | *Concentration (Retentate)* | *Benefits of Membrane* |
|---|---|---|---|
| | | ***Reverse Osmosis*** | |
| Brackish water | Water with low salinity. | Water with higher salt content | Cost-effective alternative to ion exchange and distillation. |
| Potable water | Water that meets WHO guide lines for dissolved solids | Water with higher salt concentration | Cost effective production of potable water from seawater and brackish water sources |
| Water reuse | Water with low salinity, organics and bacteria | Water with higher salt concentration | Cost-effective method of producing low salinity recycled water for industrial applications and indirect potable reuse |
| Whey | Low BOD and salinity permeate | Whey concentrate | Recovery of whey protein. |
| Dyeing effluent | Clean water | Water, BOD, salts, chemicals, etc. | Produces clean water suitable for plant recycle or discharge. |
| Polishing wastewater UF permeate | Water with low salinity, organics and bacteria | Water with higher salt concentration | Cost effective method of producing water to meet POTW standards or for water recycle |
| | | ***Nanofiltration*** | |
| Whey/Whey Permeate | Salty wastewater | Desalted whey concentrate | Allows the recovery of lactose and whey protein concentrate with reduced salt content |
| Textile | Dyes | Water, salts, BOD, COD and color | NF is used to desalt dyes resulting in a higher value product |
| Caustic cleaning solutions | Caustic cleaning solution | BOD, COD, suspended solids, caustic cleaner | Allows caustic cleaning solution to be recycled resulting in reduced cleaning chemical costs |
| Recycle of acid solutions | Acid solution | BOD, COD, calcium, suspended solids, acidic water | Allows acid solution to be recycled resulting in reduced cleaning chemical costs |
| Water | Softened water | Desalted, concentrated Antibiotics | Potable water production. Softened water reduces scaling on equipment and heat exchange surfaces |
| Antibiotics | Salty waste product | Desalted, concentrated Antibiotics | NF produces high value pharmaceutical products |

1.9.10.2.2. *Characteristics of the pervaporation process include:*

1. Low energy consumption
2. No entrainer required, no contamination
3. Permeate must be volatile at operating conditions
4. Functions independent of vapor/liquid equilibrium.

It is well known that the phase change from liquid to vapor takes place in pervaporation. Processes involving phase changes are generally energy-intensive, and distillation is a notorious example of them.

Pervaporation cleverly survives the challenge of phase change by two features.

(1) Pervaporation deals only with the minor components (usually less than 10 wt.%) of the liquid mixtures, and (2) pervaporation uses the most selective membranes. The first feature effectively reduces the energy consumption of the pervaporation process. Compared with the distillation, because of the characteristics of pervaporation operation, it is essentially true that only the minor component in the feed consumes the latent heat. The second feature generally allows pervaporation the most efficient liquid-separating technology.

For dehydration of solvents by pervaporation the performance of a membrane is usually expressed in terms of water flux, or permeance, and selectivity α. The latter is defined as:

$$\alpha = \frac{y_{water}/x_{water}}{y_j/x_j} \qquad (1.9.13)$$

where $y$ and $x$ are the molar fractions in the permeate and retentate streams, respectively. Permeance (ρ) is defined as the flux divided by the partial pressure difference over the membrane. The partial pressure of component $i$ at the retentate side is related to the mole fraction $x$ and activity coefficient $I\gamma$ in the liquid mixture

$$p_i^* = \gamma_i x_i p_i^o \qquad (1.9.14)$$

Where and $p_i^o$ is the vapor pressure of the pure component $i$. When the pressure at the permeate side is small compared to the equilibrium vapor pressure at the retentate side, permeance can be expressed as:

$$\Gamma_w = \frac{N_i}{\gamma_i x_i p_i^o} \qquad (1.9.15)$$

Process batch pervaporation is a simple system with great flexibility; however a buffer tank is required for batch operation. Continuous pervaporation consumes very little energy, operates best with low impurities in the feed, and is best for larger capacities. Vapor phase permeation is preferred for direct feeds from distillation columns or for streams with dissolved solids.

**1.9.10.1.** ***Introduction:*** Pervaporation, Vapour Permeation and Gas Permeation are very closely related processes. In all three cases the driving force for the transport of matter through the membrane is a gradient in the chemical potential which can best be described by a gradient in partial vapour pressure of the components. The separation is governed by the physical- chemical affinity between the membrane material and the species to pass through and thus by sorption and solubility phenomena. The transport through the membrane is affected by diffusion and the differences in the diffusivities of the different components in the membrane are important for the separation efficiency, too. The differences between the three processes are found in the phase and the thermodynamic conditions of the feed mixture:

Pervaporation involves the separation of two or more components across a membrane by differing rates of diffusion through a thin polymer and an evaporative phase change. A concentrate and vapor pressure gradient is used to allow one component to preferentially permeate across the membrane. A vacuum applied to the permeate side is coupled with the immediate condensation of the permeated vapors. Pervaporation is typically suited to separating a minor component of a liquid mixture, thus high selectivity through the membrane is essential. Separation by pervaporation depends on the difference in partial vapor pressure between the two sides of a membrane and the selective sorption properties of the membrane with respect to the components in the mixture. The pressure difference is created by applying a lower pressure at the permeate side. Because the separation is not driven by the liquid-vapor equilibrium, separation of azeotropic mixtures is also feasible. Methanol is rarely treated by pervaporation as it does not form an azeotrope with water and can easily be purified by distillation. Selectivity and flux of the polymeric membranes are generally not in favour for the dehydration of methanol. Other solvents are esters like ethyl- and butylacetate, ketones like acetone, butanone or methyl isobutyl ketone, ethers like tetrahydrofurane or methyl tertiary butyl ether, or acetonitrile, or mixtures of these solvents. The final water concentrations to be reached vary between 1% to below 500 ppm for the alcohols to below 100 ppm for THF. Take the separation of isopropanol/water mixtures for example, if the water content in the feed is 10 wt.%, the maximum single plate separation factor (isopropanol to water) in distillation is about 2, however, a pervaporation membrane can normally offer an one-through separation factor (water over isopropanol) of 2000–10,000. Furthermore, combination of these two features ranks pervaporation the most cost-effective liquid separation technology. In addition, pervaporation also demonstrates incomparable advantages in separating heatsensitive, close-boiling, and azeotropic mixtures due to its mild operating conditions, no emission to the environment, and no involvement of additional species into the feed stream. Unlike other membrane processes, also temperature polarization occurs in pervaporation processes. The energy transported through the membrane is high due to the vaporization and expansion of the permeating components.

Pervaporation is a unique process as not only matter, but also heat is transported through the membrane. The retentate leaving the membrane not only differs in concentration but also in temperature from the original feed.

This temperature reduction in the Pervaporation effect has two different disadvantages: The partial vapor pressure of the better permeable component is not only reduced by its reduced concentration in the retentate, but also by the decrease in temperature. For the retentate the driving force is lower than for the feed. The transmembrane flux decreases with decreasing temperature. In order to avoid too low temperatures it has become common practice to split the total membrane area into a number of stages, as shown schematically in Fig. 1.9.4. The preheated feed stream enters the first stage, where heat is lost for the evaporation of the permeate. At a lower temperature the retentate is passed through an intermediate heat exchanger where it is reheated to the original feed temperature, and fed to a second membrane stage. This procedure is repeated as many times as are necessary to reach the specified composition of the final retentate. Size and number of stages as well as allowed temperature drop over a stage are a matter of optimization in the design of any Pervaporation plant. The Pervaporation membranes used today in industrial applications are of the hydrophilic type. They preferentially permeate water but retain nearly all organic molecules. The main application of these membranes is therefore the removal of water from its mixtures with organic liquids.

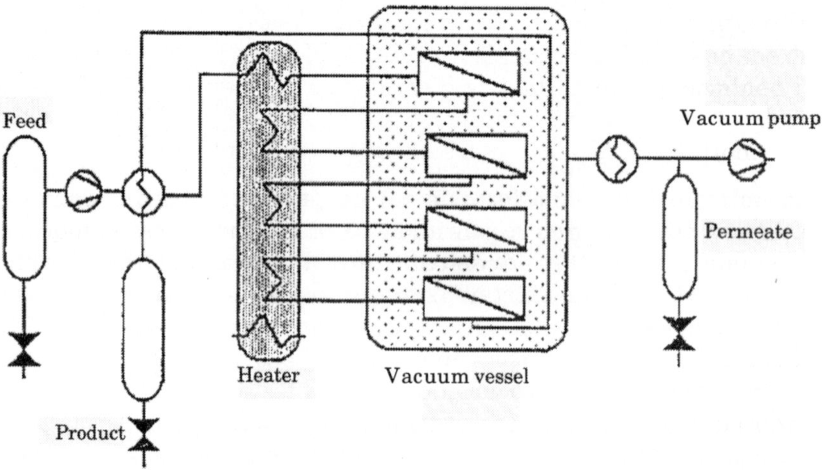


**Fig. 1.9.4:** Pervaporation process

Pervaporation is a membrane process for liquid separation, a polymeric or zeolite membrane usually serves the separating barrier for the process. When a membrane is in contact with a liquid mixture, one of the components can be preferentially removed from the mixture due to its higher affinity with, and/or quicker diffusivity in the membrane. As a result, both the more permeable species in the permeate, and the less permeable species in the feed, can be concentrated. In order to ensure the continuous mass transport, very low absolute pressures (*e.g.*, 133.3–400.0 Pa (1–3 mmHg)) are usually maintained at the downstream side of the membrane, removing all the molecules migrating to the face, and thus rendering a concentration difference across the membrane. As a variant, the use of a sweeping gas in the downstream side of the membrane is also a feasible alternative for the generally used vacuum operation.

**1.9.10.3. *Gas permeation*:** The process of gas permeation is a relatively simple process which has expanded in use rapidly since it was introduced commercially in 1979. Since that time approximately 20 or more companies have come to the market place with equipment. The range of applications covers the supply of pure or enriched gases such as He, $N_2$ and $O_2$ from air, the separation of acid gases such as $CO_2$ and $H_2S$, the separation of $H_2$ in the petrochemical and chemical industries and a variety of smaller applications. A simple phenomenological approach is sufficient in this study for the assessment of membrane quality. For single gas permeation of permanent gases through amorphous microporous silica membranes, at sufficiently high temperatures and low pressures, transport is activated and permeance is independent of pressure. Hence, permeance is described by:

$$P \equiv \frac{N}{\Delta p} = (H_o D_o)\exp\left(\frac{Q - E_D}{RT}\right) \tag{1.9.16}$$

Where $N$ is the molar flux, $H_o$ and $D_o$ are pre-exponential factors related to the Henry and diffusion coefficients, respectively, and $R$ and $T$ have their usual meaning. The overall thermally activated nature of transport arises from the simultaneous occurrence of diffusion ($E_D$) and sorption ($Q$).

Single gas permeance of a single fibre was measured in a pressure controlled deadend set-up. Viton O-rings were used for sealing, limiting the operating temperature to 210°C. The effective membrane length was 17 cm leading to an effective membrane area of 0.0017 m$^2$. Prior to measurements, the membranes were pre-treated by permeating He at 180°C for two days. Measurements were performed with He and $N_2$ (>99% pure) at temperatures ranging from 25°C to 200°C. The pressure at the permeate side (tube side) was kept slightly above ambient pressure, while the pressure at the feed side (shell-side) was varied in the range 150–350 kPa. The flow required to maintain the pressure drop over the membrane was measured by a mass flow indicator.

## 1.9.11. *Membrane modules*

Design of modules for pervaporation and vapour permeation processes had been based on the experience gained in those for water treatment by membranes, like Ultrafiltration and Reverse Osmosis. However, significant modifications had to be made due to the specific requirements of pervaporation and vapour permeation processes. Whereas in the water treatment the portion of the feed volume passing into the permeate is small, in pervaporation and vapour permeation the volume of the permeate is much larger than that of the feed. Pressure losses at the feed side have to be reduced to a minimum in vapour permeation. Otherwise the process would no longer operate at constant pressure, but the feed vapour could reach a region where superheated conditions would exist. Consequently pressure losses in vapour permeation modules have to be as low as several millibar only. In pervaporation feed side pressure losses are not that important, but in multistage arrangements will eventually limit the number of applicable stages. The partial vapour pressure at the permeate side has to be reduced in both processes to fairly low values, especially when low final concentrations of the critical component have to be reached in the retentate. Therefore any pressure losses, even in the range of a few millibar have to be avoided at the permeate side. As any feed mixture will contain organic components at high concentration, mostly at elevated temperatures, chemical and mechanical stability of all module components like spacer, gaskets, potting material and glues is critical. So far mainly four different types of modules are in use on an industrial scale. Over the last 3 years, the designs, the microporous membrane contactors for gas-transfer applications have rapidly developed. These newly designed contactors have moved the technology from small laboratory-scale devices, limited to 5 to 10 gallons per minute (gpm) capacity, to large-scale industrial devices designed for water treatment systems operating at 1000 to more than 2000 gpm. Since these devices are relatively new, the technology is often misunderstood. Membrane contactors are devices that allow a gaseous phase and a liquid phase to come into direct contact with each other, for the purpose of mass transfer between the phases, without dispersing one phase into the other. A typical use for these devices is the removal or dissolution of gases in water. The concept of using membranes to bring two phases into contact with one another is not new. However, recent developments in the design of the contactors have greatly increased their efficiency and capacity. These developments have brought membrane contactors out of the laboratory and made them economical for medium- and large-scale industrial uses. For removal of dissolved gases from an aqueous stream, membrane an aqueous stream, membrane contactors are operated with the aqueous fluid flow on one side of a hydrophobic membrane and a sweep gas and/or a vaccum applied to the other side of the membrane. Since the microporous membrane is hydrophobic, the membrane will not allow liquid water to pass through the pore into the gas side of the membrane. The membrane essentially acts as a support between the gas and liquid phases that allows them to interface at the pore. By adjusting the partial pressure of the gas in contact with the water, gases can be selectively removed or dissolved into the water.

The module is the housing that contains the membrane. With regard to failure analyses, module configuration is important because some types of modules are more reliable than others. Membrane modules are commercially available in four configurations:

- Plate-and-Frame
- Spiral-Wound
- Hollow-Fiber
- Tubular

**1.9.10.1.** ***Theory of operation:*** Membrane contactors are typically fabricated with hydrophobic hollow-fiber microporous membranes. Since the membranes are hydrophobic and have small pores, water will not easily pass through the pore. The pressure required to force water to enter the pore can be calculated by the Yong-Lapace equation modified for use with hydrophobic membrane. This pressure is often called the breakthough pressure

$$P = -2\sigma \cos\theta / r \tag{1.9.17}$$

Where, P = breakthrough pressure, θ = contact angle, σ = surface tension of the water, and r= radius of pore in microporous membrane.

A hydrophobic polypropylene membrane with a pore size of 0.05 micron (μm) has a breakthrough pressure that exceeds 150 pounds per square inch gauge (psig). This high breakthrough pressure prevents water from passing through the membrane at pressures below 150 psig. At the pore, the gas phase will be in direct contact with the liquid phase. It is at this interface that gases can be removed or dissolved into the water, depending on the equilibrium condition established.

***Principles of operation*:** To explain the contactors ability to remove dissolved gases from water, it is important to discuss the driving force for mass transfer. Henry's law states that the amount of gas that will dissolve into the water at equilibrium is proportional to its partial pressure in the vapor-phase in contact with water.

$$P = Hx$$

Where P = gas partial pressure, H = Henry's law coefficient, a function of water temperature, and x = concentration of dissolved solute at equilibrium.

1.9.11.2. *Types of Membranes and its modules*

**1.9.11.2.1. *Spiral Membrane*:** Due to their compact layout and large amount of membrane area per element, spirals are typically used for high-flow applications with minimal or no suspended solids. Their advantage is low hardware and operating costs.

Spiral wound modules with stainless steel central tubes, but otherwise similar to those known from the conventional membrane processes ultrafiltration or Reverse osmosis, are in use, mainly for organophilic membranes. Due to the larger molecular weight of the substances removed through organophilic membranes the volume of the vaporous permeate is much smaller, even at the same permeate side pressure, and the total permeate side pressure can be usually higher than in dehydration applications. Thus pressure losses in the permeate channels are less critical than in water removal. As organophilic applications operate at lower temperatures and low concentrations of organic solvents in the feed, polymers materials can be used as spacer or glue. One or several of the spiral wound modules are housed inside a pressure tube and assembled in conventional skids, very similar as in water treatment. Similar considerations are valid for organic – organic separation. Spiral wound modules have thus bee used in pilot plants for the removal of methanol and ethanol from dry organic mixtures or for the removal of aromatic from aliphatic components. Stability of the material for the feed side spacer and the glue are problems still to be solved.

There has been a development on spiral wound modules for dehydration applications, too. So far this did not lead to applications in industrial plants. Chemical stability problems of the components and too high pressure losses in the permeate side spacer could not be solved satisfactorily, and the costs of the modules and for the installation in a plant were not really lower than those for plate modules. Spiral-wound modules use a sandwich of flat sheet membranes and supports, wrapped spirally around a collection tube (see Fig. 1.9.5). The feed flows in against one end of the rolled spiral and along one side of the membrane sandwich. The support layers are designed to minimize pressure drop and allow a high packing density. Additionally, the spiral-wound modules can be designed by equipment suppliers to promote turbulence and therefore increase the mass transfer across the membrane or to provide an uninterrupted flow path to decrease membrane fouling. Spiralwound modules offer greater packing densities, but maintenance is difficult.

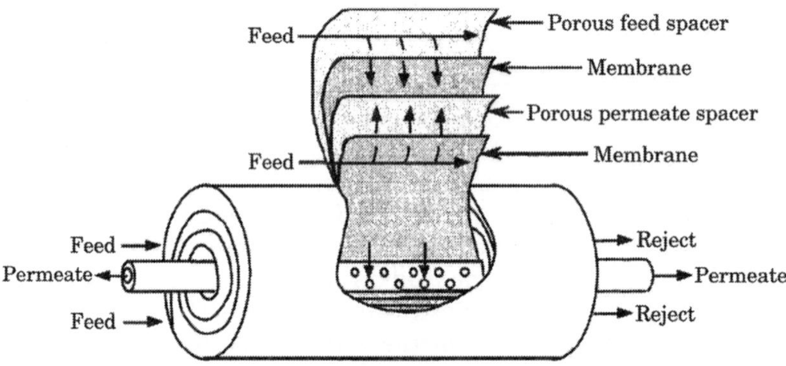


**Fig. 1.9.5:** Spiral-wound modules

**1.9.11.2.2** ***Tubular Membrane:*** As stated above, inorganic (ceramic) membranes are produced mainly as tubes. The obvious module is therefore a tube bundle very similar to a tubular heat exchanger. The detailed arrangement depends on the fact whether the separating layer of the membrane is on the inside or on the outside of the tube. In the first case a bundle of membrane tubes is connected on both ends into tube sheets, each individual tube sealed and fixed. The feed is more or less evenly distributes and directed into the inner lumen of the tubes. Well defined flow regimes and high Reynolds numbers can be obtained at the feed side, controlling polarisation effects like in water treatment application. Depending on the inner diameter of the tube the ratio of feed volume to membrane surface is rather high, and the feed stream cannot be heated inside the module. At high linear velocities this may require partial recirculation of the feed or very small modules in series with the respective intermediate heat exchangers. The permeate flow from the outside surface of the tubes is not obstructed, the permeate vapour can be condensed inside the module shell which has to be kept under vacuum. Sealing of the ceramic tubes to the metal tube sheets and the different thermal expansion of metal and ceramic in the module are problematic. In the second case, with the separating layer on the outside of the tube, the tubes need to be fixed on one side only to a tube sheet only, On this side the inner lumen of the tubes is open to the permeate volume. Baffle plates are required over the outside of the tube bundle in order to achieve good flow distributions and high Reynolds numbers in the feed. The flow regime is not that well defined as in the first case and maldistribution and dead ends may occur. Heating of the feed in the module through additional heat exchanger tubes or through the shell of the module is feasible. The flow through the membrane and the inner diameter of the tubes limit the length of the module, as otherwise the pressure drop at the open permeate side will become too high. In a more recent development each tubular membrane is again fixed and sealed to a tube sheet, with the inner lumen on one side open to the permeate compartment. Additionally each membranes tube is housed inside a heat exchanger tube. The feed flows through the annulus gap formed by the outside surface of the membrane and the inner surface of the heat exchanger tube (Fig. 1.9.6). The feed volume per surface area of membrane can now be controlled by adjusting the height of the annular gap. High linear velocities and thus very high Reynolds numbers can be achieved by this arrangement, without too high volume to surface ratios. From the outside of the heat exchanger tube the feed can be directly heated and the heat lost by the evaporation of the permeate reintroduced. By specific means the direction of the feed flow can be reversed at the end of the annular gap, and the membranes can be arranged for serial or parallel feed flow, or any combination thereof. Tubular membranes are highly resistant to plugging when processing streams with large amounts of suspended solids or fibrous compounds.

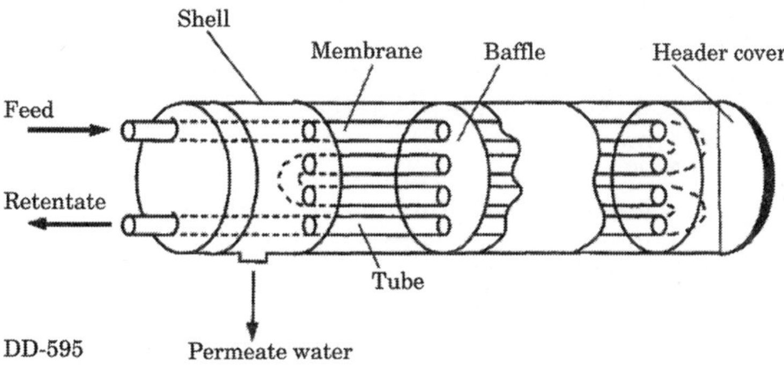


**Fig. 1.9.6:** Tubular membrane contactor

***Tubular Modules:*** Tubular modules have membranes supported within the inner part of tubes. The operator can easily service feed and permeate channels to remove fouling layers. Tubular modules are somewhat resistant to fouling when operated with a turbulent feed flow. The drawbacks of tubular modules are their high energy requirements for pumping large volumes of water, high capital costs, and low membrane surface area per unit volume of module (see Fig. 1.9.6).

**1.9.11.2.3 *Hollow Fiber Membrane Modules:*** As shown in Fig. 1.9.7, hollow-fiber modules consist of small diameter membrane fibers bundled within cylindrical pressure vessels. The fibers are pressurized from the outside. The permeate flows to the interior bore or lumen of the fiber and down the length of the fiber to the product header. Fibers can also be pressurized from the inside, but greater mechanical strength of the fibers is necessary to prevent fiber rupture.

By feeding on the shell side of the fibers, a lower pressure drop is encountered down the bore of the fiber since the permeate flow rate is less than the feed flow rate. Hollow-fiber modules offer the greatest packing densities of the configurations described. Hollow fibres or capillary modules have not yet found an industrial application in pervaporation or vapour permeation processes. A few data have been reported were organic capillary structures with an outside diameter of 0.5 to 1 mm have been coated with silicon and used in organophilic separation. With the flow on the shell side permeate pressure losses inside the bore of the fibre control the process. For specific organophilic applications, these pressure losses may be tolerable. For hydrophilic applications the modules, however, even at an inner diameter of the capillary of 1 mm the useful length of a module would be in the order of 20 to 30 cm only. Such a module, including housing and connection in any industrial application, is more costly than a

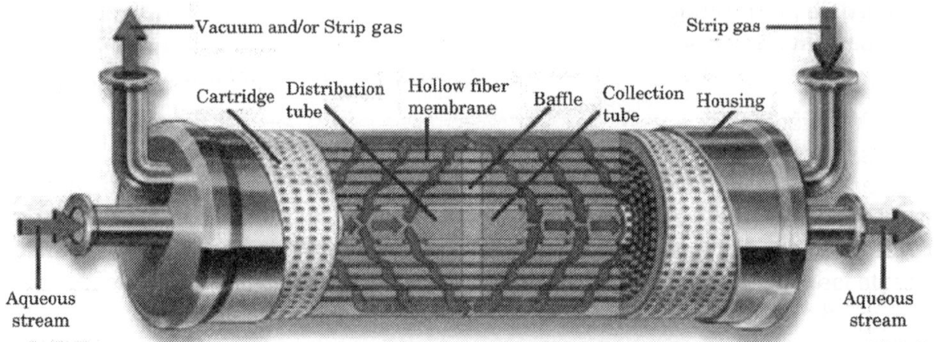


**Fig. 1.9.7:** Hollow fiber membrane contactor

plate module. So far no potting material is available which combines the necessary chemical and mechanical stability at the operation temperature and pressure of a dehydration plant. Microfibres with an inner diameter of 20 μ and a wall thickness of 10 to 20 μ have been proposed, too. Specific flux through such a homogeneous membrane would be low, but outbalanced by high packing density and low membrane costs. Fibre length would be in the order of 20 cm, arranged in a modified module, in which the fibres would be potted in axial direction into the wall of a tube. So far no reports are known on any application of such a module. Extremely high packing density and open channel design; offers the possibility of backwashing from the permeate side, particularly suited for low solids liquid streams.

**1.9.11.2.4 *Plate-and-Frame Modules*:** As shown in Fig. 1.9.8, plate-and frame modules use flat sheet membranes that are layered between spacers and supports. The supports also form a flow channel for the permeate water. The feed water flows across the flat sheets and from one layer to the next. Recent innovations have increased the packing densities for new design of plate-and-frame modules. Maintenance on plate-and frame modules is possible due to the nature of their assembly. They offer high recoveries with their long feed channels and are used to treat feed streams that often cause fouling problems. Only recently advanced designs of plate-and-frame modules capable of operating up to 25% dissolved solids and operating pressures up to 4500 psia have been placed in operation in Germany (Stanford and Miller, 1994). This development opens new opportunities for the use of reverse osmosis for concentration of metal finishing wastewaters. Plate modules are mainly used for dehydration applications; with permeate channels as open as applicable. A rectangular support

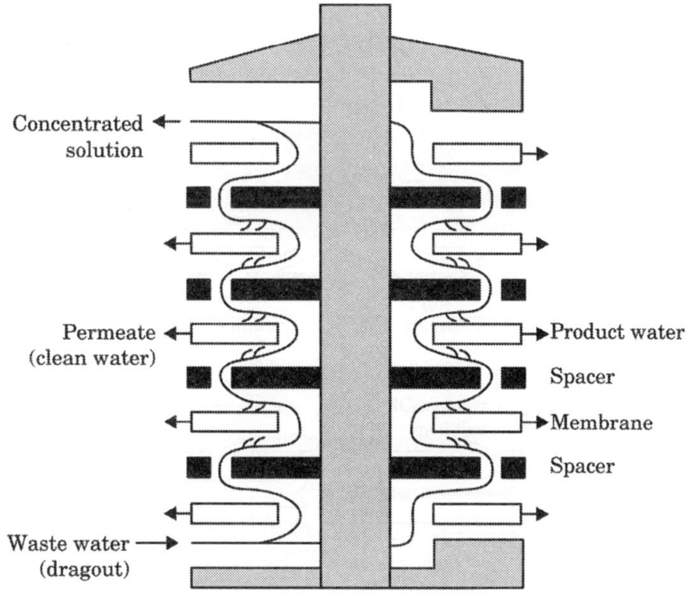


**Fig. 1.9.8:** Plate-and-Frame membrane contactor

plate is provided on both sides with gaskets, which partially cover slots in the plate, acting as distribution channels. On each gasket a membrane is placed, its feed side facing the plate. The permeate side of each membrane is supported by a perforated plate, between two perforated plates a grid or spacer is placed. A membrane, one side of the support plate, and a gasket form a feed chamber, two perforated plates and the space between them a permeate chamber. Each feed chamber is thus adjacent to a permeate chamber, each permeate chamber has a feed chamber at each side. Alternatingly feed and permeate chambers are arranged in a module. The module package is held together by means of flanges and bolts. Thickness and weight of bolts and flanges limits the maximum internal pressure for such modules to 6 to 10 bar. In order to keep weight and handability and weight of the modules within a reasonable range, the maximum size of these modules does not exceed 30 to 40 m of membrane area or less than 100 support plates. Stainless steel is used as a construction material for support plates for the membranes and for spacers. Chemically stable elastomers, like EPDM or perfluorinated polymers are used as gasket material, more widely used is expanded graphite, due to its excellent chemical and thermal resistance. Preferentially the permeate channels are open over the circumference of the module which are assembled inside special vacuum vessel. Intermediate heat exchangers and the permeate condenser are sometimes installed inside the vacuum vessel, mostly these items are installed outside for easier access and maintenance. Usually all membranes in a module are arranged for parallel flow of the feed. The feed channel, between membrane and supporting plate, has a height between 0.5 to 1 mm, linear flow rates are in the order of cm per minute. Serial flow would be desirable in order to allow for higher linear flow velocities and higher Reynolds numbers, but then feed side pressure losses will become too high.

Alternative designs are very similar to plate heat exchangers, in which the supported membrane replaces the heat exchanger plates. These modules may be open or closed to the outside at the permeate side, with internal ducts for feed and retentate, and, when closed, for permeate removal. It has been proposed to integrate plate heat exchangers as preheater and permeate condenser into such modules. Open channel design allows it to be used for products with very high viscosity, particularly suited for high solids pharmaceutical and food applications.

### 1.9.11.4. *Review and evaluation of contactor designs and operation*

In evaluating and describing contactor designs, mass transfer coefficients are very important. For mass transfer equipment in generally and membrane contactors in particular, mass transfer coefficients can be expressed using correlations of the form:

$$\mathrm{Sh} \sim \mathrm{Re}^{\alpha} Sc^{\beta} f(geometry) \tag{1.9.18}$$

The functions in equation (1.9.18) are changing for different membrane processes and the special values are characteristic for a module design and mode of operation, *i.e.* feed flow inside or outside tubular modules, then co-current, counter-current and crossflow modes and so forth. There are many examples of correlations that are given for both tube and shell side flow based on equation (1.9.18) to estimate the mass transfer coefficient for an appropriate unit design.

**1.9.11.4.1. *Tube side flow*:** Using high-speed photography and dye tracer studies, several researchers showed that tube side distribution is often not uniform. They determined that the distribution depends on the inlet manifold type (cylindrical or conical), manifold height, tube length, fiber inner diameter, shell diameter, fiber packing density and Reynolds number. With certain corrections in shell and tube geometry and when was using long manifolds at low Reynolds numbers, a nearly uniform flow was achieved. Now, if the distribution of fiber diameters leads to low flows in some fibers then the mass transfer coefficient will be reduced by polydispersity. This is showed experimentally using microporous polypropylene hollow fibers for stripping oxygen from water into water-saturated nitrogen.

**1.9.11.4.2. *Shell side flow parallel to the fibers*:** Shell side flow is not enough investigated in contrast to tube side. There are correlations, that do not give well enough experimental results on mass transfer in membrane contactors because of the lack of dependency on geometry, but a number of researchers obtained equations very similar to the equation (1.9.18) for different operations.

In the case of gas absorption and stripping, it was obtained by following correlation (Gabelman and Hwang, 1999):

$$Sh = 1{,}25\,(Re.\,d_e/L)^{0.93}\,Sc^{0.33} \tag{1.9.19}$$

The next correlation is obtained for gas absorption and stripping using close-packed fibers:

$$Sh = 0.019\ Gz \tag{1.9.20}$$

How does the shell side residence time distribution (RTD) influence on mass transfer coefficient? For an ideal contactor, plug flow would be obtained. In real shell side flow, distribution of fluid across the fiber bundle and along it leads to broaden the RTD. Mass transfer is more efficient for fluid parts with a shorter residence time because their concentration driving force is higher than that for fluid particles which spend more time in the shell. Investigators showed that higher local mass transfer coefficients are obtained for fluid elements in cross flow than those in parallel flow to the fibers. The effect of fiber packing density on shell side mass transfer was studied. The mass transfer coefficients were higher for loosely packed fibers in contrast to closely in the absorption of oxygen into water with water flowing on the shell side. Most of the published correlations for the shell side mass transfer coefficient were developed using contactors with low fiber packing densities (3-40%). The mass transfer coefficient increases steadily with increasing packing density until a maximum is enhanced on value of 65-70%. Then the fibers are very close together and there are "dead zones" which have influence on decline of mass transfer coefficient. There are differences between values which are obtained by theory and experiments. For packing density less than 65%, experimental coefficients are higher than those based on the theory because the theory assumes the fibers are packed uniformly, with axial laminar flow, whereas in the real systems there is non-uniformity which provides turbulence and consequently higher mass transfer coefficients.

**1.9.11.4.3.** ***Alternative hollow fiber geometries*:** A number of investigations led to the development of contactor designs that offer higher mass transfer coefficients than conventional parallel flow. For stripping oxygen from water into nitrogen, investigators used four module designs which were a cylindrical tube bundle, helically wound bundle, a rectangular bed of fibers and a crimped flat membrane, each with water flowing outside and across the microporous polypropylene hollow fibers. Mass transfer coefficients for the first three modules are given as:

$$Sh = 0.15\,Re^{0.8}Sc^{0.33} \qquad (Re > 2.5) \tag{1.9.21}$$

$$Sh = 0.15\,ReSc^{0.33} \qquad (Re < 2.5) \tag{1.9.22}$$

Sherwood numbers obtained at flow rates were lower than those, which are calculated by the correlation for high flow rates because of uneven flows caused by polydisperse channels between fibers. The following correlations are obtained for crimped flat membranes:

$$Sh = 1.25\ Gz \qquad (Gz<11)\ (1.9.23)$$

$$Sh = 6.0\ Gz^{0,35} \qquad (Gz>11)\ (1.9.24)$$

Researchers compared features of different cross flow modules as a function of oxygen removal efficiency on the basis of equal flow per unit membrane area and equal flow per unit membrane volume as well. Similar results were found. The rectangular bundle showed the best results, 98% oxygen was stripped from water, while the cylindrical bundle, helical bundle, and crimped flat plate removed 82%, 86% and 72%, respectively. These designs were more efficient than a parallel flow cylindrical module, with 7% of removed oxygen from the water. A rectangular design in gas absorption applications with gas flow around and perpendicular to the fibers has a number of advantages like *high mass transfer, low pressure drop, well-defined flow conditions on both sides of membrane, ability to stack modules.* The mass transfer coefficient is higher with fibers arrangement inline than when they are staggered. There are similar correlations for describing mass transfer coefficients developed in different operations like removal of oxygen or $CO_2$ from water into nitrogen, oxygenation of water using microporous polypropylene hollow fibers, bubbleless aeration of water using a sealed hollow fiber bundle unconfined in a jet stream. The mass transfer correlation:

$$Sh = 0.24\ (Re.\ d_e/L)^{0.59} Sc^{0.33} \qquad (1.9.25)$$

1.9.11.6. *Review process designs for membrane contactors applications*

There are many different designs of membrane contactors because of a large number of applications and module configurations. The module is the central part of a membrane installation and can be defined as separation unit. The main task for every engineer is to arrange the modules so that the optimal design is obtained at the lowest product cost. One of very important question that have to be considered is that of where the feed fluid phase should flow, on the shell side or tube side. In gas stripping, the feed fluid phase is liquid. The feed fluid phase is gaseous in *gas absorption*. After the selection of tube side or shell side flow of the feed fluid phase, it should be choose process design. In microfiltration is very often applied the concept of dead-end operation. All the feed is forced through the membrane, which implies that the concentration of rejected components in the feed increases and the quality of the permeate decreases with time because of fouling in dead-end operations. For industrial applications, a cross-flow operation is better because of the lower fouling in contrast to the dead-end mode. In dead-end filtration, the cake grows with time and the flux decreases with time . In cross-flow operation, flux decline is smaller and can be adjusted by cross-flow velocity. There are different cross-flow operations like: co-current, counter-current, cross-flow with perfect permeate mixing, perfect mixing. The contactors are used in the wastewater treatment and for aeration in purification of potable water. For instance, aeration in connection with biological treatment, removal of VOC's and gas stripping. Some application areas in membrane gas absorption and membrane gas stripping will be described in text below. A new application of gas liquid contacting using microporous membranes is an intra vascular oxygenator, wherein the hollow-fiber bundle is plunged through a vein in the neck or the leg into the body's main vein to provide $O_2$ and remove $CO_2$. From acidic wines produced in some unsuccessful fermentation, acetic acid can be removed through a gas membrane into a caustic solution on the other side of a hydrophobic microporous hollow fiber. Several investigators studied removal of ethanol from a fermentation blend through air seated in polytetrafluoroethylene membrane pores into an aqueous solution of ethylene or propylene glycol that does not wet the membrane.

A number of design characteristics must be taken into account to obtain the required number of transfer units and capacity. These are tube diameter, wall thickness, porosity, packing density, flow rates, distribution coefficient, fluid physical properties, and inlet concentration. Module length is a key design specification because it is a function of the number of transfer units. Longer modules provide higher efficiency and increasing pressure drop as well. Consequently, the pressure of the no wetting fluid can be lower than that of the wetting fluid, so that bulk flow of the wetting may penetrate into non-wetting. Several investigators described a design procedure, which gives the required number of fibers, length for a existing design problem and module diameter. The design objective in mass transfer with membrane contactors depends on the application. For industrial applications, the objective is to minimize the cost per amount of mass transferred, while in small installation, for example in blood oxygenations, the objective is to maximize the amount of oxygen transferred per unit volume.

**1.9.11.7. *Water and wastewater treatment*:** Nowadays, membrane technologies are becoming more frequently used for separation of wide varying mixtures in the petrochemical-related industries and can compete successfully with traditional schemes: The technology behind membrane gas separation is potentially an energy-saving one, because the separation process takes place without phase transition. It is also better for the environment, since the membrane approach requires the use of relatively simple and non-harmful materials. The recovery of minor but valuable components from a main stream using membranes can be done without substantial additional energy costs. Compared with conventional techniques, membranes can offer a simple, easy-to-operate, low-maintenance process option. The development of novel materials for gas membrane manufacturing such as organic polymeric, hybrid organic-inorganic and inorganic will expand the use of membrane technology into new fields of applications in the petrochemical industry. The process is simple. There are diverse applications which can be studied by the same basic formulations. The process is generally carried out at atmospheric conditions which, besides being energy efficient, can be important for sensitive applications encountered in pharmaceutical and food industry. Modules can be added and optimized in a process design to achieve the desired separation. Their systems have a low capital cost, compact size, modular configuration, and low specific power consumption, which reduce the production cost. It is a clean process and requires simple and inexpensive filtration. The process is continuous and the membranes do not require regeneration, unlike the adsorption or the absorption processes, which require a regeneration step leading to the use of two solid beds or a solvent regeneration unit. In the petrochemical industry, olefins such as ethylene and propylene are the most important chemicals used for the production of polyolefins such as polyethylene, polypropylene, styrene, ethyl benzene, ethylene dichloride, acrylonitrile, and isopropanol. An important step in the manufacture of olefins is large-scale separation of the olefin from the corresponding paraffin [6]. Furthermore, dehydrogenation, oxidative coupling of methane, steam reforming of methane and water gas shift reaction are some important reactions in petrochemical industry. Membrane gas separation is attractive because of its simplicity and low energy cost, but it has one major drawback and that is a reverse relationship between selectivity and permeability. Nano composite membranes, in which selectivity and permeability can simultaneously be improved, solve this problem. Petrochemical waste streams may contain phenolic compounds or aromatic amines. They are highly toxic and at high concentrations are inhibitory to biological treatment. Membrane aromatic recovery system is a relatively new process for recovery of aromatic acids and bases. Wastewater in petrochemical industry is currently treated by activated sludge process with pretreatment of oil/water separation. Tightening effluent regulations and increasing need for reuse of treated water have generated interest in the treatment of petrochemical wastewater with the advanced membrane bioreactor process.

#### 1.9.11.9. *Membrane bioreactors*

Membrane processes are increasingly being used in reaction, clarification, and recovery schemes for the production of molecules, emulsions and particles. Membrane systems take advantage of their selectivity, high surface-area-per-unit-volume, and their potential for controlling the level of contact and/or mixing between two phases. They are very well suited to the processing of biological molecules since they operate at relatively low temperatures and pressures and involve no phase changes or chemical additives, thereby minimizing the extent of denaturation, deactivation, and/or degradation of biological products. Here the various membrane processes, the established ultrafiltration and microfiltration, and emerging processes such as membrane bioreactors, membrane chromatography, and membrane contactors for the preparation of emulsions and particles is being presented. The biological membrane consists of a double layer of lipid, organised such that the non-polar regions of the phospholipid are towards the centre of the bilayer with the polar moieties Oriented towards the aqueous solvent. The protein molecules are situated either on the surfaces of this bilayer (extrinsic *or* peripheral species) or traverse from one surface to the other (intrinsic *or* integral proteins). This basic structure generates a barrier which is impermeable to polar substances unless particular proteins are present which can specifia mediate the transfer of information (ie receptors) or material (ie transporterdchannels) across the bilayer. They are alternative approaches to classical methods of immobilizing biocatalysts such as enzymes, microorganisms and antibodies, which are suspended in solution and compartmentalized by a membrane in a reaction vessel or immobilized within the membrane matrix itself. In the first method, the system might consist of a traditional stirred tank reactor combined with a membrane-separation unit; in the second method, the membrane acts as a support for the catalyst and as a separation unit (Fig. 1.9.9). The biocatalyst can be flushed along a membrane module, segregated within a membrane module, or immobilized in or on the membrane by entrapment, gelification, physical adsorption, ionic binding, covalent binding or cross-linking. The advantages of immobilizing enzymes are reported to include increased reactor stability and productivity, improved product purity and quality, and reduction in waste. Many studies are oriented to the investigation of operating conditions and optimization of the various properties of these membrane bioreactors. The efficiency of the overall system depends on the biochemical (*e.g.*, catalytic activity, reaction kinetics, concentration, viscosity of substrate and product, immobilization stability), geometric parameters (*e.g.*, membrane configuration, morphology and pore size distribution) and hydrodynamics parameters (such as transmembrane pressure and flow velocity).

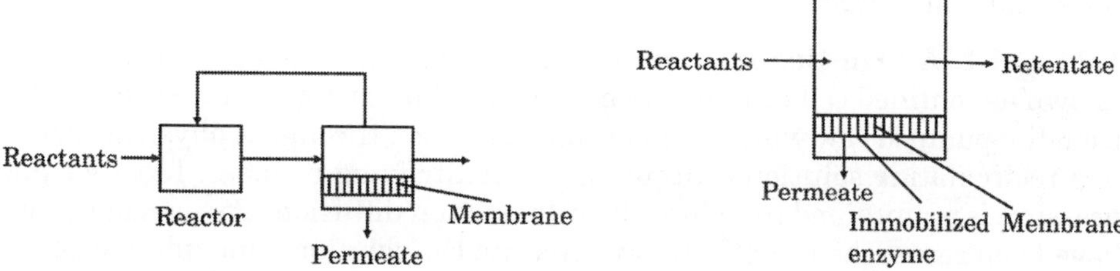


**Fig. 1.9.9.** Membrane bioreactor configurations: (a) reactor combined with a membrane operation unit, (b) reactor with the membrane active as a catalytic and separation unit.

Immobilized biocatalyst membrane reactors are frequently used in a hollow-fiber configuration because of their high packing density (large surface area per unit volume of reactor space). Membrane bioreactors have been used for the production of aminoacids, antibiotics, anti-inflammatories, anticancer drugs, vitamins, optically pure enantiomers and isomers, etc. For example, membrane bioreactors have been reported for the synthesis of lovastatin with immobilized *Candida rugosa* lipase on a nylon support, the production of diltiazem chiral intermediate with a multiphase/extractive enzyme membrane reactor, the synthesis of isomalto oligosaccharides and oligodextrans in a recycle membrane bioreactor by the combined use of dextransucrase and dextranase, the production of a derivative of kyotorphin (analgesic) in solvent media using $\alpha$-chymotrypsin as catalyst and $\alpha$- alumina mesoporous tubular support, and biodegradation of high-strength phenol solutions by *Pseudomonas putida* using microporous hollow fibers. By attaching the enzymes to the porous membrane surface, the mass transfer was improved by utilizing convection rather than diffusion, and the convection was not limited by the significant pressure drops found in enzyme-immobilized bead-filled column reactors. Membrane bioreactors for immobilized whole cells provide an environment for increased cell densities so as to produce higher product titre. The cells are perfused via a membrane with a steady continuous flow of medium and supplied with oxygen and nutrients while wastes and desired products are removed. The cells are usually retained in the bioreactor through a membrane barrier. Many membrane configurations have been tested (such as flat sheet and rotating bioreactors), although the hollow fiber configuration is a particularly interesting one. Cells are either grown in the extracapillary space with medium flow through the fibers, or grown within the fibers with medium flow outside or across the fibers. It was shown that a mass transfer limitation for oxygen and glucose was obtained and that even, well-defined spacing between the fibers was a means of reducing this phenomenon. Membrane bioreactors have been introduced over 30 years ago, and until now their main industrial applications have been for wastewater treatment (*e.g.*, industrial, domestic and municipal). In the biotechnology field, membrane bioreactors have obtained only limited success. The main technological difficulties in using membrane bioreactors on an industrial level are related with rate-limiting aspects and scale-up difficulties of this technology, together with the life-time of the enzyme, the availability of pure enzyme at an acceptable cost, the necessity for biocatalysts to operate at low substrate concentrations and microbial contamination.

**1.9.11.11.1. *Extractive membrane bioreactor:*** In this configuration also known as confined cell or fixed membrane biofilm reactor, nitrate is extracted from the pumped raw water by molecular diffusion through a physical barrier to a recirculating solution containing the denitrifying biomass. Ideally equal pressure is maintained to reduce the influence on diffusion. Various materials have been researched to effectively separate the solutions, including calcium alginate gel, polyacrylamide/alginate copolymer, an agar/microporous membrane composite structure and various microporous membranes. Membrane configurations have typically consisted of either flat sheet or tubular type. The advantage of this process is that both the electron donor and the denitrifying biomass are separated from the product water. Whilst the membrane can permit electron donor transport, biofilm formation should theoretically aid donor retention.

**1.9.11.11.2. *Ion-exchange membrane bioreactor:*** This configuration is identical to the extractive process except microporous membrane technology is replaced by dense ion-exchange membrane technology, the advantage of which is that the nonporous membrane facilitates more specific extraction of nitrate from raw water and in principle hinders the transfer of organic and inorganic pollutants present in the biomedium . The process relies on the concentration and charge gradients of the ion species present as driving forces.

**1.9.11.11.3. *Pressure driven membrane bioreactor:*** This process relies primarily on a suspended denitrifying biomass rather than biofilm development and is thought to have the advantage of extended contact between the denitrifying culture and nitrate in the reactor medium, limited only by mixing rate . Rather than supplying the electron donor or acting as a channel for nitrate laden water, the membrane is placed within or external to the bioreactor, physically rejecting the biomass (hence retaining active denitrifiers) and more refined species. External (sidestream) MBRs are considered to be more suitable for wastewater streams characterised by high temperature, high organic strength, extreme pH, high toxicity and low filterability and have been more commonly investigated in drinking water. Submerged membranes have more recently been considered for large scale. MBR applications on the basis of their lower power costs and improved mass transfer manifested as higher permeabilities. Permeate is subsequently extracted under an applied hydraulic or mechanical pressure and as permeate extraction continues, denitrifying biomass accumulates on the membrane surface in the form of a filter cake allowing further denitrification to take place while the product water passes through the membrane. Consequently this process is the only non counter ion configuration, with both nitrate and electron donor entering the developed biofilm in the same direction.

#### 1.9.11.12. *Membrane bioreactors for waste gas treatment*

The application of membranes has been proposed for a variety of purposes in waste management, as there are the separation of solids, biomass retention, aeration of bioreactors, and extraction of pollutants from wastewater. In bioreactors for waste gas treatment, organic pollutants diffuse into the liquid phase, where micro-organisms degrade the pollutants into harmless products like $CO_2$, $H_2O$ and minerals. Membrane bioreactors were designed as an alternative for conventional types of bioreactors for waste gas treatment, like the biofilter. In a biofilter waste gas is blown through a bed of compost or soil in which micro-organisms consume the gaseous organic pollutants. In such a biofilter no separate water phase is present. An advantage of the membrane bioreactor over the biofilter is the presence of a discrete water phase allowing optimal humidification of the biomass and removal of degradation products, thus avoiding inactivation of the biomass. In a membrane bioreactor for waste gas treatment the membrane serves as the interface between the gas phase and the liquid phase. The gas–liquid interface that can thus be created *e.g.* in hollow fibre reactors, is larger than in other types of gas–liquid contactors. Moreover, membrane reactors do not contain moving parts, are easy to scale up, and the flows of gas and liquid can be varied independently, without the problems of flooding, loading, or foaming commonly encountered in bubble columns. Disadvantages of membrane bioreactors are the high construction costs and the fact that their long-term operational stability still has to be demonstrated.

#### 1.9.11.13. *Liquid membrane*

Since the invention of liquid surfactant membranes by Li in 1968, metal ion extraction using liquid membranes has received considerable interest. The process is capable of achieving a higher concentration of metals in fewer stages while maintaining the high selectivity of solvent extraction. Liquid membranes can be prepared in two different configurations (see Fig. 1.9.10). A liquid can be impregnated in the pores of a porous solid for mechanical support. This form is commonly known as an immobilized liquid membrane. Several kinds of liquid membranes have been used in various disciplines, such as chemical engineering, inorganic chemistry, analytical chemistry, physiology, biotechnoloy, and biomedical engineering. Within these disciplines, the method has been applied to a wide variety of uses, such as gas separation, organic compound removal, metal ion recovery, pollutant removal and bio-separation. The technology is particularly attractive when very dilute solutions are involved since the treating ratio between receiving phase and source phase can be drastically reduced. Liquid membrane extraction is a three-phase process; two of these are miscible with each other but separated by the third phase which is the membrane phase. This membrane can be supported or emulsified and it immiscible with the other two. Solute extraction is achieved by mass transfer between the two miscible phases through the membrane phase. For water/oil/water system, it is the immiscible oil phase separating the two aqueous phases. In liquid, different solutes have different solubility and diffusion coefficients. The product of these two terms is a measure of the permeability. A liquid can yield selective permeability and therefore separation occurred. Because the diffusion coefficients in liquids are typically orders of magnitude higher than in polymers, a larger flux can be obtained. Liquid membranes can be prepared in two different configurations. A liquid can be impregnated in the pores of a porous solid for mechanical support. This form is commonly known as an immobilized liquid membrane. In the alternate configuration, the receiving phase is emulsified in an immiscible liquid membrane. This type of liquid membrane is known as a liquid surfactant or emulsion liquid membrane. Liquid membranes also can be performed in two modes, with or without a chemical carrier. With a chemical carrier, diffusion of permeate species is increased by diffusion of the reaction product. In process without an active carrier, the liquid membrane relies on solubility differences and /or diffusion coefficient differences to separate components. This will be discussed more in related section. In emulsion liquid membrane application, the emulsion must be demulsified into two immiscible

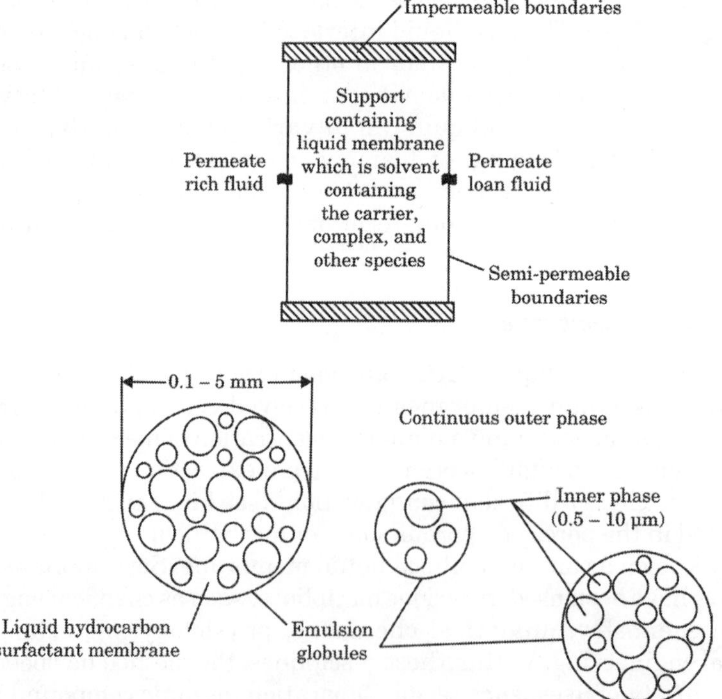


**Fig. 1.9.10:** Immobilized and emulsion liquid membrane

phases after the extraction step of the process. This is accomplished by heating, application of electric field, or centrifugation. The liquid membrane phase containing the surfactant and carrier will be recycled to the emulsion preparation step while the internal phase of the emulsion containing the concentrated solute will undergo further purification in the recovery process or treatment and disposal in the purification process. A membrane has been defined as an inter-phase between two phases. The inter-phase may be solid or liquid. In liquid membrane process, the liquid film will separates two phase from each other. The separation process occurs because of differences in solubility and diffusivity in the liquid film. There are two types of liquid membrane; Immobilized Liquid Membrane or Supported Liquid Membrane and Emulsion Liquid Membrane or Liquid Surfactant Membranes. The liquid film is immobilized within the pores of porous membrane. The porous membrane serves only as framework or supporting layer for the liquid film. Such membrane can be easily prepared by impregnating a (hydrophobic) porous membrane with a suitable organic solvent. An ILM can be made in at least three different geometries. The first one is planar or flat geometry which is very useful for laboratory purposes and for industrial purposes. However a planar geometry is not very effective since the ratio of surface area to volume is too low. Therefore, hollow fiber and spiral wound modules can be used to provide high surface area to volume ratio, which approach 10000 $m^2/m^3$ for hollow fiber and 1000 $m^2/m^3$ for spiral wound modules. There are two primary problems associated with the use of immobilized liquid membrane, which are solvent and carrier loss. Solvent loss occurred because of evaporation, dissolution, or large pressure differences forcing solvent out of the pore support structure. Carrier lost is due to irreversible side reactions or solvent condensation on one side of the membrane. To overcome the problems, ion exchange membranes have recently been studied by many researchers. Emulsion liquid membranes are known as double emulsions system. The primary emulsions were prepared by emulsifying two immiscible phases of the stripping solution and organic phase with a surfactant to produce an emulsion. This primary emulsion is then dispersed in the solution to be treated. Mass transfer takes place between the feed phase and the internal phase through the membrane phase. For example, two immiscible phases, water and oil are mixed vigorously and emulsion droplets are formed, which are stabilized by the addition of surfactant. This emulsion (water/oil emulsion) is added to a large vessel containing an aqueous phase where a water/oil/water emulsion is now formed, the oil phase being the liquid membrane in this concept. The major problem associated with ELMs is emulsion stability. If the emulsion globules break and the inner droplet phase spill into the continuous phase the separation is lost. Interfacial shear between the continuous phase and membrane phase causes the liquid membrane to thin and in some cases, rapture. Another problem is osmotic swelling although it is rarely mentioned in the literature. This phenomenon occurs when water in external phases diffuses through the organic membrane phase and swells the inner aqueous droplet phase. The increased volume of the internal phase leads to increased breakage and dilution of the concentrated droplet phase. Under agitation the emulsion liquid membrane phase separates into spherical globules of emulsion, which have typical diameters of 10 μm to 1 mm. Each globule contains many droplets of encapsulated inner or receiving phase with a typical size of 1 to 10 μm in diameter. Therefore, emulsion will produce large surface area/volume ratios for very rapid mass transfer. The emulsion liquid membrane process consists of three important components, they are: i) Internal phase (receiving phase): an aqueous phase, actually made up of encapsulated droplets, which will receive the solute. ii) External phase (source phase or feed phase): a continuous aqueous phase of feed/waste water which is the source of the metal ions. iii) Membrane phase: an organic phase that separates the internal droplets in the emulsion from the external phase. It usually consists of three components. a) Diluents/organic solvent, b) Surfactant/stabilizer of the system, c) Extractant/ carrier to transfers the metal ion from external to internal phase. An immobilized hollow spherical globule model approach is applied to the extraction process of silver ion through the ELM system. In order to satisfy the experimental finding some assumption were made based on the parameters that affect the emulsion liquid membrane extraction. Several external consideration such as emulsions distribution, uniformity of emulsion size and boundary conditions were also assumed. Finally, this model was verified to perform the validity and stability of model based on existing models. In case of membrane emulsification, droplets grow at the pore outlets until, upon reaching a certain size, they detach. This is determined by the balance between the drag force on the droplet from the flowing continuous phase, the buoyancy of the droplet, the interfacial tension forces and the driving pressure. The final droplet size and size distribution are determined by the pore size and size distribution of the membrane and also by the degree of coalescence, both at the membrane surface and in the bulk solution.

A typical experimental set-up for a membrane emulsification process includes a tubular microfiltration membrane, a pump, a feed vessel, and a pressurized ($N_2$) oil container. The oil phase (to be dispersed) is pumped under gas pressure through the membrane pores into the aqueous continuous phase which circulates tangentially to the membrane surface. Membrane emulsification has been developed during the past 25 years. The distinguishing feature is that the resulting droplet size is controlled primarily by the choice of the membrane and not by the generation of turbulent droplet break-up. The apparent shear stress is lower than the conventional emulsification systems, because small droplets are directly formed by permeation of the dispersed phase through the micropores, instead of disruption of large droplets in zones of high-energy density. Besides the possibility of using shear-sensitive ingredients, emulsions with narrow droplet size distributions can be produced. Furthermore, membrane emulsification processes allow production of emulsions at lower energy input ($10^4$–$10^6$ J/m$^3$) compared to conventional mechanical methods ($10^6$–$10^8$ J/m$^3$). Liquid membrane technology is introduced and is identified as a subset of membrane science. This technology has attracted increasing attention for its potential capability in the field of separation and has been demonstrated as an effective tool in many applications.

#### 1.9.11.13.1. *Ionic liquid as a liquid membrane*

It is very important from both the environmental and the economical point of view to find a simple and cheap way of separating gases. The use of supported ionic liquid membranes, porous supports whose pores are filled with an ionic liquid (IL), has received growing attention during recent years. These membranes have been shown to be a very attractive way for the highly selective transport of organic compounds involved in the synthesis of pharmaceutical and fine chemicals, such as alcohols, esters organic acids and amino acids. They can also be used for sea water desalination, analytical instrumentation, metal ion separation from aqueous solutions, and for gas separation. ILs are usually composed of heterocyclic organic cations and various anions. They have negligible vapor pressure and possess unique properties such as non-volatility, non-flammability, and a wide temperature range for liquid phase. The reality is that ionic liquids can be liquid at temperatures as low as –96°C and some are liquid at over 400°C. Furthermore, room-temperature ionic liquids (RTILs) are frequently colorless, fluid, and easy to handle. In the patent and academic literature, the term 'ionic liquids' now refers to liquids composed entirely of ions that are fluid around or below 100°C. Most of the ILs investigated so far comprise the organic 1-alkyl-3-methylimidazolium ([CnMIM]), N-alkylpyridinium, tetraalkylammonium or tetraalkylphosphonium cations. The anions are either organic or inorganic, including: hexafluorophosphate [PF6]; tetrafluoroborate [BF4]; trifluoromethylsulfonate [CF3$SO_3$]; bis[(trifluoromethyl)sulfonyl]amide [($CF_3SO_2$)2N]; trifluoroethanoate [CF3$CO_2$]; acetate [$CH_3CO_2$]; nitrate, and halide. ILs have many advantages, such as, they have no detectable vapor pressure, and therefore contribute no volatile organic compounds (VOCs) to the atmosphere. Another reason for booming research into ILs is that at least a million binary ionic liquids, and 1018 ternary ionic liquids, are potentially possible having their own beauty in different applications in the ecofriendly way. For comparison, more than about 600 molecular solvents are in use today. This diversity enables the solvent to be designed and fine-tuned to optimize yield, selectivity, substrate solubility, product separation, and even enantioselectivity. ILs can be highly conducting, dissolve enzymes, form versatile biphasic systems for separations, can form both polymers and gels for device applications, are media for a wide range of organic and inorganic reactions. In the past few years significant work has been carried out in the area of preparation, characterization and application of ILs for synthesis, catalysis and separation. Different ILs can be synthesized having a wide range of physical and chemical properties that can be fine-tuned by using different cations and anions to meet the requirement of specific applications. ILs are good solvents for a wide range of both inorganic and organic materials, and unusual combinations of reagents can be brought into the same phase. They are often composed of poorly coordinating ions, so they have the potential to be highly polar yet noncoordinating solvents. They are immiscible with a number of organic solvents and provide a nonaqueous, polar alternative for two-phase systems. Hydrophobic ionic liquids can also be used as immiscible polar phases with water. Ionic liquids are nonvolatile, hence they may be used in high-vacuum systems and eliminate many contaminant problems.

#### 1.9.11.13.2. *General Characteristics of ILs*

The physicochemical properties of ILs depend on the nature and size of both their cation and anion constituents. One can tuned fine the hydrophobisity, viscosity, density, and solvation ability of an IL by varying the anion, cation or alkyl chain of the cation. In context of hydrolysis stability, the lower homologues of alkylsulfate anions, namely methanesulfate and ethanesulfate may have sensitive towards hydrolysis. In the presence of water these anions may form the corresponding alcohol and hydrogensulfate at elevated temperatures. But for alkylsulfates with longer alkyl groups the sensitivity towards hydrolysis is much lower. Table 1 illustrates some physicochemical properties of the ILs commonly used in analytical chemistry (see Fig. 1.9.11). The contamination of ILs by impurities such as water, halides, and metal ions affects the physical properties. ILs generally have higher density than of water. They are often viscous and have a broad range of viscosity from 0.035-0.500 Ns/ $m^2$, compared to other molecular solvents. They have negligible vapor pressure and are in molten state at room temperature. Some ILs are fluid even at temperatures below –80°C. The melting point of ILs generally increases with increase in the substituent chain length. The temperature dependence of thermal conductivity (and heat capacity) is very weak for ILs. The heat of fusion of ILs is in the range of 1/8$^{th}$ times of that of water. The capacity of ILs for storing heat is high enough which has a crucial quality for their application in solar energy collection. ILs can be made miscible or immiscible (hydrophobic or hydrophilic) with hydrocarbons, ethers and a number of organic solvents by varying the constituent ions. They are highly polar and noncoordinating, which makes them ideal for catalytic reactions. ILs are usually composed of large asymmetric organic cations and inorganic or organic anions. They exhibit unique properties compared to other solvents. First, ILs are generally colorless liquids with relatively high viscosities. Second, they exhibit very low vapor pressures under ambient conditions and thus are effectively non-volatile. Third, ILs are good solvents for a broad spectrum of inorganic, organic and polymeric materials and are immiscible with numerous organic solvents. Some properties, such as the thermal stability and miscibility, mainly depend on the anion, while others, such as the viscosity, surface tension and density, depend on the length of the alkyl chain in the cation and/or shape or symmetry. The most commonly employed IL anions are polyatomic inorganic species, such as PF6- and BF4-. In addition to the interactions that exist in conventional organic solvents (hydrogen bonding, dipole–dipole and van der Waals interactions). ILs also have ionic interactions (mutual electrostatic attraction or repulsion of charged particles), which makes them very miscible with polar substances. At the same time, the presence of the alkyl chain on the cation determines the solubility of the ILs in less polar fluids. Hydrogen bonding is thought to exist between the oxygen or halide atoms on the anion and the hydrogen atoms on the imidazolium or pyridinium ring of the cation in ILs. The properties of the ILs can significantly change by varying the length of the alkyl groups that are incorporated into the cation and the types of anions. For example, for the 1-alkyl-3-alkylimadazolium cation, replacing the ($PF^{6-}$) anion with ($BF^{4-}$) dramatically increases the solubility of the IL in water, whereas replacing the anion with the (Tf2N") ion decreases the water solubility. The change in the length of the 1-alkyl chain from 1 to 9 on 1-alkyl-3-methylimadazolium hexafluorophosphate [CnMIm] [PF6] can make the usually soluble ILs very immiscible in water. In addition to these applications of ionic liquids, it can be found that these noble solvents are very useful in many other fields with its recycle ability and without any impact on environment which is the most impressive factor in the present era of environmental concern.

1-octyl-3-methyl imidazolium chloride ([OMIM][Cl])

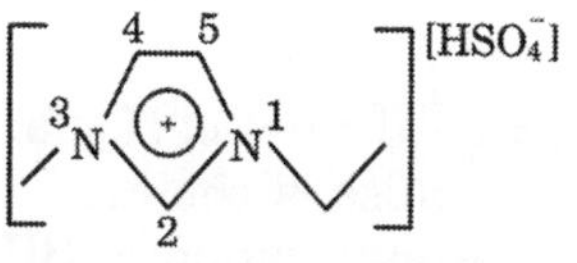

1-ethyl-3-methyl imidazolium hydrogen sulphate ([EMIM][HSO4])

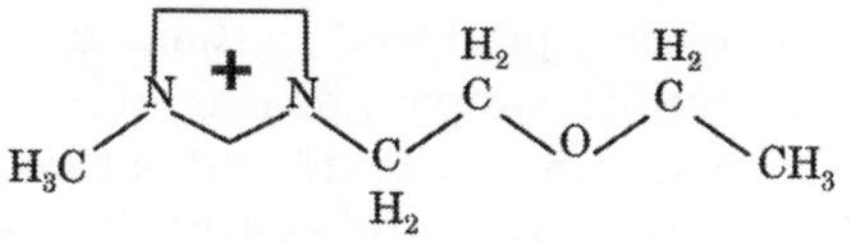

1-(2-etoxy-ethyl)-3-methylimidazolium hexafluorophosphate

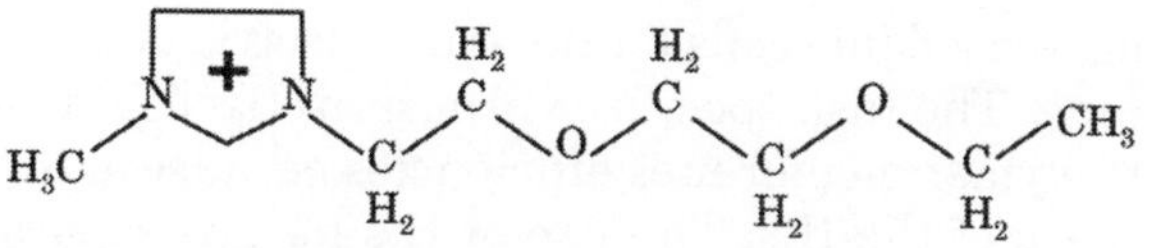

1-{2-[2-(etoxy)-etoxy]- ethyl}-3-methylimidazolium hexafluorophosphate

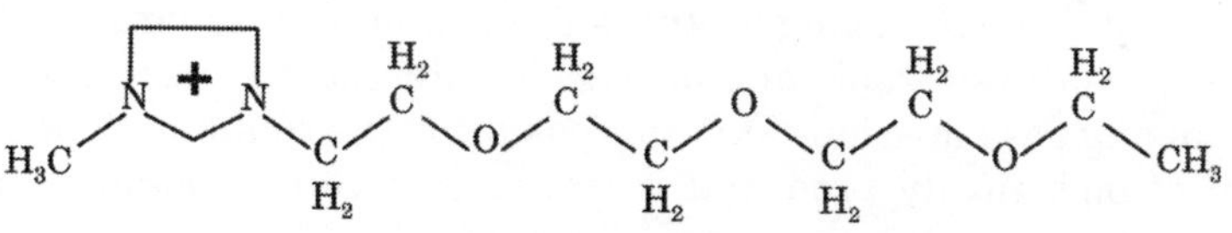

1-[2-{2-[2-(2-etoxy)-etoxy]-etoxy}-ethyl]-3-Methylimidazolium hexafluorophosphate

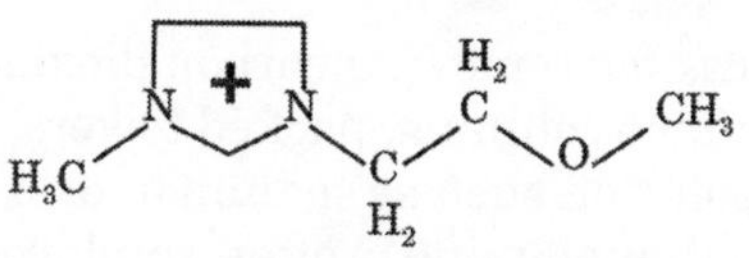

1-(2-methoxy-ethyl)-3-methylimidazolium hexafluorophosphate

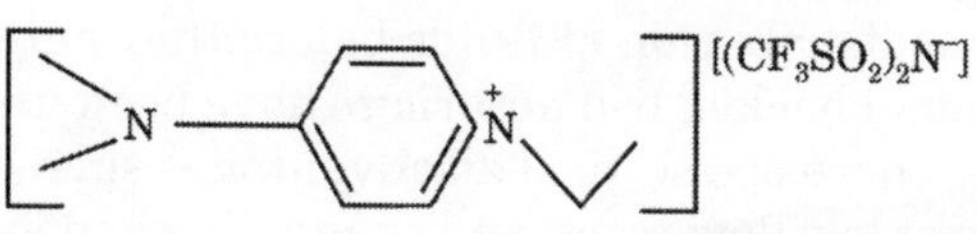

N-ethyl-4-(N'-N'-dimethyl ammonium) pyridinium bis (triflouro methyl sulfonyl) imid ([EAPY][BTI])

**Fig. 1.9.11:** Few typical structure of ILs

***Advantages of Ionic Liquids:*** Nonvolatile; Good solvent power for both polar and nonpolar compounds; Suitable density and viscosity; thermally stable; Nonflammable; Wide liquidus range; Possible to engineer ionic liquids of suitable properties by combining appropriate anions, cations, and varying their chain lengths.

Recently, the use of ILs for gas separation processes has attracted much attention because of the high solubility of different gaseous species in ionic liquids. The preparation of SILMs has also been reported for the selective separation of different organic or gaseous compounds. The $CO_2$ separation from gas streams is a worldwide topic of interest for decreasing the global warming causing gas emissions. Additionally research has focused on the refinement of biogas for the feed-in of biologically produced methane into conventional natural gas distribution networks. Sulfur compounds are interesting contaminants in many industrial gas streams. $H_2S$ is a toxic and corrosive substance that is found in nearly all natural gas sources and synthesis gas and biogas streams. $SO_2$ emissions are generated wherever fossil fuels are combusted. Other sulfur compounds, like tetrahydrothiophene (THT), are added to natural gas as odorants, making it impossible to directly use natural gas in peripheral catalytic processes like home-based fuel cells for heat and electricity generation because of catalyst poisoning. The task specific synthesis of ILs could potential exceed the conventional polymer membranes efficiencies because of the solubility and diffusivity properties of the ILs. The use of ILs for gas separation processes recently has attracted much attention due to high solubilities of different gaseous species in ionic liquids. The preparation of supported ionic liquid membranes is also reported for selective separation of different organic or gaseous compounds. Research activities have amongst others focused on $CO_2$ separation from gas streams, representing a topic of world wide interest, the decrease of global warming causing gas emissions and, increasingly, the refinement of biogas for the feed-in of biologically produced methane into conventional natural gas distribution networks.

### 1.9.11.14 *Hybrid system*

The gas absorption process for removing carbon dioxide can be carried out in many reactors such as bubble columns, packed towers, venturi scrubbers and sieve trays. An aqueous solution such as sodium hydroxide, sodium carbonate, monoethanolamine or diethanolamine is often employed as an absorbent. The main effort in designing and operating the absorber is to maximize the mass transfer rate by producing as much interfacial area as possible. In the case of the packed column, this is made possible by reasonable selection of packing material and uniform distribution of fluids before they enter the packed bed. Although the traditional packed bed absorbers have been used in the chemical industry for decades, there are several disadvantages such as flooding at high flow rates, unloading at low flow rates, and channeling and foaming, which lead to difficulties in mass transfer between gas and liquid. An alternative technology that overcomes these disadvantages and also offers more interfacial area than the conventional approaches is a non-dispersive contactor via a microporous membrane. When a suitable membrane configuration such as a hollow fiber is used, the fluids to be contacted flow on the opposite side of the membrane, and the gas–liquid interface forms at the mouth of each membrane pore. The available contact area remains undisturbed even at a high or low flow rate because the two fluid flows are independent. This porous hollow fiber membrane contactor can be an attractive alternative to packed towers for gas absorption process. The membrane contactor increased the available gas–liquid contactor area and mass transfer coefficient, and thus the hybrid process showed a higher $CO_2$ removal efficiency than the conventional absorption tower. PVDF module with an asymmetric pore structure showed a stable gas–liquid interface, and the $CO_2$ absorption rate per unit volume of the membrane contactor was 2.7 times higher than that of the packed column. In addition, the membrane contactor hybrid process was successfully operated for 80 h maintaining a greater $CO_2$ removal efficiency of above 90%. The gas absorption systems were evaluated from the viewpoint of electricity. $CO_2$ Membrane Gas Absorption Technology: Using hollow fiber membrane contactors to absorb $CO_2$ from flue gas into solvent. A hybrid process exploits the advantages of pervaporation and distillation while the negative aspects are minimized. Several processes have been presented in the literature and are applied in the industry such as for the dehydration of alcohols, aprotic solvents, and esters, as well as for the removal of VOCs from aqueous streams3. The role and application range of pervaporation in stand-alone applications and in hybrid processes can be expanded if the involved capital cost of the pervaporation unit is reduced. The stability of the pervaporation membranes, the concentration and temperature polarization and the temperature drop that occurs in the liquid are factors that increase the required membrane area, the amount of auxiliary equipment and the related capital and operating cost. In the next sections some of these drawbacks and also some characteristics of hybrid distillation-pervaporation processes are described.

Some basic hybrid configurations are presented in Fig. 1.9.12. The first configuration (Fig. 1.9.12a) is used for systems with minimum-boiling azeotrope or when a large number of trays are required in the rectifying section. Similarly, the membrane can be placed in the bottom stream for systems with a maximum-boiling azeotrope or when a considerable amount of trays are required in the stripping section. Fig. 1.9.12b is used for systems with an azeotrope at an intermediate concentration. The last system shown in the Fig. is used for systems with a low relative volatility in the whole range of concentrations where the number of trays and the reflux ratio can significantly be reduced.

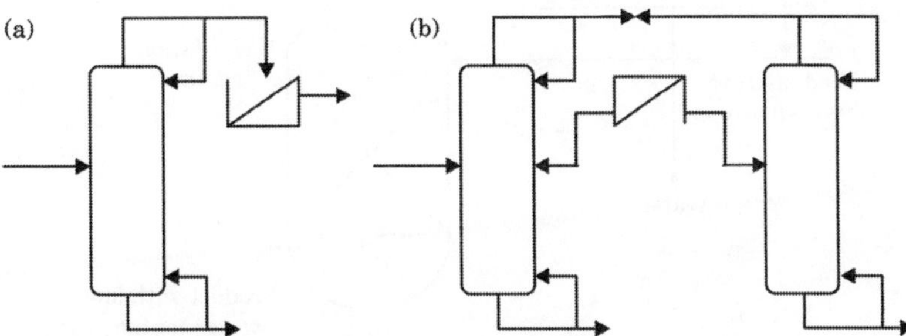


**Fig. 1.9.12:** Basic configuration for hybrid distillation-pervaporation processes

Fig. 1.9.13. A similar arrangement, however, here two organic components, an alcohol and an ester have to be separated and purified from their ternary mixture with water. The ternary vapour mixture from the first column is passed over the membrane of the vapour permeation unit, and nearly all water is removed. The permeate is recycled to the inlet of the first column. The two organic components can now be separated in the second column, which any residual water leaving together with the alcohol.

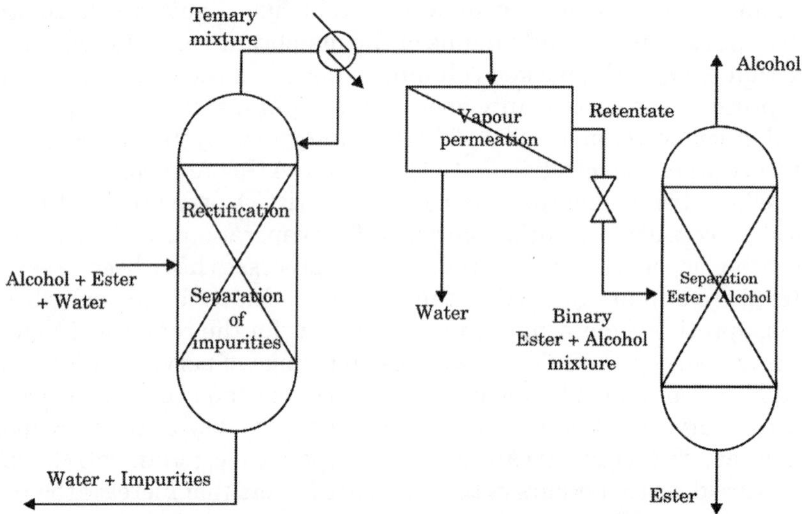


**Fig. 1.9.13:** Hybrid system for the separation of a ternary mixture

When the original feed composition is on the organic side of the azeotrope an arrangement as in Fig. 1.9.14 may have its advantages. The column separates the feed into the high boiling organic at the bottom and a low boiling mixture

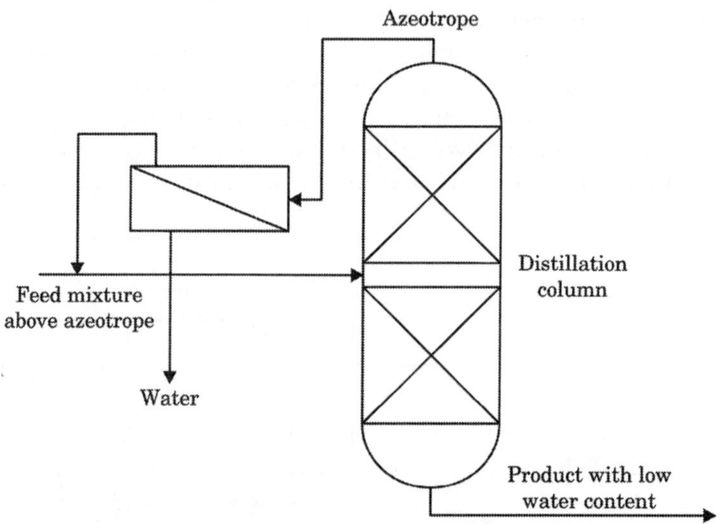


**Fig. 1.9.14:** Hybrid process – feed above azeotrope

close to the azeotrope at the top. This vapour from the top is passed through a vapour permeation plant which removes water, preferentially to a residual concentration close to that of the original feed. This retentate is recycled to the inlet of the column. All the water from the feed has to permeate through the membrane, but the most economical concentration range can be chosen. A certain drawback of this arrangement is the fact that any impurities of the original feed will remain in the purified organic stream. Existing azeotropic distillation plants for the dewatering of organic solvents can simply be revamped by the addition of a vapour permeation or pervaporation unit. Such plants comprise usually a first distillation column by which the feed is distilled close to the binary aqueous – organic azeotrope. In a second column the entrainer is added, which forms a ternary azeotrope with water and the organic. At the bottom of the second column the dry organic component is obtained, whereas the ternary azeotrope from the top of this column is condensed and split into two phases, an aqueous one, which is fed to a third column for further purification and entrainer recovery, and an organic one, which is returned to the second column. The upper part of this second column is limiting the capacity of the plant as all water has to pass through it as the ternary azeotrope. When the stream from the first column is increased, the additional water can be removed by a pervaporation – vapour permeation system, without overloading the second column. Calculations have proven that the capacity of an existing azeotropic distillation can be increased by up to 40% by reducing the reflux ratio of the first column and removal of the excess water by pervaporation. Any of the commonly used chemical engineering programs (like Aspen, Chemcad, PROII) can be used for calculation and simulation of these hybrid systems. An additional modulus has to be introduced into the program which in rather simple terms describes the membrane system.

The simplest arrangement of a pervaporation system coupled to a reactor is schematically shown in Fig. 1.9.15. A batch reactor is filled with the reaction mixture, one of the educts at precalculated stoichiometric surplus. The mixture is passed continuously over the pervaporation membrane, until the water introduced with the raw materials and freed in the reaction has been removed

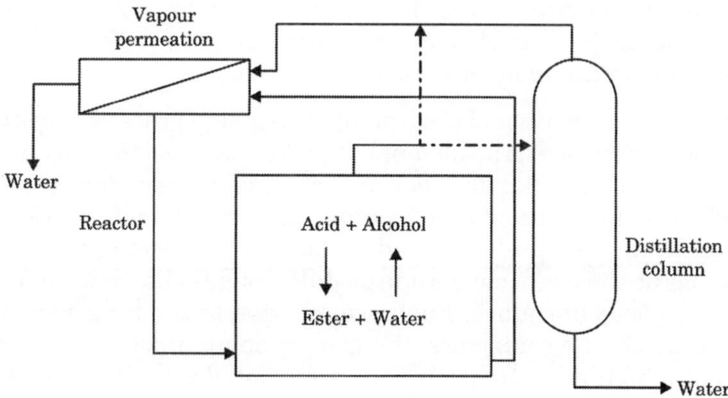


**Fig. 1.9.15:** Reaction, coupled with pervaporation/vapour permeation

to the wanted extend. As indicated in Fig. 1.9.15 the more volatile portion, *e.g.* an aqueous azeotrope can be alternatively evaporated from the reactor and passed through a vapour permeation unit. By arranging several reactors and membrane systems in a cascade and passing a bleed stream downstream of the first membrane system to a second reactor, a continuous operation is possible. Depending on the type and nature of the reaction reactors, membrane systems, and distillation columns can be combined in different arrangements for an optimum yield and downstream purification. One of the first industrial plants, combining pervaporation and an esterification reaction, is operating continuously with a cascade of reactors and pervaporation units.

## AMMONIA STRIPPING IN FLAT CIRCULAR POLYPROPYLENE HYDROPHOBIC MEMBRANE CONTACTOR

## 1.9.15. CASE STUDY

### 1.9.15.1. *Introduction*

Hollow fibre membrane contactors have been used to remove compounds having volatile nature called VCs such as $NH_3$, $H_2S$, $SO_2$, HCN, etc., from water and wastewater. The hollow fibers used are commonly microporous and hydrophobic. In this process, the feed and stripping solution are flowing on either side of a membrane. Since the membrane is hydrophobic, it prevents aqueous solution, which has higher surface tension, to penetrate the gas-filled pores. The VCs will volatilize from the feed, diffuse through the gas-filled membrane pores, and react with the stripping solution. Comparing with conventional techniques for removing of VCs, hydrophobic hollow fiber membrane contactor has provided a number of advantages. Large surface area per volume which provides fast removal of the volatile contaminants; membrane pores filled with gas rarely dramatically impede transport of VCs; no secondary pollutants are produced and the VCs can be recovered and reused if it is necessary.

Ammonia is a common and undesirable contaminant in wastewater and biomass cultivation medium. The accumulation of ammonia in water results in eutrophication and depletion of oxygen due to nitrification and thus harms the water-borne organisms such as fish. Recent development of hollow fibre membrane contactors provides an promising alternative for the removal of various volatile contaminants including ammonia.

Compared to conventional absorption or stripping processes such as bubble columns and packed beds, the membrane processes provide many advantages including larger interfacial area per unit volume, independent control of gas and liquid flow-rates without any flooding, loading, foaming, and known gas–liquid interfacial area, etc. These advantages are reflected by a number of investigations on the use of membrane modules for gas absorption and stripping reported in various literature. Furthermore, due to the tangential nature of the flow in membrane contactors, the gas stripping process does not require operation at a high-pressure drop. This also contributes to lower capital cost and ease of operation.

### 1.9.15.2. *Modeling equations*

The principle of ammonia removal through hydrophobic porous membrane is schematically shown in Fig. 1.9.16. The buffered ammonia solution flows in the fibre lumen while the acidic stripping solution in the shell side counter currently. The asymmetric PP hollow fibre membranes act as a fixed interface and keep the two streams separate. Following assumptions have been made to formulate the stripping process:

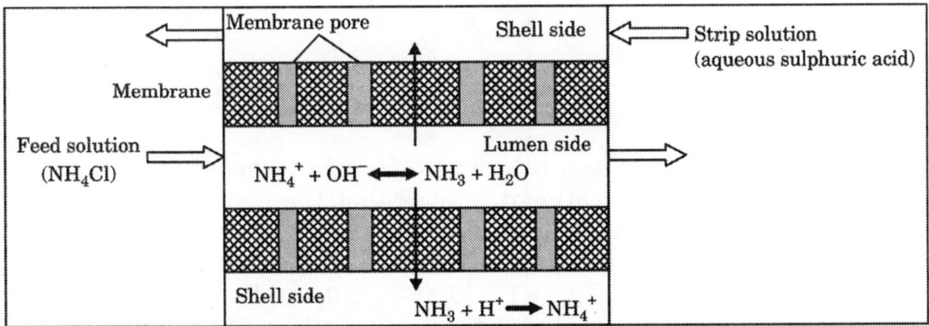


**Fig. 1.9.16:** Mechanism of ammonia removal in membrane process

*Assumptions*

1. Plug flow is applicable to the wastewater stream, that is, axial dispersion is negligible. 2. Feed flow rates of both the wastewater and the strip solution are constant and do not change with the ammonia removal. 3. The volume of the membrane module is negligible compared with the total volume of cycled wastewater. 4. Solutions in the stirred reservoirs are completely mixed.

In the membrane module, the material balance equation for ammonia is given by:

$$\frac{\partial c}{\partial t} = -\frac{Q}{n\pi R_{in}^2}\cdot\frac{\partial c}{\partial z} - \frac{2}{R_{in}}k_A c \tag{1.9.26}$$

Where $c$ is the total concentration of ammonia in solution (mol/m$^3$), $t$ the time variable (s), $Q$ the volumetric feed flow rate (m$^3$/s), $n$ the number of hollow fibres, $R_{in}$ the radius of the hollow fibre membrane (m) and $k$A is the overall mass transfer coefficient for ammonia removal (m/s). In the stirred ammonia reservoir, the material conservation equation is given by:

$$V\frac{dc_b}{dt} = -Qc_b + Q_c\big|_{z=L} \tag{1.9.27}$$

Where $V$ is the volume of wastewater in the stirred tank (m$^3$) and $c_b$ is the total concentration of ammonia in the wastewater reservoir (mol/m$^3$).

With the initial and boundary conditions as:

$t = 0, c = 0, c_b = c\ 0; t > 0, z = 0, c = c_b$

In order to solve the above governing equations, Eq. (1.9.26) is transformed into a group of ordinary differential equations by the back-difference with respect to the length variable, that is,

$$\frac{dc_1}{dt} = \frac{Q}{n\pi R_{in}^2} \cdot \frac{c_b - c_1}{\Delta z} - \frac{2k_A}{R_{in}} . c_1$$
$$\frac{dc_j}{dt} = \frac{Q}{n\pi R_{in}^2} \cdot \frac{c_{j-1} - c_j}{\Delta z} - \frac{2k_A}{R_{in}} . c_j \tag{1.9.28}$$

(j = 2, 3……n)

$$\frac{dc_b}{dt} = -\frac{Q}{V} . (c_b - c_N) \tag{1.9.29}$$

Where $\Delta z$ is given by: $\Delta z = L/N$, where $N$ is the segment number of the membrane module. This treatment implies that the membrane module is segmented equally into $N$ parts in series, each of which is considered to be a CSTR where a "first-order reaction" takes place. The initial condition for the above equations is: $[c_1, c_2 . . . c_N, c_b] = [0, 0, . . ., 0, c_0]$. By applying Laplace transform to the above differential-difference equations, the solution may be given as:

$$c_b = c_0 \sum_{k=1}^{N+1} \frac{\left( ak + \dfrac{Q}{n\pi R_{in}^2 \Delta z} + \dfrac{2k}{R_{in}} \right) \exp(a_k t)}{N\left( a_k + \dfrac{Q}{V} \right) + \left( a_k + \dfrac{Q}{n\pi R_{in}^2 \Delta z} + \dfrac{2k}{R_{in}} \right)} \tag{1.9.30}$$

in which $a_k$ ($k$ = 1, 2, . . ., $N$+ 1) are the roots of the following equation:

$$\left( s + \frac{Q}{V} \right)\left( s + \frac{Q}{n\pi R_{in}^2 \Delta z} + \frac{2k}{R_{in}} \right) = \frac{Q}{V}\left( \frac{Q}{n\pi R_{in}^2 \Delta z} \right)^N \tag{1.9.31}$$

### 1.9.15.3. *Membrane's coefficient*

Resistance in the membrane is due to the membrane structure and the presence of trapped phase within the pores, which is governed by the hydrophobic property of the PP membrane, the physico-chemical properties of absorbing liquid and the operating pressures employed. For non-wetted microporous membranes, *i.e.* gas-filled pores, the membrane's mass transfer coefficient, $k_{Am}$ can be evaluated independently by the structure properties of the pores in the membranes:

$$k_{Am} = D_{Am}\, \varepsilon p / Lp$$

Where $\varepsilon$ and $Lp$ (m) are, respectively, the surface porosity and effective pore length of the membrane and $D_{Am}$ is the diffusivity of ammonia in the membrane ($m^2/s$), which can be described by the Knudsen and the molecular diffusion models. Considering that the real pore size distribution of porous membranes is represented by the log-normal distribution, $D_{Am}$ can be expressed as:

$$D_{Am} = D_{Ag} \cdot \frac{\int_0^{\infty} r^2 / (\alpha + r) \exp[-(\ln(r / r_m)(1 + \sigma^2)^{0.5})^2 / (2 \ln(1 + \sigma^2))] dr}{\int_0^{\infty} r \exp[-(\ln(r / r_m)(1 + \sigma^2)^{0.5})^2 / (2 \ln(1 + \sigma^2))] dr} \quad (1.9.32)$$

where $D_{Ag}$ is molecular diffusion coefficient in gas phase or vapor (cm²/s) and $r_m$ (m) and $\sigma$ are the mean pore radius and standard deviation of the membrane, respectively. (cm) is defined as:

$$\alpha = \frac{3}{200} \cdot \frac{D_{Ag}}{\sqrt{8RT / \pi M_A}} \quad (1.9.33)$$

in which $M_A$ is the molecular weight of ammonia (kg/mol).

The microporous hollow fiber membranes emphasized here, developed in the early 70's, consists of small tubes with internal diameters ranging from 50-400, µm. The fibers are made of the same polymeric materials commonly used for microporous flat membranes, materials like polyolefins and cellulose esters. Properties like membrane thickness and porosity are similar to those of the corresponding flat membranes.

Many current microporous hollow fiber modules are designed like small shell and- tube heat exchangers. Each consists of a glass or plastic shell containing a bundle of fibers (Fig. 1.9.17). At each end of the shell, the fibers are embedded in a potting material, often polyurethane or an epoxy. Two distinct phases flow independently in the module. One phase flows inside the lumen of the fibers, and the other flows outside the fibers. Hollow fiber modules usually have a parallel flow configuration, but some experiments using a cross flow configuration have also been seen recently creating possibilities of development of a new configuration.

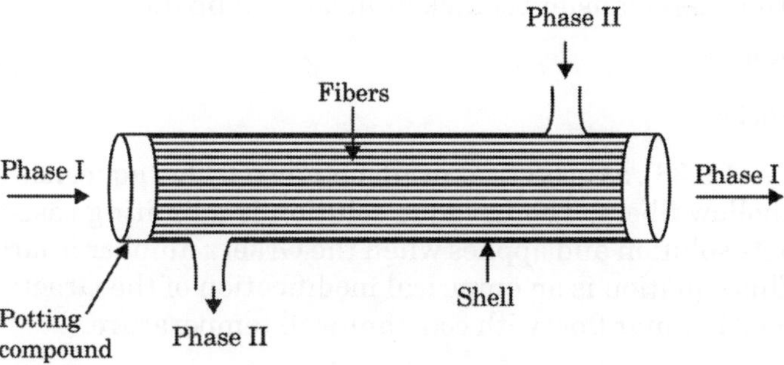


**Fig. 1.9.17.** Hollow fiber module. In this figure the shell side fluid is shown flowing parallel to the tube side fluid

We will emphasize the design of hollow fiber modules for gas stripping in this case study. Thus we are concerned with the removal of a volatile solute from a liquid into a gas. In comparing hollow fiber gas stripping with conventional equipment, we must consider the following things:

(1) The increased mass transfer due to the larger fiber surface area per unit volume;
(2) The greater range in flow rates without channeling and flooding;
(3) The decreased mass transfer due to diffusion across the membrane;
(4) The cost of the fibers; and
(5) The larger pressure drop in the modules.

### 1.9.15.4. *Mass Transfer Correlations*

The decreased mass transfer rate due to the membrane diffusion is buried in the usual expression for mass transfer:

$$j = K\Delta C \tag{1.9.34}$$

Where j is the solute flux, dc is the overall solute concentration difference, and K is the overall mass transfer coefficient. The overall coefficient K, which is usually independent of concentration, results from three contributions, one from each of the individual mass transfer coefficients (see Fig. 1.9.17). If the pores are filled, the solute concentration is a continuous function with respect to distance from the bulk phase I solution to the right hand wall of the membrane. The concentrations across the fluid-fluid interface at the right hand wall of the membrane are assumed to be in equilibrium. The membrane mass transfer coefficient $k_M$ is not a function of flow rates in either of the bulk phases, but depends upon the membrane material and the membrane thickness. The mass transfer coefficients $k_I$ and it, are functions of flow, and can be estimated using mass transfer correlations. For example, a hollow fiber with $k_I$ being the coefficient for mass transfer through the boundary layer inside the fibers, then $k_{II}$ is the coefficient for mass transfer through the boundary layer outside the fibers. To design a hollow fiber module, we first estimate the three individual mass transfer coefficients to determine the overall mass transfer coefficient and hence the flux across the membrane. Two correlations are commonly used to describe the mass transfer coefficient inside the fibers:

$$Sh = 1.62Gr^{0.33} \tag{1.9.35}$$

$$Sh = 1.86Gr^{0.33} \tag{1.9.36}$$

Equation (1.9.36), a theoretical result obtained by Levgque, has been widely verified for hollow fibers. The Levgque solution is a limiting case of the more general Graetz solution and applies when the Graetz number is large, as is the case here. This equation is an empirical modification of the Graetz solution in fully developed laminar flow with constant wall temperature.

#### 1.9.15.4.1. *Mass transfer coefficient inside the fibers dominates*

The Leveque solution is the mass transfer correlation that describes the mass transfer coefficient inside the fibers. When the solution is dilute, yb = yl ℑ 0 for gas flow rates that are much larger than the liquid flow rate. Further, since the solute is volatile, H is large. Thus HG/L >> 1. Thus eqn. (1.9.36) can be approximated by

$$\frac{K_L a c_L l}{L} = \ln\left[\frac{x_o}{x_1}\right] \tag{1.9.37}$$

The overall mass transfer coefficient, $K_L$, is related to the individual mass transfer coefficients by

$$1/K_L = 1/k_L + 1/k_M + C_L/(k_G H_{CG}) \tag{1.9.38}$$

Since the solute is volatile, $l/k_o$ is negligible compared to $l/k_L$. Equation (1.9.38) can then be simplified to

$$K_L = \frac{k_m k_L}{k_M + k_L} \tag{1.9.39}$$

where $k_m$, the membrane mass transfer coefficient is given by,

$$k_m = \frac{\varepsilon D}{d_M \tau} \tag{1.9.40}$$

Using eqns. (1.9.36), (1.9.37), (1.9.38) one can obtain

$$\alpha n^{2/3} - \beta n^{-1/3} - \gamma = 0$$
$$\alpha = 1.62 \frac{\varepsilon D_L}{0.1 \nu \tau} D_L^{2/3} \left[\frac{4Q_L}{\pi l}\right]^{1/3} \pi l, \tag{1.9.41}$$

1.9.15.4.2. *Mass transfer coefficient outside the fibers*

This case is a complete parallel to the above, differing only in the location of the limiting mass transfer coefficient and the correlation used. For an excess gas flow rate, $H_{G/L}$> 1 even though H will be small. Further, since the gas feed is solute free, $y_0$=0; due to the excess gas flow rate, yl will be small. Thus,

$$\frac{K_L a c_L l}{L} = \ln\left[\frac{x_o}{x_1}\right] \tag{1.9.42}$$

as before. The overall mass transfer coefficient is again related to the individual mass transfer coefficients by,

$$1/K_L = 1/k_L + 1/k_M + C_L/(k_G H_{CG}) \tag{1.9.43}$$

Since the solute is semi-volatile $l/k_L$ is negligible compared to $l/k_o$. This leads to,

$$K_L = \frac{k_G k_M H_{c_G}}{k_G H_{c_G} + c_L k_M} \tag{1.9.44}$$

1.9.15.5. *Experimental procedure*

1.9.15.5.1. *Ammonia removal in the PP flat circular membrane module*

The experimental setup for the removal of ammonia from synthetic effluent is schematically shown in Fig 1.9.18. The feed solution containing ammonia is pumped by a peristaltic pump (PP-10, Miclines, Peristaltic Pump, India) through the lumen of the flat circular membrane, while the stripping solution containing sulfuric acid is also pumped into the shell side (or) one side of the flat circular membrane by another peristaltic pump. Both solutions are recycled to their respective reservoirs. A pH meter is used to measure the feed pH during the experimental operation. The simulated/synthetic effluent wastewater, aqueous ammonium chloride was pumped at 25 ± 1°C and at various flow rates into the fibre lumen while the stripping solution, 0.05M or 0.1N aqueous sulphuric acid solution were (Counter and Co) currently fed into the shell side of membrane module. The flow rate of the lumen side and the shell side were controlled by two peristaltic pumps.

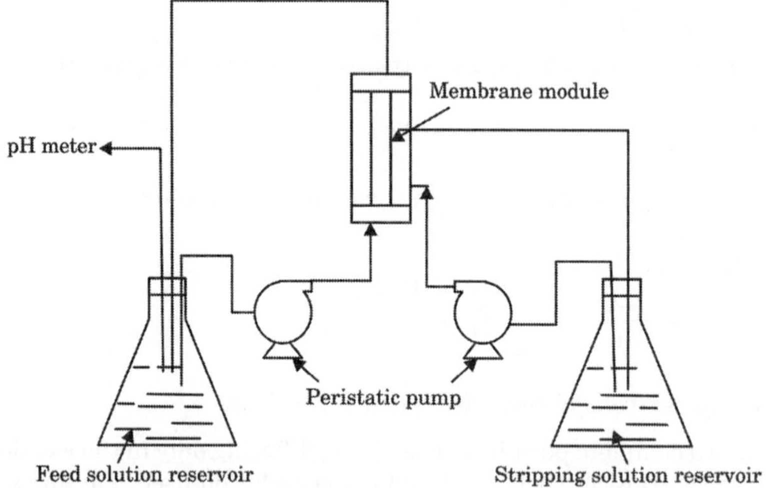


**Fig. 1.9.18:** Experimental setup for ammonia stripping in flat circular polypropylene hydrophobic membrane contactor (Suresh, unpublished work)

Fig. 1.9.19 shows the SEM image of the cross-sectional structure of PP flat circular membrane from an industrial area, consisting of 14 wt% PP, 76.87% DMAc, 5.22% LiCl and 1.95% $H_2O$. It can be seen that the membrane possesses a well-integrated asymmetric structure, where large finger pores are formed on the inner side of the fibre while smaller macro voids are formed on the outer side of the fibre. On the outer surface of the fibre membrane, a skin layer with an estimated thickness of 1 μm is present, as shown in Fig. 1.9.19. The skin layer determines the separation performance of the membrane because it provides much larger resistance to mass transfer than the porous substrate, which may only acts as a supporting layer and provides mechanical strength for the membrane.

1.9.15.5.2. *Results and Discussion*

Depletion curves of the total ammonia concentration in the synthetic effluent tank as a function of time at various operating conditions have been employed to investigate the performances of the PP hollow fibre membrane module for ammonia stripping and are presented in the following section.

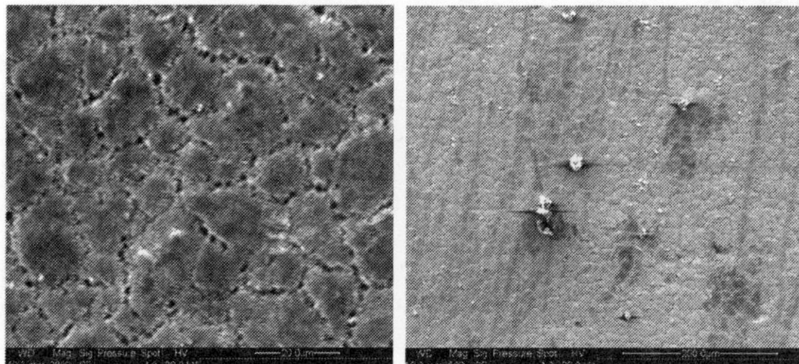

**Fig. 1.9.19:** SEM of flat circular polypropylene hydrophobic membrane before and after ammonia recovery. (a). Blank (b). After ammonia recovery (Suresh, unpublished work).

PP hydrophobic microporous membranes have been characterized and applied for the ammonia removal from water. Post-treatment with ethanol is useful to improve both the hydrophobility and the effective surface porosity of the resulting PP hollow fibre membranes, and subsequently favor the ammonia removal (Fig. 1.9.20). The initial concentration of ammonia and the flowrate of the acid stripping solution have no effect on the ammonia removal. The ammonia-stripping rate increases with increasing the feed flowrate up to 400 ml/h and further increase in the feed flowrate would give no further enhancement, which indicates that membrane resistance becomes the dominating effect. All the experimental data are in excellent agreement with the literature results. These studies are barrier /induce for further research in this same model system.

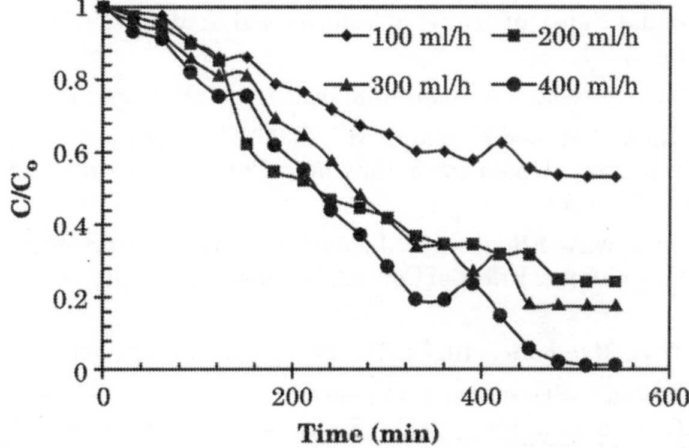


**Fig. 1.9.20:** Total ammonia decrease in feed solution with various times at various flowrate for Counter-Current flow. Fig (3) shows the change of ammonia concentration in the reservoir (dimensionless term, $c_b/c_0$) with time at conditions of pH 8 ± 1.5, T = 25 ± 1°C, $C_0$ = 9 mg/l, while the both feed flowrate were kept constant at 100 ml/h, 200 ml/h, 300 ml/h and 400 ml/h in Counter-Current flow(Suresh, unpublished work)

## REFERENCES


Barrer, R.M. 1939. Activated diffusion in membranes. *Trans. Faraday. Soc.*, 35: 644–656.

Belfort, G. and Heath, C.A. 1993. New developments in membrane bioreactors. *In*: Crespo, J.G. and Böddeker, K.W. (*Eds.*). Membrane processes in separation and purification. NATO ASI Series E: Applied Science Dordrecht: Kluwer Academic Publishers; pp. 127–49.

Belfort, G. 1989. Membranes and bioreactors: a technical challenge in biotechnology. *Biotechnol Bioeng* 33: 1047–66.

Benadda, B., Kafoufi, K., Monkam, P. and Otterbein, M. 2000. Hydrodynamics and mass transfer phenomena in counter-current packed column at elevated pressures, *Chem. Eng. Sci.* 55: 6251.

Billet, R. and Schulets, M. 1991. *Chem. Eng. Technol.*, 14: 89–95.

Brandt, S., Goffe, R.A., Kessler, S.B., O'Connor, J.L. and Zale, S.E. 1988. Membranebased affinity technology for commercial scale purifications. *Biotechnology* 6: 779–82.

Brindle, K. and Stephenson, T. 1996. The application of membrane biological reactors for the treatment of wastewaters. *Biotechnol. Bioeng.* 49: 601–610.

Deckwer, W-D. 1991. *Bubble Column Reactors,* John Wiley & Sons.

Dejanovic, I., Matijasevic, L. and Olujic, Z. 2008. Dividing wall column–A breakthrough towards sustainable distilling, *Chemical Engineering and Processing*, doi:10.1016/j.cep.2010.04.001

Delhaye, J.M. 1969. "Hot Film Anemometry in Two Phase Flows," *Two Phase Flow Instrumentation, National ASMErAIChE Heat Transfer Conf.*, Minneapolis, p. 58.

Douglas, Way, J., Richard, D.N., Thomas, M.F. and Dendy, S.E. 1982. Liquid Membrane Transport: A Survey. *J. Membr. Sci.* 12: 239–259.

Fan, L.-S. 1989. *Gas-Liquid-Solid Fluidization Engineering*, Series in Chemical Engineering, Butterworths, Boston.

Gabelman, A. and Hwang, S.-T. 1999. Hollow fiber membrane contactors, *J. Membr. Sci.* 159: 61–106.

Halwachs, W. and Schugerl, R. 1980. The Liquid Membrane Technique –A Promising Extraction Process. *Int. Chem. Eng.* 20: 519–528.

Hewitt, G., Quarini, J. and Morrell, M. 1999. More efficient distillation. *Chem Eng.*, 16–19.

Hills, J.H. 1974. "Radial Non-Uniformity of Velocity and Voidage in a Bubble Column," *Trans. Inst. Chem. Eng.*, 52: 1.

Huang, R.Y.M. 1991. *Pervaporation membrane separation processes*; Elsevier: New york.

Krishna, R., Van Baten, J.M. and Urseanu, M.I. 2000. "Three Phase Eulerian Simulations of Bubble Column Reactors Operating in the Churn Turbulent Regime: A Scale-Up Strategy," *Chem. Eng. Sci.*, 55: 3275.

Krishna, R., de Swart, J.W.A., Ellenberger, J., Martina, G.B. and Maretto, C. 1997. "Gas Holdup in Slurry Bubble Columns: Effect of Column Diameter and Slurry Concentrations," *AIChE J.*, 33: 311.

Lonsdale, H.K. 1982. *J. Memb. Sci.* 10: 81. The growth of membrane technology (review).

Luyben, W.L. 1975. "Steady-State Energy Conservation Aspects of Distillation Column Control Design", I&E.C., Fundam., 14(4): 321–325.

Matsuura, T. 1993. Synthetic Membrane and Membrane Separation Processes, CRC Press.

McHugh, M.A. and Krukonis, V.L. 1986. *Supercritical fluid extraction: principles and practice*; Butterworth: Stoneham, MA, pp. 13–22.

Noble, R.D. and Stern, S.A. 1995. *Membrane Separations Technology. Principles and Applications,* Membrane Sciences and Technology Series, 2, Elsevier.

Nottenkamper, R., Steiff, A. and Weinspach, P.-M. 1983. "Experimental Investigation of Hydrodynamics of Bubble Columns," *Ger. Chem. Eng.*, 6: 147.

Petyluk, F.B., Platonov, V.M. and Slavinskii, D.M. 1965. Thermodynamically optimal method of separating multicomponent mixtures. *Int. Chem. Eng.*, 5(3): 555–561.

Premkumar, R. and Rangaiah, G.P. 2009. Retrofitting conventional column systems to dividing-Wall Columns. *Chemical Engineering Research and Design* 87: 47–60.

Ranade, V.V. 1992. "Flow in Bubble Columns: Some Numerical Experiments," *Chem. Eng. Sci.*, 47: 1857.

Reid, C. 1966. "Principles of Reverse Osmosis", in *Desalination by Reverse Osmosis,* Merten, U., *Ed.*, pp. 1–14, M.I.T. Press, Cambridge, Mass.

Rogers, R.D. and Seddon, K.R., 2003. *Science* 302(5646): 792.

Shah, Y.T., Kelkar, B.G., Godbole, S.P. and Deckwer, W.D. 1982. "Design Parameter Estimations for Bubble Column Reactors" *AIChE J.*, 28: 353.

Sirkar, K. and Winston Ho, W.S., 1992. *Membrane Handbook*, Van Norstrand Reinhold, New York.

Skogestad, S. "Dynamics and Control of Distillation Columns - A Critical Survey", Preprints IFAC Symposium, DYCORD + 92, College Park, MD, USA, pp. 1–25.

Spillman, R. 1995. Economics of gas separation membrane processes, in Noble, R.D. and Stern, S.A. *Eds*., Membrane Separation Technology: Principles and Applications, Elsevier.

Strathmann, H. 1981. *J. Memb. Sci.* 9: 121. Membrane separation processes (review).

Tolliver, T.L. and McCune, L.C. 1978. "Distillation Column Control Design Based On Steady State Simulation", ISA Transactions, 17(3): 3–10.

Tzeng, Y.B., Chen, R.C. and Fan, L.-S. 1993. "Visualization of Flow Characteristics in a 2-D Bubble Column and Three-Phase Fluidized Bed," *AIChE J.*, 39: 733.

Van den Broeke, L.J.P. and Bakker, W.J.W. 1999. Binary permeation through a silicalite-1 membrane. *AIChE J.* 45: 976–985.

Wright, R.O. 1949. Fractionation apparatus. U.S. Patent 2, 471, 134.

# Chapter 2

# Reactive Extraction Processes

## 2.1 INTRODUCTION

Competition among various industries such as chemical, software and hardware etc. have created in the last few decades desire to produce the product which is low in cost and high in quality. Market changes every minute and ask for the quick respond. Investment into the industry shall come only if profits are visible to the share holders. Industries needs manufacturing processes which are cheaper to construct and more economic and flexible to operate so as to remain competitive in the global trade. For this an integrated system approach of complex multidisciplinary, non linear, non equilibrium process occurring on different lengths and time scale is required, which can be obtained by molecular modeling, instrumentation and powerful computational tools (CFD etc.). The solution of the problems, like large working capital, higher capital investment, long gestation period and high cost of utility, which always surrounds chemical plants, lies in the order of magnitude reduction.

A revolutionary approach toward order of magnitude reduction comes with the term coming into market "Process intensification (PI)" basically defined as, smaller, safer, and cheaper, with improved performance. Chemical industry has to adopt PI to increase energy efficiency and profitability through its introduction in process and plant design, development and implementation. In many case its implication could lead to think the process totally on new perspectives, which till now have not been though off due to safety, control, economic or political reasons. Process control and safety got its new dimensions by the introduction of PI. Its time for PI to come out of laboratory and implement itself in designing an efficient and cleaner plant. PI was introduced in 1983 by Ramshaw (1983) according to whom PI is simply equivalent to process improvement and is defined as devising an exceedingly compact plant which reduces both the main plant items and the installation cost. Heggs in 1983 gives another definition of PI: PI is concerned with order of magnitude reduction in both the process plant and equipment. Douglas (1988), interpreted process intensification as the novel equipment, processing techniques and process development methods that compared to the conventional ones, offers a substantial improvements in chemical manufacturing and processing. Process intensification offers the benefits of smaller, safer and cheaper unit; small time to the market, less waste generation and low energy consumption, thus leading to a better company image. Process intensification is an interdisciplinary field of research that needs an integrated approach. In growing number of areas, the limitations of traditional batch processing are being overcome by moving to continuous process through the application of intensification techniques. To sum-up, process intensification is a rapidly developing field inspiring ideas of modeling and new equipment designing and operating modes, which's potential are still to be explored.

The current world wide market of the chemicals is estimated to be 1.6 trillion dollars annually. This includes commodity chemicals (56%) and specialty chemicals (44%). Specialty chemicals include pharmaceuticals, agro chemicals, performance chemicals, which includes water treatment chemicals, catalysts, textile auxiliaries etc. Indian contribution to world chemical industry is only 1.75%, which is indeed small. Nevertheless there are sectors in chemical industries such as fine and specialty chemicals including bulk drugs and dyes, where Indian share of world market of 8.5 and 6% respectively, which is 4–6 times the average of 1.75% share of the overall world chemical market. The difference in contribution of India to the bulk and specialty is the matter to look for.

The reason of the difference lies in the scale of operation. The specialty chemicals are mostly produced on small/medium scale and the ingenuity and brilliance of the individual/group of individuals is the reason of the success. Large pool of scientifically trained people experiments on small and medium scale, optimize the process, develop new chemistry routes and implement it to benefit the productivity and profitability of these small and medium scale industries. Further small scale industries were the first who could easily bring PI into their plant as change is easy to bring there. On the contrary in the bulk chemical industries, the source and cost of finance for such mega-structures coupled with the inadequate infrastructure have coupled the problem for chemical industries producing them. This hindrance has always lowered the development of these industries. Large scale mostly opts for the bottlenecking process to lower the production cost and hence increase the market share values. Bottlenecking usually includes, lowering the working staffs or installing a certain unit with a modified technology such as increasing the height of prilling tower in urea manufacture to improve the quality of urea pellets etc. However, bottlenecking is not a stable solution because if a certain chemical company is adopting it other such as software can also do and hence could even lower the price much below it.

Trial and error procedure cannot be employed in mega scale industries. Even the problem of outdated technologies and poor infrastructures cannot be solved overnight. So the only cheaper alternative is process intensification *i.e.* getting more from the existing. Judicious application of process intensification in the operating plants can result into enhancement in the capacities ranging from 20 to 100% with simultaneous improvement in the selectivity's (Joshi

2001). Improvement in an operating plant can be realized through various stages of implementation.

Process Intensifying Methods, includes new or hybrid separations, integration of reaction and separation, heat exchange or phase transition (in so-called multifunctional reactors), techniques using alternative energy sources (light, ultrasound etc.) and new process-control methods (like intentional unsteady-state operation). Examples are: Multifunctional reactors: reverse-flow reactor (Stankiewicz and Moulijn, 2000), reactive (catalytic) distillation (DeGarmo *et al.*, 1992), reactive extraction (King *et al.* (1986), Wasewar *et al.* (2005); Lewis and Yang (2002)), Fuel cell (Tagawa *et al.*, 1999) etc.; membrane reactors (Stankiewicz and Moulijn, 2000); Hybrid separations: Membrane absorption and stripping (Jansen, 1995), Membrane distillation and Adsorptive distillation (Stankiewicz and Moulijn, 2000); use of alternative forms and sources of energy: Ultrasonic intensification and microwave heating (Gedye, 1988), electric field intensification (Akay and Wakeman, 1996) etc.; and other: Supercritical fluids (Goetheer, 1999), novel gas liquid reactor (Hills, 1996), microemulsion (Bhagwat and Sharma, 1988) etc.

In all the above PI methods the integrations were aiming for the better performance outputs of the process. Two basic chemical operations: reaction and separation are the key area for intensification since they are almost found in all chemical industries. The integration of these unit operations give born to subject of reactive separation processes. Presence of reaction with separation improves separations owing to enhanced mass transfer rates. On the other hand presence of separation with reactions drives the reaction to higher conversions (as if the product is removed from the system the reaction progresses in forward direction). Another advantage could be simplicity and novelty, the approach brings to the process flow sheet. These operations are also coveted for the investment and operating cost savings garnered on successful scale up to commercial operations.

Reactive extraction is a process intensification method coming under reactive separations and is gaining great attention particularly in terms of its comparison with the conventional extraction methods which have posed a great economic pressure to the industries carrying out them. Other reasons may be the emergence of new industries, demand of low cost, environmentally safe and high purity products. Reactive extraction links chemical sources and sinks to enhance reaction rates, conversions and selectivity's. Most of chemical processes are equilibrium driven and removal of product as soon as it is produced would lead to enhanced reaction rates and increased feed conversions reduce reaction severity and provide operation under milder conditions. Along with these the duo would lead to high local driving forces for separation leading to reduction in equipment size, elimination of recyclable streams and reductions in utility costs. Reduction in working inventory would definitely lead to safer units as if there is no such chemical there in any chance of its leakage, spills and environmental releases. Selectivity is also major issue in choosing the reaction-extraction duo. A selective combination could lower the byproduct formation which could lower the purity of the product. The combination of two unit operations into one would definitely lower the cost and also the smaller unit is easy to be control if any disaster is predicted to happen. The challenge however, is to commercialize this process which could be possible by the mutual working of chemists and chemical engineers.

For the last few decades, there have been intensive works on the reactive extraction for the recovery of carboxylic acid, metals, rare earth elements, amino acids, pharmaceuticals, pesticides etc. from waste water solutions or from their production sources. Reactive extraction being a clean process, since the solvent system can be completely recovered and reuse, along with simple and cheap operation had created a lot of scope of it as an efficient recovery process. The primary hinderence in the establishment of the reactive extraction process for the recovery of the acid is the search for an efficient and selective solvent. When this could be accomplished, the other task like carrying out the process is relatively simple.

To discuss the application and generate the interest of reader for reactive extraction process let us consider a case study for the recovery of carboxylic acids from fermentation broths. In fermentation processes, product inhibition leads to low productivities. So the desire is to remove the acid as soon as it is produced so that it could not affect that process. Energy studies and economic analysis have indicated that product separation is the most energy-intensive and expensive stage in fermentation processes for manufacture of carboxylic acid. Further, the heterogeneous nature of the broth, low volatility of the product and presence of product in dilute forms has caused difficulties in product separation. Distillation cannot be used for the separation of acid because of its low difference in volatility of product in comparison to water. Conventionally, recovery of acid is carried out by method of precipitation. In this recovery scheme, calcium carboxylate is precipitated, recovered by filtration and converted to carboxylic acid by addition of sulfuric acid. The dilute carboxylic acid product is then sequentially purified using activated carbon, evaporation and crystallization. These separation and final purification stages account for up to 50% of the production costs. Thus, this method of recovery is expensive and unfriendly to the environment as it consumes lime and sulphuric acid and also produces a large quantity of calcium sulphate sludge as solid waste. Thus, a process for recovery of acid is desired that should be selective and should have high capacity for the acid. Many separation processes are available like, stripping, adsorption, elctrodialysis, liquid-liquid extraction, pertraction, pervaporation, and membrane solvent extraction. All these processes have their own advantages and disadvantages. Adsorption or ion exchange process requires regeneration of ion exchange resin and adjustment of feed pH to increase the sorption efficiency requiring large amount of chemicals. Electrodialysis and dialysis have good potential and have advantage of simultaneous separation and concentration. Yet they suffer from problems of membrane fouling which requires frequent cleaning of the dialyzer. Distillation is the well-established and reliable technology but it has drawback that there is formation of high boiling esters and dimmers. In hollow fiber membrane extraction process, membrane has tendency to form emulsion but has advantage of large interfacial surface area for mass transfer in a compact unit. The drawbacks of reverse osmosis are potential formation of emulsion and complexity of operation. Interest in liquid surfactant membranes for biochemical separations has focused on their potential advantages. The main advantage of liquid surfactant membranes over other separation techniques is the large surface area available for mass transfer, which results in a fast rate of separation. In spite of these apparent advantages, very few industrial applications have been reported so far. Several drawbacks were shown to hinder implementation, mainly complexity of operation and swelling in liquid surfactant membrane. The use of supported liquid membranes for the recovery of carboxylic acid offers unique advantages. Some of the advantages are lower energy consumption, higher separation factors in a single stage, and the ability to concentrate acid during the separation. However, supported liquid membrane often suffers from membrane instability. Membrane bioreactors can be used to avoid toxicity due to extractant by immobilization of biocatalyst in membrane for the production and recovery of carboxylic acid. The major drawback of the membrane bioreactor is difficulties in cleaning and sterilization. (Wasewar *et al.*, 2005) The numerous advantages of reactive extraction, as described in the section of introduction definitely justifies its usage for the recovery of carboxylic acid from dilute solutions. The reactive substance forms the reversible complex with acid, which promotes the transfer of the polar solute from the aqueous phase to the nonpolar organic phase. Results have shown that process intensification is a vital tool to realize an improved process. Further the reactive extraction process does not affect the thermal stability of bioproducts, and energy demand is substantially low. In the remaining sections of the Chapter the process intensification using reactive extraction is talked by taking case studies on its use in the recovery of carboxylic acids from aqueous solutions.

## 2.2 SOLVENT EXTRACTION

### 2.2.1 *Solute-solute and solute-solvent interaction*

*Aqueous phase*

In order to apply reactive extraction certain desirable properties of solutes are recommended so that complexation reactions are favored, such as

(a) *Lewis acid or Lewis base function groups*: The solute of interest should have one or more function group that can participate in moderately strong complexation. Since most complexation processes involve interaction of Lewis acid with Lewis bases, acidic and basic functional groups are useful.

(b) *Low solute concentration*: Since complexing agents provide particularly high equilibrium distribution for low solute concentrations and tend to saturate stoichiometrically at high concentrations, complexation is most attractive for relatively dilute aqueous solutions.

(c) *Low activity coefficients in water*: Complexation separation processes are useful for hydrophilic solutes, because these solutes are difficult to recover from aqueous solution by most other means. Complexation is capable of providing a very low organic phase activity coefficient and hence an acceptable phase distribution coefficient, even though the activity coefficient in aqueous phase is low.

(d) *Low solute volatility*: Solutes that are less volatile than water are good candidates for separation by complexation, because they cannot be separated by stream stripping.

If all the features are available in solute it shall be attractive to employ reactive extraction for it.

*Organic phase*

In the process of extraction, separation is obtained by the partition of solute between two immiscible phases. The two phases should have sufficient density difference and surface tension so that there is a clear phase separation obtained once the phases are made to settle. If a chemical step is incorporated into the physical step it is called as reactive extraction or basically chemical reaction accompanied mass transfer. Mostly a reversible reaction is considered in the process. Separations involving reversible complexations are gaining wide importance now a days for removal of solutes from very dilute mixtures with high capacity and selectivity. The component involved in the reversible complexation is termed as a complexing agent, which reacts reversibly with the solute of interest, and is named as extractant. The choice of complexing agent is based on its strong, specific, yet reversible reaction with the solute. Choice of solvent for these agents is extremely important to work out since these extractants are either solid or highly viscous. These solvents are named as diluents and the total combination of diluent with extractant is now the solvent system for liquid-liquid extraction. Usually a low molecular weight and low viscosity diluents are employed so that they could improve the equilibrium through solvation of the complexes and can affect the process in other ways.

Chemical reactions are usually less employed in many separation, the thinking of common man are the process could not be reversed back and we are left with no option other than creating a byproduct which needs to be dumped. Further, chemical reaction often consumes large amount of expensive reactants and the reaction product is usually less valuable than the original compound. These disadvantages are overcome by using a chemical reactant that can reversible complex the component of interest in one step and the reaction can be reversed to regenerated the reactant in second step. The parameter for such type of complexations is the bond energy. Bond energy should be relatively low so that an economical process is obtained. Usually the complexation reactions with bond energy less than 50 kJ/mol, are similar to ordinary associations by van der Waals forces in the condensed state and can readily be reversed, whereas those having bond energies above 50 kJ/mol are difficult to reverse and requires undue expenses.

A simple complexation reaction gives as equilibrium of the form

$$\text{Solute} + n.\ \text{Extractant} \leftrightarrow \text{complex} \tag{2.2.1.1}$$

described by an equilibrium constant

$$K_E = \frac{[\text{Complex}]}{[\text{Solute}][\text{Extractant}]^n} \qquad (2.2.1.2)$$

In the above equation it could be seen that the attractiveness of the process lies in removal of solute from relatively low solute concentration, since it could lead to higher complexations constant ($K_E$). Another benefit of these complexations reaction are that they are selective for a certain group of solutes and hence these processes have the potential to separate only certain solutes from complex mixtures. Reactive extraction employing specific extractant also can give less co extraction of water than separations with more conventional solvents. Some important characteristics of complexing agents were also given like:

1. The complexing bond energy should be great enough to give a substantial improvement over conventional mass separating agents, which are less expansive
2. The complexation bond energy should be low enough so that the complex can be regenerated easily and completely
3. Complexing agent should minimize co-extraction of water
4. There should be no side reactions and complexation should be thermally stable, to avoid irreversible loss
5. The complexing reaction should have sufficiently fast kinetics in both directions so that equipment sizes should not become prohibitively large.

Reactive extraction has become very effective, simple, rapid, selective and sensitive method of separation. The process of reactive extraction usually involves removal by complex formation, ion exchange or chelation. The wide varieties of solvents available for reactive extraction make the subject now a day very attractive. Removal of components from dilute solution by ion-exchange mechanism is very effective and the parameter controlling in the case of liquid-liquid extraction is multiple contact such as cross current or counter current. Based on these desirable complexing agents properties, extractant are classified as:

(1) *Chelation*: Extraction by chelation is one of the successful extraction procedures. Chelate forming extractants exchange the cation with the solute and form coordinate binding. Acid alkylphosphoric acid derivatives (*i.e.* liquid cation exchangers), dithizone, cupferron, dimethylglyoxime etc are included in this group of extractants. On lab scale, some chelating agents have permitted simultaneous extraction and direct photometric determination in one single operation.

(2) *Solvation*: Tri-n-butylphosphate (TBP), tri-n-octylphoshphine oxide (TOPO), tributyl phosphine oxide (TBPO), trihexylphosphate (THP), diethyl ether, ethyl acelate or MIBK are some of the solvents that extract the solute using solvation. The first three of these compounds are Lewis bases and form non stoichimetric compounds with neutral solutes. Higher extractions were obtained if the solvent is more basic. Apart from basicity, the electron density surrounding the donor atom is also of importance.

Presence of large electron density in the compound owing to the presence of element such as oxygen, repulsive forces retards the process of extraction. Another phenomenon that can effect the extraction is the salting out of the solute when these extractant are used since salting out enhances extraction.

(3) *Ion Pair Formation*: The kind of acidic or basic solvents decides the extraction behavior. Extraction with dinonylnapthenic sulphonic acid is acidic extraction while extraction with high molecular weight amines (HMWA) is known as basic extraction. Acidic or basic extraction proceeds by ion pair formation.

(4) *Cation exchange extractants*: These are extractants which exchange proton against the cation *e.g.* phosphine and phosphoric acid. All acidic alkyl phosphoric acids, naphthenic acid, caporic acid or fatty acids are examples of liquid cation exchangers.

(5) *Anion exchange extractants*: Anion exchange extractants (*e.g.* aliphatic primary, secondary and tertiary amines) form ion pairs (salts) in acidic medium with the solute. All high molecular weight amines (HMWA) were termed as liquid anion exchangers. These amines might be primary, secondary tertiary or quaternary. These amine are formed from ammonia. Amine react with the acidic compounds by replacing one, two, three or four hydrogen atoms by anionic species to form the complexes. Since amine is in liquid form and reaction proceeds with the exchange of anion, the whole product was called 'liquid anion exchangers'. There are many type of amines such as tri-n-octylamine (TOA), tri-iso-octylamine (TIOA), Diisotridecylamine (DITDA), Alamine 308; 304, phenylben-zylamine, Alamine 336, Aliquat 336 etc. The extractability varies with rise in extraction from primary>secondary>tertiary> quaternary amines.

## *Merit and demerit of the extractant*

*Merit*

(i) The most significant advantage was separation of element from dilute solutions.

(ii) Solvent extraction allows separation of element in two different oxidation states.

(iii) The flexibility of adjustment of pH or controlling reagent concentration or choice of diluents or organic carboxylic acids facilitated the selective separation of large number of elements.

(iv) Further purification was not necessary because the product is obtained in pure form.

(v) The extraction equilibria were attained in a few seconds.

*Demerit*

(i) Formation of third phase during extraction.

(ii) Cost of the extractant is higher compared to conventional solvents. Though the process of reactive extraction is completely reversible and solvent could be regeneration without any loss still if due to some accidents these extractants are lost, the cost factor may play its role.

(iii) If the extraction of acid is carried out from bioreactors, the toxicity of solvent system is matter of concern.

### 2.2.2 *Ternary and distribution diagrams*

In reactive extraction the extractant and diluents have negligible solubility in the aqueous phase. Table 2.2.2.1 shows the solubility of some of the extractants and diluents employed in reactive extraction studies

**Table 2.2.2.1:** Solubility data for different extractant and diluents

| *S. No.* | *Extractant/diluents* | *Water solubility* |
|---|---|---|
| 1. | TOA | 0.4 g/l |
| 2. | Aliquat 336 | 0.12/100g |
| 3. | TBP | 0.422 mg/l |
| 4. | Heptane | None |
| 5. | Hexane | 0.0013/100g |
| 6. | Octane | None |
| 7. | Toluene | None |
| 8. | 1-decanol | 0.37/100g |
| 9. | 1-octanol | 0.3 mg/l |
| 10. | 2-octanol | 0.0096ml/100ml |
| 11. | Benzene | 0.18g/100ml |
| 12. | MIBK | 17g/l |
| 13. | Oleyl alcohol | None |
| 14. | Hexanol | 0.59g/100ml |

Since most of these solvent are not soluble in water, use of ternary diagram is not feasible. Distribution diagrams are used for studying the distribution of solute between the two phases. The parameter ascertaining the success of a particular solvent system is defined as distribution coefficient and is defined in the section 2.2.2. As in distillation higher is the difference between the 45 degree line and the equilibrium curve better is distillation obtained, similar is the case with liquid liquid extraction and it is preferable to have the equilibrium curve as far as possible from the 45 degree line.

## 2.3 INTENSIFICATION USING REACTIVE EXTRACTION

### 2.3.1 *Complexation*

Extraction involves the separation of the constituents of a liquid solution by contact with another insoluble liquid. Although, conventional method is the most used technique on the large scale processes, solvent extraction has been developed for the separation of the solutes from dilute solutions owing to its number of advantages over the conventional methods. Many investigators have investigated the reactive extraction of the solutes into an immiscible extractant/

solvent phase. In such process, solute is first extracted from the source by extractant and then recovered from the solvent by back extraction into another solvent. It is unlikely that a solute ion will leave an aqueous phase and enters a nonpolar organic phase unless some additional driving force is present. The reaction of solute to form organic complexes enables them to be transported from the aqueous phase to the organic phase. Reactive extraction relies on a solvent composed of a diluent that contributes the bulk physical properties and at least one chemical capable of reversibly complexing with a solute, known as an extractant. The advantage of extractive reactions include increased reactor productivity, ease in reactor pH control without requiring base addition and use of high concentration substrate as the process feed to reduce process waste and production cost. Extractive reactions also may allow the process to produce and recover the product in one continuous step and reduce the downstream processing load and the recovery cost.

The development of a new extractive recovery process for solute extraction requires the identification of a novel, more powerful extractant. The extraction system must fulfill high distribution coefficient. Further requirements of an extractant or a system of extractants are (i) low viscosity, (ii) higher density difference between the extractant and raffinate, (iii) a medium interfacial tension, (iv) thermal stability, (v) low enthalpy of vaporization, (vi) low melting points, (vii) no reaction between the extractant and raffinate, (viii) low solubility of the extractant in the raffinate, (ix) low toxicity and good biological degradability, and (x) low price and good availability. The reaction mechanism in the cases varies from proton transfer (ion pair or hydrogen bond formation) to ion exchange. In addition, the high affinity of the organic base for the solute gives selectivity for the solute over other components in the mixture.

In spite of the extensive research in the solute extraction using reactive extraction, two problems remains unsolved. The first one is the toxicity of the extractant and diluent particularly when they are used to recover solute from bioprocess unit. The second problem is the difference in the optimum pH values for extraction and the source availability. It is well known that the extraction of acid, as solute, with tertiary amine or solvating extractants is more appropriate at low pH values whereas with ion exchange extractants at high pH value. During the bio-production of these acids with the accumulation of the acid, the pH value of the system decreases. Applying a solvating extractant or amine for removal of the accumulated acid, the pH value can be maintained to the desired level, but as it was mentioned these extractants are not effective at high pH. The use of quaternary ammonium salt seems to be more appropriate at these conditions. Because of extraction, the pH value decreases, since the hydrogen ion remain in the system, while the carboxylate ions are replaced by the corresponding anions of the quaternary ammonium salt.

To overcome the above problem, mixed extractants have been tried. The presumption was that the use of such mixtures should lead to the condition where the presence of the one extractant should favor the solute extraction by the other. The proposed method gives certain advantage over the tertiary amine and quaternary ammonium salt, such as better performance in the medium pH region. Another way to solve the problem is by using modified extractants.

To improve the extraction ability of extractant, extractants are dissolved in diluents. Diluents also affect the physical properties of the extractant *e.g.* improving their viscosity etc. Two factors need to be accounted to show the influence of diluents on the extraction: (a) partial dissociation of the solute in aqueous phase and (b) dimerization in the hydrocarbon phase. Another important parameter particularly is the water of hydration. For the extraction of carboxylic acid, high attraction of binding of the acid with the water molecules requires large amount of solvent molecules so that they can compete with the water molecules that hydrate the acid at the interface. There are number of ways by which solvents for physical extraction can be classified. On the basis of molecular structure, they are classified as polar protic, dipolar aprotic and non-polar solvents. On the basis of nature, solvents are classified as inorganic (liquid anhydrous ammonia ($NH_3$), concentrated sulfuric acid ($H_2SO_4$), sulfuryl chloride fluoride ($SO_2ClF$)) and organic solvents (oxygenated such as alcohols, glycol ethers, methyl acetate, ethyl acetate, ketones, esters, and glycol esters or hydrocarbon solvent such as aliphatic solvents, aromatic solvent, halogenated solvent and natural solvent).

### 2.3.2 *Equilibria*

The ratio of total (analytical) concentration of solute in all its forms (by partition, dimmers and as complexes) in organic phase and total (analytical) concentration of all its existing forms (dissociated and undissociated) in aqueous raffinate is used as the parameter to ascertain the success of a particular solvent and is defined as distribution coefficient as

$$K_D = \frac{[A]_{org}}{[A]_{aq}} \tag{2.3.2.1}$$

Where,

$K_D$ = Distribution coefficient

$[A]_{org}$ = Concentration of solute A in organic form

$[A]_{aq}$ = Concentration of solute A in aqueous form

Usually the diluent also extract a part of acid by solvation and this should be subtracted to define the distribution coefficient of solute by chemical extraction and this can be defined as:

$$K_D^{Chem} = \frac{[A]_{org} - \nu [A]_{org}^{diluent}}{[A]_{aq}} \tag{2.3.2.2}$$

Where n is volume fraction of diluent and $[A]_{org}^{diluent}$ is the solute extracted into the organic phase by diluent alone. The overall distribution coefficient was obtained as the sum of total contribution by chemical and physical extraction and can be represented as $K_D$. Degree of extraction is defined as the ratio of solute concentration in organic phase to the sum of acid concentration in organic and aqueous phase and is defined in term of $K_D$ as

$$E\ \% = \frac{K_D \times 100}{1 + K_D} \qquad (2.3.2.3)$$

Where,

E = Degree of extraction

Physical extraction in the natural diluents, aliphatic and aromatic hydrocarbons is very poor, thus these categories of diluents behaves very poorly for the physical extraction of solute from dilute solution and hence are unacceptable for commercialization of the physical process involving these. These apolar aprotic solvents and are characterized by low relative permittivity (ε), low dipole moment, and are unable to act as an hydrogen bond donor. So these solvents interact only slightly with the acid since only the non specific directional, induction and dispersion forces are operating giving very low diluent $K_D$ values.

Dipolar aprotic solvents have large relative permittivity (ε) and dipole moment, have the presence of ion electron pairs and could solvate anion by ion-dipole and ion-induced dipole forces. Alcohols, considered as protic solvents, contain hydrogen atoms bound to electronegative element O (-O-H), are hydrogen bond donors. Their relative permittivity (ε), indicate them to be strongly polar. Thus they provide high diluent $K_D$ values since they are good anion solvators due to their hydrogen bonding ability. The tendency becomes more pronounced as the charge density (ratio of charge to volume) of the anion to be solvated increases. Further alcohols are capable of being both acceptors and donors, as are the acids; giving the highest diluent $K_D$ followed by purely basic solvents such as ethers and ketones.

Mutual solubility between an aqueous solution and a given solvent at fixed temperature is affected by the nature of distributed and its total concentration in the system. Solutes mutual solubility's can cause dramatic volume changes particularly at higher acid concentrations. Mutual solubility increases the volume of organic layer at the expense of that of aqueous phase in equilibrium and it increases with increases in solute concentration. Another important thing to be mentioned here is that the values of partition (P) and dimerization (D) are not the values obtained by pure water and solvent phases. These are obtained by mutual saturation of the two phases, obtained by equilibrium between two mutually saturated phases rather that between pure water and solvent. P and D values for monocarboxylic acid extraction using conventional solvents can be found from following equation:

$$[A]_{org} = P[A]_{aq} + 2P^2D\,[A]_{aq}^{\ 2} \qquad (2.3.2.4)$$

Where,

P = Partition Value

D = Dimerization Value

### 2.3.3 *Models for complexation*

2.3.3.1 *Physical equilibria models*

*Physical equilibria of mono-carboxylic acid*

The distribution of carboxylic acid between water and a non polar hydrocarbon is relatively free of complexities and can be regarded to be a physical distribution provided one can account for partial dissociation of the acid in the aqueous phase and its usually extensive dimerization in the hydrocarbon phase. The model equations for the physical equilibrium based on the mass action law are given below (Kertes and King, 1986):

(i) Ionization of the acid (HA) in the aqueous solution:

$$HA \leftrightarrow H^+ + A^- \quad (2.3.3.1.1)$$

$$K_{HA} = [H^+][A^-]/[HA] \quad (2.3.3.1.2)$$

(ii) Partition of the undissociated molecular acid between the two phases, aqueous (aq) and organic (o):

$$HA_{(aq)} \leftrightarrow HA_{org} \quad (2.2.3.1.3)$$

$$P = [HA_{org}]/[HA_{aq}] \quad (2.2.3.1.4)$$

(iii) Dimerization of the acid in the organic phase:

$$2HA_{org} \leftrightarrow (HA)_{2;org} \quad (2.3.3.1.5)$$

$$D = [(HA)_{2;org}]/[HA_{org}]^2 \quad (2.3.3.1.6)$$

Overall distribution coefficient is defined as the ratio of total (analytical) concentration of acid in all its forms (by partition, dimmers and as complexes) in organic phase and total (analytical) concentration of all its existing forms (dissociated and undissociated) in aqueous raffinate. It includes the effects like ionic strength, nature of ion concentration of H+ etc. of solution constituents.

$$K_D = \frac{[HA]_{org}}{[HA]_{aq}} = \frac{[HA]_{org}P + 2[(HA)_2]_{org}}{[HA]_{(aq)} + [A^-]_{(aq)}} \quad (2.3.3.1.7)$$

$$= \frac{[HA]_{aq}P + 2D[(HA)]^2{}_{org}}{[HA]_{aq} + K_{HA}[HA]_{aq}/[H^+]_{aq}}$$

$$= \frac{[HA]_{aq}P + 2DP^2[HA]^2{}_{aq}}{[HA]_{aq}(1 + K_{HA}/[H^+]_{aq})}$$

$$= \frac{P + 2P^2D[HA]_{aq}}{1 + K_{HA}/[H^+]_{aq}} \quad (2.3.3.1.8)$$

For dilute solutions the term $K_{HA}/[H^+]_{(aq)}$ is negligibly small.

$$K_D = P + 2P^2D[HA]_{aq} \qquad (2.3.3.1.9)$$

The above equation allows P and D to be determined from the plot of $K_D$ versus $[HA]_{aq}$.

*Physical equilibria of di-carboxylic acid*

Ionization of the acid ($H_2A$) in the aqueous solution is given as:

$$H_2A = HA^- + H^+ \quad k_{21} = \frac{HA^- + H^+}{H_2A} \qquad (2.3.3.1.10)$$

$$HA^- = A^{-2} + H^+ \quad k_{22} = \frac{A^{-2} + H^+}{HA^-} \qquad (2.3.3.1.11)$$

The overall dissociation constant ($K_{ai}$) can be represented as:

$$Kai = \frac{[H^+][H^{3-i}A^{i-}]}{[H^{4-i}A^{i-1}]} \qquad i= 1\text{-}2\ (2.3.3.1.12)$$

Overall distribution coefficient is defined as

$$K_D = \frac{[H_3A]_{org} + [H_3A]_{2org}}{[H_2A^-]_{aq} + [HA^{-2}]_{aq} + [A^{-3}]_{aq} + [H_3A]_{aq}} \qquad (2.3.3.1.13)$$

Where,

$K_D$ = Overall distribution coefficient

*Physical equilibria of tri-carboxylic acid*

For the ionization of the acid ($H_3A$) in the aqueous solution, dissociations reactions and dissociation constants can be represented as

$$H_3A = H_2A^- + H^+ \qquad k_{a1} = \frac{H_2A^- + H^+}{H_3A} \qquad (2.3.3.1.14)$$

$$H_2A^- = HA^{2-} + H^+ \qquad k_{a2} = \frac{HA^{2-} + H^+}{H_2A^-} \qquad (2.3.3.1.15)$$

$$HA^{2-} = A^{3-} + H^+ \qquad k_{a3} = \frac{A^{3-} + H^+}{H_2A^-} \qquad (2.3.3.1.16)$$

Where,

$k_{a1}$ = Dissociation constant for first reaction

$k_{a2}$ = Dissociation constant for second reaction

$k_{a3}$ = Dissociation constant for first reaction

The overall dissociation constant ($K_{ai}$) can be represented as:

$$Kai = \frac{[H^+][H^{3-i}A^{i-}]}{[H^{4-i}A^{i-1}]} \qquad i = 1\text{-}3 \quad (2.3.3.1.17)$$

Overall distribution coefficient is defined as

$$K_D = \frac{[H_3A]_{org} + [H_3A]_{2org}}{[H_2A^-]_{aq} + [HA^{-2}]_{aq} + [A^{-3}]_{aq} + [H_3A]_{aq}} \qquad (2.3.3.1.18)$$

### 2.3.3.2 *Chemical Equilibria*

*Chemical equilibria of monocarboxylic acids*

*Monocarboxylic Acid – Tertiary Amine*

The reactive liquid- liquid extraction of carboxylic acid (HA) with tertiary amine (B), gives a reaction complex (BHA), which remains largely in the organic phase and may be represented by

$$HA_{aq} + B_{org} \leftrightarrow BHA_{org} \qquad (2.3.3.2.1)$$

The distribution coefficient, $K_D$ is defined by:

$$K_D = [HA]_{org}/[HA]_{aq} \qquad (2.3.3.2.2)$$

The equilibrium complexation constant, $K_E$ is defined as:

$$K_E = [BHA]_{org}/[HA]_{aq}[B]_{org} \qquad (2.3.3.2.3)$$

The concentration of free amine in the organic phase would be:

$$[B]_{org} = [B]_{i,org} - [BHA]_{org} \qquad (2.3.3.2..4)$$

If (1:1) carboxylic acid – amine complex is formed then $[HA]_{org} = [BHA]_{org}$,

$$\log K_D = \log K_E + \log [B]_{org} \qquad (2.3.3.2.5)$$

If the former assumptions are valid, a plot of log $K_D$ vs log $[B]_{org}$ should yield a straight line with a slope of unity.

If the slope of distribution coefficient as a function of free amine concentration in organic is less than unity which implies that the organic phase extracts more acid than would be expected on the basis of (1:1) complex. Hence, the extraction equilibrium of the carboxylic acid is not adequately represented by equation (2.3.3.2.1).

Though the exact nature of the chemistry involved in the uptake of extra acid is unknown, and, in spite of the obvious non-ideality of the organic phase, distribution data can be interpreted by a set of equilibria involving the formation of complexes with n acid molecules and one amine molecule:

$$HA_{aq} + B_{org} \xleftrightarrow{K_{11}} BHA_{org} \qquad (2.3.3.2.6)$$

$$HA_{aq} + BHA_{org} \xleftrightarrow{K_{21}} B(HA)_{2,org} \qquad (2.3.3.2.7)$$

:

:

$$HA_{aq} + B(HA)_{n-1,org} \xleftrightarrow{K_{n1}} B(HA)_{n,org} \tag{2.3.3.2.8}$$

The equilibrium complexation constant for the reaction represented by above equations is:

$$K_E = \frac{[B(HA)_n]_{org}}{[B]_{org}[HA]_{aq}^{\ n}} \tag{2.3.3.2.9}$$

The extent to which the organic phase (Amine + Diluent) can be loaded with carboxylic acid is expressed as the loading ratio, z;

$$z = \frac{[HA]_{org}}{[B]_{i,org}} \tag{2.3.3.2.10}$$

'i' represents initial value. The value of z depends on the extractability of the acid (strength of the acid-base interaction) and its aqueous concentration. The stoichiometry of the overall extraction reaction depends on the loading ratio in the organic phase, *z*. If the organic phase is not highly concentrated *i.e.* at very low loading ratios ($z < 0.5$) the (1:1) complex is formed and the following equation holds

$$\frac{z}{1-z} = K_{E(1:1)}[HA]_{aq} \tag{2.3.3.2.11}$$

For higher loading ratios, the (2:1) complex is formed, and a plot of $z/(2-z)$ vs $[HA]_{aq}^{\ 2}$ should yield a straight line, whose slope gives the complexation constant $K_{E(2:1)}$ for the (2:1) complex.

$$\frac{z}{2-z} = K_{E(2:1)}[HA]_{aq}^{\ 2} \tag{2.3.3.2.12}$$

If the acid concentration is high enough, the (3:1) complex may be formed and a plot of $z/(3-z)$ vs $[HA]_{aq}^{\ 3}$ should yield a straight line, whose slope gives the complexation constant $K_{E(3:1)}$ for the (3:1) complex.

$$\frac{z}{3-z} = K_{E(3:1)}[HA]_{aq}^{\ 3} \tag{2.3.3.2.13}$$

*Monocarboxylic Acid – Quaternary Amine*

*Extraction with quaternary ammonium chloride*

The quaternary ammonium chloride extracts both dissociated and undissociated forms of acid. When acid anions ($A^-$) are extracted by chemical interaction with quaternary ammonium chloride ($R_4N^-Cl^-$) according to the reaction

$$[R_4N^+Cl^-]_{org} + [A^-]_{aq} \leftrightarrow [R_4N^+A^-]_{org} + [Cl^-]_{org} \tag{2.3.3.2.14}$$

The equilibrium extraction constant is (Kyuchou kov *et al.*, 2005)

$$K_{E(A-)} = \frac{[R_4N^+A^-]_{org}[Cl^-]_{org}}{[A^-]_{aq}[R_4N^+Cl^-]_{org}} = \frac{[R_4N^+A^-]_{org}[Cl^-]_{org}(K_a + [H^+])}{C_{HA}K_a[R_4N^+Cl^-]_{org}}$$

$$= \frac{m_{(A^-)}[Cl^-]_{org}}{\alpha[R_4N^+Cl^-]_{org}} \qquad (2.3.3.2.15)$$

$K_a$ is the dissociation constant of the acid and $\alpha = K_a/(K_a + [H^+])$ represents the part of dissociated form of acid. The partial distribution coefficient of acid anions ($m_{(A-)}$) can be expressed as follows

$$m_{(A-)} = \frac{\alpha K_{E(A^-)}[R_4N^+Cl^-]_{org}}{[Cl^-]_{org}} \qquad (2.3.3.2.16)$$

When undissociated molecules are assumed to be removed by chemical interaction

$$[R_4N^+Cl^-]_{org} + [HA]_{aq} \leftrightarrow [R_4N^+ Cl^- :HA]_{org} \qquad (2.3.3.2.17)$$

With the extraction constant

$$K_{E(HA)} = \frac{[R_4N^+Cl^- : HA]_{org}}{[HA][R_4N^+Cl^-]_{org}} = \frac{[R_4N^+Cl^-:HA]_{org}(K_a + [H^+])}{[H^+]C_{HA}[R_4N^+Cl^-]_{org}}$$

$$= \frac{m_{(HA)}}{[R_4N^+Cl^-]_{org}\ (1 - \alpha)} \qquad (2.3.3.2.18)$$

The following expression takes place:

$$m_{(HA)} = \frac{[R_4N^+Cl^-:HA]_{org}}{C_{HA}} = K_{E(HA)}(1- \alpha)\ [R_4N^+Cl^-]_{org} \qquad (2.3.3.2.19)$$

if the physical extraction by the diluent and modifier is negligible and can be neglected, the overall distribution coefficient $m_{(C(HA))}$ can be presented as:

$$m_{(C(HA))} = \frac{[R_4N^+Cl^-:HA]_{org} + [R_4N^+A^-]_{org}}{C_{HA}} = m_{(A-)} + m_{(HA)} \qquad (2.3.3.2.20)$$

when the organic phase consists of extractant, diluent and modifier and the physical extraction by diluent and/or modifier cannot be neglected, it must be taken into account

$$m_{(C(HA))} = \frac{[R_4N^+Cl^-:HA]_{org} + [R_4N^+A^-]_{org} + \beta_d[HA]_{m,org} + \beta_m[HA]_{m,org}}{C_{HA}}$$

(2.3.3.2.21)

$$= m_{(A-)} + m_{(HA)} + \beta_d m_d + \beta_m m_m \qquad (2.3.3.2.22)$$

where $\beta_d = \overline{V_d}/\overline{V}$ and $\beta_m = \overline{V_m}/\overline{V}$ are the volume fractions of diluent and modifier, respectively. If the undissociated molecules are extracted with extract by physical extraction (*i.e.*, chemical interaction does not occur),

$$m_{(C(HA))} = \frac{(1-\beta_d-\beta_m)[HA]_{e,org} + [R_4N^+Cl^-]_{org} + \beta_d[HA]_{m,org} + \beta_m[HA]_{m,org}}{C_{HA}}$$

$$= m_{(A-)} + (1-\beta_d-\beta_m)\, m_{(e)} + \beta_d m_d + \beta_m m_m \qquad (2.3.3.2.23)$$

where $m_e = [HA]_{e,org}/C_{HA} = p_e(1-\alpha)$ is the partial distribution coefficient of extracted undissociated molecules by physical extraction with extractant and $p_e = [HA]_{e,org}/[HA]$ is the particular distribution coefficient.

*Monocarboxylic Acid – Tertiary Amine (Based on Relative Basicity)*

Shan *et al.* (2006) described the extraction equilibrium of monocarboxylic acid from aqueous solutions on the relative basicity of the extractant. Three main factors have been found to influence the extraction equilibrium of carboxylic acid with amines from aqueous solutions, namely the acid hydrophobicity (logP), the dissociated equilibrium constant of the acid (pKa) and the apparent basicity of the extractant to HCl (pKa,s). Li *et al.* (2002; 2004a,b) and Shan (2003) studied the extraction equilibria of mono carboxylic acid by TOA and TRPO and found that the apparent extraction equilibrium constant of 1:1 complex $K_{11}$ mainly depend on the nature of solute and extractant ie logP and pKa,B. the correlated equation of $K_{11}$ was obtained with log P, pKa, pKa,BS as follows:

For TOA systems:

$$\log K_{E(1:1)} = 1.3164[(pKa,B - pKa) + \log(93.47P)] \qquad (2.3.3.2.24)$$

For TRPO systems:

$$\log K_{E(1:1)} = 1.3111 + 0.1778 \log P - 0.9124pKa + 1.2532pKa,B \qquad (2.3.3.2.25)$$

$K_{E(1:1)}$ in Eq. (2.3.3.2.24) represents extraction capacity of the acid by forming the complex of ion-pair, H-bond association and solvating power of the complex. The solvating power is a complicated H-bonding association between the complex and the diluent, which depends on the nature of the solute, extractant and diluent. Therefore, the semi-empirical correlated Eqs. (2.3.3.2.24 - 2.3.3.2.25) are limited for interpreting the extraction mechanism and predicting the extraction equilibrium behavior for other extractants. Furthermore, p*K*a,B in Eq. (2.3.3.2.24) is the relative basicity of the extractant mixture to HCl, excluding the nature of solute. If the basicity of the extractant mixture is relative to the solute, this relative basicity of the extractant can represent all of the nature of the solute, diluent and extractant, as well as special associations, *e.g.* solvating power.

Generally, the basicity of a compound represents the power of releasing hydrogen ion in the aqueous solution. The relative basicity of the extractant to the solute ($pK_{a,BS}$) is defined as follows:

$$B{\cdot}HA\ (org) = B_{(org)} + H^+_{(aq)} + A^-_{(aq)},\ Ka,\ BS \tag{2.3.3.2.26}$$

$$K_{a,BS} = [B]_{(org)}\,[H^+]_{(aq)}\,[A^-]_{(aq)}/[B{\cdot}HA]_{(org)} \tag{2.3.3.2.27}$$

where *B* and *HA* stand for extractant and carboxylic acid, respectively.

The extraction equilibrium is described by a set of equilibrium expressions involving the dissociation equilibrium of carboxylic acid in water, the association of TOA or TRPO with carboxylic acid, and physical extraction of carboxylic acid by diluent. It was assumed that only 1:1 complex was formed and that the extraction reaction between extractant and carboxylic acid took place at the organic–aqueous interface.

The reactive extraction equilibrium

$$B_{org} + HA_{aq} = [B.HA]_{org}\ K_{E(1:1)} \tag{2.3.3.2.28}$$

$$[HA]_{aq.} = [HA]_{org}\ m \tag{2.3.3.2.29}$$

$$K_{E(1:1)} = \frac{[H^+]_{(aq)}[A^-]_{(aq)}}{[HA]_{(aq)}K_{a,BS}} \tag{2.3.3.2.30}$$

$$\text{Then, } K_{E(1:1)} = K_a/K_{a,BS} \tag{2.3.3.2.31}$$

$$\log K_{E(1:1)} = pK_{a,BS} - pK_a \tag{2.3.3.2.32}$$

Since equation (2.3.4.2.32), derived from the reactive extraction mechanism and is independent of the solute and extractant, it can be used to predict the extraction equilibrium for the other new extractant system and new solute.

*Monocarboxylic Acid – Organophosphorous extractants*

Kertes and King (1986) illustrated the mass action equilibrium equations describing the transfer of a weak (monobasic) organic acid from water into an organophophorous compounds are given as:

(i) Ionization of the acid in the aqueous solution:

$$HA \leftrightarrow H^+ + A\surd\, K_{HA} = [H^+][A^-]/[HA] \tag{2.3.3.2.33}$$

(ii) Formation of the acid solvates in the organic phase:

$$HA_{(aq)} + pS_{(o)} \leftrightarrow HA.S_{p(org)} \tag{2.3.3.2.34}$$

$$K_s = [HA.S_p]_{(org)}/[HA]_{(aq)}\ [S]^p_{(org)} \tag{2.3.3.2.35}$$

The experimentally accessible distribution ratio $K_D$ is given as:

$$K_D = \frac{C_{HA(org)}}{C_{HA(aq)}}\ \frac{[HA \bullet S_p]_{(org)}}{[HA]_{(aq)} + [A^-]_{(aq)}} = \frac{K_s\,[HA]_{(aq)}\,[S]^p_{(org)}}{[HA]_{(aq)} + K_{HA}\,[HA]_{(aq)}/\left[H^+\right]_{(aq)}}$$

$$= \frac{K_s\,[S]^p_{(org)}}{1 + K_{HA}/\left[H^+\right]_{(aq)}} \tag{2.3.3.2.36}$$

the square brackets denotes the concentration in mol/l and

$$[S]_{(org)} = S_{(org)} - p\,[HA]_{(org)} \tag{2.3.3.2.37}$$

$[S]_{(org)}$ represents the concentration of the free, unbound extractant at equilibrium. If the ratio $K_{HA}/[H^+]_{(aq)}$ is constant or has negligibly small, $[H^+] << K_{HA}$, the solvation number p can be obtained by the partial differentiation of $K_D$ with respect to $[S]_{(org)}$, a plot of

$$\log K_D = \log K_s + p \log[S]_{(org)} \tag{2.3.3.2.38}$$

yields the slope p and the intercept $K_s$. The solvation number of the acids appears to be same as the number of COOH groups on the acid molecule. This indicates a stoichiometry association between individual phosphoryl group and individual carboxylic group and displays the strong affect of the extractant concentration on the experimentally determined distribution ratios.

### *Chemical equilibria of di-carboxylic acids*

*Dicarboxylic Acid – Tertiary Amine (Langmuir Type)*

The equilibrium of the active liquid-liquid extraction of carboxylic acid from aqueous solution of tertiary amine in diluent, was investigated and mathematically modeled by Poposka *et al.* (2000). Since tertiary amines extract only the undissociated molecules of carboxylic acids, the extraction of acids with amine, as a reaction of the acid base type by which acid amine complexes are formed, can be expressed by the following stoichiometric relation:

$$p[H_2A]_{aq} + q\,[R_3N]_{org} = [R_3N]_{q,\,org}\,[H_2A]_{p,\,org},\quad K_{pq} = \frac{c_{pq}}{c^p_{AA} \bullet c^q_B} \tag{2.3.3.2.39}$$

where p, q are the number of acid, amine molecules in the complex, $K_{pq}$ is the extraction equilibrium constant of (p,q) acid amine complex formation on a molarity scale, and $c_{pq}$ is the molar concentration of the complex in the organic phase.

For the interpretation of the equilibrium experimental data, two approaches have been used.

(a) *Langmuir type of the extraction isotherm*

The distribution of the undissociated acid between the aqueous and organic phases, where the acid is bound to the amine molecule/molecules can be described by the reaction of complex formation as,

$$a[H_2A]_{aq} + [NR_3]_{org} = \{[H_2A]_a\,[NR_3]\}_{org}\,.\ K = \frac{c_{complex}}{c^a_{AA} \bullet c_B} \tag{2.3.3.2.40}$$

where, a = no. of acid molecules in complex in contrast to Eq. (2.3.3.2.39), representing the formation reaction of an individual complex, Eq. (2.3.3.2.40) is conceived as an overall reaction, representing several parallel reactions. Therefore, the stoichiometric coefficients p.q are now defined as rational numbers and a = p/q.

If a is assumed to be a constant in the concentration range considered, and taking into account that in the present case the loading has been found independent of $c_{Bo}$, the shape of the isotherm $c_{Ao} = f(c_{AA})$ can be described by the modified Langmuir isotherm, independent of the initial amine concentration:

$$\frac{c_{AO}}{c_{AO,max}} = \frac{Z}{Z_{max}} = \frac{Kc_{AA}^{a}}{1 + Kc_{AA}^{a}} \tag{2.3.3.2.41}$$

Where

$C_{AA}$ = equilibrium concentration of acid in the aqueous phase
$C_{Ao}$ = equilibrium concentration of acid in the organic phase
$C_{Ao,max}$ = equilibrium concentration of acid in the organic phase at the plateau
$Z_{max}$ = maximum loading of amine
K = extraction equilibrium constant for the overall reaction
$c_{complex}$ = equilibrium concentration of the overall acid-amine complex

(b) *Mathematical model based on a proposed reaction mechanism*

The model assumes simultaneous formation of three complexes: (1, 1), (1, 2) and (2, 1) (Tomovska *et al.*, 1998). Thus according to Eq. (2.3.3.2.39), the stoichiometery of tartaric acid extraction with TOA dissolved in an alcoholic solvent, can be written as

$$[H_2A]_{aq} + [NR_3]_{org} = \{[H_2A][NR_3]\}_{org} \tag{2.3.3.2.42}$$

$$[H_2A]_{aq} + 2[NR_3]_{org} = \{[H_2A][NR_3]_2\}_{org} \tag{2.3.3.2.43}$$

$$[H_2A]_{aq} + \{[H_2A]_2[NR_3]\}_{org} = \{[H_2A]_2[NR_3]\}_{org} \tag{2.3.3.2.44}$$

Corresponding thermodynamic extraction constants $\overline{K}_{11}$ (true equilibrium constants) of the reactions are

$$\overline{K}_{11} = \frac{\left\{[H_2A][NR_3]\right\}_{org}}{(H_2A)_{aq}\,(NR_3)_{org}} = c_{11}/c_{AA}\,c_B \tag{2.3.3.2.45}$$

$$\overline{K}_{12} = \frac{\left\{[H_2A][NR_3]_2\right\}_{org}}{(H_2A)_{aq}\,(NR_3)^2_{org}} = c_{12}/c_{AA}\,c^2{}_B \tag{2.3.3.2.46}$$

$$\overline{K}_{21} = \frac{\left\{[H_2A]_2[NR_3]\right\}_{org}}{(H_2A)_{aq}\,\{[H_2A][NR_3]\}_{org}} = c_{21}/\,c_{AA}\,c_{11} \tag{2.3.3.2.47}$$

Combining Eqs. (2.3.3.2.45 - 2.3.3.2.47) with the mole balance equations of the acid and the amine in the organic phase, the mathematical model of equilibrium for (1, 1), (1,2) and (2,1) stoichiometry is obtained in the form

$$c_{Ao} = c_{11} + c_{12} + 2\,c_{21} = K_{11} c_{AA}\, c_B + K_{12}\, c_{AA}\, c^2{}_B + 2\,K_{21}\, c_{AA}\, c_{11} \qquad (2.3.3.2.48)$$

$$c_B = c^o{}_B - (c_{11} + c_{12} + 2c_{21}) = c^o{}_B - K_{11} c_{AA} c_B + K_{12} c_{AA} c^2{}_B + 2K_{21} c_{AA} c_{11} \qquad (2.3.3.2.49)$$

## 2.3.4 *Thermodynamics of complexation*

Effect of temperature and thermodynamic is essential in the design of a process. The study is aimed for two different prospective: a) operating temperature and b) back extraction/regeneration possibilities by temperature swing regeneration technique. Solute that needs to be recovered is available in various sources (*e.g.* in waste stream or produced biotechnologically in fermentors or obtained in industrial streams) in the temperature range of 298 to 313 K. Thus an extractant can be considered as good only if it could operate efficiently in this operating range. Usually extraction is an exothermic process and a decrease in extraction is expected as the temperature is increased. However, it can be stated that the decrease in extraction is the function of the extractant and diluent chosen *i.e.* the extracting medium. Further the variation in climate, feed and bioreactor conditions make the study of temperature effect more important.

Equilibrium state of a closed system is defined as the state for which the total energy is a minimum with respect to all possible changes at the given temperature and pressure.

$$(dG^t)_{T,P} \le 0 \qquad (2.3.4.1)$$

At equilibrium state differential variations can occur in the system at constant temperature (*T*) and pressure (*P*) without producing any change in Gibbs free energy constant $G^t$. Thus there are two distinctive features of the equilibrium state for given temperature and pressure: i) the total Gibbs energy is minimum and ii) its differential is zero. The total Gibbs energy can be defined as:

$$d(nG) = (nV)\,dP - (nS)\,dT \qquad (2.3.4.2)$$

Thus the differential yields:

$$d(nG) = (nV)\,dP - (nS)\,dT + \sum_i \mu_i dn_i \qquad (2.3.4.3)$$

Where, V = Volume and S = Entropy

if change in number of moles occurs as the result of chemical reaction in a closed system, $dn_i$ may be displaced by $\mu_i d\,\varepsilon$, where ε is called as the reaction coordinate and characterizes the extent or degree to which a reaction has taken place (the reaction coordinate has been given various other names, such as: degree of advancement, degree of reaction, extent of reaction and process variable), Eq. (2.2.4.3) becomes:

$$d(nG) = (nV)\,dP - (nS)\,dT + \sum_i \nu_i \mu_i d\varepsilon_i \qquad (2.3.4.4)$$

at equilibrium the last term of the above equation is zero.

Thus

$$\sum_i \nu_i \mu_i = 0 \tag{2.3.4.5}$$

Since fugacity is defined as

$$\mu_i = G_i^o + RT\ln \hat{a}_i \tag{2.2.4.6}$$

where, R = Universal Gas constant

Substituting in equation (2.2.4.5) and solving, we get the definition of equilibrium constant for the reaction as,

$$-RT \ln K_E = \sum_i \nu_i G_i^o \equiv \Delta G^o \tag{2.3.4.7}$$

$\Delta G^o$ is the conventional way of representing the quantity $\sum_i \nu_i G_i^o$ and is called the standard Gibbs energy change of reaction. If the enthalpy (ΔH) and entropy (ΔS) of reaction are assumed to be constant over the temperature range (Tamada and King, 1990), the equilibrium complexation constant is related to temperature as:

$$\ln K_E = \frac{-\Delta H}{RT} + \frac{\Delta S}{R} \tag{2.3.4.8}$$

Where, $K_E$ = Equilibrium complexation constant

The above expression indicates that a plot of in $K_E$ vs 1/ T gives a straight line. The more exothermic reaction, the more sensitive the equilibrium is to changes in the temperature. The slope is proportional to the enthalpy of reaction, and the intercept is proportional to the entropy.

Extraction of acid by extractant is via intermolecular hydrogen bonding or ion exchange of the extractant group with the acid. The extraction of acid by extractant - acid complexation is expected to be exothermic and makes the system more ordered. Extraction of acid using different extractant resulted in negative values of enthalpy and entropy (Keshav *et al.*, 2009). Thus the reactive extraction process is exothermic in nature and decrease in entropy is obtained. Higher the negative value of $\Delta H$, more exothermic is the extraction. The decrease in entropy may not be accepted from the definition of second law of thermodynamics because all spontaneous processes proceed in the direction of maximum probability, *i.e.* greatest randomness. However entropy may be looked upon as measure of randomness which is a minimum in systems with an orderly arrangement. Example, take the case of spontaneous conduction of heat along a metallic bar. This results in a random distribution of kinetic energy of the molecules. Rejection of heat from system decreases the disorder of the molecules, an equivalent or greater amount of disorder results in environment. The combination of two molecules into one *i.e.* the complex resulted in decrease in the degree of freedom and thus a more ordered arrangement. Thus decrease in entropy is expected. Further a more decrease in order of a particular interaction suggests the stronger interaction of acid with the extractant. The decrease in extraction with temperature suggests that a temperature control should be provided whenever these extractants were used *in - situ* with the fermentation broth as temperature increase could severely affect the process economics. On the other hand an advantage can be extracted out of this behavior on account of regeneration of the acid from loaded organic phase by temperature swing regeneration.

Effect of temperature is attributed to the effect of different parameters like the acid p*K*a, the extractant-acid interaction, the solubility of the acid in both phases, the extractant basicity and water co-extraction. These parameters were studied by Canari and Eyal (2004). They concluded that the p*K*a values of common carboxylic acids decrease only slightly when the temperature is increased. Similarly, p*K*a measurements for carboxylic acids exhibit a small decrease when the temperature is increased but only up to a given point. Above it, the p*K*a value increases. Similarly solubility of acids is affected by temperature in both the aqueous and extractant phase. Higher the –ve value of ΔH for a particular acid suggests that more exothermic the extraction. This may be accounted due to the difference in partial molar heat of mixing of the complex in the solvent and the partial molar heat of mixing of the acid in the aqueous phase, or $\Delta H_{true}$. Tamada and King (1990) evaluated organic phase heat of mixing using partition coefficient values (*P*) for extraction of acid by diluents alone using the following relation:

$$\frac{d\ln P}{d(1/T)} = -\Delta H_{transfer} \qquad (2.3.4.9)$$

where $\Delta H_{transfer}$ is the heat of transfer from organic to aqueous phase. Higher is the $\Delta H_{transfer}$ value for a particular diluent; higher is the effect of temperature.

### 2.2.5. *Effect of pH*

Study of effect of pH on the reactive extraction process is essential in view of the effectiveness of a particular extraction. pH on the other hand varies as the process of reactive extraction proceeds owing the recovery. Considering the case of reactive extraction of carboxylic acid, normal pH of solution containing acid is below pKa values of the respective acid but as the extraction was performed, due to the removal of the acid, pH approaches near 7.0. Thus for reactive extraction process an extractant-diluent combination is required which could recover the acid efficiently at all pHs.

Different extractant give different performance at various pH values. Tertiary amines and organophosphorous compound were found to extract the acid in undissociated form. With the increase in pH of the system, the fraction of dissociated forms increases over the undissociated form and hence these extractant perform poorly in neutral or basic media.

Quaternary amines such as Aliquat 336 on the other hand can be successfully be used *in situ* with the fermentation broth in mixed form since its performance is unaffected with the pH of system and it will extract the acid as it generated may it be in any form: dissociated or undissociated. Thus, pH will be maintained to 6–7 throughout. However it may be pointed out that a proper design of contacting device be made since quaternary amines are highly viscous and can be interfere with the physicochemical properties of the broth. Further it being toxic, it is recommended to be used in low concentration and with a non toxic diluent. Many authors had successful compared the performance of quaternary amine with the other categories of extractant and proved quaternary amine to be better at higher pH values. These authors have recommended the use of quaternary amine as favorable for the envisaged biotechnological application of the extraction process where the optimum pH range is between 5.5 and 5.8.

Fig. (2.3.5.1-2.3.5.3) shows extraction of 0.2 kmol/ $m^3$ propionic acid using TOA, TBP and Aliquat 336 in MIBK when pH was varied from 2.8 to 7. The results were in agreement with the discussion made above. For extraction using TOA and TBP, initially as the pH is increased, from natural pH (2.8 to 3.5), there is a gradual decrease in extraction and at pH greater than 5.5, very low $K_D$ were obtained. Thus at pH > p*K*a of acid, extraction is very low. The reason for this is that at higher pH, most of acid is present in dissociated form and since these extractants extracts only the undissociated form of acid, very low $K_D$ was obtained. So the best option is that the pH of system should be monitored and when pH of fermentation broth fall below 4 *i.e.* sufficient acid is

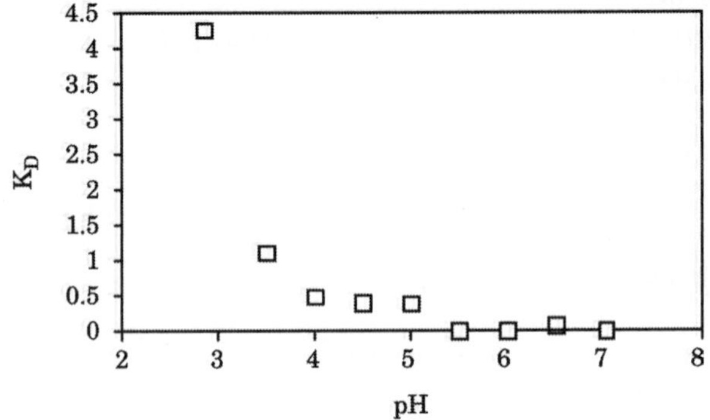


**Fig. 2.3.5.1:** Effect of pH on the reactive extraction of propionic acid (0.2mol/l) using TOA (40%) in MIBK

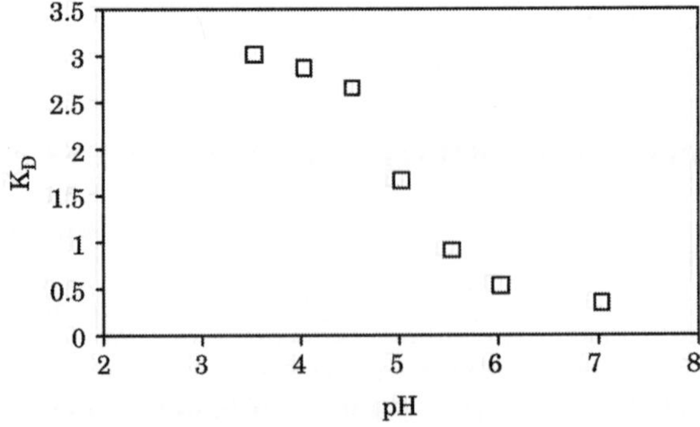


**Fig. 2.3.5.2:** Effect of pH on the reactive extraction of propionic acid (0.2 mol/l) using TBP (40%) in MIBK

generated, extractants should be injected into the system. Thus these extractant will remove the acid and raises the pH of the system, since acid will be removed. Thus a step injection mechanism should be designed, the controlled variable of which being the pH of system. On the other hand in extraction using Aliquat 336, at low pH the acid removed in undissociated form is much higher than that removed in dissociated form however at high pH the order reverses. However the total (undissociated + dissociated) acid extracted increases with increase in pH upto certain pH value, where upon it decreases with further increase of pH. The total dissociated + undissociated extracted was found to increase with increase in pH upto 5 after which it decreases. The highest value of $K_D$ was registered to be 3.3 for Aliquat 336 in MIBK which is related to practically complete deprotonation of propionic acid at pH value greater than p*K*a (4.76) of the acid. Since the optimum pH of any bio-application (bio-cultivations) is 5 to 6, Aliquat proposed to be a good extractant as it provides highest $K_D$ in that range in all diluents.

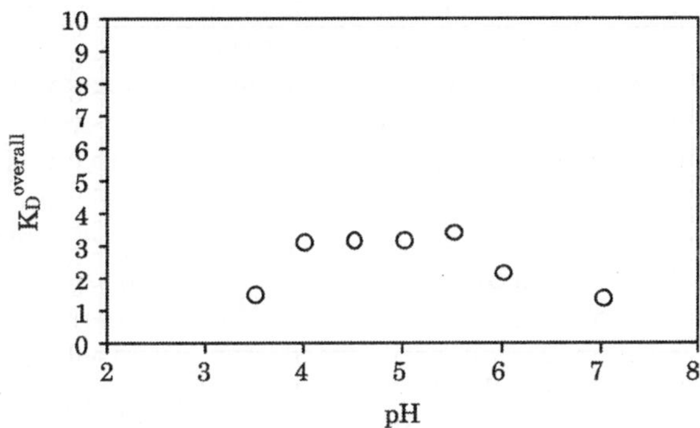


**Fig. 2.3.5.3:** Effect of pH on the reactive extraction of propionic acid (0.2 mol/l) using Aliquat 336 (30%) in MIBK

Extraction of metal in hydrometallurgical application is also effected by pH of the solution. Since cationic species is to be removed by anionic extractant, basic pH is prevailing. For the reactive extraction of lithium using extractants LIX54 + TBP it was found that negligible extraction were obtained upto pH <11.5. Increasing pH above 11.5, increases the extraction upto pH = 13.0, after which it remains constant. Natural sources are usually obtained at pH of 8-9, where as seen from above no extraction is possible, so the pH of feed solution should be increases to 12-13 before extraction of lithium should be carried out.

## 2.2.6 *Kinetics of extraction*

Kinetics of reactive extraction is important in the design and simulation of the process. Since efficiency of the reactive extraction process depends on the proper extractant diluent combination, it is the matter of utmost importance to find out the kinetic parameters of the combination to understand the nature of the phenomenon. Jet apparatus, stirred cell, mechanically agitated contactor and stirred contactor are some of the setups successfully employed to carry out kinetic studies of fluid-fluid systems. In reactive extraction systems the extractant is used in small quantities and the solvation medium is essentially the diluent and its concentration is uninfluenced in the reaction. However since it is not involved in chemical complexation, its effect does not appear in the kinetics. However in the mass transfer coefficient determination which ultimately affects the kinetics of the process, effect of diluent do appear. These mass transfer coefficient and solvation of the complexes are depended on the type of diluent employed and kinetic measurements are distinguished based on the type of diluents employed in a particular extractant – solute systems. There are some cases where the reaction occurs on both sides of the fluid. However such cases are very little and few studies are in the literature. On the other hand, number of studies could be found in literature where the reaction has been considered in one phase and that too either liquid phase in gas liquid reaction or the organic phase in liquid-liquid reactions.

As like most of the multiphase reaction systems, the reactive extraction process involves various steps such as diffusion of the reactant molecules to each other, the actual chemical transformation and the diffusion of the products away from each other. Before the substances react, they must migrate to the interface. Hence in addition to chemical reaction, physical factors which affect the mass transfer between the phases also affect the overall rate. Thus the influence of both mass transfer and chemical reaction comes in the kinetic determination of the process. Many a times the chemical step is much slower than the diffusion step and hence is the controlling, however for fast reactions diffusion may be the limiting parameter. So in order to study the influence of these two controlling parameters experimentation is desired. Visualizing the change in operating condition can also assist in detecting the dominant mechanism. *E.g.* If rate increases markedly with temperature, the chemical rate is controlling. If varying the interfacial area, by changing the diameter of the reactor, or the flow rate affect the kinetics it suggests that the reaction is mass transfer limited. Induction of turbulence by agitation is essential as it controls the amount of surface available for reaction and also eliminates the concentration gradient that otherwise hinders the diffusion. Few of fluid –fluid industrially important reactions are listed below

1. Nitration of organic compounds with aqueous nitric acid
2. Formation of soaps by action of aqueous alkalies on fats or fatty acids
3. Sulphur removal from petroleum fractions by aqueous ethanolamines
4. Treating of petroleum products with sulphuric acid
5. Hydrogenation of vegetable oils with gaseous hydrogen

2.3.6.1 *Experimental setups*

In liquid-liquid systems the kinetic is studied in two type of equipments, stirred cell or a small mechanically agitated contactor (Doraiswamy and Sharma, 1984). Figs. 2.3.6.1 and 2.3.6.2 show the diagrams for the two type of contactors.

*Stirred cell*

Stirred cell constitutes a cylindrical vessel having diameter (D) and height (H) and having a four blade dual stirrer with diameter (PD) connected to a speed

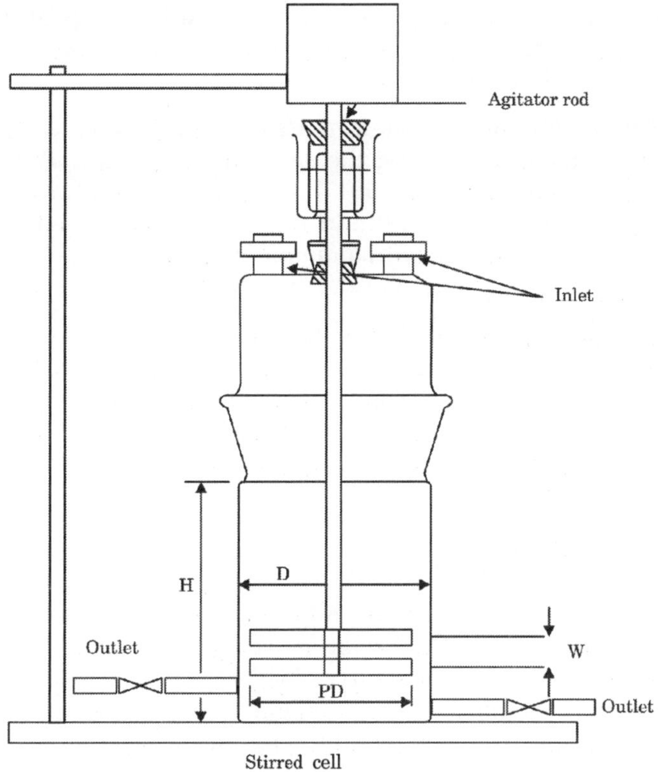


**Fig. 2.3.6.1:** Typical stirred cell

**Fig. 2.3.6.2:** Mechanically agitated stirrer

adjustable motor. Feed in entered from the top and the organic and aqueous phases were kept such that upper blades of the dual stirred are kept in one phase and the lower blades in the second phase. The concentrations of the extractant are chosen such that it does not change the viscosity of the diluent significantly. Speed of the agitators can be varied from 20-150 rev/min and volume phase ratios can be conveniently varied by the factor of 2 (as long as mass transfer coefficient value is not affected).

Effect of hydrodynamics on the kinetics of liquid-liquid reactive extraction process can be studied by varying the speed of agitation and volume phase ratios. Doraiswamy and Sharma (1984) has given the procedure for discerning the reaction mechanism for fluid-fluid reaction using stirred cell. Effect of components concentration and hydrodynamic of extraction was studied in order to predict the reaction regime.

Mass transfer accompanied chemical reactions are classified into four regimes: i) very slow reaction; ii) slow reaction; iii) fast reaction; and iv) instantaneous reactions. The classification is based on the physico-chemical and hydrodynamic parameters. A description of the regimes, rate equation for particular regime and the necessary condition for validation of experimental data are give in Section to follow.

*Mechanically agitated contactor*

Mechanically agitated contactors are high speed agitation systems which are generally baffled to avoid vortex formation. Height of liquid in these agitator is usually slightly less than the diameter of the contactor. Disk turbine having four to six blades are employed as stirrers and speed may vary as high as 2500 rpm. Table 2.3.6.1 may be used to find the regime of reaction.

**Table 2.3.6.1:** Procedure for discerning reaction mechanism: Stirred cell (Doraiswamy and Sharma, 1984)

| | ***Effect on the specific rate of absorption (kmol $m^{-2}$ $s^{-1}$)*** | | | |
|---|---|---|---|---|
| ***Regime*** | ***[A*]*** | ***[B°]*** | ***Speed of agitation*** | ***Volume of liquid (B phase)*** |
| 1 | a $[A^*]^m$ | a $[B_0]^n$ | None | α-volume |
| 2 | a $[A^*]$ | None | Increases with increase in the speed of stirring | |
| 3 | a $[A^*]^{(m+1)/2}$ | a $[B_0]^{n/2}$ | None | None |
| 4 | None[a] | a $[B_0]$ | Increases with increase in the speed of stirring in the same manner as in regime 2 | None |

[a] Provided that $[B_0]>>z[A^*]$

$[A^*]$ = equilibrium concentration of reactant 1 in the organic phase, kmol/m$^3$

$[B_0]$ = initial concentration of reactant 2 in the organic phase, kmol/m$^3$

$m$ = order of reaction with respect to A

$n$ = order of reaction with respect to B

Since the agitation is quite high the interfacial area in not predictable and is complex function of the diameter of bubble and the dispersed phase holdup. Power of agitation can be predicted using the Reynolds number based on the average density and viscosity. Dispersed phase hold up is function of various dimensionless numbers involving the various physico-chemical properties of the solvent system (extraction and diluent). In all the reaction regimes (except 1), the rate of extraction will vary with the speed of stirring whereas in regime 1 there is no influence of the speed of stirring. Critical speed of stirring is 700-900 rpm for 4-6 liter capacity contactors. The main advantage of using mechanically agitated contactor for kinetic studies is that since the large scale reactors are large scale prototype of this contactor, scale up can be easily performed. However the major limitation is the finding of the dispersed phase hold up and bubble diameter which is greatly influences by the concentrations of the organic phase components (extractant and diluent).

During the experiments the physical properties of the system can change due to the changes of the acid concentration in both phases. In particular, the interfacial tension change can exhibit effects on the dispersion interfacial area. On the other hand, the effect of the agitation level on the contacting area is very important. The stirring vessel used for the kinetic measurements can not decouple the effect of the changes of concentrations on the interfacial area. However, the stirring vessel provides agitation conditions, which allow for excluding the influence of agitation on the rate of extraction.

2.3.6.2 *Reactive extraction regimes*

Reactive extraction of solute involves reactions between the reactive component (extractant) and solute. Reaction regime could be found from the experimental results. Since the solubility of extractant (reactive component in the organic phase) is negligible in the aqueous phase, whereas solute (reactive component in aqueous phase) has significant solubility in the organic phase (diluent phase), resistance to mass transfer is confined to the organic phase. Overall extraction rate is determined by interfacial chemical kinetics and molecular diffusion of components towards and away from interface. The two resistances to transfer are by diffusion and chemical complexation which is function of mode of contact and transport properties of the components. The reaction between solute (A) in aqueous phase and extractant (S) in organic phase is represented as:

$$A_{aq} + B_{org} \leftrightarrow (A.B)_{org} \tag{2.3.6.1}$$

where subscript "aq" and "org" stand for aqueous and organic phases. Extraction rates can be calculated using the method of initial rates from the following equation

$$R_A = \frac{V_{org}}{A} \frac{d[A]}{dt}\bigg|_{t=0} \tag{2.3.6.2}$$

To ascertain the effect of reaction on the mass transfer accompanied reactions, depending upon whether the diffusion or reaction is the controlling, four different reaction regimes have been located.

*Regime 1: Very Slow Reaction*

In this regime, the reaction is very slow in comparison to the diffusion step. The bulk shall be always saturated with the A and the rate controlling is the kinetics of the homogeneous chemical reaction. The above condition is validated if volumetric rate of mass transfer is very large in comparison to the rate of homogeneous chemical reaction.

*Regime 2: Slow Reaction*

In slow reaction regime, the reaction is much faster than diffusion rate. So as soon as the reactant (A) reaches the bulk it gets consumed and the concentration of dissolved A in the bulk is zero. This also signifies that practically there is no reaction in the film and the rate of homogeneous reaction is much large than the volumetric rate of mass transfer. When the above two conditions are satisfied the reaction regime is ascertained to be slow reaction.

*Regime 3: Fast Reaction*

In fast reactions, diffusion and chemical reaction occurs simultaneously in parallel steps and the reaction occurs entirely in the film. Rate is controlled by the mass transfer as well as reaction of acid into the organic phase. The ratio of amount of A reacting in the film to that reacting in the bulk in very much greater than 1. Further since the transfer of A to the organic phase is obtained from the solubility data, and B is always in excess, the depletion of B in the film is negligible.

*Regime 4: Instantaneous Reaction*

The reaction is potentially so fast that the solute and the reactant cannot coexist. At certain distance from the interface, a reaction plane is formed at which both the solute and the reactant are instantaneously consumed by the reaction.

2.3.6.3 *Rate controlling mechanism*

*Very slow reaction*

In very slow reactions, the diffusion forces are unimportant. If increasing the speed of agitation or gas flow rate in agitated gas liquid contactors or spargers, respectively, has no effect on overall mass transfer coefficient, it suggests that the reaction is very slow reaction. Another condition to check whether the reaction belongs to this regime is interchanging the continuous and dispersed phases. If it has no effect on the rate of reaction, it concludes that the diffusional factors are unimportant. Varying the hold up is the way to vary the interfacial area as the interfacial area varies linearly with the dispersed phase hold up. Systems with very slow reaction are very easy to scale up as only the lab scale data and the information on the residence time distribution of the two phases is desired.

*Fast reactions*

For fast reactions, rate is dependent on the physicochemical properties of the system and is independent of hydrodynamics factors. Specific rate of mass transfer is proportional to (m+1)/2th power to concentration of A and n/2th power to the concentration of B (Table 2.3.6.1) and as predicted in the rate laws, square root of rate constant ($k_{mn}$). This suggests that extractant which provide higher value of rate constant is more preferable for the reactive extraction if the regime is fast reaction.

*Instantaneous reactions*

Table 2.3.6.1, suggests that the specific rate of transfer is independent of concentration of A. In this as well as in the regime of slow reaction, increasing the speed of stirring has the influence on the rate. Further it shall be seen that in both these regimes, influence of mass transfer coefficient ($k_L$) is seen in the rate equation.

### 2.3.6.4 *Rate laws*

*Very slow reaction*

Fig. 2.3.6.3 shows the flow sheet for very slow reaction. Since the kinetic is the limiting in this situation, the rate expression can be written as

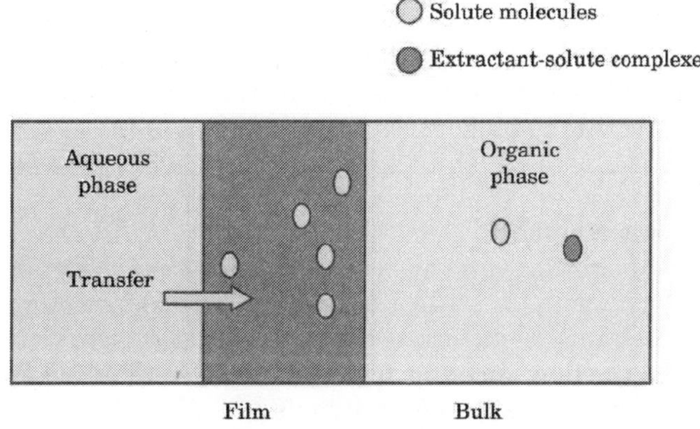


**Fig. 2.3.6.3:** Flow sheet of very slow reaction

$$R_A a = lk_{mn}[A^*]^m[B^o]^n \tag{2.3.6.3}$$

Where astrix (*) represents the interfacial value and the necessary condition for this regime is

$$k_L a[A^*] >> lk_{mn}[A^*]^m[B^o]^n \tag{2.3.6.4}$$

*Slow reaction*

Fig. 2.3.6.4 shows the flow sheet for slow reaction. Rate of reaction is faster compared to diffusion of solute; hence rate is determined from the diffusion factors. Specific rate of mass transfer is give using Higbie's theory as:

$$R_A = k_L([A^*] - [A_o]) \tag{2.3.6.5}$$

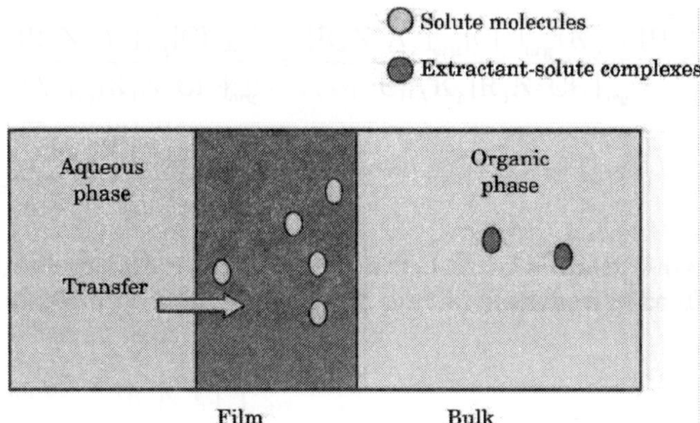


**Fig. 2.3.6.4:** Flow sheet of slow reaction

Where, $R_A$= Specific rate of mass transfer and $k_L = 2(\frac{D_A}{\pi t_E})^{1/2}$, the necessary condition for this regime is

$$k_L a[A^*] \quad << \quad l k_{mn} [A^*]^m [B^o]^n \tag{2.3.6.6}$$

Further the condition that there is practically no reaction in the film is obtained by using the factor that represents the ratio of amount of A reacting in the film to that reacting in bulk, given as square root of M

$$\sqrt{M} = \frac{\sqrt{\frac{2}{m+1} D_A k_{mn} [A^*]^{m-1} [B_o]^n}}{k_L} << 1 \tag{2.3.6.7}$$

*Fast reactions*

Fig. 2.3.6.5 shows the flow sheet for fast reaction. Reaction occurs entirely in the film and rate expression can be written as

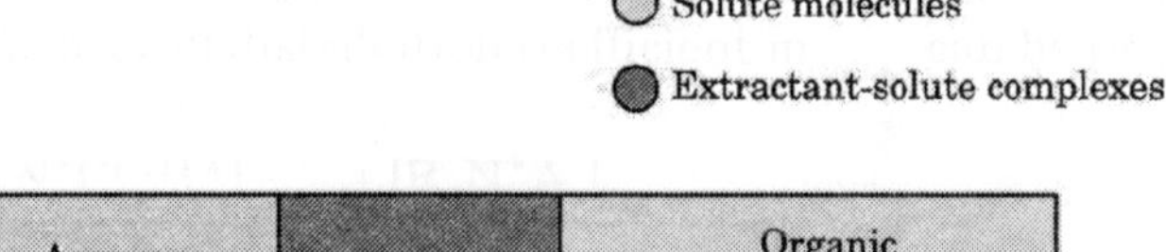


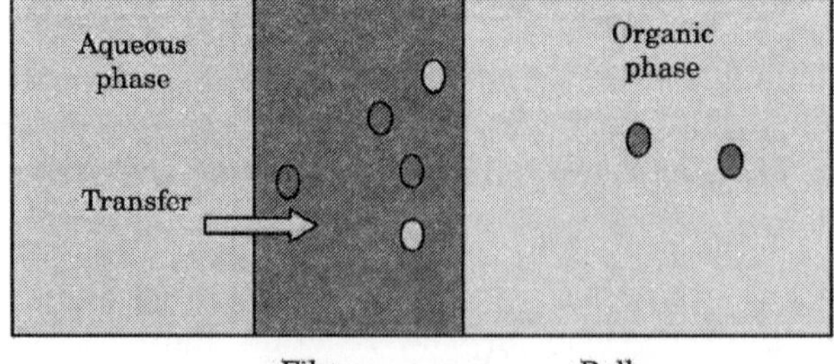


**Fig. 2.3.6.5:** Flow sheet of fast reaction

$$R_A = [A^*] \sqrt{\frac{2}{m+1} D_A k_{mn} [A^*]^{m-1} [B^o]^n} \tag{2.3.6.8}$$

And the condition for the validity is

$$\sqrt{M} = \frac{\sqrt{\frac{2}{m+1} D_A k_{mn} [A^*]^{m-1} [B_o]^n}}{k_L} << 1 \tag{2.3.6.9}$$

*Instantaneous reactions*

Fig. 2.3.6.5 shows the flow sheet for instantaneous reaction. At steady state, the rate of diffusion of dissolved solute A through the film from interface to the reaction plane will be equal to the rate of diffusion of B through the region between the reaction plane and bulk liquid phase. If $D_B$ and $D_A$ are the diffusivities of extractant and solute, the rate law can be written as

$$R_A = k_L [A^*] \left[ 1 + \frac{[B^o]}{[A^*]} \frac{D_B}{D_A} \right] \tag{2.3.6.10}$$

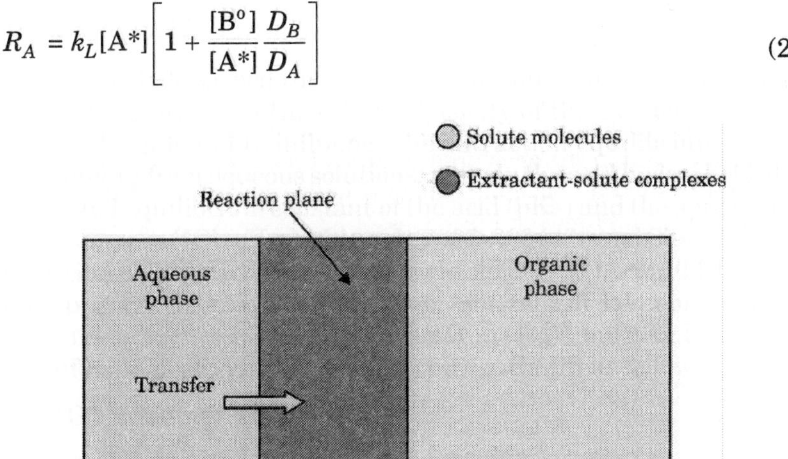


**Fig. 2.3.6.5:** shows the flow sheet for instantaneous reaction

The condition for the validity is given as

$$\frac{\sqrt{\frac{2}{m+1} D_A k_{mn} [A^*]^{m-1} [B^o]^n}}{k_L} >> \frac{[T^o]}{z[A^*]} \sqrt{\frac{D_B}{D_A}} \tag{2.3.6.11}$$

## 2.3.6.5 *Diluent effects: Effect of solvation and solvent dielectric constant on the rate of chemical reaction*

In reactive extraction using basic extractants for the recovery of acidic compounds, since the extraction is ionic in nature (acid-base neutralization), the influence of solvent on the rate of reaction can be explained using electrostatic theory, the diluent being treated as a continuum having dielectric constant å. According to the theory, the logarithm of the rate constant of reaction should vary linearly with the reciprocal of the dielectric constant

$$\ln k = \ln k_o - \frac{z_A z_B e^2}{4\pi\varepsilon_o \varepsilon d_{AB} kT} \tag{2.3.6.12}$$

Where $k_0$ is the value of $k$ in the medium of infinite dielectric constant, in which the electrostatic forces have become zero. The charges of ions are $z_A e$ and $z_B e$ where $z_A$ and $z_B$ are charge numbers and $e$ is the elementary charge ($1.602 \times 10^{-19}$ C). $d_{AB}$ is the centre to centre distance between the ions in the complex and $\mathring{a}_o$ is the permittivity of a vacuum ($8.854 \times 10^{-12}$ C$^2$/Nm$^2$).

The difference in the solvent effects for the different diluents on extraction of acids using ionic extractant can be stated in terms of the extent of solvation of the reactants and the complexes. According to the transition state theory, the rate constant in solution (for reaction say A + B $\leftrightarrow$ C) can be related to that in the gas phase by the equation

$$k = k_g \frac{\alpha_A \alpha_B}{\alpha_C} \tag{2.3.6.13}$$

The reaction between tertiary amine (pyridine) and methyl iodide is the two separate ions and in the activated complex there is partial ionization. Thus in polar diluents such as alcohols, there is better solvation of the complexes. Thus $\alpha_c$ is lowered and hence k increases.

On the other hand in some cases the reaction leads to the formation of stable substances, which decrease in polarity. A polar diluent, like here 1-decanol and 2-octanol, therefore solvates the complex less than the reactants, thus leading to the activity coefficient to be smaller for the reactants than the solvent, and such a diluents thus reduces the rate.

### 2.3.6.6 *Prediction of diffusion term*

Film theory is usually used to explain the partition of the solute to the organic phase. The rate of diffusion is proportional to area and to the driving force expressed in term of concentration difference. Thus $k_L$ can be obtained as

$$\frac{d[\mathrm{A}]_{org}}{dt} = k_L \left[S'\right]([\mathrm{A}]^*{}_{org} - [\mathrm{A}]_{org}) \tag{2.3.6.14}$$

where, $[\mathrm{A}]^*_{org}$ is the equilibrium concentration of acid in diluent only and S' is area. Integration of Eq. (2.3.6.14) yield

$$\ln\left(1 - \frac{[\mathrm{A}]_{org}}{[\mathrm{A}]^*{}_{org}}\right) = -k_L\,(S')\,t + \ln\left(1 - \frac{[\mathrm{A}]_{o;org}}{[\mathrm{A}]^*{}_{org}}\right) \tag{2.3.6.15}$$

subscript 'o' represent initial concentration. Thus the plot of $\ln\left(1 - \frac{[\mathrm{A}]_{org}}{[\mathrm{A}]^*{}_{org}}\right)$ against time $t$ enables determination of $k_L$.

For irreversible reactions, the concentration of A in the bulk of liquid phase is ordinarily zero and for reversible reaction it must be calculated from material balance and chemical equilibrium complexation constant.

$K_L$ is the function of physical properties and geometry of system. Various correlations are also available in literature in terms of dimensionless number such as Reynolds number, Schmidt number and Sherwood number. Sherwood number is expressed as function of Reynolds number and Schmidt number and this allow prediction of mass transfer coefficients. Numerous such equation are available for wetted wall towers, single droplet of liquid moving through the fluid domain and for packed beds. For wetted wall tower, Gilliland-Sherwood equation holds

$$\text{Sh} = 0.023(\text{Re})^{0.81}(\text{Sc})^{0.44} \tag{2.3.6.16}$$

Where, Sh = Sherwood's Number, Re = Reynolds Number and Sc = Schmidt Number

In packed towers correlation of Hobson and Thodos, is useful

$$\log j_d = 0.7683 - 0.9175 \log \text{Re} + 0.0817 (\log \text{Re})^2 \tag{2.3.6.17}$$

where $j_d$ for liquid is

$$j_d = \frac{k_L C_{bm}}{L_m}\left(\frac{\mu}{\rho D_v}\right) \tag{2.3.6.18}$$

2.3.7 *Regenerations*

Regeneration of solute from the loaded organic phase involves reversal of the reaction step, to recover the solute as a product phase and the solute free extractant for recycle. The solute can be back extracted from loaded organic phase using various regeneration techniques. The success of the reactive extraction technology for the recovery of solute from dilute solutions lies if complete recovery of solute from the loaded organic phase could be achieved. Forward and backward extraction steps can be achieved in a single unit by using membrane extractors. The success of a back extraction step is measured using the factor defined as level of back-extraction (LE%) as:

$$LE\% = \frac{[A]_{aq}}{[A]_{org}} * 100 \tag{2.3.7.1}$$

where [A] refers to solute concentration and subscript 'org' and 'aq' represents organic and aqueous phase concentrations, respectively. It is desirable that LE% should be high so that more of the solute is transferred from organic phase to aqueous phase in the back extraction step. A term equivalent of distribution coefficient ($K_D^{BE}$) in the forward is also used to measure the extent of back extraction and is defined as

$$K_D^{BE} = \frac{[A]_{aq}}{[A]_{org}} \tag{2.3.7.2}$$

2.3.7.1 *Various regeneration strategies*

(a) *Distillation*

Distillation is a widely used separation technology for removal of volatile compound. However as discussed in Section 1.4, it suffer from few shortcoming when applied to removal of solute from dilute systems or removal of non volatile solute. Still in many cases in order to apply distillation as the back extraction step it is preferable to use a solvent in forward step (such as hexane) which is more volatile in comparison to the solute. However, the major shortcoming in this regard is the cost factor as large amount of heat is required to removal the large quantity of solvent. Distillation cannot be utilized for regeneration when the solute is not sufficiently volatile (*e.g.* dicarboxylic and hydroxy-carboxylic acids).

(b) *Using base*

In this method of extraction, the loaded organic phase is contacted with a base such as ammonia, amine or sodium hydroxide solution. A small volume of base is usually employed so as to get the concentrated solution of solute in the back extractant (base). Laboratory studies usually employ base in stoichiometric amounts. Usually these bases results in 100% recovery of solute in all the cases. Though the method of recovery is economical and resulted in high product concentration, however, the solute is needed to be recovered. When sodium hydroxide is used for back-extraction, acid is recovered as sodium salt (*e.g.* in case of carboxylic acid extraction as sodium carboxylate). Therefore an appropriate solvent such as acid (*e.g.*, sulfuric acid) must be added to return it to the free form. This approach has the same drawbacks as the classical calcium precipitation process for direct recovery from the aqueous feed. Both sulphuric acid and NaOH are consumed, and a waste salt is formed, which requires disposal. Ammonia when used as back-extractant form amides when heated for acid removal. These amides are sufficiently stable and the process in difficult to reverse back. However, if the solute is a waste and not to be reused, the use of ammonia is still feasible as a concentrated mass of the waste solute is formed which is then disposed off.

The regeneration of organic acid has been studied in literature using a stronger volatile amine like TMA in aqueous phase (Poole and King, 1991, Keshav *et. al.*, 2010). The advantage of this regeneration technique is that it avoids consumption of chemicals and creation of salt by-product. The aqueous base, which is volatile, enables thermal decomposition of the acid–base complex in the aqueous back-extract. The decomposition forms carboxylic acid as a product and freebase as a vapor that can be reabsorbed in water and recycled for reuse in back-extraction.

(c) *Using HCl*

Mostly when a basic extractant is used in the forward step of extraction for the removal of acidic solute it can be reversed back by employing a strong acid to which the base has more attraction than the earlier one. In such cases concentrated HCl has been undertaken in the studies on back-extractions. More than stoichiometric amounts of HCl were necessary to recover most of the product from the solvent. The advantage this process is that the acid is recovered in undissociated form. It is possible to regenerate the solvent by distillation. However, the disadvantage of this technique is that aqueous HCl is highly corrosive and requires special material of construction (glass lined/ graphite). Further, in some cases the emulsion formed with the base and strong acid is very strong and cannot be reversed back to regenerate the solvent.

(d) *Temperature-swing regeneration*

Extraction was found to decrease with increase in temperature and this concept can be used to regenerate the solute from loaded organic phase. During regeneration, the extract is contacted with a fresh aqueous stream at a higher temperature to produce a solute-laden aqueous product stream and an solute-free solvent. The concentration of the solute achievable in this stream depends on the amount of change in the extraction equilibrium between temperatures and can be higher than that in the original aqueous feed stream.

(e) *pH swing*

Considering the case of recovery of acidic mass, the presence of acid in the solution as dissociated or undissociated form is function of pH of the system, at pH less than pKa of the acid, acid is present in undissociated form and as pH is raised above pKa value, various type of dissociations are encountered. Further, many extractant are effective for removal of undissociated form of acid and fails when pH of the system is above the pKa of the respective acid. This pH swing is another possibility for regeneration of acids. If the pH of the aqueous solution is increased above the pKa of the acid, back extraction is promoted. Altering the pH is usually done by adding a base. The solute is removed in the form of salt. It is required to add another acid to obtain the solute as a product from salt. This process therefore has the same drawbacks as using sodium hydroxide for back-extraction *i.e.* consumption of extra chemicals and formation of a waste salt.

(f) *Diluent swing*

Extractant employed for the forward step of reactive extraction are usually viscous or solid and their physical properties are improved using diluents which also help in solvation of the complexes formed by them with the solute. Extraction using a particular extractant is function of the diluent chosen. This concept was employed for back-extraction of solute from loaded organic phase. This technology is called as diluent swing regeneration in which, shift of the equilibrium distribution of solute from the aqueous phase to the organic phase between forward and back extraction is caused by a change in composition of the diluent with which the extractant is mixed. The composition of the solute laden organic phase leaving the extractor is altered, by either removal of the diluent or addition of another diluent, to produce a solvent system that promotes distribution of the solute to the aqueous phase. The altered organic phase is contacted with a fresh aqueous stream in the regenerator to produce the aqueous concentrated solute. The major shortcoming of this approach is that employing more than one diluent makes the designing of the back-extraction step a difficult step. This process has the disadvantage of diluting the extract stream and requiring distillation of large amounts of solvent (after the regeneration) to obtain the same shift in the active/inert diluent ratio. Expenses encountered in employing the distillation step are another factor that limits the application of this technique on large scale.

### 2.3.7.2 *Regeneration using extraction*

Using a base or HCl or sodium salts of chloride, hydroxide, sulphate, nitrate and carbonate can be utilized as the back extractants. The anion exchange reactions between the salts or base and the solute-extractant complex provide for the regeneration of the extractant in the organic phase as well as to obtain the extractant free solute in the aqueous phase.

These salts/bases are better stripping phase than ordinary deionized water since they enhances the transport and the dissociation reaction rates of the solute-extractant complex. When salt is employed, the solute (acid) is released from the complex according to following reaction:

$$Na^+ + H^+X^- \leftrightarrow Na^+X^- + H^+ \qquad (2.3.7.3)$$

The transport rate of the acid is enhanced at higher concentrations of the salts, since the concentration of the undissociated acid is lowered. Back extraction of solute acid using NaOH resulted in 100% recovery and is simple and cheap. However, the acid is recovered as sodium salt and an appropriate acid (*e.g.*, sulphuric acid) must be added to return it to the free acid form. This approach has the drawbacks of using chemical such as sulphuric acid and NaOH, and a waste salt formation, which requires disposal.

The advantage of regeneration technique using TMA is that it avoids consumption of chemicals and creation of salt by-product. The aqueous base, which is volatile, enables thermal decomposition of the acid–base complex in the aqueous back-extract. The decomposition forms acid as a product and freebase as a vapour that can be reabsorbed in water and recycled for reuse in back-extraction. However, an important matter of concern is that the organic phase, which is recycled to the extraction unit, may contain residual dissolved TMA, which can affect the performance of the unit.

In some cases, employing extraction for regeneration of solute results in emulsion formation which is to be broken using centrifugation technique. Higher concentration of stripping solution always increases the level of back-extraction. However, the solubility limitation is the factor deciding the use of particular concentration of stripping agent. Another factor which could affect the level of back-extraction is the mutual contribution of the specific diluent in the process. In some cases for a fixed solute, extractant and stripping solution, back-extraction is affected by the presence of particular diluent in the organic phase.

### 2.3.7.3 *Extraction regeneration kinetics*

In the back extraction studies, since ionic reactions occur very fast, the regeneration step should be mass transfer controlled. The kinetics studies could be performed in the similar fashion as described in section 2.3.6 and studies aims for finding the regime of reaction and rate parameters. It is desirable that the reaction should be fast so that smaller residence times are required to carry out the regeneration step in continuous towers.

## 2.3.8 *Modeling of reactive extraction process*

*Equilibrium Modeling*

The reactive liquid- liquid extraction equilibria for carboxylic acid (HA) recovery using tertiary amine, quaternary amine and organophosphorous compounds has been explained in Section 2.3.3. The complexation were dependent on number of –COOH group in a particular carboxylic acid. For 1:1 solute-extractant complexations, Eq. (2.3.3.2.1) is applicable (Kertes and King, 1986)

$$HA_{aq} + B_{org} \leftrightarrow BHA_{org} \tag{2.3.3.2.1}$$

The distribution coefficient, $K_D$ is defined by:

$$K_D = [HA]_{org}/[HA]_{aq} \tag{2.3.3.2.2}$$

The equilibrium complexation constant, $K_E$ is defined as:

$$K_E = [BHA]_{org}/[HA]_{aq}[B]_{org} \tag{2.3.3.2.3}$$

The concentration of free amine in the organic phase would be:

$$[B]_{org} = [B]_{i,org} - [BHA]_{org} \tag{2.3.3.2..4}$$

If (1:1) carboxylic acid – amine complex is formed then $[HA]_{org} = [BHA]_{org}$,

$$\log K_D = \log K_E + \log [B]_{org} \tag{2.3.3.2.5}$$

If the former assumptions are valid, a plot of log $K_D$ *vs* log $[B]_{org}$ should yield a straight line with a slope of unity.

If the exact equilibria is not known and more other than (1:1) solute-extractant complexations are formed, the equilibria could be represented as Eq. (2.3.3.2.6 - 2.3.3.2.8) as:

$$HA_{aq} + B_{org} \xleftrightarrow{K_{11}} BHA_{org} \tag{2.3.3.2.6}$$

$$HL_{aq} + BHA_{org} \xleftrightarrow{K_{21}} B(HA)_{2,org} \tag{2.3.3.2.7}$$

$$\vdots$$

$$HA_{aq} + B(HA)_{n-1,org} \xleftrightarrow{K_{n1}} B(HA)_{n,org} \tag{2.3.3.2.8}$$

The equilibrium complexation constant for the reaction represented by above equations is:

$$K_{En} = \frac{[B(HA)_n]_{org}}{[B]_{org}[HA]_{aq}^{\;n}} \tag{2.3.3.2.9}$$

$K_{En}$ = Equilibrium complexation constant

The extent to which the organic phase (Amine + Diluent) can be loaded with carboxylic acid is expressed as the loading ratio, z;

$$z = \frac{[HA]_{org}}{[B]_{i,org}} \tag{2.3.3.2.10}$$

The value of z depends on the extractability of the acid (strength of the acid-base interaction) and its aqueous concentration. The stoichiometry of the overall extraction reaction depends on the loading ratio in the organic phase, *z*. If the organic phase is not highly concentrated *i.e.* at very low loading ratios (*z* < 0.5) the (1:1) complex is formed and the following equation holds

$$\frac{z}{1-z} = K_{E(1:1)}[HA]_{aq} \tag{2.3.3.2.11}$$

For higher loading ratios, the (2:1) complex is formed, and a plot of z/(2-z) *vs* $[HA]_{aq}^{2}$ should yield a straight line, whose slope gives the complexation constant ($K_{E(2:1)}$) for the (2:1) complex.

$$\frac{z}{2-z} = K_{E(2:1)}[HA]_{aq}^{2} \tag{2.3.3.2.12}$$

If the propionic acid concentration is high enough, the (3:1) complex may be formed and a plot of z/(3-z) vs $[HA]_{aq}^{3}$ should yield a straight line, whose slope gives the complexation constant ($K_{E(3:1)}$) for the (3:1) complex.

$$\frac{z}{3-z} = K_{E(3:1)}[HA]_{aq}^{3} \tag{2.3.3.2.13}$$

Thus (1:1), (1:2) and (1:3) complexations could be predicted using Eq. (2.3.3.2.11-2.3.3.2.13). If higher complexations or aggregate formation happen, the overall equilbria can be modeled using the following equations:

Distribution of non-dissociated acid

$$nH_3A + m(B)_{org} = ((B)_mH_{3\text{-}m}A)_{org} \qquad m = 1, 2, 3 \tag{2.3.8.1}$$

Distribution of dissociated acid

$$nH^+ + H_{3\text{-}n}A^{n-} + mB = ((B)_mH_{3\text{-}m}A)_{org} \qquad m = 1, 2, 3 \tag{2.3.8.2}$$

Both sets of equations are characterized by the thermodynamic extraction constant

$$K_E = \frac{[(B)_m(H_3A)_n]_{org}}{[H_3A]_{aq}^{n}[B]_{org}^{m}} \tag{2.3.8.3}$$

Where, D = Distribution coefficient

If assuming n = 1, the above the distribution coefficient can be defined in terms of equilibrium complexation constant as

$$D = \frac{[(B)_m H_3A]_{org}}{[H_3A]_{aq}} = K_E[B]^m_{org} \quad (2.3.8.4)$$

Taking log

$$\log D \; = \; \log K_E \; + \; m[B]_{org} \quad (2.3.8.5)$$

If n is not 1, a general model could predict the various types of $K_E$ values as (Bizek *et al.*, 1992):

$$[H_3A]_{org} = \sum_{m=1}^{\alpha} \sum_{n=1}^{\beta} mK_{E(n:m)}[B]^m_{org}[H_3A]^n_{aq} \quad 1<m<\alpha; 1<n<\beta \quad (2.3.8.6)$$

The equilibrium concentration of "free" amine $R_3N$ can be calculated from the balance equation of amine:

$$[B^o]_{org} = [B]_{org} + \sum_{m=1}^{\alpha} \sum_{n=1}^{\beta} mK_{E(n:m)}[B]^m_{org}[H_3A]^n \quad (2.3.8.7)$$

*Kinetic Modeling*

*Monocarboxylic acid – Quaternary ammonium salt*

The extraction equilibria of carboxylic acids that were extracted with phosphoric and amine extractants have been reported by many researchers. However, only a few have reported studies on the extraction kinetics of these systems. Ion exchange reaction using a quaternary amine at the interface was assumed between the organic phase and aqueous phase as follows:

$$A^- + Q^+Cl^- = Cl^- + Q^+ A^- \quad (2.3.8.8)$$

$$K^* = (C^*_{aq.Cl} \; C^*_{oQA})/C^*_{aq,A} \; C^*_{oQCl} \quad (2.3.8.9)$$

Where, $Q^+Cl^-$ and $A^-$ represents the TOMAC (a quaternary amine) and carboylate and $K^*$, the extraction constant. $C_{mn}{}^*$ represents the equilibrium concentrations of each components. Mass balance in the batch operation are expressed as follows according to Eq. (3.3.8.8) because the volumes of aqueous and organic phases are identical:

$$C_{oQCL} = C_{oQCL,\,0} - C_{oQA\,=}\, C_{oQCL,\,0} - (C_{aqA,0} - C_{aqA}) \quad (2.3.8.10)$$

$$C_{aqCl} = C_{aqCl,0} + (C_{aqA,0} - C_{aqA})$$

$$C_{oQA} = C_{aqA,0} - C_{aqA} \quad (2.3.8.12)$$

The rate of acid disappearance in the aqueous phase and TOMAC in the organic phase, and rate of chloride appearance in the aqueous phase and the TOMAC - Carboxylate in the organic phase are given as follows

$$-d\, C_{aqA}/dt = k_{\,aqA}\; a\, (C_{aqA} - C_{iA}) \quad (2.3.8.13)$$

$$-d\, C_{oQCl}/dt = k_{oQCl}\; a\, (C_{oQCl} - C_{iQCl}) \quad (2.3.8.14)$$

$$-d\, C_{aqCl}/dt = k_{aqCl}\; a\, (C_{iCl} - C_{aqCl}) \quad (2.3.8.15)$$

$$-d\, C_{oQA}/dt = k_{oQA}\; a\, (C_{iQA} - C_{oQA}) \quad (2.3.8.16)$$

At quasi steady state, these rates are equal and then:

$$k_{aqA}\, a\, (C_{aqA}-C_{iA}) = k_{oQCl}\, a\, (C_{oQCl}-C_{iQCl}) = k_{aqCl}\, a\, (C_{iCl}-C_{aqCl}) = k_{oQA}\, a\, (C_{iQA}-C_{oQA}) \tag{2.3.8.16}$$

The ratio of the mass transfer coefficient in the aqueous phase and that in the organic phase are replaced with β and γ to reduce the number of parameters to be determined.

$$\beta = k_{aqA}/k_{aqCl}\ ;\ \gamma = k_{oQCl}/k_{oQA} \tag{2.3.8.17}$$

The extraction equilibrium prevails at the interface and the constant K* is expressed as:

$$K^* = \frac{C_{iCl}C_{iQA}}{C_{iA}C_{iQCl}} \tag{2.3.8.18}$$

By combining all the equations

$$dC_{aqA}\, dt = a\{-(p/2) + (\frac{p^2}{4-q})^{1/2}\} \tag{2.3.8.19}$$

where:

$$p = (k_{aqA}\, C_{aqA} + k_{oQCl}\, (1\text{-}K^*/\beta\,\gamma) + k_{aqA} C_{aqCl}\, (\gamma/K^* - 1/\beta) + k_{oQCl}\, C_{oQA}\, (\beta/K^* - 1/\gamma)$$

$$q = k_{aqA}\, k_{oQCl}\, [C_{aqA}\, C_{oQCl}(1\text{-}\ K^*/\beta\,\gamma) + C_{aqCl}\, C_{oQA}\, (1/\beta\,\gamma - 1/K^*)]$$

*Dicarboxylic Acid – Tertiary Amine*

Poposka *et al.* (2000) studied the kinetic and equilibrium data in a vigorously agitated system, neglecting the effect of transport phenomena on the process rate and obtained two types of chemical kinetic models.

(a) A formal elementary kinetic model

Considering the overall complexation reaction between the acid and amine,

$$a[H_2A]_{aq} + [NR_3]_{org} = \{[H_2A]_a[NR_3]\}_{org}\ .\ K_E = \frac{c_{complex}}{c^a{}_{AA} \bullet c_B} \tag{2.3.8.20}$$

the general form of rate equation may be written as

$$-r_A = -dc_{AA}/dt = k_1\, c^{\alpha}{}_{AA} c^{\beta}{}_{B} - k_2 c^{\gamma}{}_{C} \tag{2.3.8.21}$$

Where $k_1$,$k_2$, α, β, γ are the model parameters to be determined and $c_{AA}$, $c_B$ and $c_C$ are the actual concentrations of the acid, amine and complex at time t. For any time t of a run with initial values, the instantaneous values of $c_B$ and $c_C$ are calculated for the respective aqueous acid molarity, $c_{AA}$, from a material balance.

(b) Kinetic model of the chemical complexation as a mechanism of three reactions assuming simultaneous formation of three complexes: (1:1), (1:2) and (2:1) (Tomovska *et al.* 1998), the stoichiometry of acid extraction with extractant dissolved in an solvent, can be written as

$$[H_2A]_{aq} + [NR_3]_{org} = \{[H_2A][NR_3]\}_{org} \tag{2.3.8.22}$$

$$[H_2A]_{aq} + 2[NR_3]_{org} = \{[H_2A][NR_3]_2\}_{org} \tag{2.3.8.23}$$

$$[H_2A]_{aq} + \{[H_2A]_2[NR_3]\}_{org} = \{[H_2A]_2[NR_3]\}_{org} \tag{2.3.8.24}$$

Assuming the reaction to be elementary, the rates of the individual reactions $r_1$, $r_2$ and $r_3$ will be

$$-r_1 = -dc_{11}/dt = k_{11}\,c_{AA}c_B - k_{12}c_{11} \tag{2.3.8.25}$$

$$-r_2 = -dc_{12}/dt = k_{21}\,c_{AA}c^2_B - k_{22}c_{12} \tag{2.3.8.26}$$

$$-r_3 = -dc_{21}/dt = k_{31}\,c_{AA}c_{11} - k_{32}c_{21} \tag{2.3.8.27}$$

where $C_{AA}$ and $C_B$ is the equilibrium concentration of acid and extractant, $C_{11}$, $C_{12}$ and $C_{21}$ are concentrations of the (1, 1), (1, 2) and (2, 1) complexes.

The number of participants in the complexation is five and the number of individual reactions is three. The remaining two relations are provided for by the mole balance of the acid and amine in the organic phase.

$$c_{Ao} = c_{11} + c_{12} + 2c_{21} \tag{2.3.8.28}$$

$$c_B = c^o_B - (c_{11} + c_{12} + 2c_{21}) \tag{2.3.8.29}$$

## BACK EXTRACTION MODELING

Wasewar *et al.*, (2004), described the modeling for back-extraction of lactic acid from loaded organic phase of Alamine 336 in octanol at 30°C. The complexation of lactic acid with Alamine 336 in octanol is described as follows

$$[HA]_{aq} + [B]_{org} \leftrightarrow [HLB]_{org}; \tag{2.3.8.30}$$

$$2\,[HA]_{aq} + [B]_{org} \leftrightarrow [(HL)_2B]_{org}; \tag{2.3.8.31}$$

where HA denotes the unionized acid, B denotes the extractant (Alamine 336) and $K_{pq}$ values are the equilibrium constants for complex formation. Partition coefficient of lactic acid between water and octanol is given as

$$P = [HA]_{org}/[HA]_{aq} \tag{2.3.8.32}$$

Where, P = Partial coeffecient

The equilibria in the aqueous phase are described as follows (Poole and King, 1991):

$$HA \leftrightarrow H^+ + A^-; \tag{2.3.8.33}$$

$$R_3NH^+ \leftrightarrow R_3N + H^+; \tag{2.3.8.34}$$

$$[H^+]_{aq} + [R_3NH^+]_{aq} = [L^-]_{aq} + [OH^-]_{aq}; \tag{2.3.8.35}$$

where $R_3N$ denotes TMA. The concentration of unionized acid in the aqueous phase can be expressed as

$$[HA]_{aq} = \alpha[HA]_{aq;T}; \tag{2.3.8.36}$$

where

$$\alpha = 1/(1 + K_{a,acid}/[H^+]_{aq}) \quad (2.3.8.37)$$

Mass balance for lactic acid can be written as follows: Total acid in system = Acid in organic phase + Acid in aqueous phase

$$[HA]_T = K_{11}[B]_{org}[HA]_{aq} + 2K_{21}[B]_{org}[HA]^2_{aq} + P\Phi[HA]_{aq} + R[HA]_{aq}/\alpha \quad (2.3.8.38)$$

where $[B]_{org}$ is un-complexed Alamine 336 in organic phase, $\Phi$ is the volume fraction of diluent (octanol) in organic phase and R is the ratio of aqueous phase to organic phase used for back extraction.

Mass balance for Alamine 336 essentially present in the organic phase can be written as follows:

Total Alamine 336=Uncomplexed Alamine 336 + Complexed Alamine 336,

$$[B]_{org,T} = [B]_{org} + K_{11}[B]_{org}[HA]_{aq} + K_{21}[B]_{org}[HA]^2_{aq} \quad (2.3.8.39)$$

Substituting Eq. (2.3.8.39) into Eq. (2.3.8.38) gives a mass balance equation for lactic acid in terms of only two unknown values, the pH ($H^+$ concentration) of aqueous raffinate phase. This equation can be solved for $[HA]_{aq}$ at the set values of pH. For each pH values, the recovery of the acid into the aqueous phase can be calculated as

$$\text{Recovery} = \{R[HA]_{aq}/\alpha\}/[HA]_{org,int} \quad (2.3.8.40)$$

The ratio of moles of TMA to moles of acid originally present in the organic phase can be calculated as

$$[R_3N]_{aq,T} = [R_3N]_{aq} + [R_3NH^+]_{aq} = [R_3NH^+]_{aq}/\beta \quad (2.3.8.41)$$

where $\beta = 1/\{1 + K_{a,\ R3NH+}/[H^+]_{aq}\}$ (2.3.8.42)

*Modeling for Cascade Extraction Units*

This model was explained for the extraction of lactic acid using amine. The reaction scheme of lactic acid extraction using amine extractant can be written as:

$$[HL]_{aq} \xleftrightarrow{K_{eq}} [HL]_{org} \quad ; \quad (\text{Fast}); Keq = \frac{[HL]_{org}}{[HL]_{aq}}$$

$$[HL]_{aq} + [B]_{org} \xleftrightarrow{K_{11}} [BHL]_{org} \quad ; \quad (\text{Slow}); K_{11} = \frac{[BHL]_{org}}{[B]_{org}[HL]_{aq}}$$

$$[HL]_{aq} + [BHL]_{org} \xleftrightarrow{K_{21}} [B(HL)_2]_{org} \quad ; \quad (\text{Fast}); K_{21} = \frac{[B(HL)_2]_{org}}{[BHL]_{org}[HL]_{aq}}$$

(2.3.8.43)

The rate of complexation reaction is defined as:

$$R_F = k_1'\left([HL]_{org}[B]_{org} - \frac{[BHL]_{org}}{K_{11}'}\right) \quad (2.3.8.44)$$

$$R_F = k_1'\left(K_{eq}'[HL]_{aq}[B]_{org} - \frac{[BHL]_{org}}{K_{11}'}\right) \quad (2.3.8.45)$$

where, $k_1$' and $K_{11}$' are the concentration based forward rate and reaction equilibrium constants for the complexation reaction. The rate expression in terms of mole fraction can be written as:

$$R_F = k_1\left(\frac{x_{n,HL}}{1+10^{\left(pH-pK_A^{HL}\right)}} - y_{n,B} - \frac{y_{n,BHL}}{K_{11}}\right) \quad (2.3.8.46)$$

The xi and yi are the mole fractions of species *i* in the aqueous and the organic phases, respectively.

The steady state model for the cascade (Fig. 2.3.8.1) consisting of reactor extractors in series is shown below.

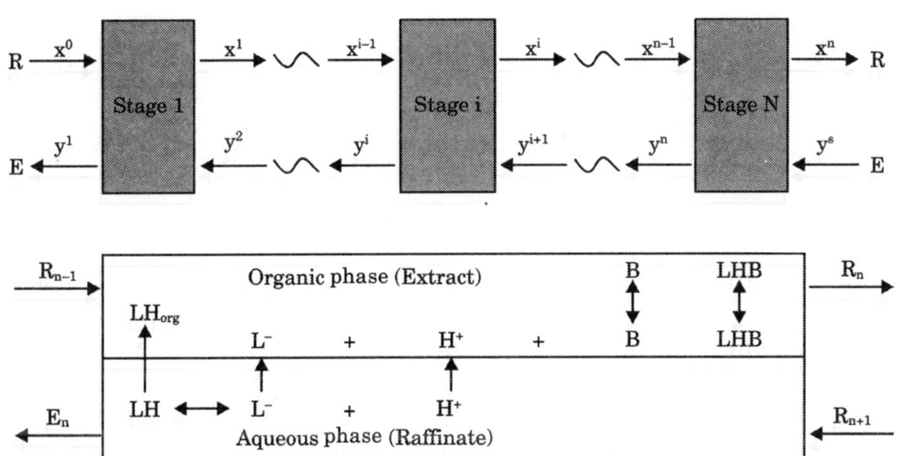


**Fig. 2.3.8.1:** Countercurrent cascade

*Assumptions*

1. The temperature and pressure of the reactors extractors are constant and the heat of the complexation reactions are neglected.
2. The dissociation of product is unaffected by the presence of other components.
3. The output streams from each reactor are in phase equilibrium.
4. The complexation reaction takes place at the interface.
5. The amine extractant is essentially insoluble in the aqueous phase.
6. The product amine complex is insoluble in the aqueous phase.
7. Undissociated product is extracted physically as by reaction into the organic phase.
8. The feed consists of product in pure water and the organic solvent enters as a mixture of an organic carrier and the amine. No product amine complex is present in the entering solve.

It is convenient to use dimensionless extract ($e_n$) and raffinate ($r_n$) flow rates.

$$e_n = \frac{\hat{E}_n}{S} = \frac{E_n + E_{n,HL} + E_{n,B} + E_{n,BHL}}{S_c + S_B} \quad (2.3.8.47)$$

$$r_n = \frac{\hat{R}_n}{F} = \frac{R_n + R_{n,HL}}{F_w + F_{HL}} \quad (2.3.8.48)$$

where $S_c$ and $F_W$ are the molar flow rates of the organic carrier and water feed, excluding lactic acid

*The phase equilibrium and component material balances*

For the n[th] reactor extractor of cascade, the component material balance is written as:

$$E_{n+1}\overline{y_{n+1}} + R_{n-1}\overline{x_{n-1}} = E_n\overline{y_n} + R_n\overline{x_n} \quad (2.3.8.49)$$

in terms of dimensionless numbers

$$\left(\frac{S}{F}\right)e_{n+1}\left\{\left(1 - y_{n+1,B} - y_{n+1,HL} - y_{n+1,BHL}\right)\vec{y}_{n+1}\right\} + r_{n-1}\left\{\left(1 - x_{n-1,HL}\right)\vec{x}_{n-1}\right\} =$$

$$\left(\frac{S}{F}\right)e_n\left\{\left(1 - y_{n,B} - y_{n,HL} - y_{n,BHL}\right)\vec{y}_n\right\} + r_n\left\{\left(1 - x_{n,HL}\right)\vec{x}_n\right\} \quad (2.3.8.50)$$

the phase equilibrium relations are

$$\overline{y_n} = f_{eq}.(\overline{x_n}) \quad (2.3.8.51)$$

The composition $\overline{y_n}$ *and* $\overline{x_n}$ describes the two components of the carrier-water system $E_n$ and $R_n$ are the molar flow rates of the extract and raffinate stream. The mole fraction of the m components in each phase sum to unity.

$$\sum_{i-1}^{m}\overline{y_{i,n}} = 1 \quad (2.3.8.51)$$

$$\sum_{i=1}^{m}\overline{x_{i,n}} = 1 \quad (2.3.8.52)$$

*Component balance for the acid*

$$E_{n+1,B} - H_n k_1 \underline{a}\left(\frac{x_{n,HL}}{1 + 10^{\left(pH - pK_A^{HL}\right)}} \cdot y_{n,B} - \frac{y_{n,BHL}}{K_{11}}\right) = E_{n,B} \quad (2.3.8.53)$$

*Component balance for the tertiary amine*

$$r_{n-1,HL} + H_n k_1 \underline{a}\left(\frac{x_{n,HL}}{1 + 10^{\left(pH - pK_A^{HL}\right)}} \cdot y_{n,B} - \frac{y_{n,BHL}}{K_{11}}\right) = R_{n,HL} + E_{n,HL} \quad (2.3.8.54)$$

*Component balance for the lactic acid-amine complex*

$$E_{n+1,BHL} - H_n k_1 \underline{a}\left(\frac{x_{n,HL}}{1+10^{\left(pH-pK_A^{HL}\right)}}\cdot y_{n,B} - \frac{y_{n,BHL}}{K_{11}}\right) = E_{n,BHL} \qquad (2.3.8.55)$$

Inserting Eq. (2.3.8.53) and (2.3.8.54) in the above equations, the component balance for the tertiary amine, lactic acid and lactic acid-amine complex can be written as

$$\left(\frac{S}{F}\right)e_{n+1}y_{n+1,B} - Da\kappa\omega_n\left(\frac{x_{n,HL}}{1+10^{\left(pH-pK_A^{HL}\right)}}y_{n,B} - \frac{y_{n,BHL}}{K_{11}}\right) = \left(\frac{S}{F}\right)e_n y_{n,B} \qquad (2.3.8.56)$$

$$r_{n-1}x_{n-1,HL} + \left(\frac{S}{F}\right)e_{n-1}y_{n+1,HL} - Da\kappa\omega_n\left(\frac{x_{n,HL}}{1+10^{\left(pH-pK_A^{HL}\right)}}\cdot y_{n,B} - \frac{y_{n,BHL}}{K_{11}}\right) =$$

$$= r_n x_{n,HL} + \left(\frac{S}{F}\right)e_n y_{n,HL} \qquad (2.3.8.57)$$

$$\left(\frac{S}{F}\right)e_{n+1}y_{n+1,BHL} - Da\kappa\omega_n\left(\frac{x_{n,HL}}{1+10^{\left(pH-pK_A^{HL}\right)}}\cdot y_{n,B} - \frac{y_{n,BHL}}{K_{11}}\right)$$

$$= r_n x_{n,HL} + \left(\frac{S}{F}\right)e_n y_{n,BHL} \qquad (2.3.8.58)$$

for n = 1,2,..n. and where Da = Domköhler number, $\omega_n$ = fraction of total hold up in the nth reactor extractor= $H_n/HT$ and $\kappa$ = the dimensional rate constant = $k_F/k_{F,ref} = 10^{(7.0-pH)}$.

The Damkohler number can be varied by changing the residence time or the specific interfacial area. The interfacial area is related to the Sauter mean diameter (d32) for droplets of the dispersed phase by

$$\underline{a} = \frac{6\varphi_{disp}}{\rho_{disp}d_{32}} \qquad (2.3.8.59)$$

Where, $\underline{a}$ = Sauter mean diameter, $\Phi_{disp}$ and $\rho_{disp}$ are the phase fraction and the molar density respectively

For physical extraction of lactic acid, phase equilibrium relationships are given as:

$$y_{n,HL} = K_{phy}x_{n,HL} \qquad (2.3.8.60)$$

where $K_{phy}$ is the distribution coefficient of lactic acid. The extract and raffinate flow rates are related to the phase fractions obtained by solving the phase equilibrium problem as

$$\frac{e_n}{r_n} = \frac{\varphi_n\left(1-x_{n,HL}\right)}{\left(\frac{S}{F}\right)\left(1-\varphi_n\right)\left(1-y_{n,B}-y_{n,HL}-y_{n,BHL}\right)} \qquad (2.3.8.61)$$

where $\Phi_n$ is the organic phase fractions; $\Phi_n = En/(E_n + R_n)$.

The fractional recovery of acid is given by

$$f_{HL} = \frac{x_{0,HL} - r_N x_{N,HL}}{x_{0,HL}} \quad (2.3.8.62)$$

**Table 2.3.8.1:** Variables and parameters

| | | |
|---|---|---|
| Parameters that are specified: | $\overline{y_{n+1}}$, $\overline{x_o}$, N, (S/F), Da, xo, HL, yN+1, HL, yN+1, B, y N+1, BHL and pH | 10 |
| Variables | $\overline{y_n}$, $\overline{x_n}$, en, rn, xn, HL, yn, HL, yn, B, yn, BHL | 2mN + 6N + 2 |
| Equations | 'm' independent material balance equations for N reactors (Eq.4), 'm' independent equilibrium equations for N reactors (Eq.5), mole fraction of m components in each phase summed to unity {2 (N+1)}, (Eq. 6,7), Individual component balances (3N), (Eq. 8-10), Phase equilibrium relationship (N) (Eq. 11) | 2mN + 6N + 2 |
| | Degree of freedom | 0 |

The number of parameters and variables are given in Table 2.3.8.1 and the degree of freedom was found to be 0. The phase equilibrium problem for the entire cascade can be solved according to the geometric method given by Minotti *et al.* (1996). The liquid-liquid equilibrium compositions and phase fractions in each of the N reactor-extractors can be calculated using the method given by Wasylkiewicz *et al.* (1996). The liquid-phase non idealities could be modeled with the UNIQUAC equation based on the parameters in Table. The equations can be solved using a nonlinear equation solver to obtain the mol fraction of the reactive components. The procedure will quantifies the viability of the reactive extraction process and estimates ranges for the design variable including the influence of the cascade configuration, number of reactor extractor stages, solvent to feed ratio, solvent composition, pH of the extraction and the Damkohler number.

## Linear Solvation Energy Relationship Modeling

The linear solvation energy relationship (LSER) that measures property XYZ, in terms of solvent properties (Kamlet *et al.*, 1983), was defined as:

$$XYZ = XYZ^0 + \text{solvent polarity/polarizability effect} + a\alpha + b\beta \quad (2.3.8.63)$$

$a$ and $b$ are the susceptibilities of XYZ to changing solvent HBD acidity and hydrogen bond acceptance (HBA) basicity, respectively and β is the HBA basicity of the solvent (Kamlet and Taft, 1976). Values of α have been determined by the use of the solvatochromic comparison method. (Taft and Kamlet, 1976; Kamlet and Taft, 1976). In this method, indicator solutes are employed for which XYZ has a negligible susceptibility to the solvent HBA basicity ($b \approx 0$) but a high one for the solvent HBD acidity (large a). The $(XYZ)_0$ + solvent polarity/polarizability effect values are obtained for the indicator in solvents devoid of HBD abilities, so that a α for HBD solvents can be calculated from the values of XYZ in the latter kind of solvents. Finally, an averaging procedure for several indicator solutes and a normalization procedure are used in order to obtain values of o~ that are in the range of 0 to 1. Many linear solvation energy correlations involving the solvent HBD acidity scale have been reported and the limitations of this scale have been discussed. (Kamlet *et al.*, 1983; Taft and Kamlet, 1976).

*Interactions of various types*: dipole-dipole and or dipole-induced dipole, hydrogen bonding etc. are responsible for the extraction of carboxylic acid from aqueous solutions. The general linear solvation energy relationship (LSER) is a comprehensive tool to include the above mentioned parameters of the diluents. Schmidt (1980) presented a systematic study on the effect of diluents on amine extraction of acids. He distinguished among three types of solvation: solvation of acid anion, alkylammonium cation or complex as a whole-and found that the solvation of cation was not common in the case of tertiary amines. The solvation of the whole amine: acid complex is based on dipole-dipole interaction and was found to play an important role in the neutralization reactions between acid and amine which are promoted by the increasing polarity of the diluent. Empirical parameters, *DP**, have been introduced to quantify this influence according to the equation based on the linear free energy principle.

$$\ln K = \ln K^{o} + a\, DP^{*} \qquad (2.3.8.64)$$

where $K^o$ and K are the extraction constants for standard and given diluent, respectively, and *a* is constant for given extractant and solute. The values of *DP** were estimated for about 25 diluents. Solvation of acid anion, which is based on specific interactions with the proton-donating diluent, is typical for the reaction of acid with alkylammonium salt. The hindering effect of the diluent can be explained as a competition between diluent and acid for the bonding with amine salt. Other empirical parameters, *DP*, were introduced to quantify this type of solvation according to the following equation, similar to equation (2.3.8.64).

$$\ln K' = \ln K'^{o} + a'\, DP \qquad (2.3.8.65)$$

The *DP* values were estimated for about 30 diluents. If both types of solvation occur simultaneously, the combined equation is to be used in the form

$$\ln K = \ln K^{o} + a\, DP^{*} + a' DP \qquad (2.3.8.66)$$

Using this concept Schmidt (1980) described the effect of diluents on the extraction of mineral acids by various tertiary amines with a comparatively good fit. The attempts were also made to correlate the values of *DP** and *DP* with various physicochemical properties of thc diluents. The main disadvantage lies in the fact that the parameters *DP** and *DP* were obtained on the base of the same data set as the extraction constants which were correlated with them. This seems to increase the empirical character of these parameters.

Bizek *et al.* (1993) modified Eq. (2.3.8.63) to adopt to describe the effect of diluents on the values of partitioning coefficients $K_D$ in the form

$$\ln K_D = \ln K_D{}^0 + P\,(\delta_h)^2/100 + s\,(\pi^* + d\delta) + b\beta + a\alpha \qquad (2.3.8.67)$$

where $\delta_h$ is the Hildebrand's solubility parameter and $\pi^*$, $d$, and $\delta$ are the solvatochromic parameters that measure the solute + solvent, dipole + dipole, and dipole + induced dipole interactions, respectively. $K_D{}^0$ represents the extraction constant for an "ideal inert" diluent. The solvatochromic parameter $\alpha$ scale of solvent HBA (hydrogen-bond donor) acidities describes the ability of the solvent to donate a proton in a solvent to solute hydrogen bond. The $\beta$ scale of HBA (hydrogen-bond acceptor) basicities provides a measure of the solvent's ability to accept a proton (donate an electron pair) in a solute to solvent hydrogen bond, respectively. The coefficients p, s, d, a and b include the properties of solute. Rather than p, s, d, a are regression coefficients. The values of solvatochromic parameters $\pi^*$, $\delta$, $\alpha$, $\beta$ have been found several hundreds of compounds. The second term of Eq. (2.3.8.67) which contains the solubility parameter $\delta_h$ does not affect the values of the objective function significantly. Thus equation (2.3.8.67) reduces to

$$\ln K_D = \ln K_D{}^0 + s\,(\pi^* + d\delta) + b\beta + a\alpha \qquad (2.3.8.68)$$

Bizek, Horacek, and Kousova (1993) calculated the solvatochromic parameters of the solvent mixtures according to:

$$SP_1 = X_1 SP_1 + (1 - X_2)\, SP_2 \qquad (2.3.8.69)$$

where $X_1$ is the mole fraction of the first solvent and $X_2$ is the mole fraction of the second solvent. $SP_1$ is the solvatochromic parameters of the first solvent and $SP_2$ is the solvatochromic parameters of the second solvent in solvent mixtures.

## 2.4 Applications and case studies

### 2.4.1 *Reactive extraction of organic acids*

- *Source*: Article: Intensification of propionic acid recovery using reactive extraction
- *Authors*: Amit Keshav, Kailas L. Wasewar and Shri Chand
- *Address*: Department of Chemical Engineering, Indian Institute of Technology (IIT) Roorkee, Uttarakhand - 247667, India
- *Conference*: 3rd Uttarakhand State Science and Technology Congress (November, 2008)

Reactive extraction is a promising technique for recovery of carboxylic acid from dilute streams. Of the different extractants used for carboxylic acids recovery, higher molecular aliphatic amines have been found to be effective. In view of this, reactive extraction of 0.05 kmol/m$^3$ propionic acid using tri-n-octyl amine (TOA), a tertiary amine, in different categories of diluents were studied and modeled. Data have been reported in terms of distribution coefficient and equilibrium complexation constant. LSER model were fitted to represent extraction mechanism for reactive extraction of propionic acid. Also various model parameters were estimated. TOA complexes were found to be function of diluents chosen, which in turn depends on polarity of the diluents. Higher extractions were found to be in alcohols with degree of extraction higher than 90%. Furthermore, LSER model has been obtained to calculate distribution coefficients of the TOA-diluent systems.

Extraction experiments involve shaking equal volumes of aqueous and organic phases for 12 hours in a constant temperature water bath, followed by settling of the mixture at a same temperature. Aqueous phase acid concentration was determined by HPLC. The acid content in the organic phase was determined with a mass balance. The reactive liquid-liquid extraction of propionic acid ([HA]) with tertiary amine TOA ([T]) gives reaction complex ($T{:}HA_n$) which remains largely in organic phase. Keeping in view that the exact chemistry involved in the uptake of extra acid is unknown, the distribution coefficient can be interpreted by the following set of equations (2.4.1.1 – 2.4.1.3),

$$[HA]_{aq} + [T]_{org} \xleftrightarrow{K_{11}} [T : HA]_{org} \tag{2.4.1.1}$$

$$[HA]_{aq} + [T : HA]_{org} \xleftrightarrow{K_{21}} [T : (HA)_2]_{org} \tag{2.4.1.2}$$

:

$$[HA]_{aq} + [T : (HA)_{n-1}]_{org} \xleftrightarrow{K_{n1}} [T : (HA)_n]_{org} \tag{2.4.1.3}$$

The equilibrium complexation constant for the reaction represented by above equations is

$$K_{En} = \frac{[T{:}(HA)_n]_{org}}{[T]_{org}[HA]_{aq}^{\ n}} \tag{2.4.1.4}$$

$K_{En}$ is expected to depend on the properties of the acid and the solvation efficiency of the diluent used. The experimentally accessible distribution ratio $K_D$ is given as:

$$K_D = \frac{[HA]_{org}}{[HA]_{aq}} \tag{2.4.1.5}$$

Degree of extraction is defined as the ratio of propionic acid concentration in organic phase to the sum of acid concentration in organic and aqueous phase and is defined in term of $K_D$ as

$$E\% = \frac{K_D \times 100}{1 + K_D} \tag{2.4.1.6}$$

The extent to which the organic phase (amine + diluent) can be loaded with carboxylic acid is expressed as the loading ratio, z;

$$z = \frac{[HA]_{org}}{[T]_{o;org}} \tag{2.4.1.7}$$

The value of $z$ in equation (2.4.1.7) depends on the extractability of the acid (strength of the acid-base interaction) and its aqueous concentration. The stoichiometry of the overall extraction reaction depends on the loading ratio in the organic phase, $z$. Extractions involving loading ratios less than 0.5, forms (1:1) complexes and the following equation holds (Kertes and King, 1986).

$$\frac{z}{1-z} = K_E\,[HA]_{aq} \tag{2.4.1.8}$$

The stoichiometric loading factor $z_s$, is the ratio of the overall complexed acid to total TOA in the organic phase. This factor includes a correction term ($\nu[HA]_{org}^{diluent}$), for the amount of acid extracted by the diluent in the solvent mixture.

$$z_s = \frac{[HA]_{org} - \nu[HA]_{org}^{diluent}}{T_{org}^{o}} \tag{2.4.1.9}$$

The linear solvation energy relationship (LSER) model by Kamlet *et al.* (1983) was modified by Bizek *et al.* (1993) to adopt to describe the effect of diluents on the values of partitioning coefficients $K_D$ in the form

$$\ln K_D = \ln K_D{}^0 + s\,(\pi^* + d\delta) + b\delta + a\alpha \tag{2.4.1.10}$$

Table 2.4.1.1 demonstrates the influence of the organic solvent on distribution coefficient of propionic acid. TOA concentration of 30% (0.686 kmol/m$^3$) in various diluents for the acid concentration of 0.05 kmol/m$^3$ was used. The function of diluent is to dilute the extractant to the desired concentration and controls the viscosity and density of the solvent phase.

**Table 2.4.1.1:** Chemical equilibria for extraction of 0.05 kmol/m$^3$ propionic acid using TOA in different diluents

| Diluent | $[HA]_{org}^{dil}$ | $[HA]_{org}$ kmol/m$^3$ | $K_D$ | $E\%$ | $z$ | $z_s$ | $K_E$ |
|---|---|---|---|---|---|---|---|
| 1-hexanol | 0.054 | 0.047 | 17.61 | 94.63 | 0.069 | 0.014 | 44.13 |
| 1-octanol | 0.030 | 0.047 | 16.68 | 94.34 | 0.069 | 0.038 | 26.11 |
| 2-octanol | 0.015 | 0.047 | 16.05 | 94.13 | 0.069 | 0.053 | 25.11 |
| 1-decanol | 0.029 | 0.048 | 19.00 | 95.00 | 0.069 | 0.040 | 29.76 |
| Oleyl alcohol | 0.023 | 0.046 | 10.11 | 91.00 | 0.066 | 0.043 | 15.79 |
| Hexane | 0.006 | 0.021 | 0.75 | 42.80 | 0.031 | 0.025 | 1.13 |
| Heptane | 0.006 | 0.015 | 0.42 | 29.44 | 0.021 | 0.015 | 0.62 |
| Kerosene | 0.002 | 0.014 | 0.37 | 27.00 | 0.020 | 0.018 | 0.55 |
| Petroleum ether | 0.002 | 0.021 | 0.75 | 42.80 | 0.031 | 0.029 | 1.13 |
| MIBK | 0.013 | 0.034 | 2.13 | 68.00 | 0.050 | 0.036 | 3.26 |
| Toluene | 0.001 | 0.030 | 1.55 | 60.80 | 0.044 | 0.043 | 2.37 |
| Butyl acetate | 0.028 | 0.041 | 4.65 | 82.32 | 0.060 | 0.031 | 7.22 |
| Ethyl acetate | 0.033 | 0.037 | 2.83 | 73.87 | 0.054 | 0.020 | 4.36 |
| TBP | 0.043 | 0.045 | 10.02 | 90.92 | 0.066 | 0.022 | 15.64 |
| Dodecane + decanol | – | 0.042 | 5.49 | 84.59 | 0.062 | – | 8.63 |
| Chloroform | – | 0.048 | 19.16 | 95.04 | 0.069 | – | 43.25 |

Tri-n-octyl amine proves out to be effective with distribution coefficient strongly depending on the nature of diluent used and its concentration in that diluent. Aliphatic and aromatic hydrocarbons (hexane, heptane, toluene, petroleum ether, kerosene) are apolar aprotic solvents and are characterized by low dipole moment, low $E_T$ value and are unable to act as an hydrogen bond donor. So these solvents interact only slightly with the acid-TOA complexes by non specific directional, induction and dispersion forces are operating. So very low $K_D$ values were obatined with E% for hexane 42.8%; heptane 29.4%; kerosene 27%; toluene 60.8%; and petroleum ether 42.8%. Esters: butyl acetate, ethyl acetate, and ketone: MIBK on the other hand are dipolar aprotic solvents and have large dipole moment, and $E_T$ values. Though these solvents do not act as hydrogen bond donors since C-H bonds are not sufficiently polarized, still they provide $K_D$ values due to presence of ion electron pairs. Anion solvation occurs mainly by ion-dipole and ion-induced dipole forces. E% for MIBK, butyl acetate and ethyl acetate were reported to be 68%, 82.32% and 73.86% respectively. Alcohols, considered as protic solvents, contains hydrogen atoms bound to electronegative element O (-O-H) are hydrogen bond donors. Their $E_T$ values are large indicating them to be strongly polar. Thus they provide high $K_D$ values since they are good acid: TOA complex solvators due to their hydrogen bonding ability. The tendency becomes more pronounced as the charge density (ratio of charge to volume) of the anion to be solvated increases. Further alcohols are capable of being both acceptors and donors, as are the acids; give the highest $K_D$ followed by purely basic solvents such as ethers and ketones. E% values were found to be between 90-95% for various alcohols used in the study.

Overall loading ratios ($z$) were obtained from Eq. (3.4.1.7). In all extraction $z$ values were found to be less than 0.5, thus only (1:1) acid: TOA complexes were assumed to be formed (Kertes and King, 11986). The stoichiometric loading factor $z_s$, which signify the loading of acid by TOA was also computed. Again TOA/diluent system favors the formation of not overloaded polar acid-TOA structures corresponding to the $z_s$ factor restricted mainly between 0.014-0.053. Since the non polar TOA by itself is a relatively poor solvating medium for the polar complexes, loading was reported in literature to decreases with increasing TOA concentration in active diluents (alcohols, esters etc), as the active solvent becomes a less favorable solvating agent. For the non polar mixture of inert diluent (aliphatic and aromatic hydrocarbons, ethers etc.) and TOA, increasing the TOA concentration, however has little effect upon loading.

Equilibrium complexation constants ($K_E$) were evaluated by Eq. (2.4.1.4) and the values were reported in Table 2.4.1.2. Again high $K_E$ values were obtained for alcohols. The order of $K_E$ for extraction of 0.05 kmol/m$^3$ lies as: TOA + 1-hexanol > TOA + chloroform > TOA + 1-decanol > TOA + 1-octanol > TOA + 2-octanol > TOA + oleyl alcohol > TOA + TBP > TOA + dodecanol + decanol > TOA + butyl acetate > TOA + ethyl acetate > TOA + MIBK > TOA + toluene > TOA + petroleum ether > TOA + hexane > TOA + heptane > TOA + kerosene.

*The solvatochromic parameters*: hydrogen bond donor acidity (*p*, d)* and hydrogen bond acceptor basicities (*a, b*)for different diluents used (only those values which can be found out from literature were presented) were obtained from Kamlet *et al.* (1983). The values of the solvatochromic parameters were fitted to the experimental data and experimental results were compared with model predictions (Table 2.4.1.2). A good description of the distribution of propionic was obtained with $R^2$=0.986 and standard error 0.33. The model equation can be written as:

$$\ln K_D = -0.567 + 8.139\,\pi^* - 3.549\delta - 3.754\,\alpha + 4.022\,\beta \qquad (2.4.1.11)$$

**Table 2.4.1.2:** Comparison between experimental and model $K_D$ values for extraction of propionic acid using TOA in different diluents.

| Diluent | $K_D^{experimental}$ | $K_D^{model}$ |
|---|---|---|
| Hexane | 0.75 | 0.67 |
| Heptanes | 0.42 | 0.51 |
| Toluene | 1.55 | 1.66 |
| Butyl acetate | 4.65 | 4.42 |
| MIBK | 2.13 | 1.98 |
| TBP | 10.02 | 11.04 |
| 1-hexanol | 17.61 | 16.41 |
| 1-octanol | 16.68 | 18.23 |
| Chloroform | 19.16 | 17.57 |

Intensification of recovery of propionic acid using reactive extraction was studied using tri-n-octyl amine in different diluents was studied to find the best diluent. Alcohols were found to be the best solvators of acid: TOA complexes, thus yielding higher distribution ($K_D$ between 10.11-17.61) and equilibrium complexation constant values ($K_E$ between 15.78-41.12 $m^3$/kmol). Loading ratios in all extraction were found to be less than 0.5, thus suggesting (1:1) acid TOA complexation. LSER model was fitted to the equilibrium data. The values of the model parameters were determined as providing best fit of the experimental results. The intensified solution (reactive extraction) to the recovery of acid from bioreactor provided advantages of higher recovery and yield, cleaner and low cost of recovery in comparison the conventional methods.

## 2.4.2 *Reactive extraction of rare earth metals*

- *Source*: Article: Reactive extraction as a tool to recover lithium
- *Authors*: Amit Keshav
- *Address*: Department of Chemical Engineering, Indian Institute of Technology (IIT) Roorkee, Uttarakhand - 247 667, India
- *Conference*: Annual research symposium in chemical engineering, (chemference-08), 5-6 july, 2008. IIT Kanpur

The design of an efficient reactive extraction process for recovery of metals from aqueous streams require the knowledge of effect of metal concentration, pH and temperature on extraction. With this aim, reactive extraction of lithium was studied using binary extractants (LIX54 + tri-n-octylphosphine oxide). The extraction using binary extractants was higher than using LIX alone and 98%

extraction was achieved. The recovery of lithium is more favorable at low temperature (< 40) and high pH (>12.5).

The breakthrough in industrial applications was achieved by LIX (liquid ion exchangers) chemicals of Henkel. With these extractants recovery from very dilutes solutions have been possible. Today cobalt, nickel, zinc, and lithium are few of the metals which are gained by reactive extraction (Marr and Bart, 1982). The process of reactive extraction constitutes an extraction, a scrub and a back-extraction step, each steps including several separation stages. The extractants, dissolved in a solvent, constitute organic phase, is recirculated counter currently to feed, scrub and strip. The extractant react with metal to form complex and thus removes the metal from dilute solution. To describe the process of reactive extraction data on effect of initial metal concentration, effect of pH, effect of temperature and effect of salts/ions etc. is required.

Lithium is the 33[rd] most abundant element on earth but due to its high reactivity only appears naturally in the form of compounds. Lithium occurs in a number of pegmatitic minerals, but is also commonly obtained from brines and clays; on a commercial scale, lithium metal is isolated electrolyrically from a mixture of lithium chloride and potassium chloride. Lithium finds increasing use in long life battery, possible fusion reactor development, refrigerants, ceramics and medicines (Averial and Olson, 1978). The main problem in lithium production is the scarcity of rich source of lithium (Chitra and Giakward, 1997). Currently the production is through mining of pegmatite ores. Recently focus has been made on recovery from sea water and other natural sources. However since the metal is present in very low concentrations (0.17 ppm), the recovery is highly uneconomical.

The reaction between LIX54 (R-H) and lithium ion ($Li^+$) involves the replacement of lithium ion with the hydrogen ion of LIX54 and can be described as:

$$[Li^+]_{\mathrm{aq}} + [R - H]_{\mathrm{org}} \leftrightarrow [Li - R]_{\mathrm{org}} + H^+; \quad \mathrm{K}_{\mathrm{s,1}} = \frac{[Li - R]_{org}[H^+]}{[Li^+]_{aq}[R - H]_{org}} \tag{2.4.2.1}$$

TOPO ([B]) is a neutral extractant which reacts with solutes by replacing water of hydration in neutral complex.

$$[Li - R]_{\mathrm{org}} + \mathrm{a}[B]_{\mathrm{org}} \leftrightarrow [Li - R - B_a]_{\mathrm{org}} \tag{2.4.2.2}$$

For synergistic extraction

$$[Li^+]_{\mathrm{aq}} + [R - H]_{\mathrm{org}} + \mathrm{a}[B]_{\mathrm{org}} \leftrightarrow [Li - R - B_a]_{\mathrm{org}} + H^+$$

$$\mathrm{K}_{\mathrm{s,2}} = \frac{[Li - R - B_{\mathrm{a}}]_{\mathrm{org}}[H^+]}{[Li^+]_{\mathrm{aq}}[R - H]_{\mathrm{org}}[B]^{\mathrm{a}}_{\mathrm{org}}} \tag{2.4.2.3}$$

Distribution coefficient can be represented as:

$$\mathrm{K}_{\mathrm{D}} = \frac{[Li+]_{\mathrm{org}}}{[Li+]_{\mathrm{aq}}} \tag{2.4.2.4}$$

Equilibrium complexation constants ($K_{s,1}$ (for extraction using LIX54 alone and $K_{s,2}$ for synergistic extraction) can be expressed in terms of distribution coefficient as:

$$\text{a)}\ K_{s,1} = \frac{K_D[H^+]}{[R-H]_{org}} \ \text{and b)}\ K_{s,2} = \frac{K_D[H^+]}{[R-H]_{org}[B]^a_{org}} \quad (2.4.2.5)$$

Taking log of above equations,

$$\log (K_D) = \log (K_{s,1}) + \log [R - H]_{org} - \log (H^+) \quad (2.4.2.6)$$

$$\text{and } \log (K_D) = \log (K_{s,1}) + \log [R - H]_{org} + a \log [B]_{org} - \log (H^+) \quad (2.4.2.7)$$

Fig. 2.4.2.1 shows the effect of LIX54 concentration of reactive extraction of lithium (20 ppm). As can be seen that increasing LIX from 5 to 20% in kerosene increase the extraction. However even at the highest concentration of LIX, $K_D$ was found to be less than 0.5. Equilibrium complexation values obtained by using LIX alone in kerosene (using Eq. 2.4.2.6) was found to be $K_{s,1}$=1.42´10$^{-12}$. On the other hand, using mixed extractants (LIX (5%) + TOPO + kerosene), increasing TOPO from 0 to 0.04 M was found to increase the extraction from 8% to 54% (Fig. 2.4.2.2). Thus synergism was observed and this can successfully be employed for recovery of lithium from dilute solutions. The distribution

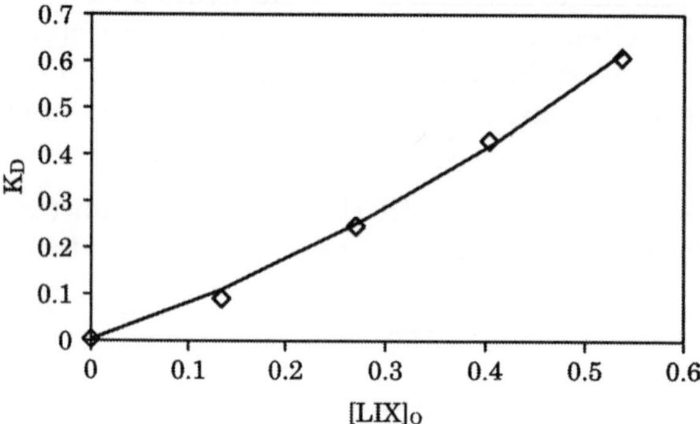


**Fig. 2.4.2.1:** Effect of LIX 54 concentration in kerosene on distribution coefficient of lithium

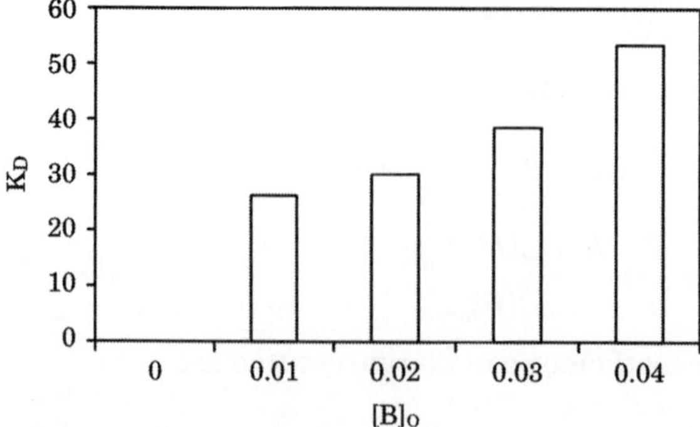


**Fig. 2.4.2.2:** Effect of TOPO addition on distribution coefficient of lithium.

coefficient obtained by synergistic extraction was 70 times (50.0) higher than what obtained by LIX54 alone. The 98% degree of extraction, clearly signify the advantage of mixed extractants.

Effect of lithium concentration ion distribution coefficient was described in Fig. 2.4.2.3. Results shows that lower lithium concentrations can be successfully extracted using the mixed extractants. Presence of lithium above 30ppm results in lower extractions. The probable reason for this may be that at those concentrations, the extractants become limiting. Since lithium concentration from natural sources is very less (< 30 ppm), higher $K_D$ would be expected.

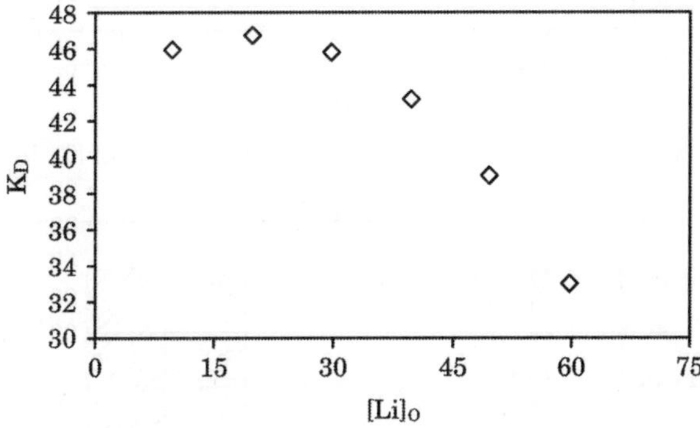


**Fig. 2.4.2.3:** Effect of initial lithium concentration on distribution coefficient of lithium

Extraction of lithium by extractant is via intermolecular hydrogen bonding or ion exchange of the extractant group with the metal. The extraction of lithium by extractant: lithium complexation is expected to be exothermic and makes the system more ordered. Thus entropy decrease is expected. Further the system becomes more ordered if the interaction of lithium: extractant is strongest. The complexation reaction in the organic phase is expected to be exothermic.

If the enthalpy and entropy of reaction are assumed to be constant over the temperature range, the expression

$$\ln \mathrm{K} = \frac{-\Delta H}{R\mathrm{T}} + \frac{\Delta S}{R} \tag{2.4.2.8}$$

indicates that a plot of $\ln K$ vs $1/T$ should gives a straight line with enthalpy as slope and entropy as intercept. The effect of temperature from 25 to 50°C on reactive extraction of lithium was studied. Plot of Equation (2.4.2.8) is shown in Fig. 2.4.2.4. Negative value of enthalpy change confirms the reaction to be exothermic (Enthalpy ($\Delta H$) = -3.14*$10^5$) and, thus extraction decreases with increase in temperature. Formation of complex makes the system more ordered and thus decreases the entropy (Entropy ($\Delta S$)= -1.05*$10^{-7}$).

The effect of pH was studied by varying pH from 3.0 to 13.5. Plot of $K_D$ versus pH, shows negligible extraction upto pH <11.5 (Fig. 2.4.2.5). Increasing pH above 11.5, increases the extraction upto pH=13.0, after which it remains constant. Natural sources are usually obtained at pH of 8-9, where as seen from above no extraction is possible, so the pH of feed solution should be increased to 12-13 before extraction of lithium is carried out.

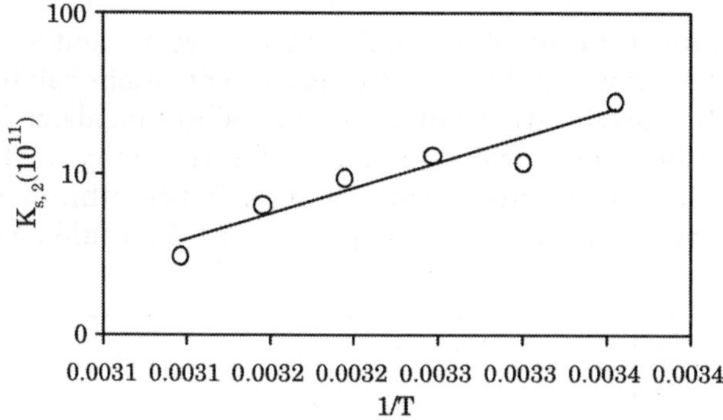


**Fig. 2.4.2.4:** Effect of temperature on extraction of lithium.

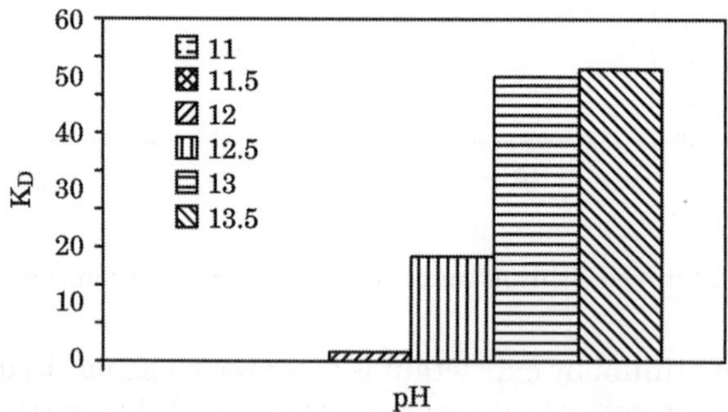


**Fig. 2.4.2.5:** Effect of pH on extraction of lithium from aqueous solutions

Reactive extraction of lithium was carried to account for parameters like, effect of lithium concentration, pH and temperature. Following conclusions can be drawn:

- Extraction using binary extractants (LIX54 + TOPO) is higher than LIX54 alone. Distribution coefficient 70 higher was obtained that cha k show the synergistic effect.
- $K_D$ is found to be independent of initial lithium concentration
- Extraction was found to decrease with increase in temperature
- Reactive extraction is effective at higher pH, thus system pH needs to be raised before extraction should be carried.
- The knowledge of effect of lithium initial concentration, pH, and temperature is important in the design of reactive extraction unit for the metal recovery.

### 2.4.3 *Reactive extraction of pharmaceutical products*

One of the most important application of reactive extraction in pharmaceutical industries in the recovery of penicillin's. Penicillin are defined as any of a large group of broad-spectrum antibiotic drugs derived directly or indirectly from molds of the genus Penicillium and other soil-inhabiting fungi grown on special culture media, which exert a bacteriocidal as well as a bacteriostatic effect on susceptible bacteria during their growth stage by the inhibition of biosynthesis of their cell wall mucopeptide. 6-Aminopenicillanic acid (6-APA) is the active core body of all penicillin's substituted at the 6-amino position resulting in a variety of antibacterial and pharmacologic characteristics.

Extraction of 6-APA using the conventional solvent such as 1-octanol and butylacetate results in degree of extraction less than 3% in both the above diluents. This conventional extraction technology has no success in recovery of 6-APA from dilute aqueous solutions. Reactive extraction of 6-APA was studied using organophosphorous compound (TBP) and quaternary amine (Aliquat 336). Fig. (2.4.3.1-2.4.3.2) shows the effect of extractant addition (in volume percentages of 10-20%) in 1-octanol and butyl acetate respectively. 1-octanol was found to be better solvation media among the two diluents chosen. This may be accounted to the hydrogen bonding ability of the 1-octanol to the acid-extractant complexes. In extraction of 6-APA using Aliquat 336 in 1-octanol and butyl acetate, the highest degree of extraction were obtained to be 45% and 16% respectively, whereas using TBP in respective diluents the highest values were 34 and 14%. Thus Aliquat 336 was more successful in recovery of 6-APA from aqueous solution and 14 times higher degree of extraction was obtained when compared with physical extraction. A higher capacity extractant such as tertiary or secondary amine could have the possibility to provide even higher extractions.

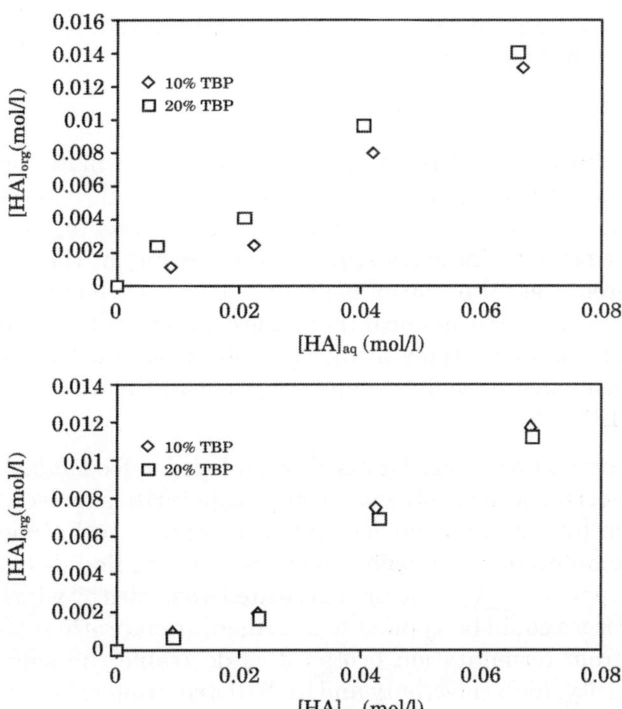


**Fig. 2.4.3.1:** Reactive extraction of 6-APA using TBP (10-20%) in a) 1-octanol and b) butyl acetate

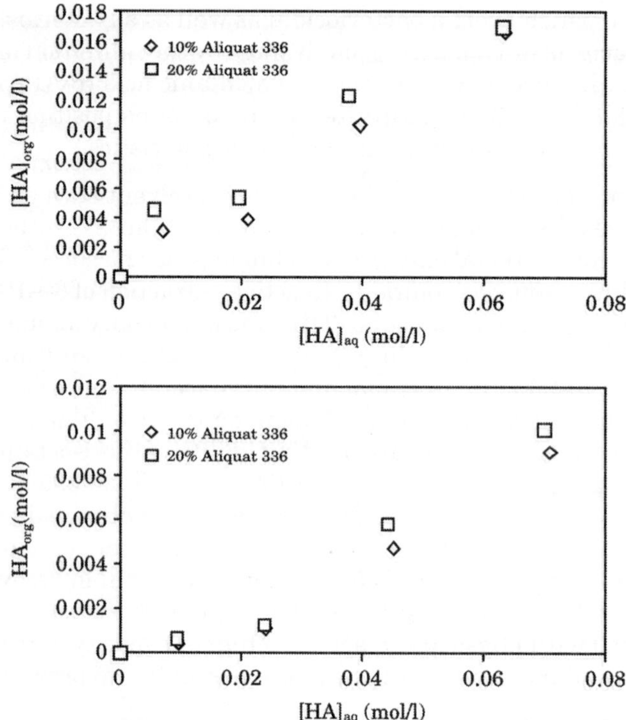


**Fig. 2.4.3.2:** Reactive extraction of 6-APA using Aliquat 336 (10-20%) in a) 1-octanol and b) butyl acetate

## 2.4.4 *Reactive extraction of other chemicals*

Extractions with chemical reactions are practiced in many other chemical operations also. Some examples include the commercialization of sulfuric acid alkylation, HF alkylation and various oxidative sweetening processes in petrochemical refining industries. LLE provides one of the most important commercial processes for performing group separations between aromatic and aliphatic species in petrochemical industries. Products that are produced by reactive extraction include polyurethanes, polycarbonates, polyamides, soaps, linear alkylbenzene sulfonates, cephalosporin antibiotic, TNT, resorcinol, gasoline and LPG etc.

Reactive extraction technology is also employed in food industries and since the major concern there is solvent toxicity, supercritical solvent such as $CO_2$ and extraction from an aqueous leachate are popular techniques. Toxicity of the solvents employed is the major factor concerning liquid-liquid extraction process to be applied to a bioreactor. There are two modes by which any reactive extraction process could be applied to a fermentation broth: *insitu* and *exsitu*. In recovery from fermentation broths it is desirable the solvent possesses minimum toxicity, high selectivity and high distribution ratios. The problem of toxicity is more of concern in *insitu* recovery of acid from fermentation broth. Toxicity is caused on two different levels: molecular level, which is due to dissolved organic solvent which can inhibit enzymes or modify cell membrane permeability and phase level, which is due to direct contact of solvent phase with the cells, which may block nutrients diffusion from medium to cells due to solvent coating and may disrupt the cell wall due to increased surface tension.

However in a particular operating system of the fermentation broth, toxicity depends on the microorganism used and widely varying results could be found depending on whether yeast or bacteria were used and on the particular strain of bacteria. There are different techniques by which the problem of toxicity could be resolved. Cell immobilization, employing membrane that can prevent cells to come in direct contact with the solvent, using microorganism strains which are unaffected by the solvent used, and employing non toxic solvents are some of the methods that could reduce problem of toxicity.

## 2.5 REFERENCES

Akay, G. and Wakeman, R. 1996. *Chem. Eng. Res. Des.*, A 74: 517–525.

Averial, W.A. and Olson, D.L. 1978. *Energy*, 3: 305–312.

Bhagwat, S.S. and Sharma, M.M. 1988. *Chem. Eng. Sci.* 43: 195–20.

Bizek, V., Horacek, J. and Kousova, M. 1993. *Chem. Eng. Sci.* 46(8): 1447–1457.

Bizek, V., Horacek, J., Kousova., M., Heyberger, A. and Prochazka, J. 1992. *Chem. Eng. Sci.* 47(6): 1433–1440.

Canari, R. and Eyal, A.M. 2004. *Ind. Eng. Chem. Res.* 43: 7608–7617.

Chitra, K.R. and Gaikward, A.G. 1997. *J. Memb. Sci.* 125: 257.

DeGarmo, J.L., Parulekar, V.N. and Pinjala, V. 1992. *Chem. Eng. Prog.* 88: 43– 50.

Doraiswamy, L.K. and Sharma, M.M. 1984. Heterogeneous reaction: Analysis, Examples, and Reactor Design, Vol. 2: Fluid-Fluid –Solid-Reactions. 1st Ed. New York, John Wiley & Sons. pp. 17–41.

Douglas, J.M. 1988. Conceptual design of chemical processes. New York: McGraw-Hill.

Gedye, R.N., Smith, F.E. and Westaway, K.C. 1988. *Can. J. of Chem.*, 66: 17–26.

Goetheer, E.L.V., Vorstman, M.A.G. and Keurentjes, J.T.F. 1999. *Chem. Eng. Sci.*, 54: 1589–1596.

Heggs, P. 1983. *Chem Engr. (London).* 13: 394–402.

Hills, J.H., Azzopardi, B.J. and Barhey, A.S. 1996. *Chem. Eng. Res. Des.* A 74: 567–574.

Jansen, A.E., Klaassen, R. and Feron, P.H.M. 1995. 1st International Conference on Process Intensification for Chemical Industry, 18, BHR Group, London, pp. 145–153.

Joshi, J.B. 2001. Process intensification, Chemical weekly, pp. 101–103.

Kamlet, M.J., Abboud, J.-L.M., Abraham, M.H. and Taft, R.W. 1983. *J. Org. Chem.* 48: 2877–2887.

Kamlet, M.J. and Taft, R.W. 1976. *J. Am. Chem. Soc.* 98: 377–383.

Kertes, A.S. and King, C.J. 1986. *Biotechnol. Bioeng.* 28: 269–282.

Keshav, A. 2008. Annual Research Symposium in Chemical Engineering (Chemference-08), 5-6 July, IIT Kanpur.

Keshav, A. and Daharwal, S. 2010. National Conference on "Emerging Trends in Chemical Engineering & Biotechnology" (BIO-CHEM 2010).

Keshav, A. and Wasewar, K.L. 2010. *Chem. Eng. Sci.* 65: 2751–2757.

Keshav, A., Wasewar, K.L. and Chand, S. 2008. 3rd Uttarakhand State Science and Technology Congress.

Keshav, A., Wasewar, K.L., Chand, S., Uslu, H. and Inci, I. 2009. *i-manager's J. Future Eng. Technol.,* 4(2): 41–49.

Kyuchoukov, G., Yankov, D., Albet, J. and Mohinier, J. 2005. *Ind. Eng. Chem.Res.*, 44: 5733–5739.

Lewis, V.P. and Yang, S.T. 1992. *Biotechnol*. 8: 104–110.

Li, Z.Y., Qin, W. and Dai, Y.Y. 2004a. *J. Chem. Ind. Eng.*, 1: 54–58.

Li, Z.Y., Qin, W. and Dai, Y.Y. 2004b. *J. Chem. Ind. Eng*. 1: 59–64.

Marr, R. and Bart, H.J. 1982. *Chem. Ing. Tech.*, 54(2): 119–125.

Minotti, M., Doherty, M.F. and Malone, M.F. 1996. *Ind. Eng. Chem. Res.*, 35: 2672–2681.

Poole, L.J. and King, C.J. 1991. *Ind. Eng. Chem. Res*. 30: 923–929.

Poposka, F.A., Prochazka, J., Tomovska, R., Nikolovski, K. and Grizo, A. 2000. *Chem. Engg. Sc.*, 55: 1591–1604.

Ramshaw, C. 1983. *The Chem. Engineer*, 389: 13–14.

Schmidt, V.S. 1980. Amine Extraction (in Russian), pp. 44–53. Atomixdat, Moscow.

Shan, X., Qin, W. and Dai, Y. 2006. *Chem. Eng. Science*, 61: 2574–2581.

Shan, X.C. 2003. Bachelor Thesis, Tsingua University.

Stankiewicz, A.I. and Moulijn, J.A. 2000. *Chem. Eng. Progr*. 96: 22–34.

Taft, R.W. and Kamlet, M.J. 1976. *J. Am. Chem. Soc*. 98(10): 2886–2894.

Taft, R.W. and Kamlet, M.J. 1979. *J. Chem. Soc., Perkin Trans*. 2: 1723–1729.

Tagawa, T., Moe, K.K., Ito, M. and Goto, S. 1999. *Chem. Eng. Sci.*, 54: 1553–1557.

Tamada, J.A. and King, C.J. 1990. *Ind. Eng. Chem. Res*. 29: 1333–1338.

Tomovska, R., Poposka, F. and Volaufova, E. 1998. Heyberger, & Prochazka, *J. Chem. Communication* pp. 305–320.

Wasewar, K.L. 2005. *Chem. Biochem. Engg.*, 19(2): 159–172.

Wasewar, K.L., Heesink, A.B.M., Versteeg, G.F., Pangarkar, V.G. 2004. *Chem. Eng. Sci*. 59: 2315–2320.

Wasylkiewicz, S.K., Sridhar, L.N., Doherty, M.F. and Malone, M.F. 1996. *Ind. Eng. Chem. Res.*, 37: 1395–1402.

# Chapter 3

# Adsorption, Ion Exchange Processes and Chromotography

## 3.1. ADSORPTION PROCESSES

### 3.1.1 *Introduction*

Adsorption (also known as adsorptive separation) can be simply defined as the concentration of a solute, which may be molecules in a gas stream or a dissolved or suspended substance in a liquid stream, on the surface of a solid. In an adsorption process, molecules or atoms or ions in a gas or liquid diffuse to the surface of a solid, where they bond with the solid surface or are held there by weak inter-molecular forces. The adsorbed solute is called the adsorbate, and the solid material is the adsorbent. Separation by adsorption depends on one component being more readily adsorbed than another. Studies and usage of adsorption dates back to eighteenth century. The uptake of gases by charcoal was studied by Scheele in 1773 and by Fontana in 1777.

#### 3.1.1.1 *Basis of Separation*

The adsorption separation is based on three distinct mechanisms: steric, equilibrium and kinetic mechanisms. In the steric separation mechanism, the porous solid has pores having dimension such that it allows small molecules to enter while excluding large molecules from entry. The equilibrium mechanism is based on the solid having different abilities to accommodate different species; that is, the stronger adsorbing species is preferentially held on its surface by the solid. The kinetic mechanism is based on the differential rates of diffusion and/ or adsorption of different species into the pore; thus by controlling the time of exposure, the faster diffusing species is preferentially removed by the solid. Based on the nature of attraction between the adsorbate and the adsorbent, the nature of adsorption may be classified as physisorption and chemisorption. In physisorption (physical adsorption), there is a weak Van-der Waals attraction of the adsorbate to the surface. The attraction to the surface is weak but long ranged and the energy released upon accommodation to the surface is of the same order of magnitude as an enthalpy of condensation (20 kJ/mol). During the process of physisorption, the chemical identity of the adsorbate remains intact, *i.e.* no breakage of the covalent structure of the adsorbate takes place. Physisorption, to be a spontaneous thermodynamic process, must have a negative $\Delta G$. Because translational degrees of freedom of the gas phase adsorbate are lost upon deposition onto the substrate. Since, $\Delta G = \Delta H - T\Delta S$, $\Delta H$ for physiosorption must be exothermic. In physiosorption, the bond is a Van-der Waals interaction and the adsorption energy is typically 5-10 kJ/mol. In chemisorption (chemical adsorption), the adsorbate sticks to the solid by the formation of a chemical bond with the surface. This interaction is much stronger than physiosorption, and, in general, chemisorption has more stringent requirements for the compatibility of the adsorbate and the surface site than physiosorption. The chemisorption may be stronger than the bonds internal to the free adsorbate which can result in the dissociation of the adsorbate upon adsorption (dissociative adsorption). In some cases $\Delta H$ for dissociative adsorption can be greater than zero, which means endothermic chemisorption, is possible. For chemisorption, the adsorption energy is comparable to the energy of a chemical bond. The molecule may chemisorb intact or it may dissociate. The chemisorption energy is 30-70 kJ/mol for molecules and 100-400 kJ/mol for atoms. A comparison between physisorption and chemisorptions for various adsorption properties is given in Table 3.1.1.

**Table. 3.1.1:** Comparison of physiosorption and chemisorption

| *Properties* | *Physisorption* | *Chemisorption* |
|---|---|---|
| Adsorption temperature | Near or below boiling point temperature of adsorbate | Virtually unlimited range |
| Adsorption enthalpy | Heat of liquifiaction (5-40 kJ/mol) | Wide range (40-800 kJ/mol) |
| Crystallographic specificity crystal planes | Independent of surface geometry | Marked difference between |
| Nature of adsorption | Non-dissociative and reversible | Often dissociative and irreversible in many cases |
| Saturation | Multiplayer occurs often | Limited to a monolyer |
| Adsorption kinetics | Fast, non-activated process | Activated process |

### 3.1.2 *Various Sorbents*

A variety of adsorbents used for removal of phenols, dyes and metals from single solute aqueous solutions include activated carbon, bentonite and perlite, rubber seed coat, hydrotalcite and its calcined product, acid-activated bituminous shale, zirconium (IV) arsenate–vanadate ion-exchanger, cross-linked polyvinyl-pyrrolidone, activated natural zeolites, water-insoluble cationic starch, and polymeric resins. Recently low cost waste materials have received considerable attention for use as adsorbents. Large numbers of researchers have carried out research using unconventional adsorbents for the removal of heavy metals like sawdust of *Pinus sylvestris*, Perlite, Pyrolyzed coffee residues and clay, low grade phosphate, $Fe_2O_3$, $Fe_3O_4$, FeS, steel wool, Mg pellets, Cu pellets, Zn pellets, Al pellets, Fe pellets, coal, goethite pretreated with phosphate, peat.

#### 3.1.2.1 *Activated carbon*

Activated carbon has many attractive properties such as, high surface area (300-1500 $m^2/g$), high adsorption capacity for large number of heavy metals.

In most instances physical adsorption has been found to prevail. Since activated carbon can be manufactured from variety of base materials (coal-bituminous to lignite, peat, wood, coconut shells, paper mill waste, molasses, petroleum coke etc.), it has varying porous structure and mechanical properties. Adsorption of organic solutes from the aqueous phase is a very important application of ACs and, it has been cited by the US Environmental Protection Agency (EPA) as one of the best available environmental control technologies. Activated carbons are profusely used as adsorbents for decontamination processes because of their extended surface area, high adsorption capacity, microporous structure, and special surface reactivity. Mean micropore size of activated carbons was prepared from solid cork wastes over the range 0.74–2.21 nm. A new and simple method of producing of activated carbons from different bituminous coals, *i.e.*, gas coal, gas-coking coal and orthocoking coal. Coal agglomerates of spherical shape obtained by oil agglomeration using rapeseed oil and linseed oil, were subjected to carbonization and activation with steam at 850°C. Activated carbons prepared from gas-coking coal with surface area BET of about 800 $m^2/g$ and pore volume of 0.40 $cm^3/g$. The adsorption capacity of the produced activated carbons was determined in terms of substituted phenolic compounds. In the case of the chemically activated samples, although they have higher total micropore volumes their mean pore size is also significantly greater and the enhancement in the amount adsorbed at low concentrations is therefore considerably reduced. Hence, the total micropore volume is no longer the controlling factor. It would be expected that the surface chemistry of the materials would play a more significant effect with adsorbates.

Activated carbon is the most effective adsorbent for organic contaminants due to its high surface area, low specificity and fast adsorption kinetics. Though activated carbon has been widely used as an adsorbent for the removal of a wide range of organic molecules and metals, high costs and regeneration difficulties hamper large-scale application of activated carbon. Adsorption onto activated carbon is often considered as the most economical and efficient process for the removal of organic compounds in dilutes aqueous solutions. Activated carbon adsorption has been cited by the USEPA as one of the best available environmental control technologies. Various adsorbents preparing from activated carbon and low-cost adsorbents used in wastewater treatment. Thus, it is used in the control of color and odors, in the removal of organic compounds or trihalomethanes precursors, to remove chlorine and in general to remove toxic compounds. Adsorption technologies for metals wastes include activated carbon and ion exchange treatment. For high strength and low volumes of wastewaters, heavy metal removal by adsorption using granular/powdered activated carbon has been widely used. Activated carbon is generally used in granular form either in batch, column (both fixed-bed and counter-current bed), or fluidized-bed operations, with fixed-bed being the most common. However, high cost of activated carbon and 10-15% loss during the regeneration are major deterrents in the utilization of activated carbon in the developing countries. Some of literature available for removal of heavy metals/dyes/phenolic compounds onto activated carbons is shown in Table 3.1.2.

3.1.2.2. *Low Cost Adsorbents*

Activated carbon is the most commonly used sorbent in wastewater treatment, due to its vast surface area and affinity for many organic chemicals. However, activated carbon is costly to regenerate; the process is energy intensive and has a high attrition rate. It also tends to sorb most organic chemicals indiscriminately, making it difficult to selectively recover certain organic chemicals for reuse. This has led to search for cheaper alternative materials as adsorbents such as bagasse fly ash, rice husk ash, lignin, bagasse pith, peat, saw dust, coal fly ash and bagasse fly ash. These substitutes have been used with varying degree of success for the removal of a wide variety of pollutants. Many agro-wastes are either arbitrarily discarded or set on fire, exploitation and utilization of these materials should bring economic and social benefits to mankind, Materials like neem leaf powder, bagasse fly ash, sugarcane dust, rice husk, banana pith, coir pith, orange peel, shells of bittim, chemically treated guava leaf powder, phoenix tree leaves, citrus grandis, unburned carbon, etc., have been studied. Thus, in the present work, adsorption potential of gulmohar plant leaf powder was investigated. Gulmohar is a flamboyant flowering tree (*Delonix regia, Caesalpiniaceae family*). For several weeks in spring and summer it is covered with exuberant clusters of flame-red flowers, 100 to 150 mm across. The delicate, fern-like leaves are composed of small individual leaflets, which fold up at the onset of dusk. Gulmohar grows 10 to 13 m tall, but its elegant wide-spreading umbrella-like canopy can be wider than its height. Gulmohar is naturalized in India and is widely cultivated as a street tree. Activated carbons which have the disordered graphitic structure, the functional groups such as carbonyl, carboxyl, alcoholic and ether groups. Activated carbon adsorption may have an edge over other methods in terms of cost especially if the starting materials for the activated carbons are biomass waste products. Oil–palm shell is one of the by-products of the palm oil industry. Malaysia, which is the world's largest palm oil producing country, generates about 4.3 million tonnes of fruit shells and 8.6 million tonnes of fibres every year. Conversion of these wastes into activated carbons would add value to this biomass and, in addition, the conversion from wastes may be able to reduce the price of commercial activated carbons. Oil–palm shell is used as a suitable precursor for preparing activated carbon because of its hardness, high carbon content and low ash content.

***Bagasses fly ash and rice husk ash:*** About 10 million tonnes of bagasse is produced annually in India. About 90% of this bagasse is used as a fuel in bagasse fired boilers. The production of bagasse fly ash is estimated to be about 3 kg/tonne of sugar produced. With the overall bagasse fly ash collection efficiency of 80% in the dust collection equipment, it is estimated that fly ash availability from the sugar industry would be of the order of 0.5 million tonnes/year. The adsorption properties of BFA are highly influenced with the firing practices of bagasse. Several investigators have recently shown interest in the use of bagasse fly ash as an adsorbent. It has been used for the COD removal from sugar mill and paper mill effluents (Srivastava *et al.*, 2005). BFA has also been used for the adsorptive removal of phenolic compounds, dyes and pyridine from wastewaters. RHA has many applications due to its attractive properties. It is an excellent insulator, so has applications in industrial processes such as steel foundries, and in the manufacture of insulation for houses and refractory bricks. RHA has also been used in the production of sodium silicate films. It is an active pozzolan and has several applications in the cement and concrete industry. Since the main components of rice husk are carbon and silica, it has the potential to be used as an adsorbent as well. Rice husk contains abundant floristic fiber, protein and some functional groups such as carboxyl, hydroxyl and amidogen, etc., which make adsorption processes possible. Secondly, the yield of rice husk obtained from agriculture as a by-product is vastness. Some of literature available for removal of heavy metals/dyes/phenolic compounds onto low cost adsorbents is shown in Table 3.1.2.

#### 3.1.2.3. *Activated Carbon Cloth / Fibres*

In many recent studies carbon fibers have been proven to display good adsorptive capabilities for organic and inorganic species. Generally, utilization of activated carbon can be in the form of powder, granular and fiber or cloth form. Activated carbon cloth having very high specific surface area and adsorption capacity, uniform pore size and mechanical strength has attracted attention in recent years. The higher adsorption kinetics and capacities of activated carbon cloths compared with granular activated carbon, due to their large external surface area and their porous specificities (high specific surface areas and a direct connection of their micropores with their external surface area). Activated carbon fibers (ACFs) are novel adsorbents of high efficiency. The raw materials of ACFs are polyacrylonitrile fibers, viscose, phenolic resin fibers or pitch fibers, etc. They are first pyrolyzed and then activated at a temperature of 700–1000°C in an atmosphere of steam or carbon dioxide. ACFs have many favorable characteristics such as high adsorption capacities and high mass transfer rates for both adsorption and desorption, and are easier to be handled in a batch adsorber than granular and powdered activated carbon. These materials in the form of felt or cloth have the advantages of having high specific surface area (as high as $2500 m^2/g$), mechanical integrity, easy handling and minimal diffusion limitation to adsorption. Some of literature available for removal of heavy metals/dyes/phenolic compounds onto activated carbon cloth/fibers is shown in Table 3.1.2.

#### 3.1.2.4. *Chitin*

Chitin, well known to consist of 2-acetamido-2-deoxy-α-d-glucose through a $(1 \leftrightarrow 4)$ linkage, is a polysaccharide that is widely spread among marine and terrestrial invertebrates and in lower forms of plant kingdom. Chitosan's properties such as non-toxicity, biodegradability, polyelectrolicity and possibility for regeneration leads to a number of applications, mostly in removing dyes, heavy metals, organic compounds and humic acid. Both hydroxyl and aminogroups found in chitosan can be modified either by using physical or chemical. Physical modification increases the sorption properties; gel formation and involves an expansion of the porous network. Functional chitosan, chemically modified by salicylaldehyde, β-cyclodextrin, and a crosslinked β-cyclodextrin polymer. Chemical modification,on the other hand, increases the sorption properties of chitosan while preventing the dissolution of chitosan in strongacids.

It also improves the mechanical strength of chitosan and improves superficial area of the beads. Most commonly used chemical modifications are cross-linking, grafting, carboxymethylation and acetylation. Some of literature available for removal of heavy metals/dyes/phenolic compounds onto chitosan is shown in Table 3.2.1.

#### 3.1.2.5. *Synthetic Polymeric Adsorbents*

Polymeric adsorbents have been increasingly regarded as an alternative to activated carbon for the selective removal of specific organic substances from contaminated water due to their wide variations in functionality, surface area, and porosity. Synthetic polymeric adsorbents are becoming alternatives because of their favorable physicochemical stability, structural diversity, and feasible regeneration. A commercial or synthetic nonionic polymeric resins like Amberlite XAD-4 resin, water-compatible hypercrosslinked polymeric adsorbent. The specific surface area of NJ-8 (906 $m^2/g$) is very close to that of Amberlite XAD-4 (880 $m^2/g$). In comparison with classical adsorbents such as silica gels, aluminas and activated carbons, macroporous polymeric adsorbents are more attractive alternatives because of their wide range of pore structures and physico-chemical characteristics, because of its high chemically stability and excellent selectivity towards aromatic solutes. Series studies have been done on the removal of phenolic compounds using Amberlite XAD-4 polymeric resin, but difficulties with extreme hydrophobic chemical structure of these polystyrene-based adsorbents have resulted in their poor contact with aqueous solutions. Some activation solvent (such as methanol, acetone, or acetonitrile) has to be used to enhance surface contact with the aqueous solutions in a pretreatment process, which limits its broad application in the chemical wastewater treatment. Adsorption onto polymeric resins has been widely studied for the treatment of effluents containing phenolic compounds. Many researchers have made more efforts on chemical modification of polymeric adsorbents with functional groups such as phenolic hydroxyl, acetyl, benzoyl and hypercrosslinked polymers. Among them, the commercial resin Amberlite XAD-4 has been considered one of the best one for the removal of phenolic compounds from wastewater. Some of literature available for removal of heavy metals/dyes/phenolic compounds onto various polymeric adsorbents is shown in Table 3.1.2.

#### 3.1.2.6. *Organo-clays*

Over the last few decades organo-clays have gained much importance for the removal of organic pollutants from aqueous solutions. Among the various types of clays, montmorillonite which is the main constituent of the low-cost and naturally abundant mineral bentonite possesses several of the properties that make it very appropriate for organo-clay preparation. Montmorillonite has a 2:1 type layer consisting of one octahedral sheet of alumina inserted in between two silica tetrahedral sheets. The hydration of exchangeable alkali and alkaline earth metal cations *e.g.*, $Na^+$, $Ca^{2+}$, $Mg^{2+}$, etc. creates a hydrophilic environment on the surface and inthe interlayer region of natural montmorillonite clay. This renders the clay relatively ineffective as an adsorbent for neutral organic contaminants (NOC) such as chlorinated phenols. The sorbent properties of montmorillonite clays for NOC can be greatly improved by replacing the natural inorganic interlayer cations with organic cations such as quaternary ammonium cations. The natural clay precursor which modified clays is commonly referred to as organo-clays. The organophilic properties of the modified clay result in part from the reduced degree of hydration of the organic cations as compared to natural inorganic exchangeable cations.

Organo-smectites, such as hexadecyl trimethylammonium-smectite, behave as a dual sorbent for organic compounds, in which the mineral fraction functions as a solid adsorbent and the organic phase as a partition medium. Organo-clays have been investigated extensively for a wide variety of environmental applications: adsorption of phenol, p-chlorophenol and 2,4-dichlorophenol with hexadecyl trimethylammonium bromide, modified bentonite like tetramethylammonium chloride, hexadecyl trimethylammonium bromide, hexadecyl benzyl dimethylammonium chloride and dioctodecyl dimethyl-ammonium bromide. Recently, there has been increasing interest to clays as useful materials for the adsorption of environmental contaminants due to their high surface areas and their less expensive prices compared to granulated activated carbon. With a chemical modification, the inorganic exchange cations of clay can be replaced by organic cations containing long chain alkyl groups. By this way, the surface nature of the clay is changed from highly hydrophilic to increasingly hydrophobic. As a result the modified organoclay becomes a powerful adsorbent for toxic organics. The adsorption of anionic dyes and non-ionic surfactants conventionally used in tanneries. Acid activation of bentonites on the removal of mimosa extract. Organoclays act as sorbents for a great variety of organic pollutants such as trichloroethylene, phenols, polycyclic aromatic hydrocarbons and some weakly acidic pesticides which have been detected in surface waters.

***Pumice treated with cationic surfactant:*** Pumice is a light, porous igneous rock. Pumice can form from any type of magma, including basalt, andesite, dacite, and rhyolite. Pumice varies in density according to the thickness of the solid material between the bubbles. It is a valuable scouring, scrubbing and polishing material in both powdered form and as pumice stone. Pumice has been found to be effective for the removal of phosphate ion from water. A modified form of pumice has been used as a sorbent material for pesticides. Immobilization or separation of contaminants in polluted waters is an objective of increasing importance. In recent years, there has been much interest in developing new, powerful sorbents to remediate contaminated environments. Previous research has found that chemically modified clay minerals (such as bentonite, montmorillonite, kaolinite and smectite) and zeolites represent an innovative and promising class of sorbent materials. Clay minerals and zeolites have permanent negative charges on their surfaces, which enable them to be modified by cationic surfactants to enhance contaminant retention and retard contaminant migration. The advantages of using such materials are their low cost and high contaminant removal capabilities. The industrial growth and population increase have resulted in the release of different pollutants in the environment and the treatment of effluents has become a challenging topic in environmental sciences. Efforts are therefore still needed to have low-cost adsorbents that are versatile *e.g.* high adsorption capacityof inorganic/organic pollutants. Alginate beads are common and well-known support materials in biosciences applications and are now used in the environmental field. Sodium alginate, a water-soluble salt of alginic acid, is a biopolymer and differs from most other polysaccharides exhibiting a sol–gel transition when simply submitted to modification of their ionic environment, *e.g.* substitution of sodium by divalent cations such as calcium. Aluminosilicates (natural or modified)are being tested for several years in the removal of toxic metals or organic pollutants from aqueous solutions due to favourable surface properties, availability, environmental and economical considerations. Some of literature available for removal of heavy metals/dyes/phenolic compounds onto various organo-clays is shown in Table 3.1.2.

### 3.1.2.7. *Zeolite*

The high cost of activated carbon has stimulated interest in examining the feasibility of using cheaper raw materials. Zeolites are hydrated aluminosilicate materials having cage like structures with internal and external surface areas of up to several hundred square metre per gram and cation exchange capacities of up to several meq./kg. Approximately 41 types of natural zeolites are known to exist, and many others have been synthesized. The cost of these adsorbent minerals indicate that cheaper and easily obtainable unconventional adsorbents such as should also be studied for the removal of pollutants from water.Suitable candidates would be the natural zeolites,with more than 40 types, forming the greatest mineral family. Among them, clinoptilolite is the most abundant one, and generally occurs as adiagenetic alteration product of volcanic glassy material and matrix in volcano sedimentary rocks.

Both natural and synthetic zeolites are used in industry as adsorbents, soil modifiers, ion exchangers, and molecular sieves. Natural zeolites possess permanent negative charges in their crystal structures, making them suitable for modification using cationic surfactants. Recent studies on the properties of surfactant-modified zeolite indicate that it is an effective sorbent for multiple types of contaminants. Natural zeolite, containing mainly clinoptilolite has been treated with a 0.02 M solution of hexadecyltrimethylammonium chloride. Surfactant modified zeolite can remove organic compounds and oxyanions from water. Surfactant modified zeolites are produced by treating zeolites with large cationic surfactants (quaternary amines). These bulky organic cations exchange selectively with the native inorganic cations from the zeolite, to form a stable, organic-rich coating on the external surface of the zeolite. Surfactant modification alters considerably the chemistry of the zeolites surface, allowing the zeolites to sorb nonpolar organic pollutants and anions, for which untreated zeolites have little affinity. Partitioning is responsible for organic sorption by surfactant-modified zeolite while transition metal cation sorption is generally unaffected. While the sorption of oxyanions by surfactant-modified zeolite was attributed to anion exchange on the positively charged surfactant bilayer, the sorption of hydrophobic organic contaminants was due to partitioning of the organics into the organic phase created by the surfactant tail groups.

*Interaction of phenols with zeolites*: The factors that affect phenols–zeolite interaction are solubility of phenols in water, repulsion between phenolate anions and negatively charged zeolite, hydrogen bonding between hydroxyl groups of phenols and zeolite surface, possible encapsulation of phenols into the pores, and complexation of phenolate anions with zeolite surface metals (Si, Al, Fe, Ti, Mn, Na, K, Ca, Mg, etc.). Some of literature available for removal of heavy metals/dyes/phenolic compounds onto various zeolite is shown in Table 3.1.2.

3.1.2.8. *Carbon Nanotubes* ($CNT_S$)

Carbon nanomaterials include fullerenes, single-walled carbon nanotubes (SWCNTs) and multi-walled carbon nanotubes (MWCNTs). The monomer structure of fullerene is a closed graphite ball, while carbon nanotubes (CNTs) are rolled-up graphite sheets forming a coaxial tube. One single rolled-up graphite sheet forms the SWCNT structure while several rolled-up graphite sheets form the MWCNT structure. They consist of sheets of carbon atoms covalently bonded in hexagonal arrays that are seamlessly rolled into a hollow cylindrical shape. The cylindrical surface geometry of CNTs has been well defined with outer diameters in the range ~ 1-100 nm and length up to several tens of μm. The CNTs because of their chemical and physical properties attract wide applications such as in the fileld of polymer composites, field emissions, energy storage and sensors. Large specific surface area of CNTs makes them suitable for adsorption of gas, metal ions and organic compounds. Recent studied suggest that CNTs can serve as good adsorbent for a variety of gas and liquid molecules, like hydrogen, butane, dioxins, dichlorobenzene and heavy metal ions [92-97]. The considerable attentions about CNTs depend on their unique structural and mechanical properties, highly thermal stability and large specific surface area. Knowledge of the adsorption behaviors of toxic chemicals by novel manufactured carbon nanotubes (CNTs) is critical for environmental application of CNTs as superior sorbents and for environmental risk assessment of both toxic chemicals and CNTs once they are released to the environment. Though activated carbons commonly have much higher surface area than multi walled carbon nanotubes. Some of literature available for removal of heavy metals/dyes/phenolic compounds onto various nanotubes is shown in Table 3.1.2.

### 3.1.3. *Adsorption kinetic study*

Kinetics studies and dynamic continuous-flow investigations, offering information on the rate of adsorption, together with the hydrodynamic parameters, are very important for adsorption process design.

***Kinetic modeling of adsorption in a batch system*:** Numerous kinetic models have been suggested to describe the reaction order of adsorption systems based on solution concentration. Kinetic models based on the capacity of the adsorbent have also been presented, such as the Lagergren's first-order equation and Ho's second-order expression (Ho, 2006). The first-order equation of Lagergren (Lagergren, 1898) and the pseudo second-order equation are the most widely used kinetic models to describe the adsorption process. The pseudo second-order equation fitted the data very well in a large quantity of literature

**Table 3.1.2:** Removal of heavy metals/dyes/phenolic compounds onto various adsorbents from aqueous solutions available in the research literatures.

| *Adsorbates* | *Adsorbents* |
|---|---|
| Cd(II) | Activated carbon, Sulphur dioxide treated activated carbon, Bagasse fly ash, Rice husk ash, Montmorillonie, Kaolin, Tobermorite, Magnetite, Silica gel and alumina, Calcium alginate beads, Goethite pretreated with phosphate, Turkish fly ashes, Surface oxidized carbon nanotubes (CNT), Activated carbon cloth, Low-grade phosphate, Coffee residues and clay, Lignite-based sorbents, Rice husks, Corncob particles, Amberlite IR-120 synthetic resin, Bone Char, Kaolinite, Rice Polish, Bentonite, Akaganeite-type nanocrystals, Palygorskite clay, Wheat Bran, |
| Cr(III), Cr(VI) | Calcium alginate beads, Goethite pretreated with phosphate, Sawdust, Phosphate treated saw dust, Photo film waste sludge, Turkish fly ashes, Surface oxidized carbon nanotubes (CNT), Bagasse fly ash, Raw Rice Bran, Bentonite, Palygorskite clay, GAC, |
| Ni(II) | Calcium alginate beads, Goethite pretreated with phosphate, Bagasse fly ash, Rice husk ash, Clinoptilolite, Phosphate treated saw dust, Raw Rice Bran, Tea Factory Waste, Crab Shell, GAC, |
| Mn(II) | Calcium alginate beads, Goethite pretreated with phosphate |
| Fe(II) | Calcium alginate beads, Goethite pretreated with phosphate |
| Zn(II) | Activated carbon, Calcium alginate beads, Goethite pretreated with phosphate, Sulphur dioxide treated activated carbon, Bagasse fly ash, Rice husk ash, Clinoptilolite, Phosphate treated saw dust, Activated carbon cloth, $Fe_2O_3$, $Fe_3O_4$, FeS; Mg, Cu, Zn, Al, and Fe pellets; coal, GAC, Low-grade phosphate, Natural zeolites, Coffee residues and clay, Rice husks, Amberlite IR-120 synthetic resin, Bentonite, Bone Char, Kaolinite, Carbon nanotubes, |
| Cu(II) | Bagasse fly ash, Rice husk ash, Calcium alginate beads, Goethite pretreated with phosphate, Clinoptilolite, Phosphate treated saw dust, Natural zeolites, Natural zeolites, Coffee residues and clay, Lignite-based sorbents, Rice husks, Coal fly ash, Corncob particles, Amberlite IR-120 synthetic resin, Bone Char, Kaolinite, |
| Pb(II) | Activated carbon, Sulphur dioxide treated activated carbon, Clinoptilolite, Bagasse fly ash, Coffee residues and clay, Lignite-based sorbents, Rice husks, Coal fly ash, Amberlite IR-120 synthetic resin, Kaolinite, palygorskite clay, GAC |
| Co(II) | Clinoptilolite, $Fe_2O_3$, $Fe_3O_4$, FeS; Mg, Cu, Zn, Al, and Fe pellets; coal, GAC, |
| Fe (II) | Lignite-based sorbents |
| Al(II) | Rice husks |
| Acrylic acid, Acrylonitrile | Powder activated carbon, Cu and Cu-Pd Catalysts, Silica gel, alumina and calcium carbonate |
| Taste and odor compounds | Powder activated carbon |
| Orange-G and Methyl Violet, Malachite green dye, Auramine-O, Congo red | Powder activated carbon, Bagasse fly ash |

**Table 3.2.2:** (*Contd...*)

**Table 3.1.2:** *(Contd...)*

| *Adsorbates* | *Adsorbents* |
|---|---|
| Phenol | Powder activated carbon, Granular activated carbon, Rice husk ash, Bagasse fly ash, Sewage sludge, Samla coal, Manganese nodule leached residue, Activated sewage sludge ($H_2SO_4$), Blast furnace sludge, Samla coal, Bentonite |
| p-methylphenol | Powder activated carbon |
| 4-chlorophenol | Granular activated carbon, Amberlite XAD-4, Rice husk char, Petroleum coke |
| 4-nitrophenol | Granular activated carbon, Bagasse fly ash, Rice husk, Coal fly ash, |
| o-cresol | Granular activated carbon, Bagasse fly ash |
| Catechol, Resorcinol, hydroquinone | Granular activated carbon , Rice husk ash, Waste Fe(III)/Cr(III) Hydroxide, $TiO_2$ surface, Carbon nano tubes, Paper mill sludge, Petroleum coke, Red mud, Activated carbon cloth, Bentonites, |
| Benzoic acid | Activated carbon |
| Aniline | Montmorillonite, Polymeric resin, Carbon nano tubes, Organo-clays, Cr-bentonite, H-Beta zeolites, Multi walled nano tubes, Ca-montmorillonite, Amberlite XAD-16 resin, Natural zeolite |

for adsorption (Ho and McKay, 1999). The Elovich equation for adsorption of gases onto a solid, and applying a second-order rate equation for gas/solid and solution/solid adsorption systems.

### 3.1.3.1. *Pseudo-First-Order Model*

The pseudo-first order model was proposed by Lagergren (1898). This model describes the rate of sorption to be proportional to the number of sites unoccupied by the solutes. The pseudo-first order model works well only in the region where biosorption process occurs rapidly. Ho (2006) noticed that the use of the Lagergren model for prediction of the biosorption kinetics is not suitable for the entire sorption period. The pseudo-second order kinetic model as developed by Ho has the following form:

The sorption of organic molecules from a liquid phase to a solid phase can be considered as a reversible process with equilibrium being established between the solution and the solid phase. Assuming a non-dissociating molecular adsorption of adsorbates onto adsorbent particles, the sorption phenomenon can be described as the diffusion controlled process (Fogler, 1998).

$$A + S \underset{k_d}{\overset{k_a}{\rightleftharpoons}} A.S \tag{3.3.1}$$

Where, A is adsorbate and S is the active site on the adsorbent and A.S is the activated complex. $k_a$ and $k_d$ are the adsorption and desorption rate constants, respectively. It can be shown using first order kinetics that with no adsorbate initially present on the adsorbent (*i.e.* $C_{AS0} = 0$ at $t = 0$), the fractional uptake of the adsorbate by the adsorbent can be expressed as

$$\frac{X_A}{X_{Ae}} = 1 - \exp\left[k_a C_S + \frac{k_a}{K_S}\right]t \tag{3.3.2}$$

where, $X_{Ae}$ = fraction of the adsorbate adsorbed on the adsorbent under equilibrium condition (*i.e.* $C_{ASe}/C_{A0}$).

$K_S = k_a/k_d$ $C_s$ = adsorbent concentration in the solution

Thus, the plot of $\ln\left(1 - X_A/X_{Ae}\right)$ versus $t$ for various values of $C_{A0}$ at a constant $C_S$ will give a straight line which will be coincident on the ordinate at $t = 0$. Constants $K_a$ and $K_d$ can be obtained using relevant relations and the slope of each curve for a given value of $C_S$. Equation (3.3.2) can be transformed as

$$\log(q_e - q_t) = \log q_e - \frac{k_f}{2.303} t \tag{3.3.3}$$

$$\text{where, } k_f = \left(k_a C_S + \frac{k_a}{k_S}\right); \tag{3.3.4}$$

$q = X_A$, and $q_e = X_{Ae}$

This equation is the so-called Lagergren equation (Lagergren, 1898). This equation is, however, valid only for the initial period of adsorption. Various investigators have erroneously fitted this equation to the adsorbate uptake data for later periods, ignoring the data of the initial period. A plot of log $(q_e - q_t)$ versus $t$ enables the determination of the kinetic constants.

Experimental results don't follow pseudo-first order model for the whole period because they differ in two important aspects: (i) $k_f(q_e - q_t)$ does not represent the number of available sites, and (ii) $\log q_e$ was not equal to the intercept of the plot of log $(q_e - q_t)$ against $t$ (Ho and Mckay, 1999).

3.1.3.2. *Pseudo-Second-Order Model*

The pseudo-second-order model can be represented in the following form (Ho and Mckay, 1999):

$$\frac{dq_t}{dt} = k_S(q_e - q_t)^2 \tag{3.3.5}$$

where, $\kappa_S$ is the pseudo-second-order rate constant (g/mg min). Integrating Eq. (3.3.5) and noting that $q_t = 0$ at t = 0, the following equation is obtained:

$$\frac{t}{q_t} = \frac{1}{k_S q_e^2} + \frac{1}{q_e} t \tag{3.3.6}$$

The initial sorption rate, h (mg/g min), as t $\rightarrow 0$ can be defined as

$$h = k_S q_e^2 \tag{3.3.7}$$

The equilibrium adsorption capacity ($q_e$) and the initial sorption rate ($h$) alongwith the pseudo-second order constant $\kappa_S$ can be determined from the slope and intercept of the plot of $t/q_t$ versus $t$.

### 3.1.3.3. *Adsorption diffusion study*

The mathematical treatment of Boyd *et al.* (1947) to distinguish between the particle, and film diffusion and mass action-controlled mechanism of exchange have laid the foundations of sorption/ion-exchange kinetics. In adsorption systems, the mass transfer of solute or adsorbate onto and within the adsorbent particle directly affects the adsorption rate. It is not only important to study the rate at which the solute is removed from aqueous solution in order to apply adsorption by solid particles to industrial uses but also it is necessary to identify the step that governs the overall removal rate in the adsorption process in order to interpret the experimental data. There are essentially four stages in the adsorption of an organic/inorganic species by a porous adsorbent

1. Transport of adsorbate from the bulk of the solution to the exterior film surrounding the adsorbent particle;
2. Movement of adsorbate across the external liquid film to the external surface sites on the adsorbent particle (film diffusion);
3. Migration of adsorbate within the pores of the adsorbent by intraparticle diffusion (pore diffusion);
4. Adsorption of adsorbate at internal surface sites.

All these processes play a role in the overall sorption within the pores of the adsorbent. In a rapidly stirred, well mixed batch adsorption, mass transport from the bulk solution to the external surface of the adsorbent is usually fast. Therefore, the resistance for the transport of the adsorbate from the bulk of the solution to the exterior film surrounding the adsorbent may be small and can be neglected. In addition, the adsorption of adsorbate at surface sites (step 4) is usually very rapid and thus offering negligible resistance in comparison to other steps, *i.e.* steps 2 and 3. Thus, these processes usually are not considered to be the rate-limiting steps in the sorption process.

In most cases, steps (2) and (3) may control the sorption phenomena. For the remaining two steps in the overall adsorbate transport, three distinct cases may occur:

| | | | |
|---|---|---|---|
| Case I: | External transport | $>$ | Internal transport. |
| Case II: | External transport | $<$ | Internal transport. |
| Case III: | External transport | $\approx$ | Internal transport. |

In cases I and II, the rate is governed by film and pore diffusion, respectively. In case III, the transport of ions to the boundary may not be possible at a significant rate, thereby, leading to the formation of a liquid film with a concentration gradient surrounding the adsorbent particles.

Usually, external transport is the rate-limiting step in systems which have (a) poor phase mixing, (b) dilute concentration of adsorbate, (c) small particle size, and (d) high affinity of the adsorbate for the adsorbent. In contrast, the intra-particle step limits the overall transfer for those systems that have (a) a high concentration of adsorbate, (b) a good phase mixing, (c) large particle size of the adsorbents, and (d) low affinity of the adsorbate for adsorbent.

*Intra-particle Diffusion Study*

A functional relationship common to most of the treatments of intraparticle transport is that uptake varies almost proportionally with the square root of time, ($t^{0.5}$), rather than time, (t), itself (Weber and Morris, 1963), *i.e.*

$$q_t = k_{id} t^{0.5} + I \tag{3.3.8}$$

where, $k_{id}$ is the intra-particle diffusion rate constant (mg/g min$^{1/2}$) and $I$ (mg/g) is a constant that gives an idea about the thickness of the boundary layer, *i.e.*, the larger the value of $I$, the greater is the boundary layer effect. The mathematical dependence of fractional uptake of adsorbate on $t^{0.5}$ is obtained if the sorption process is considered to be influenced by diffusion in the cylindrical (or spherical) particle and the convective diffusion in the adsorbate solution. It is assumed that the external resistance to mass transfer surrounding the particles is significant only in the early stages of adsorption.

Many investigators have represented the data points of q versus $t^{0.5}$ by two straight lines- the first straight portion depicting macro-pore diffusion and the second representing micro-pore diffusion (Suresh *et al.*, 1963). Extrapolation of the linear portions of the plots back to the ordinate gives the intercepts, which provide the measure of the boundary layer thickness. The deviation of straight lines from the origin is attributed to the difference in the rate of mass transfer between the initial and final stages of adsorption. Further, such deviation of straight line from the origin indicates that the pore diffusion is not the sole rate-controlling step. The slope of the Weber and Morris plots: q versus $t^{0.5}$, are defined as a rate parameter ($\kappa_{id}$), characteristic of the rate of adsorption in the region where intra-particle diffusion is rate controlling. The higher the value of $\kappa_{id}$ the higher is the intraparticle diffusion rate.

*Determination of Diffusivity*

Kinetic data could be treated by the models given by Boyd *et al.* (1947) which is valid under the experimental conditions used. With diffusion rate controlling in the adsorption particles of spherical shape, the solution of the simultaneous set of differential and algebric equations leads to (Reichenberg, 1953; Helfferich, 1962):

$$F(t) = 1 - \frac{6}{\pi^2} \sum_{z=1}^{\infty} \frac{1}{z^2} \exp\left[\frac{-z^2 \pi^2 D_e t}{R_a^2}\right] \tag{3.3.9}$$

or

$$F(t) = 1 - \frac{6}{\pi^2} \sum_{z=1}^{\infty} \frac{1}{z^2} \exp\left[-z^2 Bt\right] \tag{3.3.10}$$

where, $F(t) = q_t / q_e$ is the fractional attainment of equilibrium at time t, $D_e$ is the effective diffusion coefficient of the adsorbate in the adsorbent phase (m²/s), $R_a$ is the radius of the adsorbent particle assumed to be spherical (m), $z$ is an integer and

$$B = \frac{\pi^2 D_e}{R_a^2} \tag{3.3.11}$$

*Bt* values could be obtained for each observed value of *F(t)* from Reichenberg's Table (Reichenberg, 1953). The linearity test of *Bt* versus time plots are employed to distinguish between the film diffusion- and particle diffusion-controlled adsorption. If the plot of *Bt* versus time (having slope B) is a straight line passing through the origin, then the adsorption rate is considered to be governed by particle diffusion mechanism, otherwise it is governed by film diffusion.

For cylindrical particles, if one assumes that the diffusion takes place radially and the diffusion in the angular and axial directions are negligible, one gets the solution given by Skelland (1974), which after rearrangement leads to:

$$F(t) = 1 - \frac{4}{\pi^2}\sum_{z=1}^{\infty}\frac{1}{b_n^2}\exp\left[-D_e b_n^2 t\right] \tag{3.3.12}$$

where, $b_n$'s are roots of $J_0\left(bn_p R\right) = 0$

Vermeulen's approximation (Vermeulen, 1953) of the Eq. (3.3.2) fits the whole range $0 < F(t) < 1$, for adsorption on to spherical particles. This approximation is given as

$$F(t) = \left[1 - \exp\left(\frac{-\pi^2 D_e t}{R_a^2}\right)\right]^{1/2} \tag{3.3.13}$$

This equation could further be simplified to cover most of the data points for calculating effective particle diffusivity, *i.e.*

$$\ln\left[\frac{1}{(1 - F^2(t))}\right] = \frac{\pi^2 D_e t}{R_a^2} \tag{3.3.14}$$

The $D_e$ can be calculated from the slope of the plot of $\ln\left[1/(1 - F^2(t))\right]$ versus t.

***Bangham's Equation:*** Kinetic data can further be used to know about the step offering larger resistance in the overall sorption phenomena using Bangham's equation (Aharoni *et al.*, 1979).

$$\log\log\left(\frac{C_0}{C_0 - q_t m}\right) = \log\left(\frac{k_{0B} m}{2.303V}\right) + \alpha\log(t) \tag{3.3.15}$$

where, $C_0$ is the initial concentration of the adsorbate in the solution (mg/l), $V$ is the volume of solution (l), $m$ is the adsorbent concentration (g/l), $q_t$ (mg/g) is the amount of adsorbate retained by the adsorbent at time t, and $\alpha$ (<1) and $k_{0B}$ are constants. If the log log $(C_O/(C_O - q_t m))$ versus log($t$) plot yield very good perfect linear curves, then the diffusion of the adsorbate into the pores of the adsorbent is considered as the only rate-controlling step in the kinetic process.

Study of the adsorption equilibrium and kinetics is essential to supply the basic information required for the design and operation of adsorption equipment. Adsorption techniques have gained favor in recent years because they are considered efficient for the removal of trace organic pollutants from water that cannot be removed using other treatment processes. In addition, adsorption and desorption kinetics are technologically important, because the diffusion within solid particles is a phenomenon of great importance in catalysis, metallurgy, microelectronics, materials science, and other numerous scientific and technological applications. A mass transfer occurs during the adsorption process; the first step is the solute transfer through the adsorbent external surface film, and the others are the solute fluid diffusion into the pore holes and the adsorbed molecules' migration along the pore surfaces, if it takes place. The former is characterized by the external mass-transfer coefficient and the last ones by the internal pore and surface diffusivities. Available bulk adsorbate concentration in the liquid phase and adsorbed solute concentration on the solid phase are considered time-dependent. So, it is possible to see how the rate of adsorption changes with time by plotting the concentration decay curve in the liquid phase or the adsorbate concentration growth curve on the solid phase.

### 3.1.4. *Adsorption Equilibrium Study*

Adsorption equilibrium is established when the amount of solute being adsorbed onto the adsorbent is equal to the amount being desorbed. At this point, the equilibrium solution concentration remains constant. The relationship between the amount of adsorbate adsorbed and the adsorbate concentration remaining in solution is described by an isotherm. By plotting solid phase concentration against liquid phase concentration graphically it is possible to depict the equilibrium adsorption isotherm. To optimize the design of an adsorption system for the adsorption of adsorbates, it is important to establish the most appropriate correlation for the equilibrium curves. Equilibrium isotherms are measured to determine the capacity of the adsorbent for the adsorbate. Equilibrium isotherm models are usually classified into the empirical equations and mechanistic models, based on the mechanism of adsorption. (i) Mechanistic models can be used not only to represent, but also to explain and predict the experimental behavior. The empirical models for single solute systems used to describe the adsorption equilibrium were Freundlich, Langmuir, Brunauer–Emmett–Teller (BET) and Redlich-Peterson (R-P) have been used in the literature to describe the equilibrium characteristics of adsorption. The equilibrium of adsorption processes is usually presented using dependencies of the amounts adsorbed on the fluid phase composition for a constant temperature (Ruthven, 1984). An attempt to classify possible shapes of single solute adsorption isotherms was performed by Giles *et al.* (1960). The broad spectrum of possibilities is illustrated in Fig. 3.1.4. In general, adsorption isotherms can be determined only experimentally. Despite of the fact that there are currently a quite large number of experimental methods available, their appropriate and efficient application is still far away from being a routine job. Langmuir model (L type, based on monolayer adsorption of solute) and Freundlich model (F type, developed for heterogeneous surfaces) are the most widely accepted and used in a number of references. The BET model describes the multi-layer adsorption at the adsorbent surface, and assumes that a Langmuir isotherm applies to each layer. These models can provide information of metal-uptake capacities and differences in

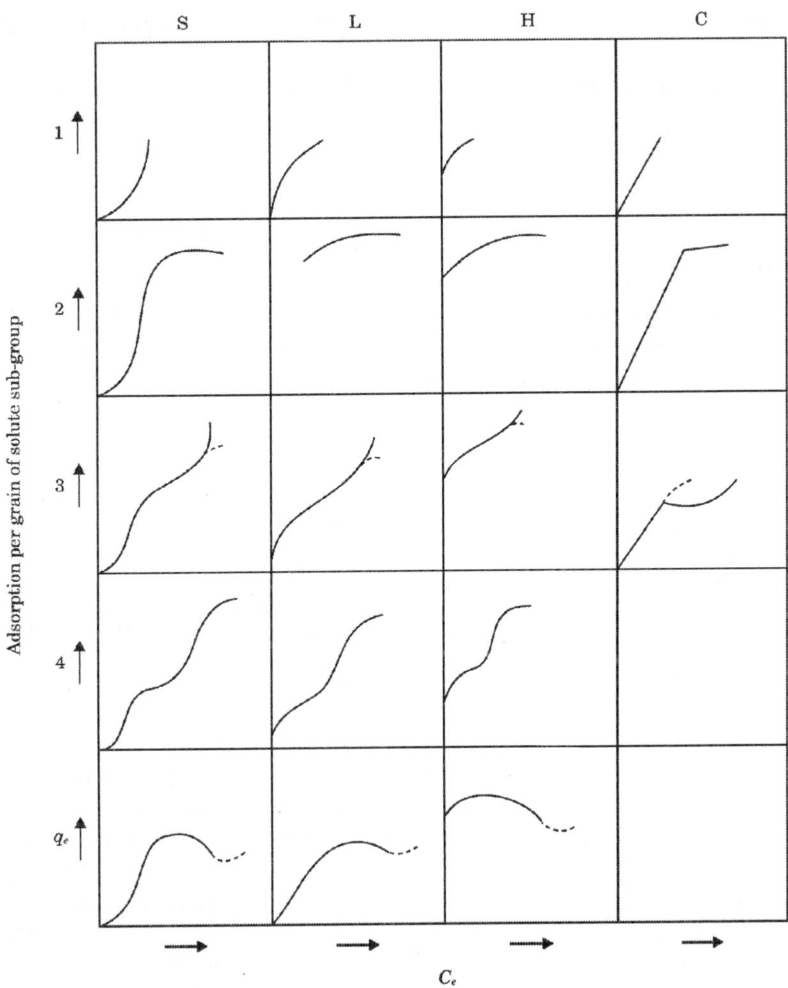


**Fig. 3.1.4:** Classification of isotherms for adsorption from solution (Giles *et al.*, 1960)

metal uptake between various species. Scatchard plots, which originally describe the attractions of proteins for small molecules and ions, were developed to describe the biosorption equilibrium. The total number of binding sites may be extrapolated from Scatchard plots if complete saturation of the biomass binding sites is not observed. To describe two- or multi-metal ion biosorption systems, various extended Langmuir models (also called comparative Langmuir model) or Freundlich type models have been developed. However, those empirical models can hardly explain the adsorption behaviour with a meaningful physical interpretation, and predictive conclusions cannot be drawn for systems operating under different conditions.

#### 3.1.4.1. *Isotherm Models for Liquid Adsorption*

The representation of the adsorption isotherms onto adsorbents can be based on models with two, three, even with more parameters. In the case of the models with two-parameter, the Langmuir and Freundlich equations are largely used. Additionally, other models such as the Temkin, Fowler–Guggenheim, Kiselev, and Hill–de Boer equations exist and allow the determination of the energy of adsorption, interaction energy between adsorbed molecules, and complex formation between adsorbate. In order to determine the best fit isotherm, the experimental equilibrium data were analyzed using thirteen adsorption isotherm models with more than two-parameter; nine three-parameter equations – the Redlich–Peterson, Sips, Langmuir–Freundlich, Fritz–Schlunder, Radke–Prasnitz (three models), Toth, and Jossens isotherms – three four-parameter equation – the Weber–van Vliet, Fritz–Schlunder, and Baudu isotherms – and one fiveparameter equation – the Fritz–Schlunder isotherm. The results reveal that the adsorption isotherm models fitted the experimental data in the order: Baudu (four-parameter) > Langmuir–Freundlich (three-parameter) > Sips (three-parameter) > Fritz–Schlunder (five-parameter) > Toth (three-parameter) > Fritz–Schlunder (four-parameter) > Redlich–Peterson (three-parameter). It is well-known that adsorption affinity is significantly greater in narrow pore materials and this leads to more rectangular isotherms due to an enhancement in adsorption at low concentrations of the adsorptive. The higher values of $K_L$, $n_F$, $E_S$ and $b_R$ obtained with the physically activated samples, in comparison to the chemically activated samples, shows that this applies equally well to solute adsorption from aqueous solutions as to, as is well-known, gas phase adsorption.

Liquid-phase adsorption, in general, is a more complex phenomena than the gas-phase adsorption [Suresh *et al.*, 2001a]. For example, although one may envision monolayer coverage in liquid adsorption, the adsorbed molecules are not necessarily tightly packed with identical orientation. Other complications include the presence of solvent molecules and the formation of micelles from adsorbed molecules. For this reason, it is prudent not to attach too great a physical significance to isotherm expression; rather they should be regarded merely as empirical expression for representing experimental data with in limited concentration ranges. Various isotherm equations like those of Freundlich, Langmuir, and Redlich-Peterson (R-P) have been used in the literature to describe the equilibrium characteristics of adsorption.

***Freundlich Equation:*** The Freundlich equation (Freundlich, 1906) is one of the earliest empirical equations used to describe equilibria data. This equation takes the following form:

$$q_e = K_F C_e^{1/n} \tag{3.4.16}$$

where, $K_F$ is the Freundlich constant (l/mg), and $1/n$ is the heterogeneity factor. The parameter n is usually greater than unity. The larger is the value of n, the more non-linear is the adsorption isotherm, and as n becomes larger than 10, the adsorption isotherm approaches a so-called rectangular isotherm (or irreversible isotherm). The term "irreversible isotherm" is normally used because the pressure (or concentration) needs to go down to an extremely low value before adsorbate molecules would desorb from the surface to the bulk fluid.

Freundlich constants, $K_F$ and $1/n$ indicate the adsorption capacity and adsorption intensity, respectively. Higher the value of $1/n$, the higher will be the affinity between the adsorbate and the adsorbent, and the heterogeneity of the adsorbent sites. The $1/n$ value indicates the relative distribution of energy sites and depends on the nature and strength of the adsorption process.

Although the Freundlich equation was proposed originally as an empirical equation, it has a sound theoretical footing. It is obtained by assuming that the surface is heterogeneous in the sense that the adsorption energy is distributed and that the surface topography is patchwise, *i.e.* the sites having the same adsorption energy are grouped together into one patch. Another assumption is that on each patch, the adsorbate molecule adsorbs onto one and only one adsorption site. These models can be applied at a constant pH. These models are used in literature for modeling of biosorption equilibrium in the presence of one metal. These values are plotted in a 2D line where the specific uptake q is reported as a function of the metal concentration $C_{eq}$, but the above said adsorption isotherms may exhibit an irregular pattern due to the complex nature of both the sorbent material and its varied multiple active sites, as well as the complex solution chemistry of some metallic compounds. Evaluation of equilibrium sorption performance needs to be supplemented by process-oriented studies of its kinetics and eventually by dynamic continuous flow tests.

***Langmuir isotherm:*** The Langmuir model assumes a monolayer adsorption of solutes onto a surface comprised of a finite number of identical sites with homogeneous adsorption energy.

Langmuir (1918) isotherm is derived with the basic assumption that the sorption takes place at specific homogeneous sites within the adsorbent. This model is expressed as follows:

$$q_e = \frac{K_L q_m C_e}{1 + K_L C_e} \tag{3.3.17}$$

where, $K_L$ is the Langmuir adsorption constant (l/mg) related to the energy of adsorption. It is a measure of the attraction forces between an adsorbate molecule and a solid adsorbent surface. $q_m$ signifies the adsorption capacity (mg/g) corresponding to complete monolayer coverage. The assumptions of the Langmuir model are that the (1) surface is homogeneous, which means that the adsorption energy is constant over all surface adsorption sites (2) adsorbed atoms or molecules are adsorbed at definite sites, and (3) that each site can accommodate only one molecule or atom.

***Redlich-Peterson (R-P) isotherm:*** Redlich and Peterson (1959) isotherm has a linear dependence on concentration in the numerator and an exponential function in the denominator. It approaches the Freundlich model at high concentrations and is in accord with the low concentration limit of the Langmuir equation. Furthermore, the R-P equation incorporates three parameters into an empirical isotherm, and therefore, can be applied either in homogenous or heterogeneous systems due to its high versatility. The R-P equation is

$$q_e = \frac{K_R C_e}{1 + a_R C_e^{\beta}} \tag{3.3.18}$$

where, $K_R$ is R-P isotherm constant (l/g), $a_R$ is also a constant $(\text{l/mg})^{1/\beta}$ and $\beta$ is the exponent which lies between 0 and 1. For $\beta = 1$, Eq. (3.3.18) gives to Langmuir equation, *i.e.*

$$q_e = \frac{K_R C_e}{1 + a_R C_e} \tag{3.3.19}$$

where $K_R = K_L q_m$ and $a_R = K_L$.

For $\beta = 0$, Eq. (3.4.1.3) gives to Henry's equation: *i.e.*

$$q_e = HC_e$$

where, $H$ is the Henry's constant and is equal to $\dfrac{K_R}{1 + a_R}$; Eq. (3.4.18) can be linearized by taking logarithms of both sides as,

$$\ln\left(K_R \frac{C_e}{q_e} - 1\right) = \ln a_R + \beta \ln C_e \tag{3.3.20}$$

A minimization procedure can be adopted to solve Eq. (3.3.19) by using a trial and error for various values $K_R$ of maximizing the correlation coefficient between the predicted data for $q_e$ from Eq. (3.3.20) and the experimental data using the solver add-in function of the MS excel.

***Temkin Isotherm:*** This isotherm assumes that (i) the heat of adsorption of all the molecules in the layer decreases linearly with coverage due to adsorbate-adsorbate interactions, and (ii) that the adsorption is characterized by a uniform distribution of binding energies up to some maximum binding energy (Temkin and Pyzhev, 1940). This isotherm contains a factor that explicitly takes into account the interactions between the adsorbing species and the adsorbent. The isotherm is represented as:

$$q_e = \frac{RT}{b} \ln(K_T C_e) \tag{3.3.21}$$

which in its linearized form as:

$$q_e = B_T \ln K_T + B_T \ln C_e \tag{3.3.22}$$

where, $B_T = \frac{RT}{b}$ (3.3.23)

$B_T$ is related to the heat of adsorption and $K_T$ is the equilibrium binding constant (l/mol) corresponding to the maximum binding energy. The isotherm constants, $K_T$ and $B_T$ can be obtained from a plot of $q_e$ versus $\ln C_e$

***Dubinin-Radushkevich (D-R) Isotherm:*** Another equation used in the analysis of isotherms was proposed by Dubinin and Radushkevich (1947).

$$q_e = q_s \exp\left(-B_{DR}\varepsilon^2\right) \tag{3.3.24}$$

where, $q_s$ is D-R constant and $\varepsilon$ can be correlated as,

$$\varepsilon = \text{RT} \ln \left(1 + \frac{1}{\text{C}_\text{e}}\right) \tag{3.3.25}$$

The constant $B_{DR}$ gives the mean free energy $E$ of sorption per molecule of sorbate when it is transferred to the surface of the solid from the infinite distance in the solution and can be computed using the following relationship:

$$E = \frac{1}{\sqrt{2B_{DR}}} \tag{3.3.26}$$

### 3.1.4.2. *Determining Best Model*

***Error analysis:*** Some of different error functions were employed in adsorption study to find out the most suitable isotherm model to represent the experimental data. These error functions are given in Table 3.1.4.2. The most commonly used one, the sum of the squares of the errors (SSE) function, has a major drawback in that it provides isotherm parameters showing a better fit at the higher end of the adsorbate concentration. This is because the magnitude of the errors and hence the square of the errors increases as the adsorbate concentration increases. Isotherm parameters determined using the sum of the absolute errors (SAE) method provide a better fit as the magnitude of the errors increases, biasing the fit towards the high concentration data. The average relative error (ARE) function attempts to minimize the fractional error distribution across the entire concentration range. The hybrid fractional error function (HYBRID) (Porter *et al.*, 1997) was developed in order to improve the fit of the SSE method at low concentration values by dividing by the measured value. In addition a divisor was included as a term for the number of degrees of freedom for the system—the number of data points (n) minus the number of parameters (p) within the isotherm equation. Marquardt's percent standard deviation (MPSD) error function (Marquardt, 1963) has been used previously by a number of researchers in the field. It is similar in some respects to a geometric mean error distribution modified according to the number of degrees of freedom of the system. A number of investigators use correlation coefficient to determine the best-fit kinetic model from the experimental data. The model whose correlation coefficient is closer to unity is considered better than the other one to represent the kinetic data. However, comparison of error between the experimental and predicted values for different models may be a better way to approximate the best-fit kinetic model. As different forms of the equation affected $R^2$ values more significantly during the linear analysis, the non-linear analysis might be a method of avoiding such errors.

**Table 3.1.4.2:** Error function generally used in the literature for kinetic/isotherm modeling in batch/column study

***Error Function***

$$HYBRID = \frac{100}{n-p}\sum_{i=1}^{n}\left[\frac{(q_{e,\exp} - q_{e,calc})}{q_{e,\exp}}\right]_i$$

HYBRID was developed to improve the fit of the square of errors function at low concentration values.

$$MPSD = 100\sqrt{\frac{1}{n-p}\sum_{i=1}^{n}\left(\frac{(q_{e,\exp} - q_{e,calc})}{q_{e,\exp}}\right)_i^2}$$

The MPSD is similar in some respects to a geometric mean error distribution modified according to the number of degrees of freedom of the system.

$$SSE = \sum_{i=1}^{n}(q_{e,\exp} - q_{e,cal})_i^2$$

The most commonly used error function, SSE, has one major drawback in that it will result in the calculated isotherm parameters providing a better fit at the higher end of the liquid phase concentration range. This is because of the magnitude of the errors, which increase as the concentration increases.

$$\text{Least square method} \left(\frac{\sum_{i}^{n}(Y_i - Y_i^*)^2}{n}\right)^{1/2}$$

$$SSWR = \sum_{i=1}^{n}\sum_{j=1}^{m}\frac{\Delta_{ij}^2}{w_j^2}$$

$$RSS = RSS_x + RSS_s + RSS_p$$

$$Q = W_X^2 \sum_{All\ times}(X - X_{\exp})^2 + W_S^2 \sum_{All\ times}(S - S_{\exp})^2 + W_P^2 \sum_{All\ times}(P - P_{\exp})^2$$

$$Q = \sqrt{\frac{\sum_{i}^{n}(S_i - S)^2}{n}}$$

$$J_K = \frac{1}{N_e}\sum_{j=1}^{N_e}\left[\frac{(X(t_j) - X_e(t_j))^2}{X_{e,\max}^2} + \frac{(S(t_j) - S_e(t_j))^2}{S_{e,\max}^2} + \frac{(P(t_j) - P_e(t_j))^2}{P_{e,\max}^2}\right]$$

**Table 3.1.4.2:** (*Contd...*)

**Table 3.1.4.2:** *(Contd...)*

$$\varphi = \sum_i \left( \frac{X_i^{\exp} - X_i^{Calc}}{X_i^{Calc}} \right)^2 + \sum_i \left( \frac{S_i^{\exp} - S_i^{Calc}}{S_i^{Calc}} \right)^2$$

$$E(\theta) = \sum_{n=1}^{np} \frac{(X_n - X_{en})^2}{Xe_{\max}^2} + \frac{(S_n - S_{en})^2}{Se_{\max}^2} + \frac{(P_n - P_{en})^2}{Pe_{\max}^2}$$

$$Z_X = \sum_{j=1}^{n} (X_{j,\exp} - X_{j,sim})^2 Z_P = \sum_{j=1}^{n} (P_{j,\exp} - P_{j,sim})^2 Z_S = \sum_{j=1}^{n} (S_{j,\exp} - S_{j,sim})^2$$

$$E = \frac{(RSS_2 - RSS_1) / (DF_2 - DF_1)}{RSS_1 / DF_1}$$

$$SS = \sqrt{\sum \frac{(y_e - y_c)^2}{N}}$$

ye and yc are the experimental value and calculated value according the model; N is the number of the experimental point.

$$ERRSQ = \sum_{i=1}^{m} (X_{cal} - X_{\exp})^2$$

### 3.1.4.3. *Multi-component Isotherm Models*

A generalized isotherm equation which was for multicomponents adsorption data has been modified for single-solute system. When a number of components are present in the solution, and are to be adsorbed simultaneously, the adsorbate-adsorbate and the adsorbate-adsorbent interactive effects play important roles in the overall sorption phenomena. In general, a mixture of components as adsorbates can produce three possible types of behavior: synergism (the effect of the mixture is greater than that of each of the individual effects of the constituents in the mixture), antagonism (the effect of the mixture is less than that of each of the individual effects of the constituents in the mixture), and non-interaction (the effect of the mixture is neither more nor less than that of each of the individual effects of the constituents in the mixture, *i.e.* the adsorbate interactions have no effects). For correlating the experimental adsorption data for multi-component (adsorbate) systems, single component isotherm equations have been modified to take into account the effects of the components on each other's sorption uptake. Some of model equations are given below (Suresh *et al.*, 2011c):

***Non-modified Competitive Langmuir Model:*** The extension of the basic Langmuir model to competitive adsorption can be formulated as follows:

$$q_{e,i} = \frac{q_{m,i} K_{L,i} C_{e,i}}{1 + \sum_{j=1}^{N} K_{L,j} C_{e,j}} \qquad (3.3.27)$$

where, the $q_{m,i}$ and $K_{Li}$ are derived from the corresponding individual Langmuir isotherm equations. When extended to binary mixtures Eq. (3.3.27) can be written as

$$q_{e,1} = \frac{q_{m,1}K_{L,1}C_{e,1}}{1 + K_{L,1}C_{e,1} + K_{L,2}C_{e,2}} \tag{3.3.28}$$

$$q_{e,2} = \frac{q_{m,1}K_{L,1}C_{e,1}}{1 + K_{L,1}C_{e,1} + K_{L,2}C_{e,2}} \tag{3.3.29}$$

where, $q_{e,1}$ and $q_{e,2}$ are the amount of adsorbate adsorbed per unit mass of sorbent for the first and the second component, respectively; $C_{e,1}$ and $C_{e,2}$ are the equilibrium concentrations, $q_{m,1}$ and $q_{m,2}$ represent the maximum adsorbate uptake and $K_{L,1}$ and $K_{L,2}$ are the Langmuir constants for the first and second components in a mixture of metal ions, respectively.

***Modified Competitive Langmuir Isotherm:*** Individual adsorption constants may not define exactly the multi-component adsorption behaviour of metal ion mixtures. For this reason, better accuracy may be achieved by using modified isotherms related to the individual isotherm parameters and to correction factors. Better accuracy may be achieved by extracting additional coefficients from experimental competitive isotherms. An interaction term, $\eta_{L,i}$, which is a characteristic of each species and depends on the concentrations of the other components, has been added in the competitive Langmuir model (Bellot and Condoret, 1993). The modified competitive Langmuir isotherm equation, then, becomes

$$q_{e,i} = \frac{q_{m,i}K_{L,i}\left(C_{e,i}/\eta_{L,i}\right)}{1 + \sum_{j=1}^{N} K_{L,j}\left(C_{e,j}/\eta_{L,j}\right)} \tag{3.3.30}$$

where, the $q_{m,i}$ and $K_{L,i}$ are derived from the corresponding individual Langmuir isotherm equations, and the $\eta_{L,i}$ are estimated from competitive adsorption data. For binary mixtures, this equation can be rewritten as for the first and the second component, respectively and these two equations can be solved simultaneously to obtain the two-component Langmuir adsorption constants of the first and the second components, respectively.

***Extended-Langmuir Isotherm:*** The major assumption of this model is that the surface sites are uniform, so that the all the adsorbates in the solution compete for the same surface sites. The final expression of the multicomponent Langmuir model can be described as follows (Yang, 1987):

$$q_{e,i} = \frac{q_{\max}K_{EL,i}C_{e,i}}{1 + \sum_{j=1}^{N} K_{EL,j}C_{e,j}} \tag{3.3.31}$$

In this model the maximal capacity $q_{max}$ is universal so that all components obey the fundamental hypothesis of the Langmuir model. The $K_{EL,i}$ values reflect the affinity between the adsorbent and the adsorbates in the multi-component systems

***Sheindorf–Rebuhn–Sheintuch (SRS) Model:*** Sheindorf *et al.* (1981) derived a Freundlich-type multi-component adsorption isotherm, the Sheindorf–Rebuhn–Sheintuch (SRS) equation, to represent experimental adsorption data. A general SRS equation for the adsorption isotherm for component $i$ in a $N$-component system is written in the form:

$$q_{e,i} = K_{F,i} C_{e,i} \left( \sum_{j=1}^{N} a_{ij} C_{e,j} \right)^{(1/n_i)-1} \tag{3.3.32}$$

The pre-exponential coefficient $K_{F,i}$ and the exponent $1/n_i$ are determined from the data of single-component systems. The competition coefficients $a_{ij}$ describe the inhibition to the adsorption of component $i$ by component $j$, and can be determined from the thermodynamic data, or more likely, from the experimental data of multicomponent systems. The assumptions incorporated in the derivation of SRS equation are: (i) each component individually obeys the Freundlich isotherm; and (ii) for each component in a multi-component adsorption system, there exists an exponential distribution of site adsorption energies, given as

$$N_i(Q) = \alpha_i \exp(-\beta_i Q/RT) \tag{3.3.33}$$

where, $\alpha_i$ and $\beta_i$ are constants; And (iii) the coverage by each adsorbate at each energy level $Q$ is given by the multicomponent Langmuir isotherm:

$$\theta_i(Q) = \frac{K_i C_{e,i}}{1 + \sum_{j=1}^{N} K_j C_{e,j}} \tag{3.3.34}$$

$$K_j = K_{0j} \exp(Q/RT) \tag{3.3.35}$$

Integration of $N_i(Q)\theta_i(Q)$ over energies in the range $-\infty$ to $+\infty$ yields and the competition coefficients are defined as $a_{ij} = K_{0j}/K_{0i}$ and thus $a_{ji} = 1/a_{ij}$. This SRS equation has been successfully applied to different types of contaminants in the multi-component systems.

***Modified Competitive Redlich-Peterson Model:*** The competitive non-modified R-P model is modified to the following modified competitive R-P Model to take the characteristic of each species into account.

$$q_{e,i} = \frac{K_{R,i}\left(C_{e,i}/\eta_{R,i}\right)}{1+\sum_{j=1}^{N} a_{R,j}\left(C_{e,j}/\eta_{R,j}\right)^{\beta,j}} \tag{3.3.36}$$

where, $\eta_{R,i}$ are estimated from the competitive adsorption data.

### 3.1.4.4. *Factors Controlling Adsorption*

The amount of adsorbate adsorbed by an adsorbent from adsorbate solution is influenced by a number of factors as discussed below.

***Nature of Adsorbent:*** The physico-chemical nature of the adsorbent is important. Adsorbents differ in their specific surface area and affinity for adsorbate. Adsorption capacity is directly proportional to the exposed surface. For the non-porous adsorbents, the adsorption capacity is inversely proportional to the particle diameter whereas for porous material it is practically independent of particle size. However, for porous substances particle size affects the rate of adsorption. For substances like granular activated carbon, the breaking of large particles to form smaller ones open up previously sealed channels making more surface accessible to adsorbent. Pore sizes are classified in accordance with the classification adopted by the International Union of Pure and Applied Chemistry (IUPAC) [IUPAC, 1982], that is, micro-pores (diameter (d) <20 Å), meso-pores (20 Å < d < 500 Å) and macro-pores (d > 500 Å). Micro-pores can be divided into ultra-micropores (d < 7 Å) and super micro-pores (7 Å < d < 20 Å). The main characteristics of phenols playing an important role in adsorption mechanisms are hydrophobicity, solubility, molecular weight and cross-sectional area, and acidity. The adsorption driving forces due to dispersion interactions between π-electrons of the aromatic ring of phenols and those of the aromatic structure of the activated carbon perfectly correlates the adsorption capacity. This mechanism can also be applied to interpret the adsorption behavior of the series of phenols. As the number of electron-withdrawing chloro substituent increases, the π-electron density in aromatic ring decreases and as a result, affinity to the π-electrons of double bonds in activated carbons is higher.

***pH of Solution:*** The surface charge as well as the degree of ionization is affected by the pH of the solution. The effects of the pH on the carbon surface are well documented in various reviews, such as the recently published one by Radovic [1999]. The adsorption capacity of the carbon for the solutes in their molecular form depends on the electron density of the solutes. This is due to the dispersive interaction between the aromatic ring of the solutes and those of the carbon surface which are the main forces involved in the adsorption process. Since the hydrogen and hydroxyl ions adsorbed readily on the adsorbent surface, the adsorption of other molecules and ions is affected by pH. It is a common observation that a surface adsorbs anions favourably at low pH and cations in high pH range.

The adsorption capacity of the activated carbons depended on the solution pH. Thus, at acid pH, the amount adsorbed remained practically constant or increased slightly with increasing pH. When increased further, there was a decrease in the amount of phenolic compounds adsorbed due to the negative charge density on the external and internal surface on the activated carbon studied at pH higher than 10. Generally, phenol has higher adsorption capacity than chlorinated phenols onto zeolites, due to benzene ring ($\pi$- electron) rather than the hydroxyl substituent of phenols interacts with synthetic zeolite surface (without pH and ionic strength adjustment) and mode of interaction, which weakens with the increase of substitution of chlorine on benzene ring.

***Contact Time:*** In physical adsorption most of the adsorbate species are adsorbed within a short interval of contact time. However, strong chemical binding of adsorbate with adsorbent requires a longer contact time for the attainment of equilibrium. Available adsorption results reveal that the uptake of adsorbate species is fast at the initial stages of the contact period, and thereafter, it becomes slower near the equilibrium. In between these two stages of the uptake, the rate of adsorption is found to be nearly constant. This may be due to the fact that a large number of active surface sites are available for adsorption at initial stages and the rate of adsorption is a function of available vacant site. Concentration of available vacant sites decreases and there is repulsion between solute molecules thereby reducing the adsorption rate. Initially, large number of vacant surface sites is available for adsorption; the adsorption rate is very fast thus increases rapidly the amount of adsorbates accumulated on the sorbent surface mainly within the first hour of adsorption. As a result, the remaining vacant surface sites are difficult to be occupied due to formation of repulsive forces between the phenolic compounds on the solid surface and the bulk phase. Most of adsorption process is normally fitted by pseudo-second-order model onto activated carbon. This suggested that the overall rate of the adsorption process was controlled by chemisorption which involved valency forces through sharing or exchange of electrons between the sorbent and sorbate. Adsorption on porous solid adsorbent from aqueous phase generally involves the transport of adsorbate in bulk phase, solid phase (by either pore diffusion and/or surface diffusion), and actual adsorption step. The last step is considered to be fast enough, thus the slowest of the transport steps would determine the overall rate of adsorption. In general, the rate of uptakes in systems with poor mixing, having adsorbate at low enough concentration, high affinity towards adsorbent and small particles are governed by external transport. In contrast, transports in systems with different characteristics than those mentioned above are limited by intra-particle diffusion step.

***Initial Concentration of Adsorbate:*** A given mass of adsorbent can adsorb only a fixed amount of adsorbate. So the initial concentration of adsorbate solution is very important. The amount adsorbed decreases with increasing adsorbate concentration as the resistance to the uptake of solute from solution of adsorbate decreases with increasing solute concentration. The rate of adsorption is increased because of the increasing driving force. A given mass of adsorbent can adsorb only a fixed amount of adsorbate. So the initial concentration of adsorbate solution is very important. The amount adsorbed decreases with an increase in the adsorbate concentration as the resistance to the uptake of solute from solution decreases with an increase in solute concentration in the solution. The rate of adsorption is thus, increased because of the increased driving force.

***Degree of Agitation:*** Agitation in batch adsorption is important to ensure proper contact between the adsorbent and the solution. At lower agitation speeds, the stationary fluid film around the particle is thicker and the process is external mass transfer controlled. With the increase in agitation this film decreases in thickness and the resistance to mass transfer due to this film is reduced and after a certain point the process becomes intraparticle diffusion controlled. Whatever is the extent of agitation, the solution inside the pores remains unaffected and hence, for intraparticle mass transfer controlled process agitation has no effect on the rate of adsorption.

***Temperature:*** Temperature dependence of adsorption is of complex nature. Adsorption processes are generally exothermic in nature and the extent and rate of adsorption in most cases decreases with increasing temperature. This trend may be explained on the basis of rapid increase in the rate of desorption or alternatively explained on the basis of Le-Chatelier's principle. Some of the adsorption studies show increased adsorption with increasing temperature. This increase in adsorption is mainly due to an increase in number of adsorption sites caused by breaking of some of the internal bonds near the edge of the active surface sites of the adsorbent. Also, if the adsorption process is controlled by the diffusion process (intraparticle transport-pore diffusion), than the sorption capacity increases with an increase in temperature due to endothermicity of the diffusion process. An increase in temperature results in an increased mobility of the metal ions and a decrease in the retarding forces acting on the diffusing ions. These result in the enhancement in the sorptive capacity of the adsorbents.

In all cases the isotherms are of class S, sub-group 4, according to the classification by Giles *et al.* (1960) based on the initial slope and the shape of the upper part of the curves. Such isotherms occur when the adsorbent–adsorbate interactions increase as the loading increases. They are characterized by low adsorption at low concentration, due to weak adsorbent–adsorbate interactions. However, once a molecule is adsorbed, adsorbent–adsorbate forces increase and promote them adsorption in a cooperative fashion. At high concentration, multilayer adsorption can take place too. The fit of the data to Freundlich isotherm indicates the heterogeneity of the sorbent surface. It has been stated by that magnitude of the exponent $1/n$ gives an indication of the favourability and capacity of the adsorbent/adsorbate system. Values $n > 1$ represent favorable adsorption conditions. In most cases, the exponent between $1 < n < 10$ shows beneficial adsorption. The three-parameter models resulted in better fittings than the two-parameter models. In this study, three parameter model, Redlich–Peterson, fitted the experimental data a little better than the Langmuir and the Toth models. Variation in the maximum adsorption capacity determined experimentally, theoretically, and from the Freundlich model in relation to five physico-chemical properties of the substrates in question. These are: molecular weight, p*K*a, solubility, octanol/water partition coefficient ($\log_{Kow}$), and cross-sectional area of adsorbed, ó. The solubility seems play a very significant role in adsorption. A decrease in solubility and p*K*a is associated to an increase in adsorption capacity. The adsorption capacity for phenol and chlorophenols are function of molecular weight and cross sectional area. Additionally, it seems that the adsorption capacity is directly proportional to the adsorbate hydrophobicity.

The shape of adsorption isotherms, according to Giles *et al.* (1960) classification, shows that phenolic compounds are adsorbed in flat position on the graphite layers, which confirms the adsorption driving forces due to $\pi$–$\pi$ dispersion interactions between the aromatic ring of phenols and the aromatic structure of the activated carbon.

### 3.1.4.5. *Thermodynamic Study*

The heat of adsorption is a significant property for the differentiation of physical and chemical adsorption. For any process to occur spontaneously, the Gibbs free energy, $\Delta G$, must decrease. Adsorption also obeys this rule. According to the second law of thermodynamics,

$$\Delta G = \Delta H - T\,\Delta S \tag{3.3.37}$$

The entropy of adsorbate decreases on going from gas phase to the adsorbed phase, therefore $\Delta S$ is negative, because the adsorbed state has fewer degrees of freedom (it is more ordered) than the gaseous state. It is generally assumed that the entropy of adsorbent is unaltered by adsorption. Hence, for the majority of adsorption, $\Delta S$ are less than zero. Therefore, according to equation 3.3.37, it is necessary that $\Delta H$ be less than zero for $\Delta G$ to be less than zero. This means that every adsorption process for which the entropy of adsorbent-adsorbate decreases during the adsorption, the adsorption process will be exothermic ($\Delta H$<0). Heat of adsorption can be calculated from adsorption isotherms using the vant Hoff equation (Ruthven, 1984):

$$\frac{\partial \ln K}{\partial T} = \frac{\Delta H}{RT^2} \tag{3.3.38}$$

Where, $\Delta H$ is the heat of adsorption

The concept of thermodynamic assumes that in an isolated system where energy cannot be gained or lost, the entropy change is the driving force. The adsorption reaction for the endothermic processes could be due to the increase in temperature increased the rate of diffusion of the adsorbate molecules across the external boundary layer and in the internal pores of the adsorbent particle, owing to the decrease in the viscosity of the solution in the other hand suggested that the increase in adsorption uptake with increase in temperature was due to the possibility of an increase in the mobility of the adsorbate molecules.

The positive values of $\Delta S^{o}$ obtained showed the affinity of the activated carbon for adsorbate and the increasing randomness at the solid–solution interface with some structural changes in the adsorbates and adsorbents during the adsorption process whereas the positive values of $\Delta G^{o}$ obtained indicated the non-spontaneous nature of the adsorption process at the range of temperatures. The negative value of $\Delta H^{o}$ suggests the exothermic nature of the adsorption. Based on the apparent enthalpy changes of adsorption (<45 kJ/mol), type of transition between physical and chemical adsorption. $\Delta G^o$ values were negative indicating that the sorption process led to a decrease in Gibbs free energy. Negative $\Delta G^o$ values indicate the feasibility and spontaneity of the adsorption process. The enthalpy changes $\Delta H^o$ were negative and the absolute values were lower than 40 kJ/mol, indicating that the adsorption is an exothermic physical process. In addition, the enthalpy change was found to be dependent of the equilibrium adsorption amount, which was due to the heterogeneity of the adsorption sites on adsorbent surface. The adsorption entropy changes ($\Delta S^o$) were less than zero, which accorded with the fact that the mobility of adsorbates on adsorbent surface becomes more restricted as compared with that of those in solution. Besides, it was shown that $\Delta S^o$ increased with the increase in equilibrium adsorption amount. The free energy changes ($\Delta G^o$) were less than zero, demonstrating that the adsorption is a spontaneous process. While the equilibrium adsorption amount increased, the absolute value of $\Delta G^o$ decreased correspondingly, which indicated the decrease in adsorption impetus. On the same equilibrium adsorption amount condition, the absolute values of $\Delta G^o$ decreased with increasing temperature. Both of the phenomena suggested that on the condition of larger equilibrium adsorption amount and higher temperature, the reverse process of adsorption–desorption could be easily carried out. The isosteric enthalpies of adsorption were calculated with a derivative Van't Hoff equation. Values of the enthalpy changes (always negative) are indicative of an exothermic process, and their magnitudes (< 43 KJ/mol) show the adsorption to be a physical adsorption.

### 3.1.5. Column Study

In order to properly design and operate fix-bed filters, the adsorption process must be understood, particularly adsorption kinetics. Adsorption in a fixed-bed is a dynamic, non-steady process, which involves the accumulation of solute molecules onto an adsorbent from a solvent crossing the column. For a sound engineering design of water treatment plants both equilibrium and kinetic data are required (Stenzel, 1993). However, most of the scientific work published on the removal of organic contaminants from water solutions has been carried out only in equilibrium conditions. There are several reasons for this, such as: (1) dynamic experiments are more expensive and time-consuming than equilibrium ones; (2) the results are more difficult to interpret; and (3) the process depends on the possible interaction of a number of variables. Powder activated carbon can be added directly to stirred reactors or fluidized-bed reactors to efficiently remove refractory organic compounds from wastewaters,but granular activated carbon packed in fixed beds is a more popular treatment process because of easier operation and less carbon loss. In order to design and operate fixed-bed carbon adsorption processes successfully, the adsorption equilibria and dynamics for specified systems must be known. Many adsorption isotherm models of single-solute and multisolute systems were reported in the literature. Schematic diagram of experimental set up is shown in Fig. 3.5.1. The columns were provided with a feed port at the bottom centre of the column. The lower portion of the column was filled with 0.2–0.3 cm diameter glass beads up to a height of 1.5 cm.

This portion was used for the uniform distribution of the solution across the whole cross-section of the column and to dampen the fluctuating flow phenomena induced by the peristaltic pump. The portion was attached to the main column through two flanges with O-ring rubber seals and a sintered metal disc in between. The column was packed with activated carbon to give different effective bed heights. The column was further loaded with aqueous solution of adsorbate using a peristaltic pump. The solution flowed upwards. Samples were withdrawn from different sampling ports at different time intervals. The column assembly was kept in a temperature controlled chamber maintained at 303 ± 1K with the help of a thermostatic heating device.

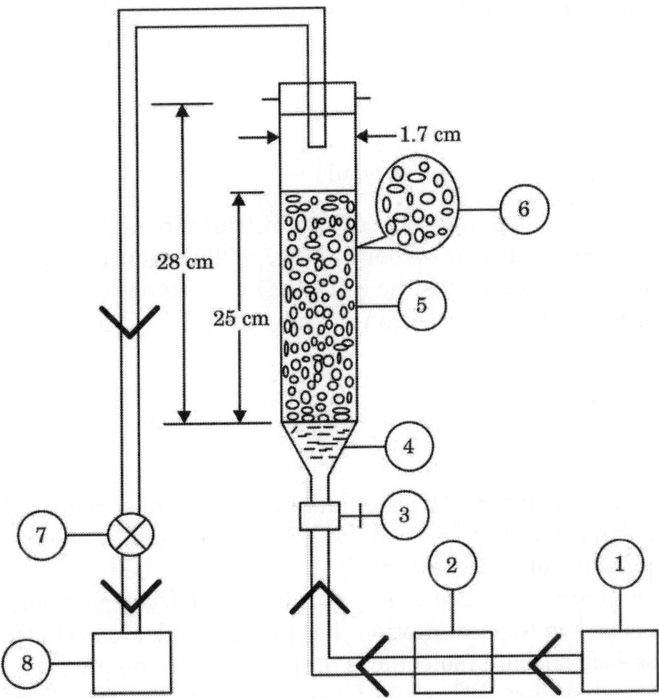


**Fig. 3.5.1:** Schematic diagram of fixed bed column bioreactor, packed with activated carbon designed to function as a continuous flow system for sorption. (1) adsorbate reservoir; (2) peristaltic pump; (3) flow control; (4) glass column reactor; (5) sorbent; (6) enlarged view of column packing; (7) extention for secondary column; and (8) effluent storage. All dimensions are referred.

***Theory:*** The most important criterion in the design of a column adsorber is the prediction of the column breakthrough or the shape of the sorption wave front, which determine the bed length ($z$) and the operating life span and regeneration time of the bed. Factors that affect the column breakthrough include both operating variables and characteristics of the sorbate and the sorbent. Several models have been used for the prediction of breakthrough times. Some of these are discussed below.

***Thomas model:*** The data obtained in column in continuous mode studies were used to calculate maximum solid phase concentration of sorbate onto sorbent and the adsorption rate constant using the kinetic model developed by Thomas (1944). The Thomas solution is one of the most general and widely used methods in column performance theory. The expression by Thomas for an adsorption column is given as follows: This model assumes plug-flow behavior in the bed and uses the Langmuir isotherm for equilibrium and second-order reversible reaction kinetics. It assumes a constant separation factor but is applicable to both favorable and unfavorable sorption conditions. Sorption is usually not limited by chemical reaction kinetics but is often controlled by interphase mass transfer. Therefore, this model is suitable for sorption processes in which external and internal diffusion limitations are absent. Application of the model can lead to some error in sorption processes that follow first-order reaction kinetics. The Thomas model is given by

$$\frac{C_t}{C_0} = \frac{1}{1 + \exp[k_{Th}(q_o x - c_0 V_{eff}) / \upsilon]} \tag{3.3.39}$$

Where $k_{Th}$ is the Thomas rate constant (ml/min mg), $q_0$ the equilibrium sorbate uptake per g of the adsorbent (mg/g), $x$ the amount of adsorbent in the column (g), $V_{eff}$ the effluent volume (ml), $C_0$ the influent sorbate concentration (mg/l), $C_t$ the effluent concentration at time $t$ (mg/l) and $\upsilon$ is the flowrate (ml/min).

$$\ln\left(\frac{C_0}{C_t} - 1\right) = \frac{k_{Th} q_0 m}{Q} - k_{Th} C_0 t \tag{3.3.40}$$

The value of $C_t/C_0$ is the ratio of effluent and influent sorbate concentrations. The value of time $t$ (min) is $t = V_{eff}/\upsilon$. The kinetic coefficient $k_{Th}$ and the adsorption capacity of the column $q_0$ can be determined from a plot of $C_t/C_0$ against $t$ at a given flow rate using non-linear regression.Yoon and Nelson (1984) model based on the assumption that the rate of decrease in the probability of sorption for each sorbate molecule is proportional to the probability of sorbate sorption and sorbate breakthrough on the sorbent. The equation for the 50% breakthrough concentration from a fixed bed of sorbent is

$$\ln\left(\frac{C}{C_0 - C}\right) = k_{YN} t - t_{0.5} k_{YN} \tag{3.3.41}$$

Where $k_{YN}$ is the Yoon-Nelson rate constant ($min^{-1}$). The values of $k_{YN}$ and $t_{0.5}$ can be obtained from the slope and intercept, respectively, of a linear plot of $\ln[C/(C_0 - C)]$ versus t. Clark (1987) model used the mass-transfer coefficient in combination with the Freundlich isotherm to define a new relation for the breakthrough curve as

$$\frac{C}{C_0} = \left(\frac{1}{1 + Ae^{-rt}}\right)^{1/n-1} \tag{3.3.42}$$

with

$$A = \left( \frac{C_0^{n-1}}{C_b^{n-1}} - 1 \right) e^{rt_b} \tag{3.3.43}$$

and

$$r = \frac{\beta}{U} \upsilon_m (n-1) \tag{3.3.44}$$

Where $n$ is the Freundlich constant, $C_b$ is the concentration of sorbate at breakthrough time $t_b$ (mg/l), and $\upsilon_m$ is the migration velocity of the concentration front in the bed (cm/min). $\upsilon_m$ can be determined from the relationship

$$\upsilon_m = \frac{UC_0}{N_0 + C_0} \tag{3.3.45}$$

Equation 3.3.42 can be rearranged to the following linear form. For a particular sorption process in a fixed bed with a chosen treatment objective, the values of $A$ and $r$ can be determined by using the above equation, thereby enabling the prediction of the breakthrough curve.

$$\ln\left[ \left( \frac{C_0}{C} \right)^{n-1} - 1 \right] = -rt + \ln A \tag{3.3.46}$$

Wolborska Model (1989) deduced the following relationship for describing the concentration distribution in a bed for the low-concentration range of the breakthrough curve

$$\ln \frac{C}{C_0} = \frac{\beta C_0}{N_0} t - \frac{\beta Z}{U} \tag{3.3.47}$$

Where $\beta$ is the kinetic coefficient of external mass transfer ($min^{-1}$) and the other symbols have their usual meanings. The values of $\beta$ and $N_0$ can be determined from a plot of $\ln(C/C_0)$ versus $t$ at a given $Z$ and $Q$.

***The bed-depth/service time analysis (BDST) model:*** The BDST model is based on physically measuring the capacity of the bed at different breakthrough values. The BDST model works well and provides useful modeling equations for the changes of system parameters. The Bohart-Adams model is used for the description of the initial part of the breakthrough curve, which relates $C/C_0$ to time, $t$, for a continuous-flow sorber column. A modified form of the equation that expresses the service time at breakthrough, $t$, as a fixed function of operation parameters is BDST model. Hutchins (1973) linearized eq 3.3.45 to give

$$t = \frac{N_0}{C_0 F} Z - \frac{1}{K_a C_0} \ln\left( \frac{C_0}{C_t} - 1 \right) \tag{3.3.48}$$

Where $C_t$ is the effluent concentration of solute in the liquid phase (mg/l), $C_0$ the initial concentration of solute in the liquid phase (mg/l), F the influent linear velocity (cm/min), $N_0$ the adsorption capacity (mg/g), $K_a$ the rate constant in BDST model (l/ mg min), t the time (min) and Z is the bed depth of column (cm).

This equation predicts linear BDST plots between t and Z. The sorptive capacity of the system, $N_0$, can be evaluated from the slope of the plot, and the rate constant, k, can be calculated from the intercept, I, where

$$I = -\frac{1}{K_a C_0} \ln\left(\frac{C_0}{C} - 1\right) \tag{3.3.49}$$

At 50% breakthrough, $C/C_0 = 0.5$ and $t = t_{0.5}$. Therefore, the last term in eq 3.5.11 becomes 0, and the half-time, $t_{0.5}$, is obtained as

$$t_{0.5} = \frac{N_0}{C_0 F} Z \tag{3.3.50}$$

Thus, a plot of $t_{0.5}$, the time at 50% breakthrough, versus $Z$ should be a straight line passing through the origin, and $N_0$ can be calculated from the slope. Where $N_0$ and $K$, the adsorption capacity and rate constant, respectively, can be evaluated.

***The effect of flow rate on breakthrough curve:*** Breakthrough generally occurred faster with higher flow rate, while reaching saturation time was increased significantly with a decreased in the flow rate. When at a low rate of influent, adsorbate had more time to contact with rice husk that resulted in higher removal of adsorbate in column. The variation in the slope of the breakthrough curve and adsorption capacity may be explained on the basis of mass transfer fundamentals. The reason is that at higher flow rate the rate of mass transfer gets increases, *i.e.* the amount of adsorbate adsorbed onto unit bed height (mass transfer zone) gets increased with increasing flow rate leading to faster saturation at higher flow rate. According to the "tailing" of a breakthrough curve (*i.e.*, a slow approach of $C/C_0$ toward 1) is commonly observed in liquid-phase sorption where intraparticle diffusion is the rate limiting transport process. As the flowrate value through the bed increased, the length of the sorption zone, Z, increased, and the time of contact of the solution with the sorption zone decreased.

***Effect of influent adsorbate concentration on breakthrough curve:*** At lower influent concentrations, breakthrough curves were dispersed and breakthrough occurred slower. As influent concentration increased, sharper breakthrough curves were obtained. This can be explained by the fact that more adsorption sites were being covered as the adsorbate concentration increasing. The larger the influent concentration, the steeper is the slope of breakthrough curve and smaller is the breakthrough time. These demonstrate that the change of concentration gradient affects the saturation rate and breakthrough time, or in other words, the diffusion process is concentration dependent. As the influent concentration increases, adsorbate loading rate increases, so does the driving force increase for mass transfer, which in a decrease in the adsorption zone length.

***Empty bed contact time (EBCT) and sorbent usage rate (Ur):*** Contact time is an important parameter in the design of sorption columns and is often expressed in terms of the empty bed contact time. EBCT affects the volume to breakthrough and the shape of the breakthrough curve. This is determined by the equation

$$EBCT = \frac{V_C}{Q} = \frac{A_C Z}{Q} \tag{3.3.51}$$

The performance of a fixed bed can be further evaluated in terms of the sorbent usage rate, $U_r$, defined as the weight of sorbent saturated per cubic decimeter of sorbate solution treated. $U_r$ is given by

$$U_r = \frac{m_C}{V_b} = \frac{V_C \rho}{V_C N_b} = \frac{\rho}{N_b} \tag{3.3.52}$$

Where $V_C$ is the volume of sorbent in the bed (l), $A_C$ is the cross-sectional area of the column (cm$^2$), $m_C$ is the mass of sorbent in the column (g), $V_b$ is the volume of solution treated at breakthrough (l), F is the apparent density of sorbent in the column (g/cm$^3$), and $N_b$ is the number of bed volumes of solution to breakthrough.

***The effect of different bed depth on breakthrough curve:*** The higher bed column results in a decrease in the solute concentration in the effluent at the same time. The slope of breakthrough curve decreased with increasing bed height, which resulted in a broadened mass transfer zone. High uptake was observed at the highest bed height due to an increase in the surface area of biosorbent, which provided more binding sites for the sorption. The BDST model constants can be helpful to scale up the process for other flow rates and concentration without further experimental run.

***Modeling of column study:*** Thomas model was applied to the experimental data with respect to flow rate, influent concentration of adsorbate and bed depth. A non-linear regression analysis was used on each set of data to determine the Thomas model parameters of $q_0$ and $k_{Th}$. The determined coefficients and the error values were also obtained using nonlinear regression analysis according Eq. (3.5.1). With flow rate increasing, the value of $q_0$ decreased but the value of $k_{Th}$ increased. As the bed depth increased, the value of $q_0$ increased significantly while the value of $k_{Th}$ decreased significantly. Thomas model is the second-best fit after only the Woolborska model. This rate controlling mechanism is not constant but is specific to feed concentration, sorbent, column design, and operation conditions. Empirical methods for determining the limiting diffusion resistance of a particular exchanger typically entail testing with varying particle sizes or 'interruption' experiments, which stop the contact between particle and fluid at different time intervals, either in batch or column contactor. When particle diffusion is the rate controlling resistance, the mass-transfer rate is proportional to the diffusivity of ions in the solid phase and inversely proportional to the square of the particle radius.

## 3.1.6. CASE STUDIES

### 3.1.6.1 *I Adsorption of Phenol onto Activated Alumina and Molecular Sieve*

*Introduction*

Phenols are commonly encountered in aqueous effluents from various manufacturing processes such as oil refineries, coke plants, and phenolic resin plants, and are toxic substances that should be removed from the aquatic environment. Phenol and chlorophenols are long-lived pollutants frequently found in industrial effluents. However, these compounds are considered as the primary pollutants in the wastewater due to their high toxicity, high oxygen demand, and low-biodegradability. Consequently, removal of them from wastewater has attracted significant environmental concerns. Phenolic compounds impart objectionable taste and odour to drinking water at concentration as low as 0.005 mg/l. Besides, they are harmful to organisms and toxic to humans. Phenols are considered as priority pollutants since they are harmful to organisms at low concentrations and many of them have been classified as hazardous pollutants because of their potential to harm human health. Because of their toxicity, the US Environmental Protection Agency (EPA) and the European Union have designated phenols as priority pollutants (USEPA, 1987). Increasing concern for public health and environmental quality has led to the establishment of rigid limits on the acceptable environmental levels of specific pollutants. Thus, the removal or destruction of phenols from process or waste streams becomes a major environmental problem. Activated alumina and molecular sieve has good adsorptive properties and has been used for the removal of organic compound from aqueous solutions; and surfactants like sodium dodecyl and octylphenol.

The present case study aims to model the isotherm of phenol adsorption onto molecular sieve by various isotherm models, available in the literature and to find out the best-fit isotherm models.

*Experimental design*

***Characterization of activated alumina (AA) and molecular sieve (MS):*** The AA and MS was supplied by S.D. fine-chem limited, (Mumbai, India) was manufactured from coconut shells. The physico-chemical characterization of AA and MS was performed using standard procedures. The particle size analysis of AA and MS was: Mesh size 3-5 mm, 5-8 mm for AA and 2.5-5mm, 2-3mm, 1-2mm for MS. The energy dispersive spectrum (EDX) of fresh AA and MS showed the presence of oxygen, aluminum, and sodium and silica oxide (Fig. 3.1.6.1). The morphologies of blank AA and blank MS have been examined by SEM is shown in Fig. 3.1.6.2. This Fig. reveals surface texture and porosity of the blank AA and MS. XRD patterns of blank AA and MS are shown in Fig. 3.1.6.3. It shows that alumina is an aggregate of small crystalline structure.

*Chemicals*

All the chemicals used in the study were of analytical reagent (AR) grade. Phenol was obtained from Ranbaxy Fine Chemicals, New Delhi, India and S.D. fine

Chemicals, Mumbai, India. NaOH, and $H_2SO_4$ or HCl were obtained from S.D. fine Chemicals, Mumbai, India. A stock solution of phenol was made by dissolving exact amount of phenol in distilled water, respectively. Test solutions were prepared by diluting 1 g/l of stock solution with double-distilled water. The range of concentration of phenol prepared from stock solution varied between 50-1000 mg/l.

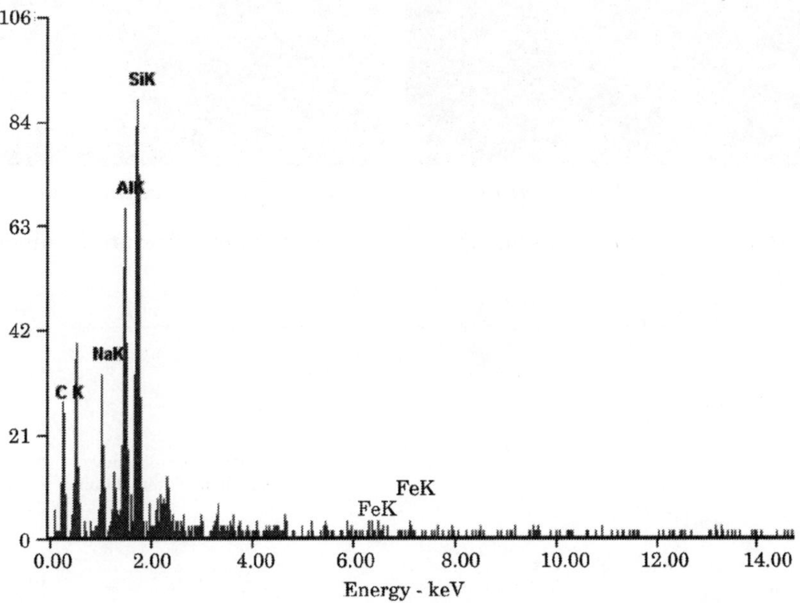


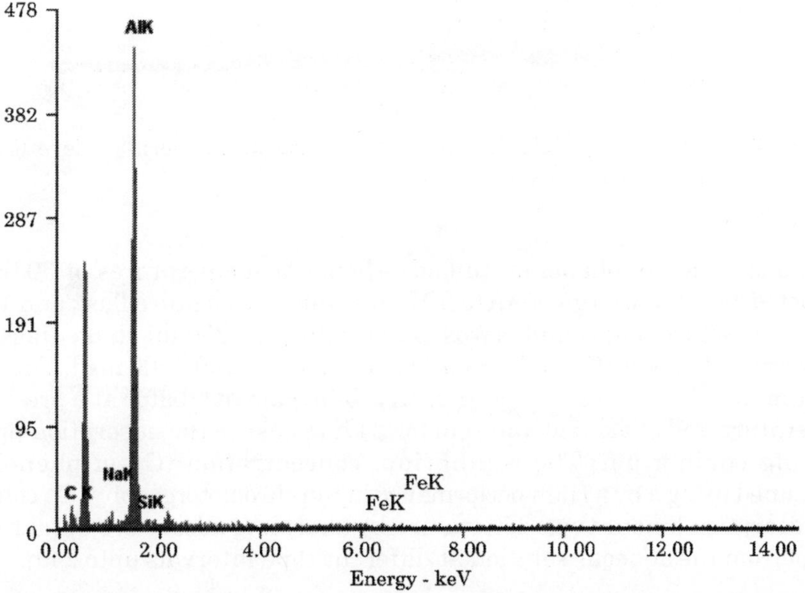


**Fig. 3.1.6.1:** EDX of activated alumina and molecular sieve (Suresh, unpublished work)

***Fresh activated alumina*** ***Fresh molecular sieve***

**Fig. 3.1.6.2:** SEM of blank-Activated alumina and Molecular sieve(Suresh, unpublished work).

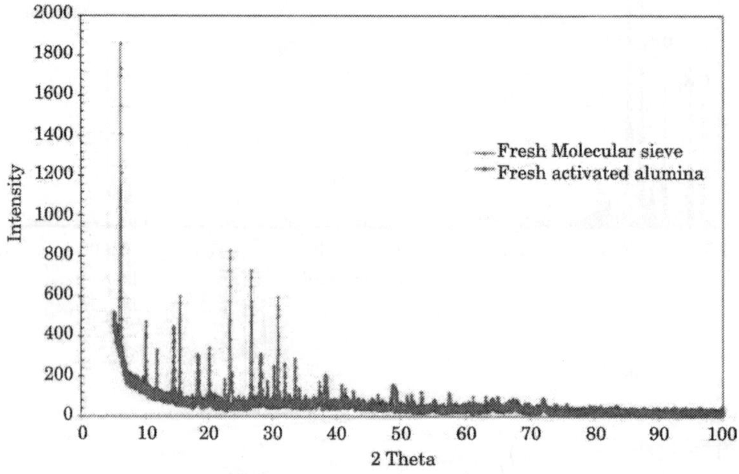


**Fig. 3.1.6.3:** XRD pattern of blank activated alumina and molecular sieve (Suresh, unpublished work).

*Batch adsorption*

Static adsorption of phenol on the adsorbents at temperatures of 303K was conducted as follows: 10 g/l of AA (or) MS was introduced into a flask and 100 ml of aqueous solution of C and R was added into each 250 ml shake flask. The initial concentrations ($C_0$) of the solutions ranged from 50-1000 mg/l. The flasks were completely sealed and placed in an orbital cum incubator at a pre-settled temperature and shaken at 150 rpm for 24 h to ensure the adsorption process reaching equilibrium. The equilibrium concentration ($C_e$) of phenol was determined using a by a High performance liquid chromotorgraphy ($C_{18}$ column). The adsorption kinetics was determined by analyzing adsorptive uptake of the phenol from the aqueous solution at different time intervals upto 24 h.

The removal of phenol from the solution and the adsorption uptake in solid phase, ($q_t$) calculated using the following relationships:

$$\text{Percentage phenol removal} = 100(C_o - C_t) / C_o \qquad (3.3.53)$$

$$\text{Amount of adsorbed phenol per g of solid, } q_t = (C_o - C_t)V / w \qquad (3.3.54)$$

Where, $C_0$ is the initial phenol concentration (mg/l), $C_t$ is the phenol concentration (mg/l) at any time, $t$; $V$ is the volume of the solution (L) and $w$ is the mass of the adsorbent (g). For equilibrium values, the corresponding values at equilibrium time were taken for calculation.

### *Results and Discussion*

*Effect of dosage (m)*

The effect of the adsorbent dosage on the adsorption of phenol on AA or MS is shown in Fig. 3.1.6.4. AA or MS dosage was varied between 1 and 50 g/l and equilibrated for 24 h. It can be revealed from Fig. 3.1.6.4. that the removal of phenol increases with increase in AA or MS dosage. The phenol removal efficiency of the AA resin was 44% at the adsorbent concentration of 10 g/l, while it is varied from 41% to 38% at adsorbent concentrations of 15–50 g/l. The removal of phenol at AA or MS dosage greater than 10 g/l remains almost constant and then decrease. Thus, the equilibrium concentrationis considered to be 10 g/l (44% removal for AA and 51% removal for MS.

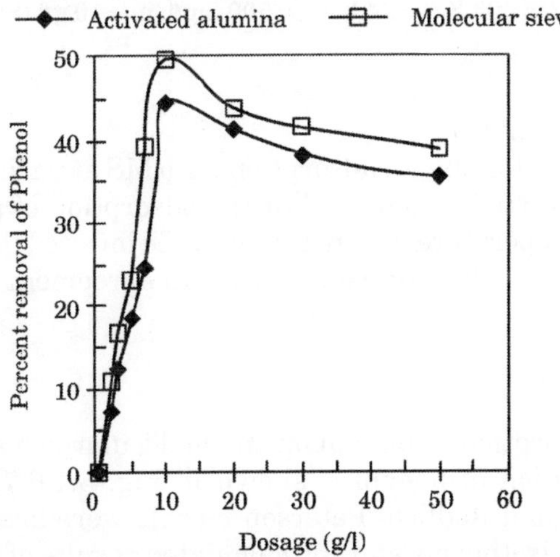


**Fig. 3.1.6.4:** Effect of adsorbent dose on the adsorption of phenol by activated alumina and molecular sieve.$T$ = 303 K, $t$ = 24 h, $C_0$ = 500 mg/l (Suresh, unpublished work).

*Effect of pH*

The initial pH of adsorption medium is one of the most important parameters affecting the adsorption process. Effect of initial pH on adsorption of phenol was studied with initial concentration of 500 mg/l and optimum dose of 10 g/l. The experimental result for the uptake of the phenol on AA or MS in the pH range of 2-12 is shown in Fig. 3.1.6.5. It was observed that the relative amount of phenol adsorbed was significantly affected by pH. The maximum adsorption occurs at pH 4 for both adsorbents. Significant decline in removal efficiency was observed for further increase in pH, which may be attributed to formation of phenolate anions. Therefore, in this work, the pH 4 of the solutions of phenol was maintained constant by using suitable buffer solutions.

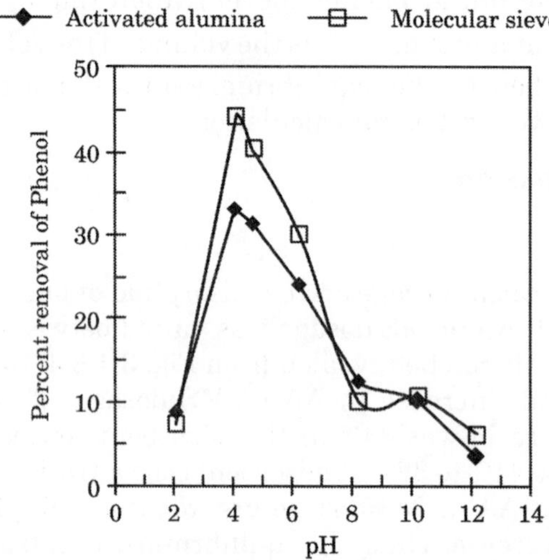


**Fig. 3.1.6.5:** Effect of *pH* on the adsorption of phenol by molecular sieve and activated alumina. $T$ = 303 K, $t$ = 24 h, $C_0$ = 500 mg/l, m = 10 g/l (Suresh, unpublished work).

*Effect of temperature*

Equilibrium isotherms for phenol adsorption onto MS at various temperatures are shown in Fig. 3.1.6.6. It is found that the adsorption of phenol decreases with an increase in temperature. It can, therefore, be inferred that the adsorption is exothermic in nature. The trend obtained is in agreement with the general adsorption processes.

*Adsorption equilibrium*

The equilibrium adsorption of phenol on the molecular sieve as a function of the initial concentration of phenol is shown in Fig. 3.1.6.7. The Langmuir, Freundlich, Temkin and Redlich–Peterson models were used to describe the equilibrium sorption isotherms and the calculated results of these models are given in Table 3.1.6.1. It is found that the adsorption of phenol on the molecular sieve was correlated well with the R-P equation as compared with other three equations under the concentration range studied (Fig. 3.1.6.6). The $R^2$ values for R-P isotherm are closer to unity in comparison to the values obtained for other isotherms, and also the SSE error values are least for the fit of R-P isotherm. Therefore, R-P the best-fit isotherm equation for the adsorption of phenol at all temperatures.

*Adsorption thermodynamics*

The temperature dependence of phenol uptake was studied by performing batch experiments at 303-318 K. According to classical thermodynamics, the Gibbs

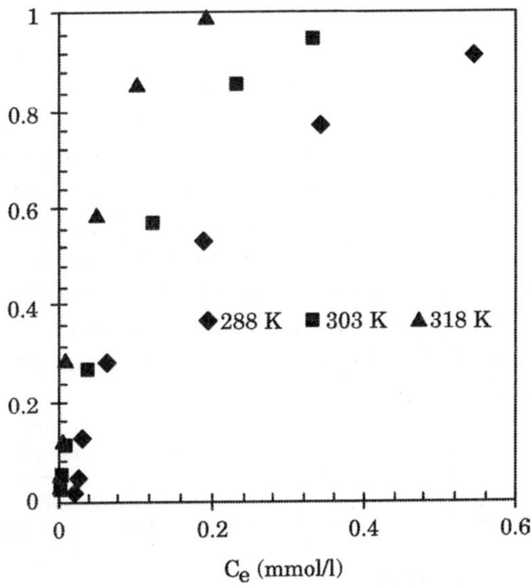


**Fig. 3.1.6.6:** Equilibrium adsorption isotherms at different temperature for Phenol-Molecular sieve system, t = 24 h, $C_0$ = 0.53–10.63 mmol/l, m = 10 g/l. Experimental best data points predicted by R-P equation (Suresh, unpublished work).

**Table 3.1.6.1:** Isotherm parameters for the removal of phenol by molecular sieve (t=24 h, $C_0$= 0.53–10.63 mmol/l, m = 10 g/l) (Suresh, unpublished work).

| | | **Freundlich** | | |
|---|---|---|---|---|
| ***Temp.(K)*** | ***$K_F$(mmol/g) (l/mmol g)$^{-1/n}$*** | ***1/ n*** | ***$R^2$*** | ***SSE*** |
| 288 | 0.52 | 0.73 | 0.991 | 0.072 |
| 303 | 0.62 | z0.70 | 0.993 | 0.081 |
| 318 | 1.04 | 0.55 | 0.967 | 0.089 |

| | | **Langmuir** | | |
|---|---|---|---|---|
| ***Temp. (K)*** | ***$q_m$ mmol/g)*** | ***$K_L$ (l/mmol)*** | ***$R^2$*** | ***SSE*** |
| 288 | 0.52 | 2.73 | 0.993 | 0.053 |
| 303 | 0.57 | 4.52 | 0.991 | 0.034 |
| 318 | 0.67 | 8.84 | 0.993 | 0.042 |

| | | **Temkin** | | |
|---|---|---|---|---|
| ***Temp. (K)*** | ***$K_T$(l/ mmol)*** | ***$B_1$*** | ***$R^2$*** | ***SSE*** |
| 288 | 58.58 | 0.08 | 0.972 | 0.039 |
| 303 | 448.53 | 0.06 | 0.890 | 0.050 |
| 318 | 551.43 | 0.07 | 0.960 | 0.032 |

| | | **R-P** | | | |
|---|---|---|---|---|---|
| ***Temp. (K)*** | ***$K_R$(l/g)*** | ***$a_R$ (l/mmol)*** | ***β*** | ***$R^2$*** | ***SSE*** |
| 288 | 1.88 | 3.33 | 0.66 | 0.996 | 0.035 |
| 303 | 39.19 | 49.36 | 0.31 | 0.997 | 0.031 |
| 318 | 19.85 | 26.79 | 0.69 | 0.999 | 0.024 |

free energy change $\Delta G_0$ is also related to the entropy change ($\Delta S_0$) and the heat of adsorption ($\Delta H_0$) at a constant temperature ($T$) by the equation is given in the theory thermodynamics section. The values of $\Delta G_0$, $\Delta H_0$ and $\Delta S_0$ for phenol-molecular sieve system is shown in Table 3.1.6.2. $\Delta G_0$ values were negative indicating that the sorption process led to a decrease in Gibbs free energy and that the adsorption process is feasible and spontaneous. The adsorption of phenol onto molecular sieve is endothermic in nature, giving a positive value of $\Delta Ho$. The value of ΔHæ% for phenol–MS, adsorption systems were found to be 51.20 kJ/mol, respectively. The values of ΔSæ% for the respective systems were found to be 231.32 kJ/mol K. The positive value of ΔSæ% suggests an increased randomness at the solid–solution interface, and an increase in the degree of freedom of the adsorbed species.

**Table 3.1.6.2:** Thermodynamic parameters of the adsorption of phenol onto MS.

| *Temp. (K)* | $\Delta G_0$*(kJ/ mol)* | $\Delta H_0$*(kJ/ mol)* | $\Delta S_0$*(J/ mol K)* |
|---|---|---|---|
| 288 | –19.03 | 51.20 | 231.32 |
| 303 | –27.05 | | |
| 318 | –31.52 | | |

***Conculsion:*** In the present case study, isotherm studies and related thermodynamics are reported for the adsorption of phenol from aqueous solution onto molecular sieve. Langmuir, Freundlich, Temkin and Redlich–Peterson isotherm models were used to illustrate the experimental isotherms and isotherms constants. Equilibrium data fitted very well with the Redlich–Peterson model. Adsorption of phenol adsorption onto molecular sieve is favourably influenced by an increase in the temperature of the operation. Adsorption was found to endothermic in nature.

### 3.1.6.2. Case Study. II. Adsorption of Colour in Activated Carbon Packed Column

#### *Introduction*

Most dyes are inert and non-toxic at the concentration discharged into the receiving water; however, they impart color undesirable to the water user. The release of these colored waste waters in the ecosystem is a dramatic source of esthetic pollution, of eutrophication and of perturbations in the aquatic life. Colored wastes in the receiving streams interfere with transmission of sunlight into streams and, therefore, reduce photosynthetic activity. Activated carbon (AC) is the most popular adsorbent, which has been used with great success. However, AC is expensive and its regeneration and reuse makes it more costly. Batch experiments are usually done to measure the effectiveness of adsorption for removing specific adsorbates as well as to determine the maximum adsorption capacity. The continuous adsorption in fixed-bed column is often desired from industrial point of view. It is simple to operate and can be scaled-up from a laboratory process. A continuous packed bed adsorber does not run under equilibrium conditions and the effect of flow condition at any cross-section in the column affects the flow behaviour downstream. The flow behaviour and mass transfer aspects become peculiar beyond a particular length to diameter ratio of the column. In order to design and operate fixed-bed adsorption process successfully, the breakthrough curves under specified operating conditions must be predictable. The shape of this curve is influenced by the individual transport process in the column and in the adsorbent. Breakthrough determines bed height (Z) and the operating life span of the bed and regeneration times.

The present case study aims to explore the possibility of utilizing AC or the adsorptive removal of colour in continuous mode using a packed bed column. In order to analyze the column dynamics in the adsorption process, the influence of the bed height (Z), on breakthrough curves have been investigated.

## *Experimental design*

***Column Studies:*** Experimental arrangement used for the dynamic column studies was for removal of colour using activated carbon. Adsorption experiments are conducted in continuous mode using the experimental setup as shown in Fig. 3.1.6.21. The experimental setup consists of a vapor generation unit and a fixed bed for adsorption. Adsorption experiments are conducted at room temperature, and the fixed bed column is a borosilicate glass of 30 cm diameter and three different lengths 1 m, 0.5 m, 0.3 m. The column was packed with activated carbon up to different lengths, 1 m, 0.5 m, 0.3 m from the bottom, and was loaded with aqueous colour solution a peristaltic pump. Samples were withdrawn from different ports at different time intervals. The column assembly was kept in a temperature-controlled chamber maintained at 303 K by use of a thermostatic heating device. The samples are analyzed in the colorimeter for

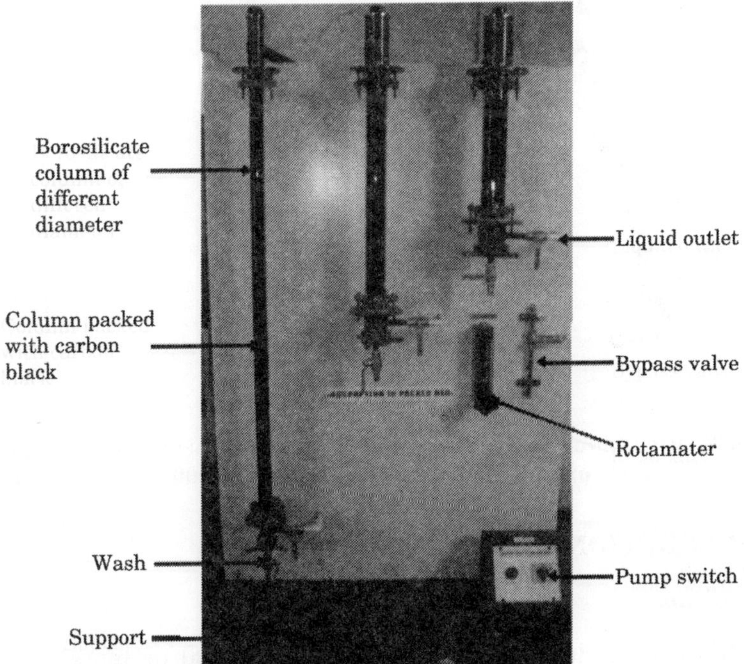


**Fig. 3.1.6.21:** Laboratory scale of adsorption in packed column (Suresh, unpublished work).

measuring colour of the solution. In most of the experimental runs, the breakthrough point was taken at $C/C_0$=0.10, *i.e.*, at $t_{0.1}$. All experimental runs were performed with initial colour = 100 PSU. The length of the bed, Z, was varied from 30 to 90 cm, and the flow rate, Q, was varied using rotatomer.

***Chemical:*** $KMNO_4$ was obtained from Ranbaxy Fine Chemicals, New Delhi, India and S.D. fine Chemicals, Mumbai, India.

### *Results and Discussion*

Operational parameters such bed length (Z), $C_o$, Q and D are important in column design. The effects of these parameters on the adsorption capacity of AC were studied for the colour reduction. The breakthrough curves were examined at different length (Z) at constant Co, Q and D (3cm) and the results are shown in Fig. 3.1.6.22 by data points. Higher reduction of colour was observed at highest Z. This was due to the increase in the number of binding sites owing to increase in adsorption surface area of the adsorbent. As the value of Z increased, colour reduction had sufficient time to diffuse into whole mass of AC that resulted in an increase in $t_b$. The slope of breakthrough curve decreased with an increase in Z due to the broadening of mass transfer zone. For three time ($t_1$=540 min, $t_2$= 660 min, $t_3$=1320 min for $Z_1$=0.3 m, $Z_2$= 0.5m, $Z_3$=1m) required for reaching saturation increased significantly with increases bed length.

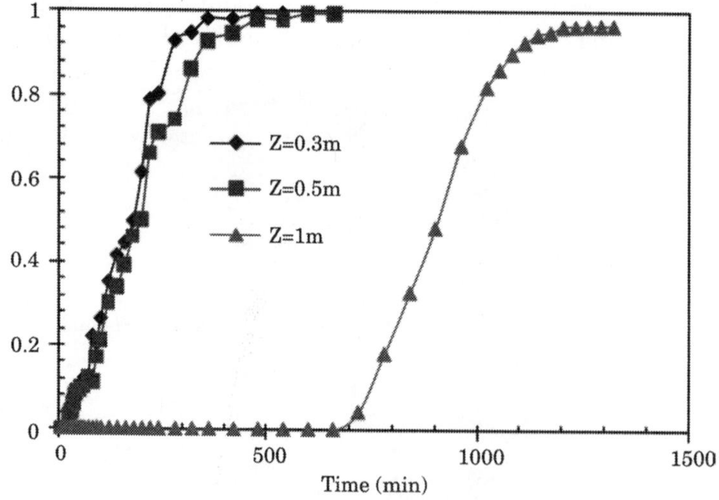


**Fig. 3.1.6.22:** Breakthrough curves for colour sorption on activated carbon column (Suresh, unpublished work). Initial colour = 100 PCU; D = 3cm; Z = 1m.

## 3.1.7. BIOSORPTION

### *Introduction*

Biosorption can be defined as the removal of metal or metalloid species, compounds and particulates from solution by biological material. A large quantity of materials has been investigated as biosorbents for the removal of metals or organics or others extensively. The biosorbents can be basically classified into the following categories: algae, fungi, and bacteria, industrial wastes (*e.g.*, *S. cerevisiae* waste biomass from fermentation and food industry), agricultural wastes (*e.g.* corn core) and other polysaccharide materials, etc. Waste materials or by-product biomass from large scale fermentation processes are the source of new family of biosorbents conveniently. In particular, some waste mycelia are available in large quantities for the removal of heavy metals. Seaweeds from the oceans produced in copious quantities are another inexpensive source of biomass. Marine algae, especially brown algae such Sargasso seaweed was investigated for metal removal. Abundant natural materials, particularly cellulosic nature, have been suggested as potential biosorbents for the removal of heavy metals. For economical reasons, other low-cost biosorbents are of interest recently, such as agricultural wastes (Bailey *et al.*, 1999). The first major challenge for the biosorption field was to select the most promising types of biomass from an extremely large pool of readily available and inexpensive biomaterials. Although many biological materials can bind heavy metals, only those with sufficiently high metal-binding capacity and selectivity for heavy metals are suitable for use in a full-scale biosorption process to improve biosorption process, including immobilization of biomaterials, improvement of regeneration and re-use, optimization of biosorption process etc.

### 3.1.7.1. *Biosorbents*

Low cost and widely available natural materials are represented by almond shell, hazelnut and walnut, birch wood, *Cassia fistula* biomass, coconut fiber, lemon peel, *Moringa oleifera* seeds, neem oil cake, olive pomace, orange peel, palm flower, pine bark, pomegranate peel, ponkan peel, reed, rice husk, rice straw, sour orange residue, and yellow passion fruit shell, and they have lucratively been applied as biosorbents to eliminate metals from aqueous media. With the aim of increasing biosorption capacity, modifications of the biosorbent surface have been carried out. Thiolation of coconut fiber, pine bark using the Fenton reagent. These modification processes tend to improve the adsorption capacity of the biosorbents (see the Table 3.1.7.1).

*Living organism biosorbents*

***Bacterial biosrobents:*** Bacteria are the most abundant and versatile of microorganisms and constitute a significant fraction of the entire living terrestrial biomass of ~$10^{18}$ g. Early 1980, some microorganisms were found to accumulate metallic elements with high capacity. Some marine microorganisms enriched Pb and Cd by factors of $1.7 \times 10^5$ and $1.0 \times 10^5$ respectively, relative to the aqueous solute concentration of these elements in ocean waters. Bacteria were used as biosorbents because of their small size, their ubiquity, their ability to grow under controlled conditions, and their resilience to a wide range of environmental. Table 3.1.7.1. summarizes some of the important results of metal biosorption using fungal biomasses.

***Fungal biosorbents:*** Table 3.1.7.1 summarizes some of metal ions/dyes/ phenolic compounds biosorption using fungal biomasses. Some of filamentous

**Table 3.1.7.1:** Some of examples of metal ions dyes and phenolic compounds sorbates onto various biosorbents (Living/Non-living) available in the reasearch literatures.

| ***Biosorbates*** | ***Biosorbents*** |
|---|---|
| ***Metals*** | ***Non-living biosorbents*** |
| As (III), As (IV) | Herbaceous plant, Tea waste, Coconut fiber, Green coconut shell powder, *Moringa oleifera* |
| Zn (II) | Eucalyptus bark, Saw dust, Mango bark, Pipeapple fruit peel, Coffee husk, Rick husk, Residual biomass, Neem leaves, Animal bones, Barley straw, Coniferous bark, Papaya wood, Palm tree leaves, Rice bran, Banana pith *(Musacea zingiberales),* Water lettuce |
| Cu (II) | Peanut shell, Coffee husk, Tartaric acid, Modified rice husk, Powdered waste sludge, Impregnated GAC, Sunflower stark, Residual biomass, Chitosan, Crab shell, Peat, Bagasse, Birch wood betula sp., Chaff, Neem oil cake, Olive pomace, Tea waste |
| Cr (III), Cr (VI) | Peanut shell, Neem bark, Mango bark, Corn cob powder, Bagasses, Rice husk, Modified rice husk ash, Waste tyres, Sawdust, Groundnut hull, Potato peel waste, Wheat shell, Orange peel, Almond shell, *Azadirachta indica* (Neem) leaf powder, Crushed cocount shell, Green coconut shell powder, Ground nut shell, Hazelnut shell, Palm flower, *Tamarindus indica* seed, Walnut shell, Yellow passion-fruit shell, *Hardwickia binata* bark, Nutshell, *Pinus pinasca* bark, Turkish Coffee, Untreated and Treated *Pinus sylvestris* bark, wallnut shell, waste tea, banana pith *(Musacea zingiberales),* water lettuce |
| Cd (II) | Black gram husk, Coffee husk, Seed coat, Mango peel waste, Neem bark, Mango bark, Chemically modified orange peel, Eucalptus bark, Maize leaves, Grape bunch, Banana peel, Unmodified and EDTA-modified maize husk, Residual biomass, Baker's yeast, Reed leaves, Vermiculite, Pomelo peel, *Azadirachta indica* (Neem) Leaf Powder, Coconut Copra meal, Green coconut shell powder, Grapefruit peels, Lemon feel, Neem oil cake, Olive pomace, Orange peels, Pine bark, Black oak bark, Formaldehyde polymerized peanut skins, *Hardwickia binata* bark, Nutshell, *Pinus pinasca* bark, Redwood bark, Turkish Coffee, Wallnut shell, Waste tea, Banana pith *(Musacea zingiberales)* |
| Pb (II) | Tartaric acid, Modified rice husk, Mango peel waste, Peepul leaves, Chemically modified orange peel, Residual biomass, Chitosan, Crab shell, Neem leaves, Bagasse ash, Groundnut hull, *Azadirachta Indica* (Neem) Leaf powder, Chaff, Coconut fiber, Lemon feel, Lemon resin, Neem oil cake, Reed (*Phragmites australis*), Tea waste, Yellow passion-fruit shell, Formaldehyde polymerized peanut skins, Redwood bark, Banana pith *(Musacea zingiberales)* |
| Hg (II) | Rice husk, Neem bark, Mango bark, Coconut fibre, Black oak bark, Doughlas fir bark, Redwood bark |
| Ni (II) | Tamarindd bark, Chemically modified orange peel, Residual biomass, Neem leaves, *Cassia fistula* Golden shower biomass, Water lettuce |
| Mn (II) | Lemon feel, Lemon resin, Rice husk, Water lettuce |
| | **Living biosorbents** |
| As (III), As (IV) | Marine green algae |
| Cu (II) | *Sargassum filipendula, Pseudomonas sp, aquatic fern, S. cerevisiae, Chlorella vulgaris, Cladonia rangiformis, Sargassum, Sargassum filipendula, Sargassum fluituns, Sargassum wighti, Sphaeroplea, Sphaeroplea (Acid treated), Spirogyra, Aspergillus niger, Cupriavidus* |

**Table 3.1.7.1:** (*Contd...*)

**Table 3.1.7.1:** (*Contd...*)

| ***Biosorbates*** <br> ***Metals*** | ***Biosorbents*** <br> ***Non-living biosorbents*** |
|---|---|
| | *taiwanensis TJ 208, Neurospora crassa, Bacillus cereus, Pseudomonas aeuginosa, Bacillus sp., Sphaerotitus natans, Streptomyces coelicolor, Penicillium* sp., *Penicillium* sp., *Penicillium* sp., *Aspergillus, Trichoderma, Rhizopus, Mucor, Enterobacter* sp., *Macro fungus (Pycnoporus sanguineus)* |
| Cr (VI) | *Pseudomonas sp, free cells/immobilized cell of seeweads, Fungi, Sargassum sp biomass, Microalgae, Marcoalgae, Marine algal mass, Chlorella miniata, Chlorella vulgaris, Aspergillus niger, Basidiomycete BDT-14 (DSM 15396), Pencicilduim janthinellum, Rhizopus arrhizus, Sacchoromyces cerevistae, Bacillus licheniformis, Pseudomonas* sp., *Staphylocococus xylosus, S. cerevisiae* |
| Cr(III) | *Pseudomonas aeuginosa* |
| Cd (II) | *Pseudomonas sp, aquatic fern, Chlamudpmonas reinhardril, Chlorella vulgaris, Fuscus spiralis, Laminaria hyperborean, Sargassum, Sargassum filipendula, Cupriavidus taiwanensis TJ 208, P. Axalicum, Tolypociadium* sp., *Escherichia coli, Pseudomonas aeuginosa, Pseudomonas sp., Staphylocococus xylosus, Ochrobactrum anthropi, Aermonas caviae, Phanerochaete chryosporium, Sphingomonas paucimobilis, Enterobacter sp., Pseudomonas putida, Streptomyces pimprina, Macro fungus (Pycnoporus sanguineus)* |
| Ni(II) | *Pseudomonas sp, Chlorella vulgaris, Sargassum filipendula, Sargassum wighti, Sphaeroplea, Sphaeroplea (Acid treated), Escherichia coli, Pseudomonas aeuginosa, Bacillus thurigiensis, Streptomyces rimosus* |
| Pb(II) | *Green algae, free cells/immobilized cell of seeweads, aquatic fern, Chlamudpmonas reinhardril, Fuscus spiralis, Laminaria hyperborean, Sargassum, Turbinaria conoides, Aspergillus niger, Cupriavidus taiwanensis TJ 208, Cephalosparium aphidicola, Neurospora crassa, P. axalicum, Tolypociadium sp., Bacillus cereus, Bacillus* sp., *pseudomonas* sp., *Penicillium sp., Aspergillus, Trichoderma, Rhizopus, Mucor, Macro fungus (Pycnoporus sanguineus)* |
| Zn (II) | *Aquatic fern, A. niger, Hyperborean, Fuscus spiralis, Laminaria hyperborean, Sargassum, Sargassum filipendula, Pseudomonas aeuginosa, Pseudomonas putida, Streptomyces rimosus, Thiobacillus ferroaxidans, S. cerevisiae, Fusarium, Saccharomyces* |
| Hg (II) | *Marine green algae, Chlamudpmonas reinhardril, Aspergillus niger, P. Axalicum, Sacchoromyces cerevistae, Tolypociadium* sp., |
| Co(II) | *Eleven kinds of Fungus, Fungal biomass* |
| Mn(II) | *Sacchoromyces cerevistae* |
| **Dyes** | **Non-living biosorbents** |
| Methylene blue | Gulmohar plant leaf powder, Rice husk |
| Basic dye | Sphagnum moss peat |
| Malachite green, | |
| Auramin O, | |
| Thioflavine T | Forest plants |
| Rhodamine dye | Jack fruit peel |

**Table 3.1.7.1:** (*Contd...*)

**Table 3.1.7.1:** *(Contd...)*

| ***Biosorbates*** ***Metals*** | ***Biosorbents*** ***Non-living biosorbents*** |
|---|---|
| Acid green dye | Brewery waste |
| Reactive dye | Bagasse pith |
| Congo red dye | Bagasse pith |
| | **Living biosorbents** |
| **Phenolic compounds** | **Non-living biosorbents** |
| P, 2CP, 4CP | Chitosan, Banana peel |
| P | Chicken feathers, Dried sewage sludge |
| | **Living biosorbents** |
| P, 2CP, 4CP | *Brown algae, Fungal biomass, $H_2SO_4$-treated Aspergillus niger* |
| Acrylic acid | *Bacillus thuringiensis, Pseudomonas putida, Comamonas testosteroni* and *Acidovorax* sp |
| 2NP, 4NP, 2,4-DNP | *Rhizopus oryzae* |

fungi (*Penicillium* spp., *Aspergillus* spp., *Rhizopus* spp. and white rot fungi) are available for biosorption. Penicillium group, cashmere-like Penicillium Asian group, the middle *P. chrysogenum* and it is typical of penicillin-producing bacteria. Fungi have smooth surface, 150–350 µm long, 3 to 3.5 µm wide, 2–3 branches, they are brush sticks asymmetry, ranging from the length of the vice sticks, based on the growing stems and small stems. Conidia is oval, (2–4) µm×(2.8–3.5) µm. They grow faster, round, tight cashmere-like or slightly tempered, blue and green (white edge), a radial grooving on the surface, often yellow-colored droplets exudative. Bright yellow colony back to a dark brown and yellow-soluble spread to the medium. They are widely distributed in the air, soil and organic matters on biodeterioration. Penicillium can remove a variety of heavy metal ions from aqueous solutions, such as Cu, Au, Zn, Cd, Mn, U and Th. For example, *P. chrysogenum* can extract gold from a cyanide solution and sorbing well for uranium and lead. *Aspergillus niger* is an important microorganism in biotechnological applications. It has been used to produce extracellular enzymes such as glucoamylas, pectinase, acidic lipase, feruloyl esterase, and xylanase and organic acids such as gluconic acid and citric acid. *A. niger* is also ecologically important in biodegradation of toxic chemicals such as hexadecane, treatment of waste beet molasses and olive mill waste, and bioconversion of wastewater sludge. Waste biomass of *A. niger* from fermentative industry, is used to remove hazardous heavy metal ions, such as cadmium, lead, chromium, and copper from aqueous solution. Non-living waste biomass consisting of *A. niger* attached to wheat bran was used as a biosorbent for the removal of copper and zinc from aqueous solution. The cell wall of *R. arrhizus* involves a high content of chitin. The ability of chitin to complex metal ions has been confirmed.

***Algae biosorbents:*** Algae are of special interest in search for and the development of new biosorbents materials due to their high sorption capacity and their ready availability in practically unlimited quantities in the seas and oceans. However, there are few research on biosorption with algae as compared to those using other biomass (mainly fungi and bacteria), and there is still fewer for multi metallic systems. According to the statistic review on biosorption, algae have been less used as biosorbent material than other kinds of biomass, especially fungi and bacteria. Brown algae among the three groups of algae (red, green, brown algae) received the most attention. Some marine micro and macro algal species as biosorbents for heavy metal ions like *Chlamydomonas reinhardtii, Chlorella salina, Chlorella sorokiniana, Chlorella vulgaris, Chlorella miniata, Chlorococcum sp, Cyclotella cryptica, Lyngbya taylorii, Phaeodactylum tricornutum* and *Ascophyllum nodosum, Ascophyllum sp., Cladophora crispata, Cladophora fascicularis, Codium fragile, Colpomenia sinuosa, Corallina officinalis, Ecklonia sp., Fucus vesiculosus, Fucus ceranoides, Fucus serratus, Fucus spiralis, Gracilaria fischeri, Gracilaria* sp., *F. spiralis, G. fischeri, Gracilaria* sp., *Jania rubrens, Laminaria digitata, Laminaria japonica, Laurencia obtuse, Padina pavonia, Padina* sp., *Palmaria palmata, Petalonia fascia, Pilayella littoralis, Porphyra columbina, Sargassum asperifolium, Sargassum hemiphyllum, Sargassum hystrix, Sargassum natans, Sargassum* sp., *Sargassum* sp., *Sargassum vulgaris*. These algae were reported to be able to adsorb one or more heavy metal ions, including K, Mg, Ca, Fe, Sr, Co, Cu, Mn, Ni, V, Zn, As, Cd, Mo, Pb, Se, Al, with good metal uptake capacity

***Modified biosorbent:*** In recent years, attention has been focused on the utilization of unmodified or modified rice husk as a sorbent for the removal of pollutants. Unmodified rice husk has been evaluated for their ability to bind metal ions. Various modifications on rice husk have been reported in order to enhance sorption capacities for metal ions and other pollutants. Rice husk contains a high proportion of cellulose (28–36%), thus it appears to be a good candidate for modification with carboxylic acids. The biosorption system consists of a bacterial biofilm supported on granular activated carbon, placed in contact with the polluted solutions. The bacteria used for the formation of the different biofilms were *Arthrobacter viscosus, Bacillus coagulans, Escherichia coli* and *Streptococcus equisimilis*. Microbial biomass consists of small particles with low density, poor mechanical strength and little rigidity. The immobilization of the biomass in solid structures qeates a material with the right size, mechanical strength and rigidity and porosity necessary for metal accumulation. Immobilisation can also yield beads and granules that can be stripped of metals, reactivated and reused in a manner similar to ion exchange resins and activated carbon. Various applications are available for biomass immobilization. The principal techniques that are available in literature for the application of biosorption are based on adsorption on inert supports, on entrapment in polymeric matrix, on covalent bonds in vector compounds, or on cell cross-linking. Support materials are introduced prior to sterilization and inoculation with starter culture and are left inside the continuous culture for a period of time, after which a film of microorganisms is apparent on the support surfaces. Entrapment in polymeric matrices: The materials obtained from immobilization in calcium alginate and polyacrylamides are in the form of gel particles. Those obtained from immobilization in polysulfone and polyethyleneimine are the strongest. Covalent bonds to vector compounds: The most common vector compound (carrier) is silica gel. The material obtained is in the form of gel particles. Cross-linking: The addition of the cross-linker leads to the formation of stable cellular aggregates. This technique was found useful for the immobilization of algae. The most common cross linkers are: formaldehyde, glutaric dialdehyde, divinylsulfone and formaldehyde – urea mixtures. Alginate beads are common and well-known support materials in biosciences applications and are now used in the environmental field. Sodium alginate, a water-soluble salt of alginic acid, is a biopolymer and differs from most other polysaccharides exhibiting a sol–gel transition when simply submitted to modification of their ionic environment, *e.g.* substitution of sodium by divalent cations such as calcium. This material is biocompatible, low cost and the technique of encapsulating is simple. Industrial processes in water treatment (*e.g.* filtration, fluidized bed) require granular materials of controlled size and the encapsulation of different components in the alginate beads to enhance the properties (*e.g.* adsorption capacity of organic/inorganic pollutants, mechanical properties) are widely investigated. For instance, aluminosilicates (natural or modified) are being tested for more than 20 years in the removal of toxic metals or organic pollutants from aqueous solutions due to favourable surface properties, availability, environmental and economical considerations.

### 3.1.7.2. *Mechanism(s) of Biosorption*

Knowledge accumulated on the molecular biology of the yeast is very helpful to identify the molecular mechanism biosorption in metal ion removal. The mechanisms involved in biosorption or metal–microbe interactions should be studied with great efforts by utilizing various techniques and the combination of them. The imprecise definition of sorption perhaps gives a clue that the mechanism(s) involved in biosorption are often difficult to characterize, except perhaps in the simplest laboratory systems. Biological material is complex and a variety of mechanisms may be operative under given conditions. The variety of structural components present in biomass means that many functional groups are able to interact with metal species, *e.g.* carboxyl, phosphate, hydroxyl, amino, thiol, etc., to varying degrees and influenced by physico-chemical factors. For biosorption, defined as a physicochemical process independent of metabolism, such mechanisms as adsorption, ion exchange and complexation/coordination may be important and, in these cases, biosorption can be rapid and reversible with biomass properties analogous to conventional ion exchange resins. Ion exchange is the replacement of an ion in a solid phase in contact with a solution by another ion. More specifically, it is the replacement of an absorbed, readily-exchangeable ion by another. Although a simple concept, in reality it can be a mechanistically highly complex process depending on the system. Several other mechanisms may occur which may complicate sorption and/or desorption. Precipitation, where bound metal/radionuclide species can act as loci for subsequent deposition, can lead very high uptake capacities but this may inhibit desorption. Extensive precipitation of actinides such as uranium and thorium on non-living fungal biomass is a good example of precipitation arising from hydrolysis product formation. It is likely that the various mechanisms involved in biosorption can operate simultaneously to varying degrees. There are several ways of classifying metals according to their chemical properties. In a biological context, relevant schemes consider ligand preferences, a property that can-underlie biological activity, and are relevant to biosorption. Some behaviour will be affected by metal concentration, as well as the relative metal concentrations in mixtures where competitive effects may occur. The hard/soft scheme predicts that bonds formed between hard acids and hard ligands will be predominantly ionic whereas soft acid–ligand complexes are more covalent in character. However, this can vary depending on the nature of the biomass: $Sr^{2+}$ binding to denatured yeast biomass was ionic but interaction with cell wall of living yeast exhibited increased covalent binding. This could be due to the involvement of amine and sulphydryl functional groups that are active on cell surfaces of living cells but which may be denatured in dead biomass, where phosphate and carboxylate groups may be more significant.

Hydrophobic compounds, *e.g.* hydrocarbons, are, by definition, not readily soluble in water. However, these substances will associate with non-polar environments such as the surface of organic particles. Such hydrophobic sorption clearly occurs when hydrophobic compounds encounter biomass in biosorption systems. Additionally, the lipophilic nature of hydrophobic compounds means that they can pass through membranes and be absorbed into the organic matrix. Absorption may be a significant component of biosorption in such cases. The mechanisms involved in dye biosorption onto chitosan are various and include surface adsorption, chemisorption, diffusion and adsorption-complexation. The most important steps are film diffusion, pore diffusion and chemical reactions like ion exchange and complexation. Amine sites appear to be the main reactive groups for dyes, though hydroxyl groups might contribute, with intermolecular interactions of the dye molecules being most probable in chitosan–dye systems.

### 3.1.7.3. *The Following Factors Affect Biosorption Process*

The type and nature of the biomass or derived product can be very important as discussed, including the nature of its applicationas, *e.g.* freely-suspended cells or biomass, immobilized preparations, living biofilms, etc. Physical and chemical treatments such as boiling, drying, autoclaving and mechanical disruption will all affect binding properties while chemical treatments such as alkali treatment often improve biosorption capacity, especially evident in some fungal systems because of deacetylation of chitin to form chitosan-glucan complexes with higher metal affinities. Growth and nutrition of the biomass, and age can also influence biosorption due to changes in cell size, wall composition, extracellular product formation, etc. The surface area to volume ratio may be important for individual cells or particles, as well as the available surface area of immobilized biofilms. In addition, the biomass concentration may also affect biosorption efficiency with a reduction in sorption per unit weight occurring with increasing biomass concentration. Apart from these, physico-chemical factors such as pH, the presence of other anions and cations, metal speciation, pollutant solubility and form, and temperature may also have an influence. With living cell systems, the provision of nutrients and optimal growth conditions is an obvious requirement. Fig. 3.1.7.1. shows general biosorption procedure for grass root researcher in the area of biosorption.

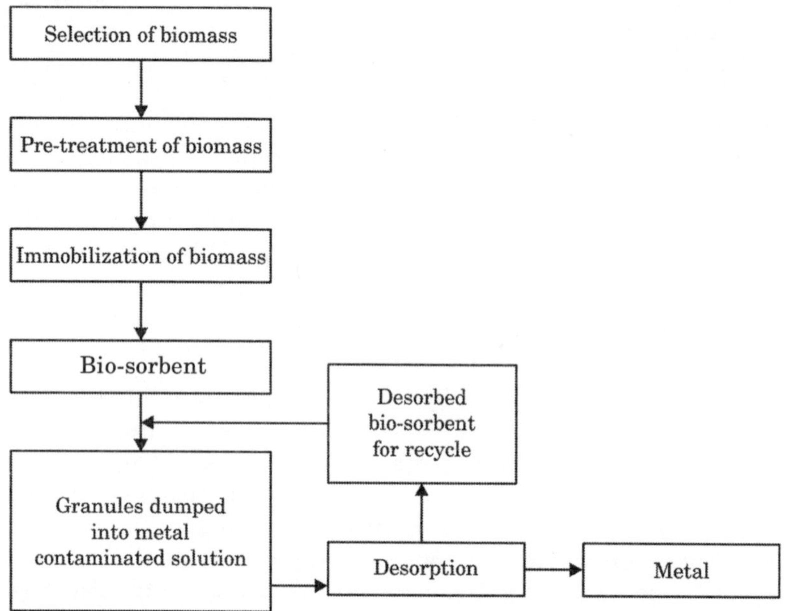


**Fig. 3.1.7.1.** Schematic representation of biosorption procedure.

Metal biosorption has frequently been shown to be strongly pH dependent in almost all systems examined, including bacteria, cyanobacteria, algae, and fungi. Competition between cations and protons for binding sites means that biosorption of metals like Cu, Cd, Ni, Co and Zn is often reduced at low pH values. Conversely for anionic metal species like $TcO^4$, $PtCl_4^3$, $CrO_4^2$, $SeO_4^2$ and $Au(CN)_2$, increased biosorption may be seen at lower pH values.145 Biosorption of some metals may be pH-independent and some examples exist for, *e.g.* $Ag^+$, $Hg^{2+}$ and$AuCl^4$, explained by the formation of covalent complexes with N and S-containing ligands.

***pH:*** pH is one of the most important parameter in the biosorptive process and environmental factors. The pH value of solution strongly influences not only the site dissociation of the biomass surface, but also the solution chemistry of the heavy metals: hydrolysis, complexation by organic and/or inorganic ligands, redox reactions, precipitation, the speciation and the biosorption availability of the heavy metals. The biosorptive capacity of metal cations increases with increasing pH of the sorption system, but not in a liner relationship. On the other hand, too high pH value can cause precipitation of metal complexes, so it should be avoided during experiments. For different biosorption system of metal ions, the optimal pH is different. It is reported that the optimal pH value is 5–9 for copper biosorption by *S. cerevisiae*, and 4–5 for uranium. When metal solution in low pH, the affinity with the proton at the binding site of yeast is much greater than that of the metal ion compared with that at higher pH. At higher pH, *e.g.* 5.0, divalent positive ions are suitable to interact with negatively charged groups in biomass. On the other hand, the outer layer of the cell wall of *S. cerevisiae* consists of a coat protein, which can cause a charge through dissociation of ionisable side groups of the amino acids. The ionic state of ligands such as carboxyl, phosphate, imidazole and amino groups will promote reaction with the positively charged metal ions. However, metal anions, such as chromium (VI), exhibit different pH features from metal cations. Generally speaking, a low pH value is favorable for biosorption. The effect of pH on biosorption of Cr(VI) has been investigated by various researchers using a variety of different types of biomass. Optimal biosorptive removal of Cr(VI) at low pH (2.0) has been reported for *Rhizopus nigricans, Bacillus* sp. and *Dunaliella* sp. Modification method for the yeast may offer a promising way to utilize the biomass at high pH for the removal of metal anions. The explanation of the pH shift in biosorption process and the pH effect is helpful to identify the mechanism of metal biosorption. The metal uptake capacity, expressed in milligrams per gram of biomass, was found to be dependent upon the following factors: the initial metal concentration (with the uptake capacity decreasing with increasing initial concentration), the biomass loading (with the uptake capacity decreasing with increasing biomass loading) and pH (with the uptake capacity increasing with increasing pH in the range of 1.5 to 6.0). The metal uptake capacity was significantly affected in the presence of the co-existed ions.

Biosorption is mainly used to treat wastewater where more than one type of metal ions would be present; the removal of one metal ion may be influenced by the presence of other metal ions. For example: Uranium uptake by biomass of bacteria, fungi and yeasts was not affected by the presence of manganese, cobalt, copper, cadmium, mercury and lead in solution. In contrast, the presence of $Fe^{2+}$ and $Zn^{2+}$ was found to influence uranium uptake by *Rhizopus arrhizus* and cobalt uptake by different microorganisms seemed to be completely inhibited by the presence of uranium, lead, mercury and copper. To understand the effects of nutritional and environmental factors on the quality of microbial biomass could help us to cultivate the desirable types of organisms on cheap nutrients, and to use it in a suitable form for direct application.

***Competing ions/co-ions:*** Real industrial effluent usually contains various ionic components, including metal cations and anions. Some studies indicated that cations and anions additional to the ions of interest have a generally detrimental impact on metal accumulation. Usually, the biosorption capacity of one metal ion is interfered and reduced by co-ions, including other metal ions and anions presenting in solution, however the gross uptake capacity of all metals in solutions remains almost unchangeable. Competitive effect occurred in a mixed solution containing lead and chromium during the biosorption process by flocculating brewer's yeast. The yeast of *S. cerevisiae* seems to have more affinity, higher selectivity and biosorption capacity to Pb(II) than to Cr(III) in aqueous solutions. The decrease of metal uptake in competitive conditions was thought to be a response to increased competition between same charged species for binding sites of the ethanol-treated yeast cells. The presence of anions also affects the biosorption of metal ions. The anions existing in the solution may form a complex with the metal ions, which would reduce the metal biosorption capacity seriously. Nowadays only limited information is available on the competition of fungal biosorption, including *S. cerevisiae.* More research is required in competitive study for the biosorption systems containing two- or multi-metal ions by utilizing multi-parameter mathematical models.

***Initial concentration of metal ions and biomass and kinetic modelling:*** The uptake rate of the metal ion will increase along with increasing the initial concentration if the amount of biomass is kept unchanged. Lower values of biomass concentrations there is an increase in the specific uptake. Contrary to that, biosorptive capacity of the metal ions is inversely proportional to the initial concentration of the biomass when the initial concentration of metal ions is kept constant. Increase of the biomass concentration of the biosorption system could result in increasing the sorption site interactions. When the biomass concentration is low, metal ions in the solution would not only be adsorbed to the surface of the biomass, but also enter into intracellular part through facilitating the concentration gradient of metal ion. As a matter of fact, biosorptive capacity of metal ions was reported to be related to the ratio of the concentration of initial metal ions to the concentration of biomass. Both aspects cannot be neglected when assessing the influence of concentration of the metal ion and biomass on biosorption, otherwise, error would occur. Generally speaking, the biosorption capacity and the removal efficiency of metal ions by *S. cerevisiae* became higher with prolonging the contact time. However, in practice, it is necessary to optimize the contact time, considering the efficiency of desorption and regeneration of the biomass. Kinetic models can be helpful to better understand the mechanisms of metal biosorption and to evaluate performance of bio-sorbents for metal removal. There are many models that allow describing the kinetics of biosorption. Models most commonly described in the literature include the pseudo-second order model describing the process kinetics. Description of the sorption process makes use of numerous models to determine the sorbent volume. The total sorption effect may be described by empirical equations with constants determined in an experimental manner. Linear regression is commonly used to determine the model parameters; however, non-linear regression was shown to allow for better accuracy and correlation of the model with the experimental results. In addition, linear regression does not permit for determination of the parameters of a three-parameter model. Pseudo-first-order and pseudo-second-order equations were used to fit the experimental data of Ni(II) uptake by the deactivated protonated yeast of *S. cerevisiae*. In comparison with the adsorption rate constants of both the models, we can see that the second order model provided the best correlation of the data.

***Temperature and isotherm modelling:*** Temperature has also an influence on the biosorption of metal ions, but to a limited extent under a certain range of temperature, which indicates that ion exchange mechanism exists in biosorption to some extent. Temperature seems not to influence the biosorption performances in the range of 20–35°C. Biosorption process is usually not operated at high temperature because it will increase the operational cost. Temperature (5–40°C) had minor effect on the accumulation level of $Cu^{2+}$, $Co^{2+}$ or $Cd^{2+}$ by free cells of *S. cerevisiae* in suspension. Adsorption reactions are normally exothermic, so biosorption capacity increases with decrease of temperature. In the range of 15–40°C, the maximum equilibrium biosorption capacity for Pb(II), Ni(II) and Cr(VI) ions by the inactive *S. cerevisiae* was reached at temperature of 25°C. The decrease in capacity at higher temperature between 25 and 40°C revealed that the processes of biosorption for these metal ions by *S. cerevisiae* are exothermic. The decrease of biosorption capacity at higher temperature may be due to the damage of active binding sites in the biomass. However, metal biosorption of Cr(VI) by *S. cerevisiae* increases with increasing temperature in the range of 25–45°C, they explained that higher temperature would lead to higher affinity of sites for metal or binding sites on the yeast. The energy of the system facilitates Cr(VI) attachment on the cell surface to some extent. When the temperature is too high, there is a decrease in metal sorption due to distortion of some sites of the cell surface available for metal biosorption. Over modest physiological-type ranges, temperature usually has little effect on biosorption, although high temperatures, *e.g.* 50°C, may increase biosorption in some cases. Low temperature will, however, affect living cell systems and any auxiliary metabolism-dependent processes that aid biosorption. Sorption capability is a very important parameter that allows for estimation of process costs, since the determined isotherms allow to predict the amount of sorbent required for effective sorption. Ability to regenerate the sorbent is very important in cyclic processes, especially when expensive and selective sorbents are used. The simplest method for determination of constants in two-parameter models is transformation of the equation describing the equilibrium to a linear equation form. Simple basic models (Langmuir model and Freundlich model) also do not incorporate the effects of any external variable environmental factors, although they are capable of describing many biosorption isotherms in most cases. The mechanistic conclusions from the good fit of the models alone should be avoided. Moreover, biosorption isotherms may exhibit an irregular pattern due to the complex nature of both the biosorbents and its varied multiple active sites, as well as the complex solution chemistry of some metallic compounds. To describe two- or multi-metal ions biosorption system, various extended Langmuir models (also called competitive Langmuir model) or Freundlich type models have been developed. These empirical models hardly reflect the sorption mechanism. The biosorption of copper and zinc conformed to the Langmuir isotherm.

### 3.1.8. Case Study

#### 3.1.8.1. *Biosorption of As(III) Onto Dried E. crassipes Biomass*

*Introduction*

Biosorption is a potential technology not only toxic metal removal, but also for precious metal recovery. In recent years, applying biotechnology in controlling and removing metal pollution has been paid much attention, and gradually becomes hot topic in the field of metal pollution control because of its potential application. Alternative process is biosorption, which utilizes various certain natural materials of biological origin, including bacteria, fungi, yeast, algae, etc. These biosorbents possess metal-sequestering property and can be used to decrease the concentration of heavy metal ions in solution from ppm to ppb level. It can effectively sequester dissolved metal ions out of dilute complex solutions with high efficiency and quickly, therefore it is an ideal candidate for the treatment of high volume and low concentration complex wastewaters. Bacteria, fungi, yeast and algae can remove heavy metals and radio nuclides from aqueous solutions in substantial quantities. The uptake of heavy metals by biomass can take place by an active mode (dependent on the metabolic activity) known as bioaccumulation or by a passive mode (sorption and/or complexation) termed as biosorption. Cadmium, mercury, copper, arsenic, zinc and lead are the potentially hazardous heavy metals,which can be of serious threat to animals and human beings for their high toxicity and tendency for bioaccumulation in the food chain even in relatively low concentrations in the environment. Biomass-based sorption (tea waste), be one of the best and low cost techniques for Cd (II) and As (III) ions removal from liquid phase (Kamsonlian *et al.*, 2011).

The objective of this case study to the efficiency of dried *E. crassipes* biomass in for removal of As (III) ion. Fig. shows photoview of batch biosorption study in the laboratory conical flask after sorption of As(III) onto dried *E. crassipes* biomass.

**Fig. 3.1.8.1:** Photoview of batch biosorption study (As (III)) onto dried *E. crassipes* biomass) (Suresh, unpublished work).

### *Materials and Methods*

***Biomass preparation:*** *E. crassipes* biomass was collected from the phytoremediation process and wet form of biomass was initially air-dried for 5 h and then kept in the oven at around 60°C for another 24 h. The dried biomass was examined to remove the presence of dust and fines.

***Reagents and equipments:*** The chemicals, used in the experimental work, were of AR grade. To prepare solutions, Milli Q water was used.

*Batch experiments*

Biosorption of As (III) from the samples were conducted by optimizing the parameters like pH, dosage, contact time and temperature. The desired amount of biomass was added and the samples were then agitated inside the orbital shaker at 150 rpm. The batch biosorption procedures were carried out at constant solution pH, temperature is 303 K and biosorption dosage of 8 g/l and contact time of 24 h. Experiment was done by drawing out the samples at regular intervals of time and filtered to separate adsorbent from liquid phase, then analyzed through the atomic absorption spectroscopy.

The details of batch experiment procedure, kinetic modeling and calculation are given in the case study of adsorption section.

## ***Results and Discussion***

### *Effect of contact time*

Effect of initial concentration on the percentage removal of As(III) on dried *E. crassipes* biomass was studied at different initial concentrations by keeping other parameters constant is shown in Fig. 3.8.5.2. The data showed that a contact time of 45 min was required to achieve an optimum biosorption and there was no significant change in concentration of the metal ion with further increase in contact time. From Fig. 3.8.5.2 it was observed that with the increase in initial substrate concentration, the pecent removal As(III) increases. Increasing the initial As(III) concentration would increase the mass transfer driving force and therefore the rate at which As(III) molecules pass from the solution to the particle surface. This would result in higher As(III) adsorption. However, some of researchers found that the percentage biosorption of As(III) decreased as the initial concentration decreases.

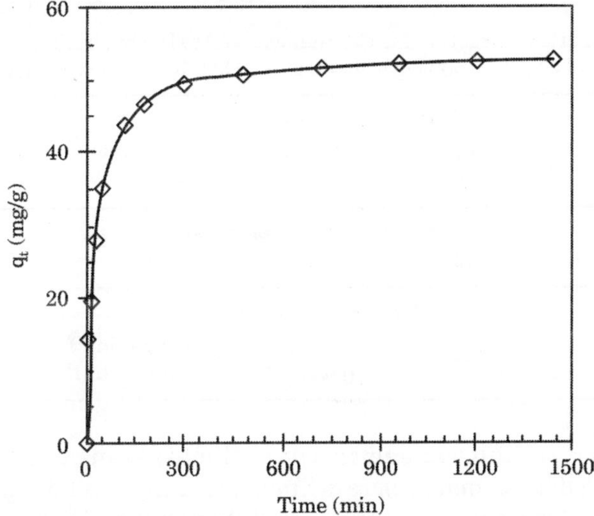


**Fig. 3.1.8.2:** Effect of contact time on the biosorption of As(III) by dried *E. crassipes* biomass. Experimental data points given by the symbols and the lines predicted by the pseudo-second-order model. $C_0$ = 500 mg/l, T = 303 K, m = 8 g/l (Suresh, unpublished work).

### *Biosorpiton kinetics*

In the present study, two kinetic models namely pseudo-first-order and the pseudo-second-order model were applied to kinetic data in order to investigate the adsorption behaviour of sorbate onto dried *E. crassipes* biomass. The results of the fit of these models are given in Table 3.1.8.1. The fit of experimental data to the pseudo first-order and the pseudo-second-order equations seemed to be quite good when correlation coefficients ($R^2$) obtained from non-linear regression analyses were examined. However, it was very difficult to decide which model represents the experimental data better, just on the basis of $R^2$. A better criterion to find the best model for the experimental data is MPSD parameter. It is known that the lower the MPSD value, the better is the fit. When the MPSD values given in Table 3.1.8.1 are examined, it can be seen that they are much smaller for the pseudo-second-order model as compared to that for the pseudo-first-order model, leading to the conclusion that the kinetic data of adsorption of As(III) onto dried *E. crassipes* biomass fit to the pseudo-second order model better than the pseudo-first-order model. Some of researchers fitted the kinetic data of As(III) adsorption onto various biomass by pseudo-first order model.

***Weber and Morris intra-particle diffusion plot:*** If the Weber and Morris (1963) plot of $q_t$ versus $t^{0.5}$ satisfies the linear relationship with the experimental data, then the sorption process is supposed be controlled by intra-particle diffusion only. However, if the data exhibit multi-linear plots, then two or more steps influence the sorption process. Fig. 3.1.8.3 presents the plots of $q_t$ versus $t^{0.5}$ for all the adsorbates and the parametric values are given in Table 3.1.8.1. In this figure, the data points are related by two straight lines. The curvature from the origin to the start of the first straight portion (not shown in figure) represents

**Table 3.1.8.1:** Kinetic parameters for the removal of As (III) by dried *E. crassipes* biomass (t = 24 h, C0 = 500 mg/l, m = 8 g/l, T= 303 K) (Suresh, unpublished work).

| | | **Pseudo 1st order model** | | | |
|---|---|---|---|---|---|
| $q_{e,exp}$ (mg/g) | $q_{e,cal}$ (mg/g) | $K_f$ (min$^{-1}$) | $R^2$ (non-linear) | MPSD | |
| 26.99 | 26.61 | 0.036 | 0.983 | 37.45 | |
| | | **Pseudo 2nd order model** | | | |
| 27.67 | 0.0016 | 1.25 | 0.997 | 18.42 | |
| | | **Weber Morris model** | | | |
| $K_{id1}$ (mg/g min$^{1/2}$) | $I_1$ | $R^2$ | $K_{id2}$ (mg/g min$^{1/2}$) | $I_2$ | $R^2$ |
| 2.96 | 9.47 | 0.966 | 0.17 | 46.7 | 0.926 |

the boundary layer diffusion and/or external mass transfer effects. The first straight portion depicts macro-pore diffusion and the second represents meso-pore diffusion. These show only the pore diffusion data. Extrapolation of the linear portions of the plots back to the y-axis gives the intercepts which provide the measure of the boundary or film layer thickness. The first linear portion is attributed to the gradual equilibrium stage with intra-particle diffusion dominating. The second portion is the final equilibrium stage for which the intra-particle diffusion starts to slowdown due to the extremely low sorbate concentration left in the solution. The deviation of straight lines from the origin indicates that the pore diffusion is not the sole rate-controlling step. Therefore, the biosorption proceeds via a complex mechanism consisting of both surface biosorption and intra-particle transport within the pores of dried *E. crassipes* biomass.

It can be inferred from the Fig. 3.1.8.3 that the diffusion of As (III) from the bulk phase to the external surface of dried *E. crassipes* biomass, which begins at the start of the biosorption process, is the fastest. The slope of the linear portions are defined as a rate parameters ($k_{id,1}$ and $k_{id,2}$) and are characteristics of the rate of biosorption in the region where intra-particle diffusion is rate controlling. It seems that the intra-particle diffusion of As (III) into meso-pores (second linear portion) is the rate-controlling step in the biosorption process. The portion of the plots are nearly parallel ($k_{id,2} \approx 0.17$ mg/g min$^{0.5}$), suggesting that the rate of biosorption As(III) into dried *E. crassipes* biomass is comparable at all $C_0$. Slopes of first portions ($k_{id,1}$) are higher for higher $C_0$ for As(III), due to the greater driving force at higher $C_0$.

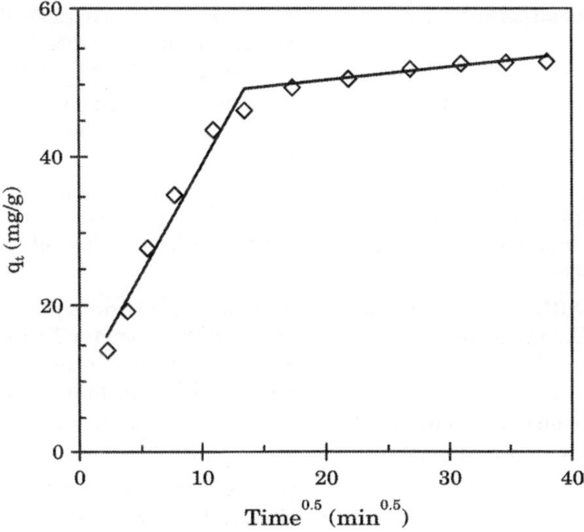


**Fig. 3.1.8.3:** Weber and Morris intra-particle diffusion plot for the removal of As(III) by dried *E. crassipes* biomass. $C_0$ = 500 mg/l, *T* = 303 K, *m* = 8 g/l (Suresh, unpublished work).

***Conculsion*:** The adsorption dynamic curves were analysed by pseudo-first-order, pseudo-second-order and intra-particle diffusion models. The rates of adsorption were found to obey the pseudo-second order model with good correlation. Diffusion study showed that the adsorption dynamics of As(III) biosorption onto dried *E. crassipes* biomass is controlled by combination of surface and pore diffusion.

*Instrumental Tools and Techniques used in Metal Adsoption / Biosorption Studies*

Various analytical techniques applied for investigation of sorption by various adsobents (Table 3.1.8.2). Different techniques provide different information on sorption, which are complementary to each other. It is necessary to combine different techniques to explore the mechanisms of sorption.

### 3.1.9. Regeneration/Desorption study

Recyclability of an adsorbent is of crucial importance in industrial practice for metal removal from wastewater. To test the suitability and stability of the

**Table 3.1.8.2:** Some techniques applied in metal adsorption/biosorption studies

| ***Main techniques*** | ***Analysis*** |
|---|---|
| CHNS analyser | Carbon, hydrogen, nitrogen and sulphur presents of before and after adsorption/ biosorption process |
| Particle size analyser | To determine the particle size distribution (undersize/oversize) of before and after adsorption/biosorption process |
| Fourier Transform Infra Red (FTIR) spectroscopy | To determine the presence of functional groups in the adsorbents before and after the sorption of pollutants at room temperature |
| Differential pulse polarography | Monitors the biosorption of metal ions from solution by live bacteria from mixed solutions, *e.g.* Cd(II), Zn(II) and Ni(II) |
| Semi-electron microscope (SEM) | To understand the morphology of the sorbents, before and after the adsorption of various pollutants |
| Atomic-force microscopy (AFM) in conjugation with transmission electron microscopy (TEM) and electron probe X-ray microanalysis | To visualize deposition of uranyl phosphate at the cell surface. To investigate the local electrostatic properties of microbial cell |
| X-ray diffraction (XRD) and proton induced X-ray emission (PIXE) analysis | The structure of sorbents with the help of a X-ray diffractometer Ni removal mechanism via intercalation to preform the crystalline $UO_2PO_4.4H_2O$, forming $Ni(UO_2PO_4)_2.7H_2O$ |
| Transmission Electron microscopy (TEM) and energy dispersive X-ray (EDAX) microanalysis | To obtain information on the localization of metals palladium, silver, nickel, in the different parts of the bacteria cells. To understand the chemical composition of the sorbents surface before and after the adsorption |
| Potentiometric titrations and time resolved laser-induced fluorescence spectroscopy | To determine the binding site of the biomass for lanthanide ions; possible metal binding groups are diverse, and carboxyl and phosphate group may be main groups.To study the acid properties of the biomass in order to identify the functional acidic groups for sorption and to determine their affinities by considering their partial or total ionization equilibrium reactions |
| The electron spin resonance (ESR) spectra of Cu(II) in bacterial cell | Revealed Cu(II) in the cells has the tetragonally distorted octahedral structure with nitrogen and oxygen as ligands atoms, indicating Cu(II) binding with amino acids residues in the cell surface proteins |
| Pore area distribution and Surface area analyser | To determine before and after adsorption of any pollutants of their surface area as well as pore area distribution |
| The differential thermal analysis (DTA), differential thermal gravimetery (DTG) and thermo gravimetric (TG) under oxidizing atmosphere at the heating rate | To deduce drying and thermal degradation characteristics like loss of moisture, exothermic or endothermic temperature etc |
| Inductively coupled plasma-mass spectrometry (ICP-MS), Atomic absorption spectroscopy (AAS), UV-spectrophotometer, High performance liquid chromatography (HPLC), Gas chromatography (GC) | To determine the concentration of metals/dyes/ phenolic compounds/others from the before and after adsorption/biosorption process |

biosorbent, it was subjected to successive biosorption and desorption cycles. As the number of adsorption cycles increases the adsorptive capacity of the adsorbent decreases because with successive adsorption the active sites of the adsorbent decreases which results in the decrease in its removal efficiency.

Most of the authors of the cited studies have indicated that P and related compounds are both physically and chemically adsorbed on the surface of an activated carbon. Typically, thermogravimetric analysis (TG) or temperature programmed desorption coupled with mass spectrometry (TPD-MS) experiments were performed to determine the conditions where phenol or phenolic compounds are desorbed and, indirectly the nature of adsorbed species onto the carbon surface. Unfortunately, in most of the MS analysis, the masses that are analyzed are corresponding to secondary molecular species (as CO, $CO_2$, $H_2O$) and not the molecular ion of the adsorbed compound. This fact can induce to losses of information that could give rise to additional erroneous conclusions.

**3.1.9.1. *Solvent desorption/regeneration:*** Solvent desorption studies help in finding the mechanism of an adsorption process. If the adsorbates adsorbed onto the adsorbent can be desorbed by water, it can be said that the attachment of the adsorbate onto the adsorbent is by weak bonds. If the strong acid, such as $HNO_3$ and HCl, or strong bases, such as NaOH can substantially desorb the adsorbent, it can be said that the attachment of the adsorbate onto the adsorbent is by ion exchange. If organic acids, for example, such as $CH_3COOH$, can desorb the adsorbent, it can be said that the adsorption of the adsorbate onto the adsorbent is by chemisorption. Various solvents, *viz.* ethanol, $HNO_3$, HCl, NaOH, $CH_3COOH$, acetone and water were literarure available for the elution of adsorbates. Phenol and 4-nitrophenol, NaOH as the optimal desorption reagent due more desorption efficiency as per literature available information. The regeneration of the saturated yeast (immobilized *S. cerevisiae* by formaldehyde) is satisfying at 4°C by the combination of reduction and desorption using formaldehyde and $HNO_3$. The usefulness of HCl as an efficient metal desorbent has been literature reported for agrowaste sorbents. The desorption efficiency for both the metals remained almost unchanged during the five sorption–desorption cycles.

**3.1.9.2. *Thermal desorption/regeneration:*** It is necessary to properly dispose the spent-sorbents and/or utilize it for some beneficial purpose if possible. The dried spent-sorbents can be used directly or by making fire-briquettes in the furnace combustors/incinerators to recover its energy value.

The process of thermal regeneration involves a pyrolytic and an oxidative stage. The pyrolytic stage occurs when the spent carbon is exposed to temperatures up to 800°C under inert conditions. This results in the elimination of any volatile compounds adsorbed in the carbon porosity, including residual moisture, and the thermal decomposition of other less-volatile compounds. A residue of carbonised char is formed from the adsorbed compounds which occupies some of the carbon porosity. The oxidative stage involves the controlled gasification of the pyrolysed carbon at temperatures usually around 800°C in the presence of a mildly oxidising atmosphere, usually steam or carbon dioxide or a mixture of both. This results in the elimination of the charred residue and the exposure of the original carbon-pore structure. After exhaustion, activated carbon must be regenerated or replaced by fresh carbon. The first option is preferred due to the high cost of this adsorbent. Moreover, from an environmental viewpoint regeneration is recommended because the exhausted spent carbon could be land filled. Although thermal regeneration is widely used, the process is not totally understood and consequently the choice of design parameters and operating conditions of regeneration are far from optimal. Moreover, operation costs are sensitive to adsorption capacity recovery and mass losses incurred during regeneration. Therefore, the fundamentals of this type of regeneration need to be studied. From an economic point of view, the gasifying agent used is very important because it determines to a great extent the processing conditions (temperature and reaction time) and the design parameters. Steam and $CO_2$ are generally used as gasifying agent; air, however, is seldom used even though it is afree, no-cost reactive and operation conditions could be more favourable (lower temperature and shorter reaction time). Air gasification is hindered by the high reactivity of oxygen, which can damage the porous structure of the carbon. The regeneration of several field-spent sorbent representatives of those typically generated by the drinking water treatment industry. Regeneration was conducted in steam at 800°C over reaction times between 0 and 60 min in order to achieve different degrees of material gasification.

The accumulation of metals affects the process of thermal regeneration as some of them have beenreported to catalyse the reaction between solid carbon and the oxidising agents. Catalytic metals chemisorb water or carbon dioxide molecules and transfer the oxygen to the carbon surface which causes enhanced carbon gasification in the area adjacent to the catalytic metal. The presence of catalytic metals affects the regeneration process in two ways. Firstly, they increase the reactivity of the carbons, which means that lower temperatures or reduced reaction times are required to achieve a certain degree of gasification. Secondly, they affect the development of the carbonporosity during the regeneration process usuallypromoting localised carbon gasification and theformation of larger pores.

**3.1.9.3. *Bioregeneration:*** Bioregeneration is defined as the renewal of the adsorptive capacity of activated carbon by microorganisms for further adsorption. Bioregeneration of activated carbon increases the service-life of activated carbons. It is traditionally believed that bioregeneration occurs as a result of a concentration gradient between the carbon and the bulk liquid. Therefore, bioregeneration can only occur for compounds that readily desorb. Some researchers suggest that also exo-enzymes act in bioregeneration. Bioregeneration has been demontrated in offline systems, which involve desorption and biological removal of adsorbed organic matter in a closed-loop recirculating batch system.

It has also been literature reported that bioregeneration of carbon occurs during the time course of wastewater treatment processes that are based on biological activated carbons. However, most of the studies aiming at quantifying bioregeneration were performed using offline systems because of difficulties encountered in quantifying bioregeneration during biological activated carbon treatments of wastewater. Bioregeneration is dependent on several factors including biodegradability, adsorbability and desorbability of sorbate, characteristics of activated carbon and process configuration, and it can be optimized by varying the operational conditions.

Bioregeneration is the renewing of activated carbon by microbial activities. Bioregeneration can be achieved either by mixing bacteria with saturated activated carbon in offline systems (Goeddertz *et al.*, 1988) or it can be achieved in the course of biological treatments as in the case of biological activated carbon treatment systems (Sublette *et al.*, 1982). Bioregeneration can be optimized by varying the nature of microorganisms, the environmental conditions and the loading on activated carbon.

***Offline bioregeneration:*** Offline bioregeneration involves removal of adsorbed organic matter from contaminated carbons through desorption and biodegradation occurring inside a closed batch system. It consists of regenerating used activated carbon in a column in which a mixture of acclimated bacteria, nutrients and dissolved oxygen are recirculated to remove adsorbed organic matter (Goeddertz *et al.*, 1988). Most of the conclusions that can be drawn on the bioregeneration process were obtained from investigations involving offline bioregeneration, because in simultaneous processes it is very difficult to differentiate between adsorption/desorption and biodegradation. Goeddertz *et al.* (1988) suggested that the offline process for activated carbon bioregeneration should be preferred to the bioregeneration process occurring during the biological activated carbon treatments based on the limitations of the biological activated carbon process for availability of nutrients and dissolved oxygen, the persistence of many organic compounds and operational difficulties including hydraulic short circuiting and excessive head loss.

### 3.1.9.10. PHYTO-SORPTION

***Definitions/Terms***

***Phytoremediation:*** Phytoremediation is the use of certain plants to clean up soil, sediment, and water contaminated with metals and other pollutants.

***Phytoextraction (or) phytoaccumulation:*** Plant roots take metal contaminants and store in stems and leaves (harvestable regions). This technique is generally used for metals/organic compounds like nickel, zinc, copper, lead, chromium and cadmium, phenol etc.

***Phytodegradation:*** Enzymatic breakdown of organic pollutants such as trichloroethylene and herbicides, both internally and through secreted plant enzymes.

### 3.1.10.1. *Introduction*

Phytoremediation can be defined as engineered or constructed systems that use aquatic macrophyte in the treatment of domestic or industrial wastewater, designed to achieve a specific treatment goal (Suresh, 2006). Such systems have been used successfully since 1970, for a variety of applications involving secondary, advanced secondary and tertiary treatment. Hydraulic retention time, hydraulic loading rate, and organic loading rate, are the most common parameters used and needed to size aquatic plant systems. Systems in which plants are cultured in shallow tanks or trenches have earlier been called by other authors as nutrient film technique (Jewell *et al.*, 1983) and thin film technique. Phytoremediation is comprised of two components, one by the root colonizing microbes and the other by plants themselves, which degrade the toxic compounds to further non-toxic metabolites. Various compounds, *viz.* organic compounds, xenobiotics, pesticides and heavy metals, are among the contaminants that can be effectively remediated by plants.

One more dimension to this challenge is the development of wastewater treatment technologies appropriate for rural and small urban localities. In such localities, the volume of wastewater generated is low. It is very expensive to treat such low volumes by conventional technology because, according to the economics of scale, lower the quantity of waste to be treated by any technology, higher is the cost of treatment per litre. This is one of the reasons sewage treatment facility. This leads to widespread water and soil pollution and frequent outbreaks of water borne diseases. Against this background, the emerging technology of using aquatic macrophytes-especially common weeds-which enable cost-effective treatment of biodegradable wastewaters, assumes greater significance and promise.

The realisation that aquatic plants can improve water quality must have come about several centuries ago, with the observation that wastewaters flowing out of channels infested with vascular plants, such as water hyacinth, seemed to be clearer than the wastewaters entering such channels. Asian farmers, down the ages, have known the advantage of culturing aquatic plants and fishes in wastewaters ponds to speed up the treatment of wastes as well as to recover nutrients in the form of live biomass. However, scientific studies to employ aquatic plants as bioagents in water purification began only during the 1985.

The macrophytes commonly found in eutrophic water bodies are classified as *free floating weeds*: their entire body except roots keep above water example *Eichhornia crassipes, Pistia stratiotes, Lemna minor*, *submerged weed*: where the whole body of the plant is submerged in water example *Hydrilla vercillate* and *emergent weeds*: which are rooted in soil but emerge significant heights above water example *Typha latifolia, Phragmites communis*.

Boyd (1970) was pioneer to demonstrate the nutrient removal potential of aquatic plants. Macrophytes play prominent role in nutrient and heavy metal recycling of many aquatic eco-system. *Eichhornia crassipes* could be used for removal of nutrients and heavy metals from sewage and sludge ponds have been recognised worldwide. The highlight of water treatment by aquatic weeds which some of perspective for developing countries have been reported by US National Academy of Science (1976).

#### 3.1.10.2. *Selection of Appropriate Macrophyte*

The following criteria have been identified over the years for selecting a plants species or a combination of species as the main bioagent(s) in water treatment systems.

(a) Adaptability to local climate
(b) High oxygen transport capability
(c) Tolerance to adverse concentration of pollutant
(d) High pollutant uptake efficiency
(e) Tolerance to adverse climate conditions
(f) Resistance to pests and diseases
(g) Easy management

The plants commonly used so far in aquatic macrophyte systems are water hyacinth, salvinia, bulrush and pennywort. The nutrient assimilation capacities of aquatic macrophtyes are closely related to growth rate, standing crop and tissue composition.

3.1.10.3. *Function of Aquatic Plants and Micro-organisms in Aquatic Macrophyte System*

Plants function in phytoremediation in two ways, the major one being facilitation of favorable conditions for microbial degradation, specifically by plant root colonizing microbes, and the second aspect is plant root itself, providing a simple and inexpensive means of accessing contaminants existing in subsurface soils and water. In the rhizosphere, microbial population is supported by plant nutrients in the form of decaying biomass and root exudates. 3 Plant root exudates also stimulate co-metabolic transformations, leading to degradation of some organic contaminants, as in the case of phenolics released by mulberry acting as co-metabolites for poly-chlorinated biphenyls degrading bacteria.

*Root and or stem in water column*

i. Provide surface on which bacteria grow
ii. Provide media for filtration and adsorption of solids
iii. In some case release natural polymers that facilitate flocculation and sedimentation.

Stems and or leaves at above water surface assist in

i. Attenuation of sunlight and thus curtail growth of algae
ii. Reduce effect of wind on water (*e.g.* rolling of settled matter)
iii. Facilitate transfer of gases and heat between atmosphere and water; some plants are very efficient' oxygen pumps' and hasten waste stabilization serving as mini aerators. Floating aquatic plants water hyacinth (*Eichhornia* sp.) acts as an excellent filter.

3.1.10.4. *Pollutant Removal Mechanism*

Vegetated or aquatic macrophyte system separate and transform contaminants by physical, chemical and biological mechanism. These mechanisms may occur simultaneously or sequentially as the wastewater flows through the system (Suresh, 2006).

***Flocculation and sedimentation:*** Aquatic macrophyte system has been very efficient in solids removal like settleable and floatable solids from the wastewater. Sedimentation is a partial separation or concentration of suspended solids particle from a liquid by gravity settling. In addition to the normal mechanisms of sedimentation which are operative in such system, the roots of aquatic plants serve as nuclei for the growth of clusters which then settle down. Nonsettling and colloidal particles are also removed, at least partially, by bacteria growth-which results in the settling of some colloidal solids and the microbial decay of others. There is also secretion of biopolymers from the plants roots which assists in flocculation. In aquatic macrophyte system, plants surface in the water column are coated with an active biofilm of periphyton. This biofilm can adsorb colloidal and supracolloidal particles as well as absorb soluble molecules. Roots, shoots and litter from the plants support the biofilm (aerobic bacteria and associated animals and plants) which treat wastewater in constructed wetland systems.

***Aeration:*** Aquatic macrophytes have remarkable abilities of transporting atmospheric oxygen to wastewater through the leaves, stems and roots. After a portion of oxygen thus transported is consumed during root respiration, the rest enters the water column and is utilized by facultative and aerobic bacteria for the oxidation of organic carbon. This ability to 'pump in' oxygen can be strong enough with some plants to enable oxidation of even strongly reduced sediments in eutrophic water bodies. The potential amount of oxygen transferred by plants roots in to the wastewater depends on many factors including dissolved oxygen concentration, root depth, air and leaf temperature and plants growth status (rapid growth Vs senescence).In vertical filters; oxygen mainly enters the soil filter by virtue of the suction effect of the water as it flows downwards. In subsurface horizontal flow systems oxygen is chiefly input by the marsh plants (helophytes). The introduction of atmospheric air into the plants interior means that under anoxic conditions a sufficient amount of oxygen is available in the rhizome and root zones which can be used for respiration however, the oxygen transported in the airflow is also vital to the plants survival in another respect. Oxygen is released in to the rhizosphere and parts of the root system, mainly around the root tips and young laterals. The release of oxygen caused the formation of an oxidative protective film directly on the root surface this film protects the sensitive root areas from being damaged by toxic components in the anoxic, usually extremely reduced rhizosphere. Submerged macrophytes increase the content of dissolved oxygen and deplete dissolved carbon dioxide in the water during the periods of high photosynthetic activity. Emergent macrophytes have large internal air space for transportation for oxygen to roots and rhizomes the oxygen thus transported is partly utilized in respiration and balance is released from the roots for stimulating growth of nitrifying bacteria in the rhizosphere. The oxygen flow rates which have been measured as 126μ mol $0_2$/ h g root dry mass for giant rush and 120-200 μ mol $0_2$/h g root dry mass for cattail.

***Transpiration:*** In addition to the above mentioned facts, the transpiration of plants also influences their technological application for wastewater treatment. In practice, it is usually evapotanspiration that is measured. Evapotranspiration is the sum of physical evaporation from water surfaces and plants transpiration and it depends on numerous factors influencing the ecosystem microclimate, as listed by Kadlec and Knight of evapotranspirated (1996).

***Removal of dissolved BOD/COD:*** Biochemical oxygen demand is removed as result of metabolic activity by microorganism. BOD removal mechanism in aquatic macrophyte system is a kin to the one that occurs in slow-rate horizontal flow trickling filters with built-in secondary clarification macrophytes replacing gravel as bacterial support structure. In Aquatic Macrophyte System, BOD is utilized by bacteria as an energy source and for cell synthesis. These bacteria inhabit microenvironments in the sediment, the root zone, and may also be dispersed throughout the water column. Anaerobic bacteria utilize oxygen as an electron acceptor in the breakdown of substrate carbon, whereas facultative anaerobic bacteria utilize oxidized inorganic compounds such as nitrate and sulphate as electron acceptors. The BOD removal rates of aquatic systems could be increased by supplement in the natural wetlands environment with additional oxygen (aeration) and bacterial support structure. *Water hyacinth* plants were found to be responsible for 90% of BOD removal, while the remaining 10% of BOD removal was due to oxygen transport directly from air.

John (1984) *Water hyacinth* at hydraulic retention time (HRT) of 25 days, reduces the biochemical oxygen demand (BOD) and Chemical oxygen demand (COD) of anaerobically and aerobically treated palm oil mill effluent by 4980mg/l to 180mg/l (96.4%) and 8850mg/l to 1120mg/l (87.3%).

***Nitrogen removal:*** Nitrogen is removed from aquatic macrophyte system by a number of mechanisms; Uptake by and subsequent harvesting of plants; Volatilization of ammonia; Bacterial nitrification/denitrification. Of these, bacterial nitrification-denitrification has the most dominant influence of nitrification if ammonium occurs in the oxidized root zone of macrophyte. This process is enhanced beneath stands of plants which transport large quantities of oxygen nitrate-N thus formed. It will diffuses into reduced microenvironments in the pond system where it is utilized as an electron acceptor by facultative anaerobic bacteria and is lost from the system as' N'gas. Both native waste water carbon and plants detritus can be utilized by these bacteria as a carbon source. More than 50% of nutrients was found to be stored in below ground portions of the plants; tissues which may be difficult to harvest to achieve effective nutrient removal. Because emergent macrophytes have more supportive tissue than floating macrophyte, they may have a greater potential for storing the nutrients over a longer period. For effective nutrients removal the biomass must be periodically harvested as the amount of nutrients assimilated by the plants would otherwise be released back to the water upon decay of the dead plants material. The uptake of nitrogen into the plants biomass is of minor importance from a technical viewpoint since harvesting the above ground biomass would remove only 5-10% of the nitrogen.

***Removal of phosphorus:*** In aquatic system, phosphorus can be removed from the wastewater by plants uptake, microbial assimilation, precipitation with cations such as Ca, Mg, Fe, Mn and adsorption onto clays and organic matter. The sorption of phosphorus on clays involves both the chemical bonding of the negatively charged phosphate with positively charged clay and the substitution of phosphate for silicates in the clay matrix. Uptake of P can also be stimulated by the presence of other micronutrients such as iron. If not harvested the macrophytes can re-release phosphorus back in to the wastewater especially from the dead plants tissues. Less than 5% of the phosphorus load in municipal wastewater is taken up by the plants, so that effect of harvesting the plants biomass is insignificant.

***Removal of heavy metals:*** In aquatic macrophyte plants removal of heavy metals occurs through: Plants uptake especially by roots; Precipitation through biogeochemical (microbial) processes; Adsorption or ion exchange to settled clay and organic compounds; Chemical precipitation. For example, mine drainage, Fe (II) is oxidized to Fe (III) by a biotic and microbial oxidation and other elements such as arsenic also precipitated. Uptakes of mercury and methyl mercury by submerged macrophytes was always taken up to a greater degree than inorganic mercury.

***Removal of pathogenic germs:*** In aquatic macrophyte system, concentration pathogenic bacteria and virus are reduced because of the long exposures - of the order of a few days-of these, organisms to physical, chemical and biological factors hostile to these organisms. This removal is though seldom complete and as consequence disinfection of the aquatic macrophyte system effluents is necessary before their final discharge. The bactericidal effect of higher plants on pathogenic germs. The elimination of pathogens cannot be solely explained by filtration effects. Whereas amoebae in a gravel filter were removed to a degree of 95% (tropical climate) or 75% (subtropical climate), the efficiency of the soil filter was only 15-17%.

The economic success of energy production and wastewater treatment using aquatic macrophytes based system for wastewater treatment and biomass production depends to a large extent on the photosynthetic activity and growth rates of the plants. Engineering analysis has shown that in many situations the cost of secondary and advanced water treatment can be reduced by utilizing aquatic macrophtes based systems rather than conventional treatment methods (Abbasi, 1987). Water hyacinth systems are very efficient in pollutant removal from wastewaters. Efficiency of nitrogen removal ranges from 10 to 90%, phosphorus removal was from 35 to 80%. The same facility successfully removed about 60% of $BOD_5$ and 43% of suspended materials from wastewater. These systems can also remove heavy metals like chrome, cadmium, copper and other effectively. John (1984) conducted laboratory scale and pilot plant trials on the treatment of effluent from palm oil mills. Water hyacinth was propagated on palm oil mills and BOD was reduced from 4980 mg/l to 180 mg/l (96.4% reduction) and COD was reduced from 8850mg/l to 1120mg/l (87.3% reduction) at hydraulic retention time (HRT) of 25 days.

### 3.1.11. *Case Study*

#### 3.1.11.1. *Treatment of Domestic Wastewater by using Eichhornia crassipes Plant*

*Introduction*

Phytoremediation is an eco-friendly approach for remediation of contaminated soil and water by natural macrophytes plants. They have the ability to assist the breakdrown of human-and animal-derived wastewater and remove disease-causing microorganisms and pollutants. Various aquatic organisms and macrophytes are known to accumulate metals to level greatly in excess of these in their environment. Liquid biowastes are associated with all aspects of human endeavor. In industries too, such wastes, in the form of sewage, are generated due to human activity in the industry. Therefore, the treatment of liquid biowastes is an essential component associated with industries. Various techniques and methodologies are available to treat the different types of wastes, but in most of these the costs are prohibitive. This makes the effective implementation of known technologies difficult. It therefore, becomes a matter of urgent importance to develop wastewater treatment systems. Biodegradable liquid wastes can be effectively treated by using aquatic macrophytes as bioagents. *Eichhornia crassipes* has been studied for its tendency to accumulate the heavy metal contaminants present in water bodies is considered as world's most troublesome aquatic weed. Its population can double in as little as 6 days. Water hyacinth is helophyte plant growing best in warm waters rich in macronutrients.Optimal water pH for growth of this aquatic plant is neutral but it can tolerate pH values from 4 to 10. This is very important fact because it points that *E. crassipes* can be used for treatment of different type of wastewater. Optimal water temperature for growth is 28-30ºC and optimal air temperature is 21-30ºC. Water hyacinth can be used as food for people because its leaves are rich in proteins and vitamin A. It can be utilized as green fertilizer or as compost and ash in regenerating degraded soils. It can easily replace straw as substrate for mushroom growing or it can be use as fodder.It is used as feed source for anaerobic digestion, biogas generation and fermentation of sugar into alcohol.

The present case study aims to wastewater treatment process is to eliminate or reduce on domestic wastewater to the level that cause no adverse effect on human or the receiving environment.

*Analytical methods*

The wastewater was analyzed for its characteristics as explained in the following sections using procedures outlined in American Public Health Association (APHA, 1995). The COD value was assayed with a COD analyzer (Aqualytic, Germany). The standard dichromate closed reflux method was used for COD determination.

*Experimental*

The experimentals were conducted in laboratory experiments with 5 l capacity plastic tubs total volume. The 2 l working volume of domestic wastewater, and 150 g of well acclimatized in tap water plants of *Eichhornia crassipes* was taken in tubs. All the experiment was set in duplicate. Pollutant was measured in terms of chemical oxygen demand (COD). The pollutant concentration (mg/l) prepared from original sewage wastewater varied between 250-500 mg/l. The initial COD (250 mg/l) was prepared by diluting domestic wastewater with tap water.

*Results and Discussion*

Fig 3.1.11.1. shows the pecent removal of COD in the sewage wastewater in the culture of *Eichhornia crassipes* and unplanted reactor for 10 days. After 7 days, the percent removal of COD was constant and trends to decrease the percent removal due to the excess of the micro-and macro nutrients present in the feed. Reduction of COD as found to be attained in *E. crassipes* reactor is mainly due to the reduction in the amount of organic solids. This is brought about by the filtration and adsorption as well as microbial mineralization of trapped solids. The results indicate that plant biomass plays a major role in the efficiency of such wastewater.

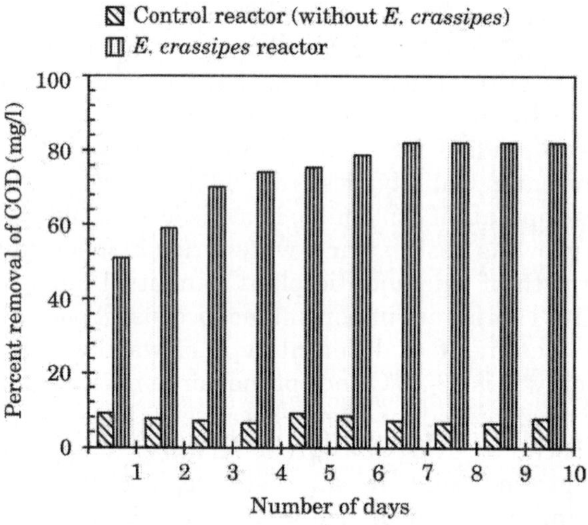


**Fig. 3.1.11.1:** Phytosorption of domestic wastewater (Suresh, unpublished work).

### 3.1.12. Design Of Experiments

Statistical methods to investigate the combination of those parameters for the best parameters and conditions for any processes; Evaluating the effects of initial concentration, pH, temperature and adsorbents loading on the percentage removal of adsorbates and statistically optimizing these variables for maximum removal of adsorbate efficiently and economically; The application of statistical experimental design in sorption techniques, results in higher percentage yields, less treatment time with minimum costs. Most optimization studies during the development of a process involve variation of one factor at a time, keeping all other factors constant. But the experiments conducted using the factorial designs, enable all factors to vary simultaneously. This helps in quantifying linear, square and interactive effects of the test variables. Another important advantage is that, the experimental designs could be changed progressively until a fitted model is found to describe the studied phenomenon.Taguchi's experiments are, generally, conducted in such a manner as to optimize 'one-factor at-a-time'. This methodology overlooks the interactive effects of various factors on the sorption process. Fractional factorial design based on Taguchi's orthogonal array can be a very effective methodology to investigate the effects of multiple factors as well as potential interactions between these factors in a time and cost effective manner. It has been used extensively in carrying out experiments and devising a strategy for quality control in the manufacturing industries. Recently, chemical and environmental engineers have started applying Taguchi's design of experimental methodology in various studies. Literature information: formation of nano-sized silver particles by chemical reduction method; separated copper ions by electrodialysis; optimized the conditions for wastewater treatment using spiral-wound reverse osmosis element; recovering heavy metals from acidified aqueous solutions for different process parameters as metallic ion concentration, current density and bed expansion. Response surface methodology (RSM) is a collection of mathematical and statistical techniques useful for designing experiments, building models and analyzing the effects of the several independent variables (factors). The main advantage of RSM is the reduced number of experimental trials needed to evaluate multiple factors and their interactions. The study of the individual and interactive effects of these factors will be helpful in efforts to find the target value. Hence, RSM provides an effective tool for investigating the aspects affecting desired response if there are many factors and interactions in the experiment. In order to determine a suitable polynomial equation for describing the response surface, RSM can be employed to optimize the process. Box–Behnken is a spherical, revolving design; it consists of a central point and the middle points of the edges of the cube circumscribed on the sphere. This procedure decreases the number of actual experiments to be performed, while allowing thorough observation of the possible interactions between input parameters studied and their effect on the removal of adsorbates and pollutants.

## 3.2. Ion-exchange Technique

### 3.2.1. *Introduction*

Ion exchange is an exchange of *ions* between two *electrolytes* or between an electrolyte *solution* and a *complex*. In most cases the term is used to denote the processes of purification, separation, and decontamination of aqueous and other ion-containing solutions with solid *polymeric* or *mineralic* 'ion exchangers'. Ion exchange is a unit operation in its own right, often sharing theory with adsorption or chromatography, although it has its own special areas of application. The oldest and most enduring application of ion exchange is in water treatment, to soften or demineralise water before industrial use, to recover components from an aqueous effluent before it is discharged or recirculated (Arden, 1968). Ion exchange may also be used to separate ionic species in various liquids as discussed by Helffrich (1962) and Schweitzer (1988). Applications of ion exchange membranes are discussed in Chapter 1. Thompson (1850) studied the exchange between ammonium ions in fertilisers and calcium ions in soil. Ion exchange is a versatile separation process used to remove metal contaminants from aqueous waste streams and to recycle or discharge the treated solution. Ion exchange techniques involve the use of an ion selective resin to remove ionic contaminants (metals) from solution. Three basic types of resins are employed; cation exchange resins, anion exchange resins, and metal selective chelating resins. A typical heavy metal ion exchange system has a fixed bed of resin with the ability to exchange or remove cations or anions. In general, divalent and trivalent ions are easier to remove than monovalent ions using ion exchange. Ion exchange is probably the most frequently used chromatographic technique for the separation and purification of proteins, polypeptides, nucleic acids, polynucleotides, and other charged biomolecules. Typical ion exchangers are *ion exchange resins* (functionalized porous or *gel* polymer), *zeolites, montmorillonite, clay*, and *soil humus*. Ion exchangers are either cation exchangers that exchange positively *charged* ions *i.e* cations or anion exchangers that exchange negatively charged ions *i.e* anions. There are also *amphoteric* exchangers that are able to exchange both cations and anions simultaneously. However, the simultaneous exchange of cations and anions can be more efficiently performed in *mixed beds* that contain a mixture of anion and cation exchange resins, or passing the treated solution through several different ion exchange materials. Ion exchangers can be unselective or have binding preferences for certain ions or classes of ions, depending on their *chemical structure*. This can be dependent on the size of the ions, their charge, or their structure. Typical examples of ions that can bind to ion exchangers are:

- $H^+$ (*proton*) and $OH^-$ (*hydroxide*)
- Single charged monoatomic ions like $Na^+$, $K^+$, or Cl,
- Double charged monoatomic ions like $Ca^{2+}$ or $Mg^{2+}$
- Polyatomic *inorganic* ions like $SO_4^{2}$ or $PO_4^{3}$
- *Organic bases*, usually *molecules* containing the *amino functional group* $-NR_2H^+$
- *Organic acids*, often *molecules* containing -COO (*carboxylic acid*) functional groups
- *Biomolecules* which can be ionized: *amino acids, peptides, proteins*, etc

Ion exchange is a reversible process and the ion exchanger can be *regenerated* or *loaded* with desirable ions by washing with an excess of these ions.

**3.2.2. *Matrix:*** An ion exchanger consists of an insoluble matrix to which charged groups have been covalently bound. The charged groups are associated with mobile counter ions. These counter-ions can be reversibly exchanged with other ions of the same charge without altering the matrix. It is possible to have both positively and negatively charged exchangers. Positively charged exchangers have negatively charged counter-ions (anions) available for exchange and are called anion exchangers. Negatively charged exchangers have positively charged counter-ions (cations) and are termed cation exchangers. The matrix may be based on inorganic compounds, synthetic resins or polysaccharides. The characteristics of the matrix determine its chromatographic properties such as efficiency, capacity and recovery as well as its chemical stability, mechanical strength and flow properties. The nature of the matrix will also affect its behaviour towards biological substances and the maintenance of biological activity. The first ion exchangers were synthetic resins designed for applications such as demineralisation, water treatment, and recovery of ions from wastes. Such ion exchangers consist of hydrophobic polymer matrices highly substituted with ionic groups, and have very high capacities for small ions. Due to their low permeability these matrices have low capacities for proteins and other macromolecules. In addition, the extremely high charge density gives very strong binding and the hydrophobic matrix tends to denature labile biological materials. Thus despite their excellent flow properties and capacities for small ions, these types of ion exchanger are unsuitable for use with biological samples. Ion exchangers based on dextran (Sephadex), followed by those based on agarose and cross-linked cellulose were the first ion exchange matrices to combine a spherical form with high porosity, leading to improved flow properties and high capacities for macromolecules.

***Charged groups:*** The presence of charged groups is a fundamental property of an ion exchanger. The type of group determines the type and strength of the ion exchanger; their total number and availability determines the capacity. There is a variety of groups which have been chosen for use in ion exchangers; some of these are shown in Table 3.2.1.1. Sulphonic and quaternary amino groups are used to form strong ion exchangers; the other groups form weak ion exchangers. The terms strong and weak refer to the extent of variation of ionization with pH and not the strength of binding. Strong ion exchangers are completely ionized over a wide pH range whereas with weak ion exchangers, the degree of dissociation and thus exchange capacity varies much more markedly with pH.

**Table 3.2.1.1:** Fucntional groups used on ion exchangers

| ***Anion exchangers*** | ***Functional group*** |
|---|---|
| Diethylaminoethyl | $-O-CH_2-CH_2-N^+H(CH_2CH_3)_2$ |
| Quaternary aminoethyl | $-O-CH_2-CH_2-N^+(C_2H_5)_2-CH_2-CHOH-CH_3$ |
| Quaternary ammonium | $-O-CH_2-CHOH-CH_2-O-CH_2-CHOH-CH_2-N^+(CH_3)_3$ |
| ***Cation exchangers*** | ***Functional group*** |
| Carboxymethyl | $-O-CH_2-COO^-$ |
| Sulphopropyl | $-O-CH_2-CHOH-CH_2-O-CH_2-CH_2-CH_2SO_3^-$ |
| Methyl sulphonate | $-O-CH_2-CHOH-CH_2-O-CH_2-CHOH- CH_2SO_3^-$ |

***Some properties of strong ion exchangers:*** (1) Sample loading capacity does not decrease at high or low pH values due to loss of charge from the ion exchanger. (2) A very simple mechanism of interaction exists between the ion exchanger and the solute. (3) Ion exchange experiments are more controllable since the charge characteristics of the media do not change with changes in pH. This makes strong exchangers ideal for working with data derived from electrophoretic titration curves.

### 3.2.3. *Application of Ion Exchange Resin*

The ion exchange method involves the exchange of ions in the solution with ions of similar charge on the resin. Ion exchange process has received considerable attention for separation and concentration of a variety of metals from waste effluents. It is generally used for effective recovery of silver from rinse water or other dilute solutions of silver. The soluble silver thiosulfate complex is exchanged with the anion on the resin. This is the exhaustion step and is accomplished by running the solution through a column containing the resin. For large operation, the next step is the regeneration step in which the silver is removed from the resin column with a silver complexing agent such as ammonium thiosulphate. This method are considered uneconomical for silver recovery from photographic waste but are feasible for low silver concentration in separating waste streams where the total ionic strength is not much greater than the silver salt concentration (Smith and Bonnie, 1977). In consequence, ion exchange equipment more resembles a manufacturing plant than a waste water treatment facility. As a result, ion exchange has been regarded more as a method to produce pure water than as a method to purify waste.

***Oil removal:*** The presence of oil in wastewater is a major concern in a number of industries such as food industry, petroleum refinery, vegetable oil and hydrogenated oil manufacturing, petrochemical industry, metal processing, and finishing and metallurgical industries, etc. oil contamination is generally of two types: immiscible mixture and miscible solution. Immiscible mixture may be in the form of free oil or emulsions – which may be unstable or stable emulsion. Stable oil-in-water or water-in-oil emulsions and miscible solutions are very difficult to be treated for removing oil or water, as the case may be, from one another. Oil emulsions are deleterious to environment as they impede penetration of sunlight and absorption of oxygen from air by the water bodies contaminated by oil, and in many cases are toxic to the aquatic life. Because of these reasons, the Ministry of Environment and Forests, Government of India (CPCB, 2008) has prescribed the discharge wastewater with oil concentration not exceeding 5 mg/l. Chinese Government has also set the discharge limit for oil as 5 mg/l. USEPA has prescribed a limit of 3 mg/l for oil in wastewater for discharge. Since the concentration of oil in water, either in miscible or immiscible form, generally is very large in waste water emanating from the petroleum refineries, metal processing and vegetable oil processing industries, it is necessary to use such treatment methods as are capable of reducing oil concentration less than 5 mg/l. Immiscible oil in water is generally in free form or unstable dispersed form. However, emulsions are very difficult to deal with as they are in stable form, finely dispersed in the mixture, and the presence of a surfactant and co-surfactant makes them more difficult to treat them. Such kinds of wastewaters generally have interfacial films and electrostatic repulsion forces operating on fine droplets to sustain stability. Emulsion stability with respect to film/layer formation and coalescence/aggregation of the droplets and the rheology of the aqueous dispersion medium are very important aspects in the emulsion treatment. The droplets size distribution is determined mainly by the amount of the surfactants and the energy expended during emulsification.

Conventional oil removal methods which include such methods as gravity separation and tilted plate interception, API separator, floatation and chemical coagulation do not meet the statutory requirements of oil concentration in the treated water. The commercially available technology for treatment of emulsion involves chemical de-emulsification followed by air floatation and gravity separation. But, these processes require the use of expensive specific chemicals and generate excess volume of sludge and corrosive environment in the handling equipments. Recently a number of researchers have focused their attention on coalescence methods for separation of oil from emulsion. Coalescers are basically improved filtration systems using designed filtration medium, usually modified polymeric resins, which provide high adsorption capacity and oil-removal efficiency. These characteristics are important due to hydrogen bonding between oil molecules and free hydrophobic part of the fixed surfactant. Hydrogen bonding and hydrophilic interaction are considered to be the principal driving forces for the adhesion of oil onto polymeric resins. The adsorption process is found to be reversible because of the much higher ionic bond energy of the surfactant fixation than the hydrogen bond energy. This opens up the possibility of breaking the hydrogen bond while maintaining the ionic bond intact, and subsequent separation and recovery of oil. A coalescer is found to be effective in reducing oil in an appropriate droplet size range while treating synthetic and actual industrial effluent resulting in effective oil separation from emulsion. Different sorbent materials can be used for removal of oil from wastewater. Two types of sorbents were investigated organic (loose natural wool fibers (NWF) and recycled wool based non-woven material (RWNM) and inorganic (sepiolite). Sorption was carried out in continuous tubular contractor (initial oil concentration of 1511 mg/l) and batch tank (initial oil concentration of 5066 mg/l). Wool-based sorbents showed higher sorption capacity (5.56 g/g for NWF and 5.48 g/g for RWNM) compared to sepiolite (0.19 g/g) in case of sorption in batch tank.

***Treatment of oil-in-water emulsions by a sawdust bed filter:*** The effect of operating conditions on the performance of a sawdust bed filter used for the treatment of an oil-in-water emulsion: A metalworking fluid (3 vol.% oil) was used as oil-in-water emulsion and sawdust as filter medium and sorbent. Because of the high stability of the emulsion, small amounts of inorganic salt (calcium sulphate) were mixed with the sorbent, acting as coagulant to achieve the emulsion breakdown. The influence of flow rate, bed height, temperature and the amount of coagulant salt added was studied. Experimental results show that several processes are involved in oil removal from oil-in-water emulsions, *i.e.* coagulation, coalescence, adsorption or straining. More than 99% of oil content in the influent stream was removed. Experimental results show that low-cost sorbents like sawdust are feasible to be used in the treatment of oil-in-water emulsions if small amounts of coagulant salts are added to the filter media. Application of commonly used filter mediums such as activated carbon and polypropylene is limited due to their high cost and inefficient regeneration, which needs further improvement and modification of adsorption medium. This can be overcome by coalescence which is an effective process used to destabilize and remove finely dispersed particles from wastewaters. The studies have shown that a coalescer is effective in reducing high levels of oil and grease in appropriate particle size range in the treatment of synthetic and actual industrial effluent. When oil in water emulsion is passed through a porous medium (resin bed) the micro droplets of oil are captured by the medium forming a film of hydrocarbon around the bed. The micro droplets combine together and form macro droplets, which is discharged continuously from the bed.

**3.2.4. *Darcy's law for fluid flow through porous medium:*** Darcy's law is a simple proportional relationship between the instantaneous discharge rate through a porous medium, the viscosity of the fluid and the pressure drop over a given distance.

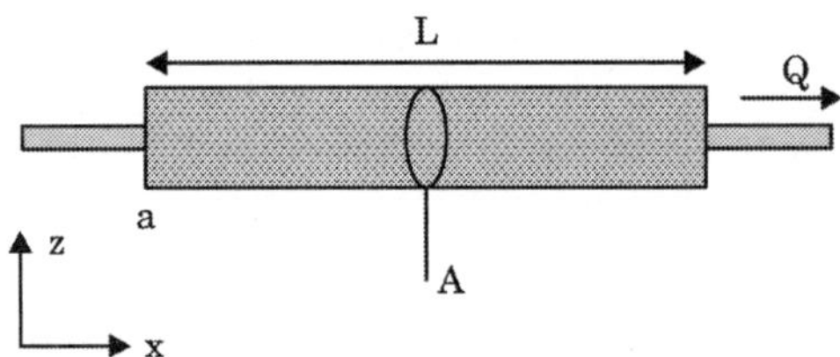


**Fig. 3.2.2.1:** Fluid flow through a porous medium (McCabe, 2001).

The total discharge, $Q$ (volume per time, m$^3$/s) is equal to the product of the permeability (ê units of area, m$^2$) of the medium, the cross-sectional area (A) to flow, and the pressure drop ($P_b$–$P_a$), all divided by the dynamic viscosity ì (Pa·s), and the length $L$ the pressure drop is taking place over.

$$Q = \frac{-kA}{\mu} \frac{(P_b - P_a)}{L} \tag{3.3.53}$$

The negative sign is needed because fluids flow from high pressure to low pressure. So if the change in pressure is negative (in the $x$-direction) then the flow will be positive (in the $x$-direction). Dividing both sides of the equation by the area and using more general notation leads to

$$q = \frac{-k}{\mu} \nabla P \tag{3.3.54}$$

Where $q$ is the flux (discharge per unit area, with units of length per time, m/s) and "P is the pressure gradient vector (McCabe, 2001). The most commonly used sorbent for removal of organic compounds from water is granulated activated carbon because of its highly porous structure that provides large internal surfaces for adsorbed molecules to reside. However, it has been reported that hydrophobic $CF_3$ silica aerogels have a much larger absorption capacity than granulated activated carbon for removing oil and other organic contaminants.

## 3.3. Chromatographic Separations

### *Terms/Definitions*

***Analytical technique:*** A technique is a fundamental scientific phenomenon that has proved useful for providing information on the composition of substances. Infrared spectrophometry is an example of an analytical technique.

***Analytical method:*** A method is a specific application of a technique to solve an analytical problem. The infrared analysis of styrene-acrylonitrile copolymers is an example of an instrumental method.

***Procedure:*** A procedure is the written instructions for carrying out a method. The 'standard method' developed by the American Society for Testing Materials and Assocation of Official Analytical Chemists are, in reality, standardized procedures. The procedure for infrated analysis of styrene-acrylonitrile copolymers involves the extraction of the residual styrene and acylonitrile monomers from the polymer into carbon disulfide.

***Protocol:*** The sample absorbances are compated with those of standards of known concentration. In contrast, the most specific decription of a method is known as a protocol. The detailed directions must be followed, without exception, if the analytical results are to be accepted for a given purpose, such as an environmental analysis to meet Environmental Protection Agency requirements or blood alcohol determinations for legal proceedings.

### 3.3.1. *Introduction*

Analytical instrumentation plays an important role in the production and evaluation of new products and in the protection of consumers and the environment. This instrumentation provides the lower detection limits required to assure safe foods, drugs, water and air.

***Classification of instrumental techniques:*** Most instrumental techniques fit into one of three principal areas: spectroscopy electrochemistry, and chromatography (Table 3.13.1). Although several important techniques, including mass spectrometry and thermal analysis, do not fit conveniently into these classifications, these three areas do provide a basis for a systematic study of chemical instrumentation. Advances in both chemistry and technology are making new techniques available and expanding the use of existing ones. Photoacoustic spectroscopy is an example of an emerging analytical technique. A number of exisiting techniques have been combined to expand the utility of the component methods. Gas chromatography-mass spectrometry and inductively coupled plasma-mass spectrometry are examples of successful 'hyphenated' methods (Table 3.3.1). The distribution of computer power to individual instruments has led to the widespread use of methods such as the Fourier

**Table 3.3.1:** Classificaiton of chemical instrumentation

| ***Principal Types*** | ***Techniques*** |
|---|---|
| Spectroscopic techniques | Ultraviolet and visible spectrophotometry |
| | Fluorescence and phosphorescence spectrophotometry |
| | Atomic spectrometry (emission and absorption) |
| | Infrared spectrophotometry |
| | Raman spectroscopy |
| | X-ray spectroscopy |
| | Radiochemical techniques including activation analysis |
| | Nuclear magnetic resonance spectroscopy |
| | Electron spin resonance spectroscopy |
| Eletrochemical techniques | Potentiometry (pH and ion selective electrodes) |
| | Voltammetry |
| | Voltametric techniques |
| | Stripping techniques |
| | Amperometric techniques |
| | Coulometry |
| | Electrogravimetry |
| | Conductance techniques |

**Table 3.3.1:** (*Contd...*)

**Table 3.3.1:** *(Contd...)*

| ***Principal Types*** | ***Techniques*** |
|---|---|
| Chromatographic techniques | Gas chromatography<br>High-performance liquid chromatographic techniques |
| Miscellaneous techniques | Thermal analysis<br>Mass spectrometry<br>Kinetic techniques |
| Hyphenated techniques | Gas chromatography-mass spectrometry<br>Inductively coupled plasma-mass spectrometry<br>Gas chromatography-infrared spectroscopy<br>Mass spectrometry-mass spectrometry |

transform to produce new techniques: Fourier transform infrared and pulsed nuclear magnetic resonance spectroscopy.

***Important considerations in analytical methods:*** Qualitative information may include elemental composition, oxidation states, functional groups, major components, minor components, and complete identification of all species present in the sample. Quantitative data include the required accuracy and precision, range of expected analyte (substance being determined) concentrations, and detection limits for the analyte. Other considerations are the unique physical and chemical properties of the analyte, properties of the sample matrix, the presence of interferences that are likely to eliminate the use of certain analyte properties of measurement indicators, and finally an estimated cost of the analysis.

### *Classification of Chromatographic Methods*

The mobile phase can be a gas or liquid, whereas the stationary phase can be only a liquid or a solid (see Table 3.3.2). When the separation involves predominantly a simple partitioning between two immiscible liquid phases, one stationary and the other mobile, the process is called liquid liquid chroma-tography. When physical surface forces are mainly involved in the retentive ability of the stationary phase, the process is denoted liquid-solid chromatography.

**Table 3.3.2:** Chromatographic methods

| | |
|---|---|
| Gas chromatography | Gas-liquid chromatography<br>Gas-solid chromatography |
| Liquid chromatography | Liquid-liquid chromatography<br>Liquid-solid chromatography<br>Ion-exchange chromatography<br>Exclusion chromatography |

## 3.3.2. *Chromatography*

Chromatographic methods of separation are distinguished by their high *selectivity* that is their ability to separate components of closely similar physical and chemical properties. Many mixtures which are difficult to separate by other methods may be separated by chromatography. The range of materials which can be processed covers the entire spectrum of molecular weights, from hydrogen to proteins. Chromatography also plays two other roles in process engineering. First, the detailed design of each unit operation in a process requires knowledge of a variety of physical and chemical properties of the materials involved. Chromatographic techniques can provide rapid and accurate methods of measuring a great variety of thermodynamic, kinetic and other physico-chemical properties. Secondly, chromatography is widely used for routine chemical analysis in quality control and for automated analysis of process streams in process control (process chromatography). This chapter, however, is concerned mainly with production chromatography as a unit operation whose use is increasing as the demand for high purity materials grows.

The design of a new plant begins with a proposed chemical route from starting materials to finished product, already tested at least on a small laboratory scale. In many cases, *analytical chromatography* will have been used in the laboratory to separate and identify the products in the mixtures produced by the proposed chemical process. In the biotechnology area, for example, the original starting material may be a natural product or a complex synthetic mixture which has been analysed by chromatography. Information on the choice of chromatographic stationary and mobile phases and operating conditions is therefore often already available and provides a basis for initial scale–up to an intermediate laboratory scale, *preparative chromatography*, and for the subsequent design of the commercial separation process, *production* or *large-scale chromatography*.

### 3.3.2.1 *Gas Chromatography*

Gas chromatography is the technique of choice for the separation of thermally stable and volatile organic and inorganic compounds. Gas-liquid chromatography accomplishes the separation by partitioning the components of a chemical mixture between a moving (mobile) gas phase and a stationary liquid phase held on a solid support. Gas-solid chromatography uses a solid adsorbent as the stationary phase. The availability of versatile and specific detectors and the possibility of coupling the gas chromatograph to a mass spectrometer or an infrared spectrophotometer further enhance the usefulness of gas chromatography.

***Gas Chromatographs:*** A gas chromatograph consists of several basic modules joined together to (1) provide a constant flow of carrier (mobile phase) gas, (2) permit the introduction of sample vapors into the flowing gas stream, (3) contain the appropriate length of stationary phase. (4) maintain the column at the appropriate temperature (or temperature-program sequence), (5) detect the sample components as they elute from the column, and (6) provide a readable signal proportional in magnitude to the amount of each component.

***Sample Injection System:*** The standard mode, suitable for approximately 95% of packed column applications is, direct injection. The sample is injected by a hypodermic syringe through a self-sealing silicone rubber septum onto a glass liner within a metal block, where it is sealing silicone rubber septum onto a glass liner within a metal block, where it is vaporized and swept onto the column. Reduction in sample volume is necessary when working with capillary columns. This is accomplished by an injector-splitter, where typically 1µl sample is injected but only 0.01 µl enters the capillary; the remainder is vented. This technique prevents column overload but wastes a significant portion of the sample. When analyses are performed on small amounts of sample with some component concentrations in the parts-per-billion range, too little material is placed on the column if splitting is used for these samples. Splitless injection is required for these samples. The entire sample, including any solvent, is injected onto the open tubular column through a modified heated flash vaporizer. A large solvent tail is avoided by venting the injection part to the atmosphere at that time. When most of the solvent and essentially all the sample have entered the column. The proper time until venting is critical; too long a time causes a solvent peak large than necessary which might bury some of the peaks of interest.

***Automatic Sampler:*** An automatic sampler duplicates manual sample measurements and injection. Sample vials are a glass, throw-away type with vapor-tight septum caps. The sampler flushes the syringe with a new sample to remove traces of the previous sample, pumps new sample to wet the syringe completely and eliminate any bubbles, takes in a precisely measured amount of sample, and injects the sample into the gas chromatograph. The automatic samplers are machine reproducible and consistently more precise than a skilled chromatographer; also unattended operation releases the operator for other duties.

***Derivative formation:*** Derivatization prior to gas chromatography is often desirable to (1) improve the thermal stability of compounds, particularly compounds that contain polar function groups. (2) change the separation properties of compounds by the purposeful adjustment of their volatility, and (3) introduce a detector-oriented tag into a molecule. In an ideal derivatization, all the relevant functional groups should be derivatized quantitatively and quickly, preferably in less than 10 min. Then, without further processing, the reaction mixture can be injected directly into the gas chromatograph. Separation of the derivatives should be complete and excess reagent should elute with the solvent peak. A few typical examples with be mentioned).

A derivatization method makes feasible the quantitative and qualitative analysis of amino acids. After they are dry, amino acids are converted to butyl ester hydrochlorides by treatment with 1-butanol and hydrogen chloride at 100 °C for 30 min:

$$\underset{\displaystyle NH_2}{RCHCOOH} + C_4H_9OH + HC1 \text{——} \underset{\displaystyle NH^+_3\ \ CI^-}{RCHCOOC\,H} + H_2O$$

The butyl esters are converted to volatile N-trifluoroacetyl butyl esters in sealed tubes:

$$(CF_3CO)_2O + \underset{\displaystyle NH^+_3CI^-}{RCHCOOC_4H_9} \text{ —— } \underset{\displaystyle NHCOCF_3}{RCHCOOC_4H_9} + HCI + CF_3COOH$$

Subsequent chromatographic resolution into single peaks by temperature programming is then easily accomplished.

***Gas chromatographic columns:*** Two basic types of columns are in general use: the packed column and the open tubular or capillary column. Packed columns are constructed from tubing of stainless steel, nickel, or glass. Inner diameters may range from 1.6 to 9.5 mm and length is often 3 m. These columns are packed with an inert support, usually a derived from firebrick or filter aids. The internal column diameter should be at least eight times derived from filter aids. The internal column diameter should be at least eight times the diameter of the support particles. Especially with low liquid loadings (less than 10%) and nonpolar phases, the support must be thoroughly inactivated when polar compounds are to be analyzed. Surface mineral impurities, which can serve as adsorption sites, can be removed by acid washing. If the support in not deactivated at all, it is called nonacid washed. The acid-washed grade performs quite well for the analysis of relatively nonpolar samples. However, the silanol (Si-OH) groups that cover the surface also tend to adsorb solutes. These surface silanol groups can cause peak tailing through hydrogen bonding with polar samples. The problem can be minimized by converting the silanol groups to silyl ethers by treating the column packing with dimethyldichlorosilane (unused silylating agent is removed with methanol). When silanized supports are used, the loading of the stationary phase should be limited to maximum of 10%. Special packing material may be needed for particular applications. Very lightly loaded (roughened or texturized) glass beads are used for very rapid analyses well below the boiling point of the sample components. Sieved Teflon supports are used when corrosive substances are handled. In gas-solid chromatography the packing material is an adsorbent, such as silica gel, a bonded-phase support, or a molecular sieve.

***Capillary Columns:*** Capillary columns have an internal diameter of 1 mm or less. These are usually constructed of fused silica (a very high-purity glass), which has much higher degree of cross linking within the silicon-oxygen matrix than does ordinary glass. The high tensile strength of the silica tubing permits the construction of thin-walled, flexible columns. To protect the thin wall against scratches, a protective coating of polyimide is applied to the outer wall. There are two main types of capillary columns: (1) packed columns with solid particles over the whole diameter of the column (micropacked) and (2) open tubular columns with an open and unrestricted flow path through the middle of the column. The latter are divided into wall-coated open tubular columns, and porous-layer open tubular columns. The inner surface of the wide-bore capillary column is coated with the stationary liquid phase. Uniform coating of the surface was difficult to obtain until bonded-phase (polymer) columns were introduced. The polymer wets the surface of a fused silica column well, with the result that a very uniform film covers the column walls. A variety of functional groups can be blended into the polysiloxane chain to provide stationary phases of different polarity or selectivity. The result is a film that is both thermostable and nonextractable. The columns can be flushed with pure solvents to remove contaminants—nonvolatile and pyrolysis products. The major advantage of wide-bore capillaries regardless of the separation is an increase in the speed of analysis. Using packed column flowrates, wide-bore column produces separations with packed column efficiencies but with a total run time roughly three times faster. However, when the flow rate is optimized for the tubing diameter, the wide-bore column produces far superior efficiencies and therefore superior separations, with analysis times approximately equal to those for packed columns to wide-bore capillaries: shorter retention times, greater inertness, longer life, lower bleed, higher efficiencies, and greater reproducibility. The ability to manipulate the film thickness reproducibly has led to columns that can be tailored for the analysis of very volatile or very high-boiling mixtures. The customary thickness of the bonded liquid phase is 1-2 μm. Thicker films (3-5 μm) permit the analysis of volatile materials without the need for subambient column temperatures. Films less than 1μm thick are valuable for analyzing high-molecular-weight compounds with reasonable analysis times.

***Ovens:*** Chromatographic columns are coiled and held in a basket that is mounted inside an oven. The column oven must be able to be rapidly heated and cooled. This requires a well-designed and adequate system of air flow. In most designs the air is blown past the heating coils, then through baffles that make up the inner wall of the oven, past the column, and back to the blower to be reheated and recirculated. Ovens are usually constructed of low-mass stainless steel. For temperature programming it is desirable to have a range of temperature program rates from 0.1 to 50°C for isotherm runs and within ± 2°C of the desired temperature during programming.

***Detectors for gas chromatography:*** A detector, located at the exit of the separation column, senses the presence of the individual components as they leve the column. The detector volume must be small to prevent the remixing of components separated on the column. The electrical analog output of the detector is amplified and then sent directly to a strip chart recorder or converted to a digital signal and sent to a microcomputer system. The computer system can process the data, store them, and display the chromatogram with analytical results on a video screen or recorder. Such systems provide a continuous record of the mass of solute eluted or concentration profiles of eluted solute bands-that is, a chromatogram of concentration versus time.

***Flame Ionization Detector:*** The flame ionization detector (FID) adds hydrogen to the column effluent. Subsequently, the mixture is passed through a jet where it is mixed external air and burned. The flame jet and a cylinder positioned above the tip of the flame constitute the twin electrodes. In another arrangement two parallel plates are mounted above the flame tip. A potential is applied across the two electrodes, which lowers the resistance between the electrodes and causes a current to flow. The current arises from the ions and free electrons generated in a pure hydrogen/air flame. When ionizable material from the column effluent enters the flame and is burned, the current markedly increases. The current flows through an output device, a recorder or microcomputer. An opposing voltage equal to the signal from the hydrogen/air flame when only pure carrier gas is passing permits the baseline to be adjusted. The FID is enclosed within a chimney so that it is unaffected by drafts and can be heated sufficiently to avoid condensation of water droplets from the combustion process. An ignitor coil and flame-out sensor are placed above the jet to reignite the flame if it becomes extinguished. The FID responds proportionately to the number of $-CH_2-$ groups introduced into the flame. A portable gas chromatograph that can be carried by a shoulder strap has been built around the FID as detector. The dual-mode instrument provides either continuous direct readout of total organic vapor concentration (bypassing the chromatographic column) for screening or survey purposes, or qualitative and quantitative analyses using the GC option. A portion of the sample air is directed through an integral charcoal filter to provide the detector with a supply of pure air. The portable unit carries an 8hr supply of hydrogen, a small diaphragm air pump, and a readout meter. An isothermal pack offers temperature control of the separation column; the pack encloses a supercooled mixture that, when seeded, provides a constant temperature during its solidification process. Fig. 3.3.2.1 shows fermented product, is then separated from the sediment by using suitable filter medium and then ethanol is separated by simple distillation method (Suresh, 2004). Ethanol content is a distilled product and is thus analyzed by gas chromatography with flame ionization detector. Though the result was

CLASS–GC10 Ver.–2.00 04/03/30 17 : 52 : 52

Column : BP--624, 50mt:., 0.53, 3U
Ovan : 60C-8 min-15C/min-220C
Inj./FID : 200C/260C
Flow/Split : 4.0ml/min./12
Sample : EcOH 100 PPM Std
Analyst : ASA

*** Chromatogram *** Filename : MAR30.002

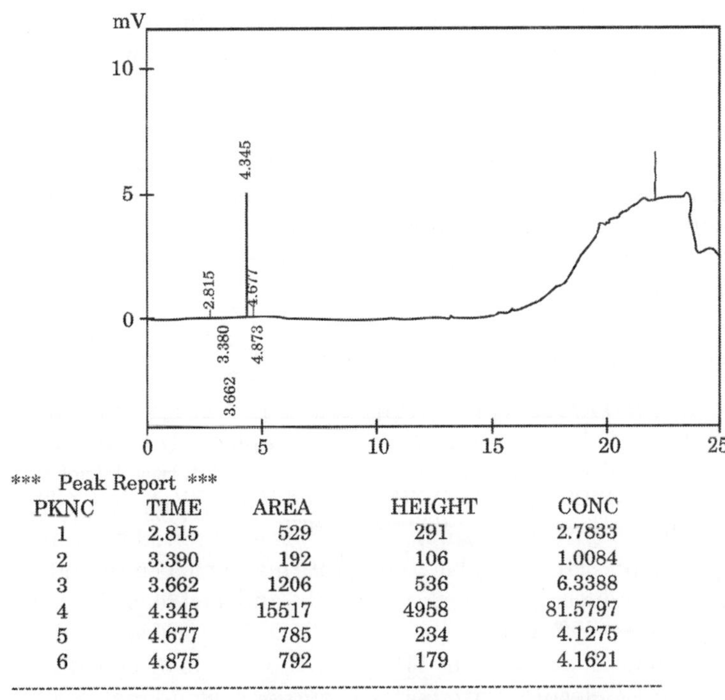


*** Peak Report ***

| PKNC | TIME | AREA | HEIGHT | CONC |
|---|---|---|---|---|
| 1 | 2.815 | 529 | 291 | 2.7833 |
| 2 | 3.390 | 192 | 106 | 1.0084 |
| 3 | 3.662 | 1206 | 536 | 6.3388 |
| 4 | 4.345 | 15517 | 4958 | 81.5797 |
| 5 | 4.677 | 785 | 234 | 4.1275 |
| 6 | 4.875 | 792 | 179 | 4.1621 |
| | | 19021 | 6303 | 100.0000 |

**Fig. 3.3.2.1:** Chromatography for evaluating of ethanol in fermented cashew fruit sample (Suresh, unpublished work)

expected to be greater than 10% (weight) of ethanol, it was only 19,969 ppm [~4% (weight)] by gas chromatography.

***Electron Capture Detector:*** In the electron capture detector (ECD) the column effluent passes between two electrodes. One of the electrodes has on its surface a radioisotope that emits high-energy electrons, (beta particles) as it decays. These electrons bombard the carrier gas (nitrogen), resulting in the formation of a plasma of positive ions, radicals, and thermal electrons by a series of elastic and inelastic collisions. The application of a potential difference to the electron-capture cell allows the collection of the thermal electrons that constitute the detector standing current, or baseline signal, when only carrier gas is passing through the detector. Electron-absorbing compounds in the carrier gas stream react with the thermal electrons to produce negative ions of larger mass. The rate of recombination between negative and positive ions is many times faster than between thermal electrons and positive ions. The decrease in detector current due to removal of thermal electrons by recombination in the presence of electron capturing compounds forms the quantitative basis of the detector operation. Efficient electron collection with nitrogen as the carrier gas is achived with a displaced coaxial cylindrical cell configuration. During most of the time no field is applied; this enables the free electrons to reach thermal equilibrium with the gas molecules. The opportunity for electron capture is maximized, and stable response factors are thereby attained. Negative ion formation occurs in a region where positive ions are also present and recombination can efficiently take place. There is on collection of negative ions. The formation of a space charge is prevented by the brevity of the pulse interval. The linearity of ECD is a function of the cell design, the composition of the carrier gas, and the method of applying the potential difference to collect the thermal electrons.

Steroids, biological amines, amino acids, and various drug metabolites can be converted to perfluoro derivatives prior to electron capture detection. Other applications involve halogenated anesthetics, polynuclear carcinogens, and sulfur hexafluoride. The ECD also responds to anhydrides, peroxides, conjugated carbonyls, nitriles and nitrated, plus ozone and organometallics and sulfur-containing compounds.

### 3.3.2.2. *High-perfomance Liquid Chromatography*

***Theory and instrumentation:*** Only about 20% of known compounds lend themselves to analysis by gas chromatography either because they are insufficiently volatile or cannot pass through the column or because they are thermally unstable and decompose under the conditions of separation. High-perfomance liquid chromatography (HPLC) is not limited by sample volatility or thermal stability. HPLC is able to separate macromolecules and ionic species, labile natural products, polymeric materials, and a wide variety of other high-molecular-weight polyfunctional groups. With an interactive liquid mobile phase. Chromatographic separation in HPLC is the result of specific interactions between sample molecules with both the stationary and mobile phases. These interactions are essentially absent in the mobile phase of gas chromatography.

HPLC offers a greater variety of stationary phases, which allows a greater variety of these selective interactions and more possibilities for separation. Sample recovery is easy in HPLC. Separated fractions are easily collected by placing an open vessel at the end of the column. Recovery is usually quantitative (barring irreversible adsorption on a column), and separated sample components are readily isolated from the mobile-phase solvent. In addition to the usual type of organic compounds, liquid column chromatography can handle separations of ionic compounds, labile naturally occurring products, polymeric materials, and high-molecular-weight polyfunctional compounds.

***Derivatization:*** The most common derivatization techniques in HPLC are intended to enhance detectability by ultraviolet absorption, fluorescence, or electrochemistry, Precolumn derivatization is carried out before separation. The derivatized sample is then injected into the column. Postcolumn reactions are carried out after separation. Reaction can be run off-line for long periods, if necessary. The product need not have different detection properties from the reactant or analyte so long as each has different separation properties. In cases where the detection sensitivity to the 254-nm wavelength in ultraviolet absorption is zero or very low, detection can be enhanced by attaching to the solute a chromophore with high absorption at 254 nm. This might be accomplished, for example, with the following reagents: All these reagents make reductive electrochemical detection and also fluorescence feasible. The formation of fluorescent derivatives serves two purposes: it allows the sensitive detection of otherwise nonfluorescent molecules, and it exploits the selectivity of fluorescence by allowing the detection of all compounds with a particular functional group in a sample after derivatization. In postcolumn reactions, effluent from an HPLC column is mixed with a regent before it enters the detector. The equipment needed to perform postcolumn rections is relatively simple. A column filled with glass beads mixes the effluent and reagent (added through a tee valve) before the mixture passes into the detector. Reaction times must be short, although a delay tube can be added to the system to allow slow reactions to take place. In postcolumn work there is much less need to convert quantitatively each analyte into a single product; reproducibility is the only requirement. Postcolumn addition of an alkaline buffer (pH 10.4) increases 20-fold the sensitivity of HPLC analysis for barbiturates. As little as 2 μg of drug can be detected by using a postcolumn reaction system because at alkaline pHs the uv absorption maxima for barbiturates shift to longer wavelengths 240 nm). Simply adjusting the mobile phase to an alkaline pH would dissolve gradually a silica-based column packing.

***Separation columns:*** Columns are constructed of heavy-wall, glass-lined metal tubing or stainless steel tubing to withstand high pressures (up to 680 atm) and the chemical action of the mobile phase. The interior of the tubing must be smooth with a very uniform bore diameter. Straight columns are preferred and are operated in the vertical position. Column end fittings and connectors must be designed with zero void volume to avoid unswept corners or stagnant pockets of mobile phase that can contribute significantly to extra-column band broadening. Packing is usually retained by inserting stainless steel frits into the end of the column. Most column lengths range from 10 to 30 cm; short, fast columns are 3 to 8 cm long. For exclusion chromatography, columns are 50 to 100 cm long. Many HPLC separations are done on columns with an internal diameter of 4 to 5 mm, such columns provide a good compromise between efficiency, sample capacity, and the amount of packing and solvent required. Column packings feature particles that are uniformly sized and mechanically stable. Particle diameters lie in the range 3-5 µm, occasionally up to 10 µm or higher for preparative chromatography.

***Fixed-Wavelength Detectors:*** A fixed-wavelength detector uses a light source that emits maximum light intensity at one or several discrete wavelengths that are isolated by appropriate filters. Using a medium-pressure mercury lamp, wavelengths of 254, 280, 313, 334, and 365 nm can be selected by the use of narrow-bandpass interference filters. Operation in the visible region can be accomplished by using a quartz iodine lamp and appropriate interference filters. Many compounds, including the nucleic acids, have absorption bands, of which portions encompass the 254 nm wavelength.

***Variable-Wavelength Detector:*** A variable-wavelength detector is a relatively wide-bandpass ultraviolet-visible spectrophotometer coupled to a chromatographic system. It offers a wide selection of uv and visible wavelengths, but at an increased cost. To obtain a complete spectrum, the eluent flow must be stopped to trap the component of interest in the detector cell while the uv-visible spectral region is scanned.

***Comparison of HPLC detectors:*** Detectors based on the absorption of visible-ultraviolet radiation are virtually insensitive to temperature variations of the lest liquid. They may be used with gradients. The advantages of a filter photometer are its relatively low cost and the high sensitivity (nanogram level) achieved for many compounds of chemical and biological interest that absorb ultraviolet or visible light at a fixed wavelength. A variable-wavelength detector offers a range of wavelengths from 190 to 600 nm, which permits a wavelength to be chosen where the solute absorbance is maximum. It is also possible to select wavelength that suppresses the absorption of an interfering solute or the mobile phase, yet where the solute has some absorbance. The detector noise levels are greater, usually by a factor of five to ten, than with filter photometers, since the light intensity for each individual wavelength is less. The simultaneous of radiation at many wavelengths and the acquisition of the resulting data mean that spectra are available for visual presentation as three-dimensional chromatograms (time vs. wavelength vs. absorbance) or storage at any time without disrupting the chromatographic signals. With a single wavelength detector, after one has obtained a chromatogram at the fixed wavelength, one cannot go back and look for other bits of information at other wavelengths. With diode arrays it is possible later to extract data at other wavelengths from the memory. Comparison of absorption spectra with spectra in a user-generated library often gives positive identification of sample components. It is now possible to evaluate a peak for purity by software data manipulation rather than by iteration and refinement of the chromatographic separation. Data manipulation can be a cost-effective substitute for higher chromatographic resolution and can perhaps justify the cost differential between diode arrays and a conventional scanning instrument. Fluorescence is a means of increasing both the selectivity and the sensitivity of an HPLC analysis. Selectivity is enhanced, since not all compounds that absorb radiation fluoresce. Of course, this limits the applicability, although fluorescent derivatives of many nonfluorescing substances can be prepared. Typical fluorescing compounds are polynuclear aromatics, steroids, plant pigments, vitamins, alkaloids, catecholamines, aflatoxines, and porphyrins. Sensitivity is improved, since the fluorescence signal is measured against a low background, assuming that the mobile phase does not fluoresce. Fluorescence is frequently 100 to 1000 times more sensitive than absorbance detection (although the two methods are frequently comparable if derivatization is needed) and is approximately 1 ng/ml for strongly fluorescing compounds. The fluorescence detector is not for general use. It is a powerful tool for specific application with the selective detection of trace components. It may be used with gradients.

***Reverse-phase chromatography:*** Reverse-phase chromatography uses a hydrophobic bonded packing usually with an octadecyl (C-18) or octyl (C-8) functional group and a polar mobile phase, often a partially or fully aqueous mobile phase. Polar substances prefer the mobile phase and elute first. As the hydrophobic character of the solutes increases, retention increases. Generally, the lower the polarity of the mobile phase, the higher is its eluent strength. Hydrocarbons are retained more strongly than alcohols. Water is the weakest eluent. Methanol and acetonitrile are popular solvents because they have low viscosity and are readily available with excellent purity. Eluents intermediate in strength between these solvents and water are usually obtained by preparing mixtures. Since the optimum composition of the mobile phase must generally be found by trial and error, it is convenient to start with a 1:1 water/methanol mixture. If the sample components elute at or near the transit time of a nonretained solute, lower concentration of methanol is indicated. Changing to acetonitrile, dioxane, or mixtures of 1,4-dioxane/methanol or acetonitrile/2-propanol can often improve selectivity. In reverse-phase chromatography, solvent gradients are generated by a continuous decrease in the polarity of the eluent during the separation -for example, by gradually increasing the organic solvent content in water/ methanol or water / acetonitrile mixtures. The reverse-phase technique in its various forms is the most widely used mode in HPLC and comprise nearly half of all the liquid chromatographic methods described in the literature. This techniques is the most likely to provide optimum retention and selectivity when compounds have no hydrogen-bonding groups or have a predominant aliphatic or aromatic character. The method is well suited for separating solutes based on the size and structure of alkyl groups. In clinical chemistry the quantitative analysis of drugs of abuse is increasingly carried out by this technique. Pharmaceuticals routinely analyzed include barbiturates, antiepileptic drugs, analgesics, and sedatives. Food preservatives, herbicides, and sugars are additional applications for reverse-phase methods. In the pharmaceutical field the use of this technique has been increasing at the expense of adsorption chromatography. Because the water content of the mobile phase can range from 100% to quite low percentages or none at all, a broad spectrum of biomolecules can be chromatographed, lipophilic or ionic, small or large.

Lipophilic compounds such as triglycerides, which have very poor solubility in aqueous reverse-phase solvents, often can be separated by a nonaqueous reverse phase using an octadecyl-packed column with polar organic solvents such as acetonitrile or tetrahydrofuran. The chromatographic behavior of a solute can be described in several ways. For column chromatography, the retention volume, (or corresponding retention time) and the partition ratio are the terms most frequently used. By varing the stationary-mobile- phase combinations and various operating parameters, the degree of retention can be varied from nearly total retention to a state of free migration. The retention behavior reflects the distribution of a solute between the mobile and stationary phases. Fig. 3.3.2.2 and Fig. 3.3.2.3 shows the separation of three phenolic compounds from the two different columns. The volume of mobile phase necessary to convey a solute band from the point of injection, through the column, and to the detector (to the apex of the solute peak) is defined as the retention volume. It may be obtained directly from the corresponding retention time, on the chromatogram by multiplying the later by the volumentric flowrate.

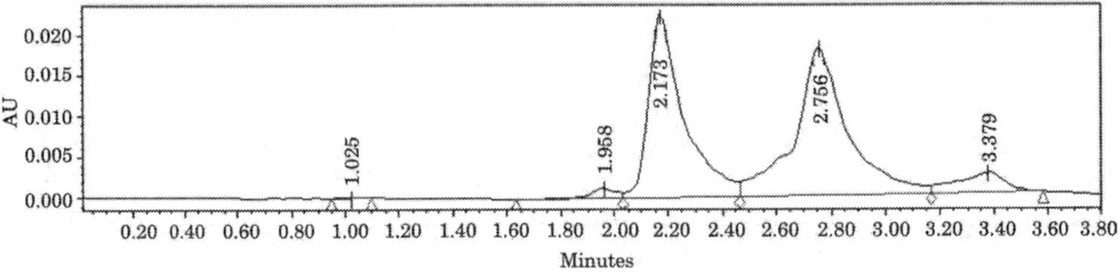


**Fig. 3.3.2.2:** Shows the separation of phenolic compounds. Column: $C_{18}$ Nova-pak type, Column pressure=1340 psi, flowrate=1ml/min (a) Phenol ($C_0$= 500 mg/l; retention time= ~2.17 min) (b) 4-nitrophenol ($C_0$=1000mg/l; retention time=~2.75min) (c) 4-chlorophenol ($C_0$=500mg/l; retention time= ~ 3.37 min) (Suresh, unpublished material)

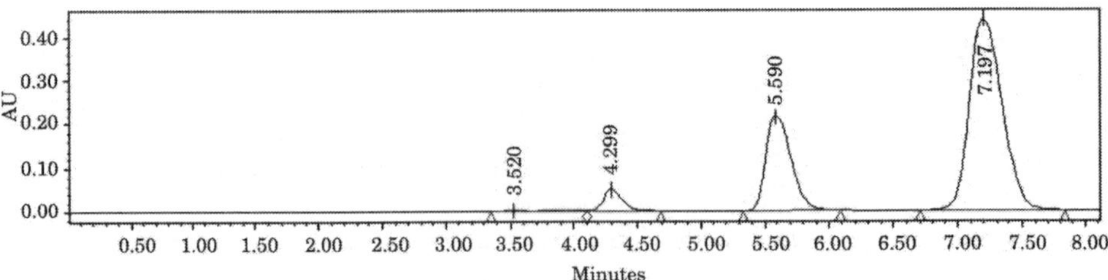


**Fig. 3.3.2.3:** Shows the separation of phenolic compounds. Column: $C_{18}$ (3.9 × 150 mm) type, Column pressure=1210 psi, flowrate=0.5ml/min (a) Hydroquinone ($C_0$= 100 mg/l; retention time=~4.29 min) (b) Resorcinol ($C_0$=1000mg/l; retention time=~5.59min) (c) Catechol ($C_0$=1000mg/l; retention time= ~7.19 min) (Suresh, unpublished material).

**3.3.2.3. *Ion-pair chromatography*:** Ion-pair chromatography, which can be considered a subset of reverse-phase chromatography, can deal with ionized or ionizable species on reverse-phase columns. The method overcomes difficulty in handling certain samples by the other liquid chromatography methods; these are samples that are very polar, multiply ionized, and/or strongly basic. In ordinary reverse-pahse HPLC, organic ions show poor peak shapes and inadequate retention. The ion suppression method is also limited to the pH range 2.0-7.5 by the instability of stationary bonded phases outside this pH range. Ion-exchange packings offer limited choice with little ability to vary selectivity by changing the column packing. In ion-pair chromatography an ion-pair reagent is added at low concentration to the mobile phase. The ion-pair reagent is itself ionized.

**3.3.2.4. *Exclusion chromatography*:** Exclusion chromatography, also called gel permeation chromatography, is a non interactive mode of separation. Essentially a maze for molecules, the particles of the column packing have various size pores and pore networks, so that solute molecules are retained or excluded on the basis of their hydrodynamic molecular volume-that is, their size and shape. Strictly speaking, separation in exclusion chromatography is not based on molecular weight. As the sample passes through the column, the solute molecules are sorted. Very large molecules cannot enter many of the pores, and they also penetrate less into the comparatively open regions of the packing. Thus exluded, they travel mostly around the exterior of the packing and elute at the bed void volume of the mobile phase. Very small molecules diffuse into all or many of the pores accessible to them. With a larger column volume at their disposal, small molecules exit the column last. Between these two extremes, intermediate-size molecules can penetrate some passages but not others and, consequently, suffer retardations in their progress down the column and exit at intermediate times.

**3.3.2.5. *Ion-Exclusion Chromatography:*** Ion-exclusion chromatography is somewhat of a hybrid method. The principal mechanism involved in these separations is ion exclusion, although steric exclusion and reverse-phase partitioning effects have been observed. As a result, separations are dependent on such factors as resin particle size and even isomer distribution in addition to the obvious variables such as cross linking and ionic form. The fact that multiple modes are involved is advantageous, often permitting the separation of species that would otherwise coelute. Separations are carried out on hight-capacity polystyrene-based exchange resins using dilute mineral acid eluents. Acetonitrile may be used as an organic modifier to decrease the retention time of relatively nonpolar compounds. For most applications the column must be maintained at elevated temperatures, so a column heating device must be used. Much work in the food and beverage industry, or physiological samples in which most of the tricarboxylic acid cycle are present, is easily done with most of the tricarboxylic acid cycle are present, is easily done with an anion exclusion column. Wine beer, fruit juice, and many dairy products are quickly analyzed with minimal sample preparation (usually only filtration or centrifugation). Silica columns have been only partially successful in separating organic acids. Another application is the separation of oligosaccharides and sugar alcohols using an anion exclusion column in the calcium ionic form with water as eluent and a temperature of 90°C. Maltotriose and higher oligosaccharides are separated from mono- and disaccharides by steric exclusion effects.

**3.3.2.6. *Affinity chromatography:*** The general scheme of affinity chromatography involves the covalent attachment of an immobilized biochemical (called an affinity ligand) to a solid support. When a sample is passed through the column, only the solute(s) that selectively binds to the complementary ligand is retained; the other sample components elute without retention. The separations exploit the 'lock and key' binding that is prevalent in biological systems. The retained solute(s) can be eluted from the column by changing the mobile-phase conditions. The major advantage of this technique is its tremendous specificity, which permits rapid isolation (in preparative work) with a good yield in a singles step. Column matrices encompass a wide variety of materials: organic gels such as beaded agarose (a purified form of the polysaccharide agar), cellulose, dextran, polyacrylamide, combinations of these polymers, and microporous glass beads. The matrix should be sufficiently hydrophilic to avoid the nonspecific binding of solutes and stable to most water - compatible organic solvents. A porous matrix permits a high degree of ligand substitution and is more accessible to larger molecules. Preactivated agarose supports are available commercially or can be prepared as outlined: Agarose is a very popular matrix; its porous meshwork can be strengthened by cross-linking it with 1-chloro-2,3-epoxypropane. Other commercial supports emphasize active groups such as -CO-NHR, - CH(OH) $CH_2$ NHR, - $CH_2$ NHR, -S-S-R, and - $C_6H_4$-N=N-$C_6H_3$ (OH)R.

## REFERENCES

Abbasi, S.A. 1987. Aquatic plant-based water treatment systems in Asia. Reddy, K.R. and Smith, W.H. (*Eds.*). Aquatic plants for water treatment and resource recovery. Magnolia Publishing Inc., Orlando. pp. 175–98.

Aharoni, C., Sideman, S. and Hoffer, E. 1979. Adsorption of phosphate ions by colloid ion-coated alumina, *J. Chem. Technol. Biotechnol.*, 29: 404–412.

APHA (American Public Health Association), AWWA (American Water Works Association), WPCF (Water Pollution Control Federation), Standard Methods for the Examination of Water and Wastewater, APHA, 1995. Washington, DC, USA.

Arden, T. V. *Water Purification by Ion Exchange* (Butterworth, London 1968).

Boyd, G.E., Adamson, A.W. and Meyers, L.S. 1947. The exchange adsorption of ions from aqueous solution by organic zeolites II Kinetics, *J. Am. Chem. Soc.*, 69: 2836–2848.

Boyd, C.E. 1970. Vascular aquatic plants for mineral nutrient removal from polluted waters. *Economic Botany* 23: 95–103.

Clark, R.M. 1987. Evaluating the cost and performance of field-scale granular activated carbon systems. *Environ. Sci. Technol.* 21: 573–580.

Dubinin, M.M. and Radushkevich, L.V. 1947. Equation of the characteristic curve of activated charcoal, *Chem. Zentr.*, 1: 875.

Fogler, H.S. 1998. Elements of Chemical Reaction Engineering, Prentice-Hall PTR, p. 3.

Freundlich, H.M.F. 1906. Uber die adsorption in losungen, *Z. Phys. Chem.* 57(A): 385–470.

Freundlich, H.M.F. Over the adsorption in solution. *J. Phys. Chem.* 57: 385–471.

Giles, C.H., MacEwan, T.H., Nakhwa, S.N. and Smith, D. 1960. Studies in adsorption. Part XI. A system of classification of solution adsorption isotherms, and its use in diagnosis of adsorption mechanisms and in measurements of specific surface areas of solids. *J. Chem. Soc.* 10: 3973–3993.

Goeddertz, J.G., Matsumoto, M.R. and Weber, A.S. 1988. Offline bioregeneration of granular activated carbon. *Journal of Environmental Engineering* 114: 1063–1076.

Helfferich, F. 1962. Ion exchange. New York: McGraw-Hill.

Helffrich, F. 1962. *Ion Exchange* (McGraw-Hill, New York).

Ho, Y.S., Wase, D.A.J. and Forster, C.F. 1996. Kinetic studies of competitive heavy metal adsorption by sphagnum moss peat, *Environ. Technol*. 17: 71–77.

Ho, Y.S. 2006. Second-order kinetic model for the sorption of cadmium on to tree fern: A comparison of linear and non-linear methods. *Water. Res.* 40: 119–125.

Ho, Y.S. and McKay, G. 1999. Pseudo-second order model for sorption processes. *Process Biochem.* 34: 451–465.

Hutchins, R.A. 1973. New method simplifies design of activated-carbon system. *Chem. Eng.*, 80: 133–138.

Jewell, W.J., Clarkson, W.W. and Madras, J.J., Delancey-Pompe, H. and Kabrick, R.M. 1983. Wastewater treatment with plants in nutrient films *EPA Project Summery*; EPA-600/S2-83-067; pp. 1–5.

John, C.K. 1984. Treatment of agro-industrial wastes using *Water hyacinth, Water Science and Technology*; 17: 781–790.

Kamsonlian, S., Balomajumder, C., Chand, S. and Suresh, S. 2011. Biosorption of Cd (II) and As (III) ions from aqueous solution by tea waste biomass. *African Journal of Environmental Science and Technology* 5(1): 1–7.

Lagergren, S. and Sven, K., 1898. Zur theorie der sogenannten adsorption gelöster stoffe, *Vetenskapsakad. Handl*. 24: 1–39.

Lagergren, S. 1898. About the theory of so called adsorption of soluble substances, *Ksver Veterskapsakad Handl*, 24: 1–6.

Langmuir, I. 1916. Constitution and fundamental properties of solids and liquids, *J. Am. Chem. Soc*. 38: 2221–2295.

Marquardt, D.W. 1963. An algorithm for least-squares estimation of nonlinear parameters, *J. Soc. Ind. Appl. Math.*, 11: 431–441

Ministry of Environment and Forest Notification. Environment (Protection) Amendment Rules, 2008. (Petroleum oil Refinery) G.S.R. 186(E), 18/03/2008, India.

Porter, J.F., Mckay, G. and Choy, K.H. 1999. The prediction of sorption from a binary mixture of acidic dyes using single-and mixed isotherm variants of the ideal adsorbed solute theory. *Chem. Eng. Sci.* 54: 5863–5885.

Redlich, O. and Peterson, D.L. 1959. A useful adsorption isotherm, *J. Phys. Chem.*, 63: 1024–1026.

Reichenberg, D. 1953. Properties of Ion-exchange resin in relation to their structure. III. Kinetics of exchange, *J. Am. Chem. Soc.*, 75: 589.

Schweitzer, P.A. (*Ed.*). 1988. *Handbook of Separation Techniques for Chemical Engineers*, 2nd edn. (McGraw- Hill, New York).

Sheindorf, C., Rebhum, M. and Sheintuch, M. 1981. A Freundlich-type multicomponent isotherm, *J. Colloid Interface Sci.*, 79: 136–142.

Skelland, A.H.P. 1974. Diffusional Mass Transfer, Wiley, New York.

Spielman, Y. and Yeang, P. 1977. Coalescence of Oil-in-Water Suspensions by Flow through Porous Media. *Ind. Eng. Chem., Fundam*, 16.

Srivastava, V.C., Mall, I.D. and Mishra, I.M. 2005. Treatment of pulp and paper mill wastewaters with poly aluminium chloride and bagasse fly ash. *Colloids and Surfaces A: Physicochem. Eng. Aspects* 260: 17–28.

Sublette, K.L., Snider, E.H. and Sylvester, N.D. 1982. A review of the mechanism of powdered activated carbon enhancement of activated sludge treatment. *Water Research* 16: 1075–1082.

Suresh, S., Srivastava, V.C. and Mishra, I.M. 2011b. Study of catechol and resorcinol adsorption mechanism through granular activated carbon characterization, pH and kinetic study, *Sep. Sci. Technol.* 46: 1750–1766.

Suresh, S., Srivastava, V.C. and Mishra, I.M. 2011a. Isotherm thermodynamics, desorption and disposal study for the adsorption of catechol and resorcinol onto granular activated carbon, *J. Chem. Eng. Data*, 56: 811–818.

Suresh, S., Srivastava, V.C. and Mishra, I.M. 2011c. Adsorptive removal of phenol from binary aqueous solution with aniline and 4-nitrophenol by granular activated carbon, *Chem. Eng. J.*, 171: 997–1003.

Suresh, S. 2006. Biological treatment of wastewater using constructed wetland. M. Tech thesis, pp. 1–103.

Suresh, S. 2004. Fermentation of cashew fruit. B.Tech report, pp. 1–30.

Taguchi, G. 1986. Introduction to Quality Engineering, Quality Resources, New York.

Tempkin, M.J. and Pyzhev, V. 1940. *Acta Physiochim, URSS*, 12: 217–222.

Thomas, H.C. 1944. Heterogeneous ion exchange in a flowing system. *J. Am. Chem. Soc.*, 66: 1664–1666.

Thompson, H.S. 1850. *J. Roy. Agr. Soc. Eng.* 11: 68. On the absorbent power of soils.

UNEP. 1989. Environmental aspects of metal finishing industry: a technical guide, United Nations Environment Programme, Industry and Environment Office, Paris. pp. 19–39.

US National Academy of Science, 1976. Making aquatic weeds useful: some perspective for developing countries.

Vermeulen, T. 1953. Theory for irreversible and constant pattern solid diffusion, *Ind. Eng. Chem.*, 45(8): 1664–1670.

Volesky, B. 1990. Biosorption of Heavy Metals, CRC Press, Boca Raton, FL.

Weber Jr. W.J., and Morris, J.C. 1963. Kinetics of adsorption on carbon from solution, *J. of the Sanitary Engg. Div. ASCE*, 89(SA2): 31–59.

Wolborska, A. 1989. Adsorption on activated carbon of *p*-nitrophenol from aqueous solution. *Water Res.,* 23: 85–91.

World Health Organisation, 1971. International Standards for Drinking Water. WHO, Geneva.

Yoon, Y.H. and Nelson, J.H. 1984. Application of gas adsorption kinetics. I. A theoretical model for respirator cartridge service time. *Am. Indian Hygiene Assoc. J.* 45: 509–516.

# Chapter 4

# Catalytic Processes

## 4.1. INTRODUCTION

Catalysis is a term coined by *Baron.* J.J. Berzelius in 1835 to describe the property of substances that facilitate chemical reactions without being consumed in them. A broad definition of catalysis also allows for materials that slow the rate of a reaction. Whereas catalysts can greately affect the rate of a reaction, the equilibrium composition of reactants and products is still determined solely by thermodynamics.

Hetrogenous catalysts are distinguished from homogenerous catalysts by the different phases present during reaction. Homogenerous catalysts are present in the same phase as reactants and products, usually liquid, while heterogenerous catalysts are present in a different phase, usually soild. The main advantage of using a heterogenerous catalyst is the relative ease of castalyst separation from the product stream that aids in the creation of continuous chemical processes. Additionally, heterogenerous catalysts are typically more tolerant of extreme operating conditions than their homogenerous analogues.

A heterogenous catalytic reaction involves adsorption of reactants from a fluid phase onto a solid surface, surface reaction of adsorbed species, and desorption of products into the fluid phase. Clearly, the presence of a catalyst provides an alternative sequence of elementary steps to accomplish the desired chemical reaction from that in its absence. If the energy barriers of the catalytic path are much lower than the barrier(s) of the noncatalytic path, significant enhancements in the reaction rate can be realized by use of a catalyst. This concenpt has already been introduced in the previous chapter with regard to the CI catalysed decomposition of ozone and enzyme-catalysed conversion of substrate. A similar reaction profile can be constructed with a heterogenous catalytic reaction. For example, plantinum-catalysed oxidation of carbon monoxide according to the overall reaction $CO + 1/2O_2 = CO_2$. The first step in the profile represents the adsorption of carbon monoxide and dioxygen onto the catalyst. In this case, adsorption of dioxygen involves dissociation into individual oxygen atoms on the Pt surface. The product is formed by addition of an adsorbed oxygen atom to an adsorbed carbon monoxide molecule. The final step in the catalytic reaction is desorption of adsorbed carbon dioxide into the gas phase. The Pt catalyst facilitates the reaction by providing a low energy path to dissociate dioxygen and form the product. The noncatalytic route depicted is extremely slow at normal temperatures due to the stability of dioxygen molecules.

## 4.2. HOMOGENOUS REACTION

In homogeneous catalysis are applied, in contrast with heterogeneous catalysis, where solid catalysts are used. One example of homogeneous catalysis, alkylation of isobutane with alkenes, with acids such as HF and $H_2SO_4$ as catalyst. In this chapter, we will confine ourselves to homogenous catalysts based on transition metals. These catalysts are becoming increasingly important, and already have found a large number of industrial applications. Homogeneous catalysis is applied in essentially all sectors of the chemical process industry, in particular in polymerization, in the synthesis of bulk chemicals (solvents, detergents, plasticizers) and in fine chemistry. It is useful to compare homogenous and heterogenous transition metal based catalysis. In homogeneous catalysis the reaction mixture contains the catalyst complex in solution. This means that all of the metal is exposed to the reaction mixture. In heterogenous catalysis on the other hand, the metal is typically applied on a carrier material or as a porous metal sponge type material, and only the surface atoms are active (Fig. 4.1).

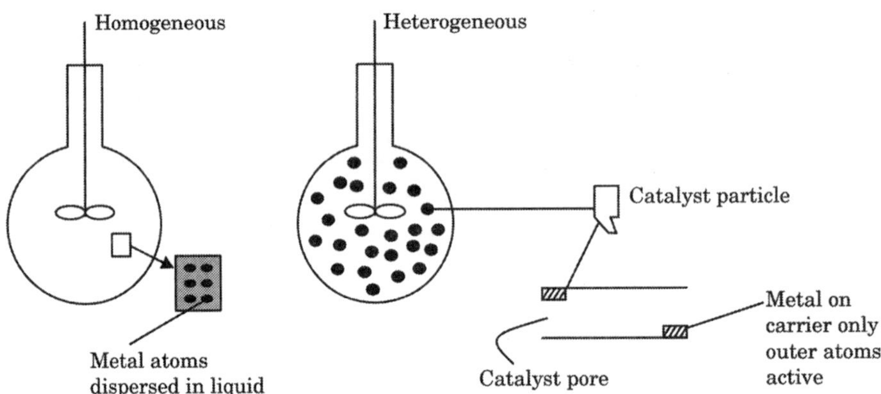


**Fig. 4.1:** Catalyst dispersion in homogeneous and heterogeneous catalysis

Thus, in terms of activity per metal center, homogenous catalysts are often more active. The high dispersion of homogenous catalysts also minimizes the effect of catalyst poisons; in homogenous catalysis one poison molecule only deactivates one metal complex, whereas in hetergenous catalysis a poison molecule can block a pore containing many active sites (pore plugging). Homogenous catalysts are also capable of being much more selective because with homogenous catalysts there is usually only one type of active site, whereas heterogenous catalysts often contain many different types of active sites, some of which could even catalyze often contain many different types of active sites, some of which could even catalyze undesired reactions. The discrete metal complexes in homogenous systems provide a well-defined catalyst system, which is easy to study by conventional techniques (infrared spectroscopy, NMR). From this another advantage of homogeneous catalysis arises, namely the possibility to follow the effect of changes in ligands and /or reaction conditions accurately. Hence, the selectivity behavior of heterogeneous catalysts, with their complex surfaces, is much more difficult to understand, and therefore, effective modification is fairly difficult. Because homogenous catalysis is usually carried out in the liquid phase, temperature control is relatively easy. Homogenous transition-metal catalysis not applied in every known process. First of all, its high specificity is not always called for. In oil refinery processes especially, the complex feeds and products do not justify the use of a highly specific catalyst. Secondly, homogeneous catalysis of course also has disadvantages. The main problem is the separation of catalyst and product. This is often only feasible for low molecular weight products. Furthermore, the use of solvents adds an additional separation step. Thirdly, the temperature window that can be applied is limited, in particular because catalyst complexes are often not resistant to high temperature. In homogenous catalysis often precious metals are used. Furthermore, the ligands are also expensive. Therefore, the catalyst productivity indeed should be high, and catalyst losses should be minimized, requiring nearly complete recovery of metal and ligands. Fouling of the reactor and other equipment is another problem specific for homogeneous catalysis. Homogenous catalysis is a highly innovative area. It also typically is an area where progress is accelerated when chemical engineers and chemists co-operate closely from the start. Novel developments in homogeneous catalysis are for a large part related to new insights in organometallic chemistry. Homogenous catalysts are not suitable for strongly endothermic reactions such as cracking of C-C bonds in Fluid Catalytic Cracking (FCC), because of their limited stability at higher temperatures. In most cases pressures are low but there are exceptions, for example the hydroformylation and carbonylation reactions with cobalt. It is fair to state that also heterogeneously catalyzed processes exist at moderate temperatures.

## 4.3. HETEROGENEOUS REACTION

Heterogeneous catalysis refers to the form of catalysis where the phase of the catalyst differs from that of the reactants. Phase here refers not only to solid, liquid, vs gas, but also immiscible liquids, *e.g.* oil and water. The great majority of practical heterogeneous catalysts are solids and the great majority of reactants are gases or liquids (Refer table 5.1). Heterogeneous catalysis is of paramount importance in many areas of the chemical and energy industries. In heterogeneous catalysis, the reactants diffuse to the catalyst surface and adsorb onto it, via the formation of chemical bonds. After reaction, the products desorb from the surface and diffuse away. Understanding the transport phenomena and surface chemistry such as dispersion is important. If diffusion rates are not taken into account, the reaction rates for various reactions on surfaces depend solely on the rate constants and reactant concentrations. For solid heterogeneous catalysts, the surface area of the catalyst is critical since it determines the availability of catalytic sites. Surface areas can be large, for example some mesoporous silicates have areas of 1000 $m^2/g$. The most common approach to

**Table 4.1:** Classification of heterogeneous catalysts

| ***Reacting phases*** | ***Examples*** | ***Comment*** |
|---|---|---|
| Solid + gas | Ammonia synthesis from $N_2$ + $H_2$ over iron catalysts | |
| Solid + solution | Hydrogenation of fatty acids with nickel | Used for the production of margarine |
| Immiscible liquid phases | Hydroformylation of propene | Catalyst in aqueous phase, reactants and products mainly in nonaqueous phase |

maximizing surface area is by the use of catalyst supports, which are the materials over which the catalysts are spread.

### 4.3.1. *Methyl ester production from vegetable on heterogeneous basic catalysts*

Heterogeneous mixed oxide catalysts; magnesium and calcium oxide on alumina and silica supports are used for transesterification of vegetable oils; canola and sunflower, to methyl esters. The primary objective is to understand whether basicity of the solid catalysts and/or its crystallite size affect the biodiesel yield during the reaction. The mixed oxide catalysts were prepared by a single step sol-gel method. MgO and CaO were chosen because of their basic property. The crystallite size does not have any direct effect on the transesterification activity. Yet, the interaction of the crystallites with the support material may change the number of the active sites and/or the creation of new "basic" sites. In fact, the basicity was found as the critical factor for the choice of the suitable mixed oxide for the transesterification reaction. The addition of CaO to alumina creates synergetic effect. But this is not observed with the catalysts: $CaO/SiO_2$ and also the $MgO/Al_2O_3$ and $MgO/SiO_2$. $CaO/Al_2O_3$ differs from the others in such a way that it contains the basic sites with the mild basic strength. Not only the crystallite size but also the type of the oxide is important to obtain the heterogeneous catalysts as active as the homogeneous catalyst. $CaO/Al_2O_3$ has the same activity as that of NaOH for the transesterification of canola and sunflower oils at 50°C. It was also found that the activity of $CaO/Al_2O_3$ was truly heterogeneous. Four mixed oxide catalysts are synthesized; $CaO/Al_2O_3$, $CaO/SiO_2$, $MgOAl_2O_3$ and $MgO/SiO_2$ with the loadings of 5, 20, 50 and 80 wt%. In the synthesis of the catalysts, magnesium nitrate hexahydrate is used as a precursor for magnesium oxide and calcium nitrate tetrahydrate as precursor for calcium oxide. Aluminum isopropoxide and tetraethyl orthosilicate are used as precursors for alumina and silica support oxides, respectively. Ethanol and deionised water are used as solvents and HCl, $HNO_3$ and $NH_4OH$ are used as peptizer in the sol-gel method. Homogeneous transesterification reaction is carried out using NaOH and MeOH. Commercially available canola and sunflower oils are used as the lipid feedstock. Acetic acid is used to remove the unreacted NaOH from the biodiesel. In this example, the effect of basicity (amount of the basic sites) of the solid catalysts and/or its crystallite size on the biodiesel yield during the transesterification of canola and sunflower vegetable oils are investigated on the heterogeneous mixed oxide catalysts: calcium oxide and magnesium oxide on silica and alumina supports. Both alumina and silica supported catalysts are prepared via the modified single step sol-gel methods and XRD results showed that the high dispersion of active phases is achieved for CaO based catalysts even with the high CaO loading of 80 wt%. Also, the effect of support material on crystallite size is observed on CaO and MgO based catalysts. $CaO/Al_2O_3$ catalysts with the high CaO loadings show the high activity even under the unfavourable reaction conditions, such as low MeOH/Oil molar ratio of 3.6 and the low reaction temperature, 50°C. This study shows that:

- The crystallite size does not have any direct effect on the transesterification activity.
- The change in the crystallite size may be responsible for the variation in the observed biodiesel yield through either directly with the changes in the number of the active sites and/or the creation of new "basic" sites due to the interaction of the crystallites with the specific support materials.
- Basicity is the main factor for the choice of the suitable oxide for the transesterification reaction, since high biodiesel yield is obtained with high basicity but the basic strength is also as critical as the basicity.
- Mild basic strength is required to achieve the high biodiesel yield in the transesterification of canola and sunflower oils.

### 4.3.2. *Heterogeneous photo catalysis*

Heterogeneous semiconductor photo catalysis using $TiO_2$ as the photo catalyst is an emerging technology with key advantages including operation at ambient conditions as well as the fact that the catalyst itself is inexpensive, commercially available at various crystalline forms and particle characteristics, non-toxic and photo chemically stable. From a mechanistic point of view, illumination of an aqueous $TiO_2$ suspension with irradiation with energy greater than the band gap energy of the semiconductor generates valence band holes and conduction band electrons. Holes and electrons may either undesirably recombine liberating heat or make their separate ways to the surface of $TiO_2$, where they can react with species adsorbed on the catalyst surface. Valence band holes can react with water and the hydroxide ion (*i.e.* under alkaline conditions) to generate hydroxyl radicals, while electrons can react with adsorbed molecular oxygen reducing it to superoxide radical anion which, in turn, reacts with protons to form peroxide radicals. Although available at various crystalline forms, a commercially available product containing 80:20 anatase: rutile shows exceptional activity and its superiority against other grades of $TiO_2$ is attributed to the morphology of its crystallites. This morphology allows an easy electron transfer from rutile to anatase, thus stabilizing charge separation and, therefore, lowering the recombination of photogenerated carriers. Besides $TiO_2$, ZnO and CdS have also been employed as photocatalysts in water treatment. In the context of pharmaceuticals treatment, the catalytic activity of ZnO and Degussa $TiO_2$ for the degradation of sulfamethazine and chloramphenicol respectively is compared and found that ZnO is slightly more effective than $TiO_2$. Photocatalytic reactions typically involve $TiO_2$ suspensions with the catalyst concentration being an important parameter that affects performance. Other parameters are the light wavelength and intensity, the solution pH which dictates the ionization state of the catalyst surface and consequently affects the extent of organics adsorption and degradation, the addition of $H_2O_2$ as an extra oxidant to promote reactions, and the water matrix (*i.e.* the presence of humic substances, bicarbonates or dissolved gases). Photocatalytic reactions usually obey to Langmuir–Hinshelwood kinetic model which is reduced to pseudofirst or -zero-order kinetics depending on the operating conditions. From an engineering point of view, the use of catalyst in slurry form requires an additional treatment step to remove it from the treated effluent. The role of the membrane is to retain the used catalyst, the unreacted pharmaceuticals and their by-products, which could then be recycled to the photoreactor. Alternatively, the catalyst may be immobilized on suitable support matrices, thus eliminating the need for post-treatment removal. However, it should be pointed out that catalyst immobilization unavoidably leads to a decrease of the surface area available for reactions compared to suspended systems. From an economic point of view, heterogeneous (as well as homogeneous) photocatalysis is likely to benefit from the use of renewable energy sources to power the process. In this direction, solar photocatalysis has gained considerable attention and several studies report the use of natural or simulated sunlight irradiation for pharmaceuticals treatment. A solar photocatalytic system comprising 100 $m^2$ of compound parabolic collectors has recently been developed to treat effluents at flow rates up to 250 l/h. This industrial-scale unit, whose application has been demonstrated for the treatment of α-methyl-phenylglycine solutions, is designed to treat various types of effluents through a combination of solar photo-Fenton pre-treatment and immobilized biomass post-treatment. Furthermore, its environmental impact was evaluated by means of life cycle analysis and compared to that of a similar system comprising $TiO_2$ photocatalysis instead of photo-Fenton pre-treatment. It was concluded that homogeneous photocatalysis involved a substantially lower environmental impact than the respective heterogeneous process.

### 4.3.3. *Ozonation*

Ozone is a strong oxidant that either decomposes in water to form hydroxyl radicals which are stronger oxidizing agents than ozone itself, thus inducing the so-called indirect oxidation or attacks selectively certain functional groups of organic molecules through an electrophilic mechanism. Depending on the type of the substrate and the operating conditions in question, ozone oxidation is usually favored at increased pH values due to the increased production of hydroxyl radicals. Moreover, treatment performance is enhanced if ozone is combined with light irradiation, hydrogen peroxide or with iron or copper complexes that act as catalysts. Ozonation has been traditionally employed in drinking water treatment for odor and taste control and disinfection, as well as (in some cases) for wastewater disinfection. Therefore, it is not surprising that several studies have been carried out onsite in drinking water plants and Wastewater treatment plants (WWTPs). Special attention is given on WWTPs since pharmaceuticals usually exit secondary treatment unaffected and, therefore, they need to be treated in subsequent stages. Moreover, they may partly be adsorbed on primary and secondary sludge although data regarding their concentration in sludge is scarce due to inherent analytical difficulties associated with such samples. Sludge preconditioning by ozonation have been studied with emphasis on the fate of adsorbed pharmaceuticals. Ozone is a powerful oxidizing agent ($E_0$ = 2.08 V), when compared to other agents, such as $H_2O_2$ ($E_0$ = 1.78 V), and can react with several classes of compounds through direct or indirect reactions. In the absence of light, ozone can react directly with an organic substrate, S, through slow and selective reactions of electrophilic substitution or dipolar cycloaddition.

$$O_3 + S \rightarrow S_{ox} \tag{4.1}$$

This reaction is important in acid media and for solutes that react very fast with ozone. The electrophilic attack may occur at atoms with a negative charge density (*e.g.* N, P, O or nucle ophilic carbons) or to double/triple bonds like carbon–carbon, carbon–nitrogen and nitrogen–nitrogen. Ozone can also react indirectly through a fast and non selective reaction by the HO• free radical generated by its decomposition, more favoured in alkaline media:

$$2O_3 + 2H_2O \rightarrow 2HO^{\bullet} + O_2 + 2HO_2^{\bullet} \tag{4.2}$$

The HO• radical is a powerful and non-selective oxidant that can react in aqueous solution through three possible mechanisms: (i) hydrogen abstraction; (ii) electron transfer; and (iii) radical addition. The secondary radicals formed during this reaction can react with ozone or other compounds. Addition of hydrogen peroxide to the ozonation system produces a sensible improvement. The process, called Perozone, combines the direct and indirect ozone oxidation of organic compounds. $H_2O_2$ initiates $O_3$ decomposition by electron transfer through a reaction that can be envisaged as the activation of $H_2O_2$ by ozone

$$\mathrm{O_3 + H_2O_2 \rightarrow HO^{\bullet} + O_2 + HO_2^{\bullet}} \tag{4.3}$$

The process is expensive but fast, and can treat organic pollutants at very low concentrations (μ/L), at pH between 7 and 8. Irradiation of ozone in water with near-UV light (wavelengths > 300 nm) produces $H_2O_2$ quantitatively:

$$O_3 + H_2O + hv \rightarrow H_2O_2 + O_2 \tag{4.4}$$

The generated hydrogen peroxide is photolyzed, producing HO• radicals:

$$H_2O_2 + hv \rightarrow 2HO^{\bullet} \tag{4.5}$$

This combination might be considered in principle only an expensive form of generating $H_2O_2$ and then HO•. Indeed, it is a combination of UV/$H_2O_2$ and $O_3$/$H_2O_2$, but the advantage is that at 254 nm ozone has a higher absorption coefficient than $H_2O_2$ (ε254 = 3300/M cm versus 18.6/M cm), and can be used to treat water with high UV absorption background. The efficiency is higher than that of $O_3$ or direct UV, and the reactor does not need to be in quartz because UV-B light (280–315 nm) can also be used. If wavelengths lower than 300 nm are employed, photolysis of $O_3$ takes place, generating additional HO• and other oxidants, with the subsequent increase of the efficiency

$$O_3 + hv \rightarrow O_2 + O \tag{4.6}$$

$$O + H_2O \rightarrow 2HO^{\bullet} \tag{4.7}$$

In conclusion, the addition of light to the $H_2O_2/O_3$ process produces a net increase on the efficiency for decontamination of groundwater and for soil remediation. On this basis, the simultaneous application of ozonation and photocatalysis, hereafter indicated as photocatalytic ozonation, has potential use for the efficient treatment of contaminated waters in a wide range of concentrations. When $O_3$ is present in a photocatalytic suspension, in addition to the homogeneous pathways of direct ozonation, the role of $O_3$ as an electron trap must be considered. FTIR studies report different modes of ozone interaction with $TiO_2$ surface, such as: (i) physical adsorption; (ii) formation of weak hydrogen bonds with surface hydroxyl groups of the catalyst; and (iii) dissociative adsorption on strong surface Lewis acid sites leading to $O_3$ decomposition with formation of a free oxygen molecule and a surface oxygen atom that remains attached to the site.

In the presence of $TiO_2$ and near-UV radiation, adsorbed ozone acts as a very strong electrophilic agent generating an ozonide radical, $O^{3-}$ whose further reactions produce eventually HO• radicals in the adsorption layer

$$O_3 + e^- \rightarrow O_3^{\bullet} \tag{4.8}$$

$$O_3^{\bullet} + H^+ \rightarrow HO_3^{\bullet} \tag{4.9}$$

$$HO_3^{\bullet} \rightarrow HO^{\bullet} + O_2 \tag{4.10}$$

The above reported mechanism gives rise to the formation of one hydroxyl radical per each trapped electron. In conventional photocatalytic processes, oxygen is used as the oxidant compound. The disadvantages of using oxygen as electron trap are the slow electron transfer from $TiO_2$ to $O_2$ and the fact that three electrons are needed for the generation of a hydroxyl radical

$$3O_2^{\bullet} + 3e^- \rightarrow 3O_2^{\bullet -} \tag{4.11}$$

$$2O_2^{\bullet -} + 2H_2O \rightarrow 2HO_2^{\bullet} + 2HO^- \tag{4.12}$$

$$2HO_2^{\bullet} \rightarrow O_2 + H_2O_2 \tag{4.13}$$

$$O_2^{\bullet -} + H_2O_2 \rightarrow HO^- + HO^{\bullet} + O_2 \tag{4.14}$$

Consequently, when the photocatalyst is irradiated in the presence of $O_3$, a greater number of HO• radicals is produced. Ozone therefore acts as a powerful oxidant in place of oxygen, mainly owing to the fact that $O_3$ is more electrophilic than $O_2$ towards electron photogenerated onto the $TiO_2$ surface. HO• generation from $O_3$ is pH dependent and increases with decreasing pH; this avoids the use of high alkaline pH to induce HO• formation from $O_3$. It is more advantageous to combine photocatalysis and ozonation for the better working of carbon black modified nano- $TiO_2$ (CB-$TiO_2$) thin films supported on Al sheets for removing catechol. The combined process considerably increased the total organic carbon removal rate, and the kinetic study showed that the rate constants of the complete mineralization of catechol with $TiO_2$/UV/$O_3$ are about two times higher than that of UV/$O_3$ with the same concentration of ozone. The CB-$TiO_2$ thin film is also used for degrading and mineralizing dibutylphtalate by photocatalytic ozonation; also in this case, the degradation activityis higher than that showed by only photocatalysis.

# Chapter 5

# Future Research in the Various Separation Processes

The various separation processes provides a wide scope of many useful research which will benefit the current industrial needs and enhance the objectives to satisfy multiple needs.

## 5.1. DISTILLATION

The steady state models continue to be powerful and efficient tools for designing control systems for distillation columns. Steady state techniques have been used for decades to develop control strategies for distillation columns. While theses techniques are effective for screening out clearly undesirable control structures and suggesting viable candidates, they provide incomplete and sometimes misleading information. To accurately assess the performance and suitability of alternative control schemes, rigorous dynamic simulation is required. There is need for rigorous dynamic simulation for final selection of a workable and robust strategy for distillation.Among the plethora of bioreactors available for aerobic culture, bubble columns, which are composed of a cylindrical vessel fitted with a gas sparger, are gaining in use. The simple construction of bubble-column reactors makes them easy to maintain. In addition, it is possible to control the degree of shear, uniformly within the reactor, which is critical to the growth of plant and animal cells in particular. The computer-automated radioactive-particle tracking technique is used to investigate liquid recirculation and turbulence in bubble columns. Experiments are conducted in air and water bubble columns of different sizes, with various gas distributors, and operated over a range of superficial gas velocities. Supplementary experiments, using the hot-probe anemometry technique, are performed at selected flow conditions. Distillation, the most common separation process in chemical process industries, requires significant energy inputs. Since first industrial application in 1985, the dividing wall column has moved from a conceptual to an industry-proven technology, steadily growing in the number and size of applications. Today numerous industrial applications are known, mainly concerning 3 components separations. Further development and implementation efforts are oriented towards separation of more than three components using a Dividing-Wall Column (DWC). In order to maximize potential gain on energy saving, multiple partition wall configurations should be employed, which however may prove to be too demanding or expensive, which means that viable designs will not necessarily be the most energy efficient ones. Similar trade-off exists between energy efficiency and controllability. On the other side, adopting non-welded partition wall design approach minimizes difficulties to be associated with the construction of complex columns, at least for those to be equipped with structured packings. DWC, which works on the basis of Fully Thermally Coupled Distillation System, has been studied due to its lower energy consumption compared to the conventional column system. Retrofitting the C2C systems to DWCs for several industrial applications has been in use, to reduce the energy consumption and hence the operating cost of distillation. There are lower energy requirements and consequently cost savings due to use of DWCs for new plants as well as there is significant potential for retrofitting existing C2C systems. An effective and efficient control design procedure for distillation columns using steady state models have been developed and successfully applied.

Compared to conventional columns-in-series and/or in-parallel configurations a DWC requires much less energy, capital and space. This makes DWC to something that corresponds with the present day idea of sustainable process technology. Dividing wall distillation columns, DWC, is used to separate azeotropic mixtures. Use the DWC for separation of azeotropic mixtures is feasible, and important energy savings, until 50%, can be obtained in comparison with conventional sequences. DWCs are being build for obtaining four or even more products containing two or more partition walls in parallel, which is something not yet attempted in industrial practice. Bubble columns have many advantageous characteristics that make them an attractive choice for the scaling-up of biological processes. Their mechanical simplicity enables a more sterile operation. Perhaps more importantly, it is possible to control levels of shear within the reactor - a crucial factor for the growth of plant and animal cells in particular. Reliable design correlations are also available for momentum, mass and heat transfer, making it possible to construct large reactors rationally. Even though the effectiveness of simple bubble columns for highly viscous liquids (viscosity ~1 Pa.s) has been questioned, recent evidence indicating that this type of reactor has certain advantages in the high viscosity. The mass-transfer rate in such liquids is, however, adversely affected. Most of the mass transfer occurs in the vicinity of the gas sparger where bubbles are relatively small; the upper sections containing very large bubbles offer negligible interfacial area. Notwithstanding some disadvantages, bubble columns will continue to find new applications in bioprocessing. Available process simulators still do not include a DWC as a separate model. Instead, modeller can use various sequences of conventional column modules to mimic a DWC. However, there are difficulties regarding initialization and convergence of these models. Based on own experience, it is generally more beneficial to use models with as little as possible column sections to reduce convergence time and increase stability, however single column pump-around model is not recommended, especially not for separation of more than three products, because it is difficult to set up. Rigorous simulation procedures are basically performance evaluation and not design oriented, and built-in optimization algorithms can not handle integer variables, *i.e.* number of stages in particular sections. Therefore, to obtain design related parameters one has to use an external optimization routine. A good design procedure that would overcome this problem could significantly contribute to further expanding the field of DWC applications. Although interactive sequential approach to performance simulation and optimization may be improved, it appears that simultaneous solving, equations-based approach, with its superior convergence stability and speed is much more effective in this respect. Namely, it allows combining rigorous simulations with powerful optimization routines. This is the key to successful simulation and design of DWC configurations for obtaining four and more products. The experimental research on DWC is presently aimed at acquiring knowledge that will support further development and implementation of reactive distillation applications. Since a reactive DWC can be associated with implementing practically an ultimate sustainability approach in chemical process industries. Regarding the number and variety of industrial applications, dividing wall column can already be considered as a distillation process intensification success story. Being a genuine representative of substantial distilling, making slowly inroad into the applications dominated by tray columns, DWC will certainly develop into a standard type of distillation column in the near future.

***Suggestions for future reseach:*** Taking heating and cooling duties into consideration, to make heat-integrated distillation column sequences, Extend the method to also consider design of batch distillation, Extend the concept of scaling factors to include, for example, a scaling of the cost of alternative separation methods/techniques, Extend the controllability analysis for various distillation column configurations.

## 5.2. SUPERCRITICAL FLUID EXTRACTION

Supercritical fluid extraction using carbon dioxide as solvent is now the on going area of research, in order to establish an efficient method to predict extraction curves on large scale. Extraction of oil and tocopherols from almond seeds using supercritical carbon dioxide is also becoming popular. It is found that extraction yield is greatly influenced by particle size. Hollow fiber dispersion-free solvent extraction of Pu(IV) from nitric acid medium using micro porous hydrophobic polypropylene hollow fiber contactor using Aliquat-336 /Solvesso-100 as an extractant is also a wide area of research. The food industry has a need for new processes to obtain valuable additives of natural origin. Supercritical extraction can be considered a viable alternative to the extraction with conventional solvents, the utilisation of which is being progressively limited in the food industry and other sectors, by the imposition of increasingly demanding regulations for the processes permitted for obtaining the additives needed for many food products. The possible utilisation of additives that not only serve as colorants but also present antioxidant characteristics, which provide health benefits to consumers, contributes to the added value of the final product.

If these additives can be obtained by processing methods that are more respectful of the natural environment, as is the case of extraction with fluids in supercritical conditions, this represents a further improvement of the sales prospects of the final product by adding another feature to its intrinsic quality. In addition, the high degree of purity of the pigments obtained by means of supercritical extraction is one of the factors that should be studied over the medium to long term, since it constitutes a factor of considerable importance, in the context of the possible implementation of this technology on an industrial scale.

## 5.3. CRYSTALLIZATION

Crystallization is one of the most powerful purification methods available in process engineering. Fractional crystallization is a stagewise separation technique that relies upon liquid-solid phase transition and enables multicomponent mixtures to be split into narrow fractions, ultimately leading to top purities of selected components, through the virtue of selectivity found in solid liquid equilibria. Fractional crystallization is most frequently encountered in the separation of organic materials ranging from isomer separations to tar chemical mixtures and from organic acids to monomers. Fractional crystallization is used to purify a wide variety of chemicals.

## 5.4. MEMBRANE PROCESSES

Membrane processes are increasingly reported for various applications in both upstream and downstream technology, such as the established ultrafiltration and microfiltration, and emerging processes as membrane bioreactors, membrane chromatography, and membrane contactors for the preparation of emulsions and particles. Membrane systems exploit the inherent properties of high selectivity, high surface-area-per-unit-volume, and their potential for controlling the level of contact and/or mixing between two phases. Flat sheet dialysis membranes are casted using phase inversion method, using cellulose acetate as the polymer and acetic acid as the solvent. Modification agents such as polyethylene glycol 400 and non-solvent swelling agent, *i.e.* distillated water are added. Recently membrane reactors are developed for biological treatment of waste gases. In this type of bioreactor gaseous pollutants are transferred through a membrane to the liquid phase, where micro-organisms degrade the pollutants. The membrane bioreactor combines the advantages of membrane devices with the clean technology of biological air purification. Two types of membrane materials can be used for gas–liquid contact: hydrophobic microporous material and dense material, such as silicone rubber. Microporous material generally has a higher permeability, but dense membranes can be advantageous in the case specific selectivity is required. Biomass is generally present as a biofilm on the membrane, but may also be suspended in the liquid phase. In a number of cases the reactor performance appears to be hampered by an unstable biofilm performance and:or by clogging of the liquid channels due to excess biomass formation. So far, membrane bioreactors for waste gas treatment have only been tested on laboratory scale. If the long-term stability of these reactors can be demonstrated, we expect membrane bioreactors to be useful tools in the treatment of gas streams containing poorly water-soluble pollutants and highly chlorinated hydrocarbons, which are difficult to treat with the conventional methods for biofiltration. Biological denitrification is a comparatively low cost option, though its poor retention of bacteria and requirement for external carbon necessitates further downstream processing. To eradicate the associated disadvantages, the biological process can be integrated with membrane technology. Further research involves nitrate removal MBR technologies. The application of membrane processes in petrochemical industry, processes such as olefin/paraffin separation, light solvent separation, solvent dewaxing, phenol and aromatic recovery, dehydrogenation, oxidative coupling of methane and steam reforming of methane is of extensive area for research. Besides, separation using polymer-inorganic nano composite membranes and wastewater treatment using membrane bio-reactors is also in practice. Membrane reactors have found utility in a broad range of applications including biochemical, chemical, environmental, and petrochemical systems. The variety of membrane separation processes, the novel characteristics of membrane structures, and the geometrical advantages offered by the membrane modules have been employed to enhance and assist reaction schemes to attain higher performance levels compared to conventional approaches. In these, membranes perform a wide variety of functions, often more than one function in a given context. An understanding of these various membrane functions will be quite useful in future development and commercialization of membrane reactors. Various functions of the membranes in a reactor can be categorized according to the essential role of the membranes. They can be employed to introduce/separate/purify reactant(s) and products, to provide the surface for reactions, to provide a structure for the reaction medium, or to retain specific catalysts. Within these broad contexts, the membranes can be catalytic/noncatalytic, polymeric/inorganic, and ionic/nonionic and have different physical/chemical structures and geometries. The functions of the membrane in a reaction can be enhanced or increased also by the use of multiple membrane-based schemes. The irreversible thermodynamic theories already developed for solid membranes with fixed charged site density have been extended to include the different physicochemical aspects of the liquid membranes. It has been found that the mobile charged site, the counterions, and the uncharged associated species contribute to the electrical membrane potential through their phenomenological coefficients. The electrical membrane potential equations have been integrated in isothermal and nonisothermal conditions for monoionic and biionic systems. PVC supported liquid membrane and carbon paste potentiometric sensors incorporating an Mn (III)-porphyrin complex as a neutral host molecule were developed for the determination of paracetamol. The use of a crossflow membrane bioreactor in treating wastewater discharged by a petroleum refinery is a lage area of research. Due to several advantages of membrane processes, researchers examined various membrane systems by which reasonable results were obtained in laboratory scale. Based upon large capital expense and energy cost of industrial distillation columns, it seems worthy to recommend membrane processes as a substitution with distillation columns in industrial applications. Of course, maintaining the system stability for long term operations is still questionable. In petrochemical industry, reaction and separation play an important role in achieving desirable products. Using a membrane reactor, these two parameters can be controlled simultaneously. Dehydrogenation, oxidative coupling of methane, steam reforming of methane and water gas shift reaction are important reactions in petrochemical industry. However, parameters such as 'long term reliability with catalyst life', 'high temperature resistance' and 'low coke formation' should be considered in applying membrane reactors in industrial cases. In case of the application of nanotechnology in the production of gas separation membranes, nano composite membranes can simultaneously improve permeability and selectivity. Producing these membranes in industrial scales, in future, refineries and petrochemical complexes will be free from huge energy-intensive distillation columns, and membrane modules can easily be replaced. Petrochemical waste streams may contain phenolic compounds or aromatic amines. They are highly toxic and at high concentrations are inhibitory to biological treatment. Membrane aromatic recovery system is a relatively new process for recovery of aromatic acids and bases. Wastewater in petrochemical industry is currently treated by activated sludge process with pretreatment of oil/water separation. Tightening effluent regulations and increasing need for reuse of treated water have generated interest in the treatment of petrochemical wastewater with the advanced membrane bioreactor process. It can be predicted that up to 2020, investments in the production of gas separation membranes will greatly increase. Even when the membrane is located outside the reactor and connected to it in a recycle mode, such coupled reaction-separation processes provide considerable improvements in the overall reaction performance. Many investigations on membrane reactors have been conducted in biochemical processing, petrochemical applications, and environmental pollution control. A few processes employing polymeric membranes have been commercialized. Utilizing more than one membrane in a reactor imparts additional functional capabilities. For effective large-scale utilization of the diverse functional capabilities of a membrane in a reactor, considerable research and development on membrane lifetime, available space time, module design, membrane fouling, membrane poisoning, and membrane cost are essential. This is especially true for inorganic membranes and higher temperature applications where the hurdles for industrial use are considerable.

Membranes have long been an integral part of biotechnology processes. The most well-known examples are ultrafiltration and microfiltration, which have become routine methods. The sterile filtration of fermentation media, purification of buffers and proteins are now standard practices. Other applications of membrane processes have been introduced more recently, as membrane bioreactors, membrane chromatography and membrane contactors. Although ultrafiltration and microfiltration are well-established processes, membrane bioreactors and membrane chromatography have certainly not obtained the expected levels of success. The reasons are related to a reticence of users for applying new technologies, and also to their potential advantages compared to other processes, such as classical reactors for biological reactions and conventional bead chromatography. In this sense, the examples of ultrafiltration and microfiltration may be advantageous, for which a continuous improvement in understanding the physicochemical phenomena governing the processes has been shown to enhance considerably the separation of proteins using their inherent electrostatic properties. In addition, the extensive study of hydrodynamics combined with both numerical simulation and experimental investigation, has resulted in considerable improvement in performance of ultrafiltration and microfiltration, both in terms of fluxes and separation. Ultrafiltration (UF) is a low pressure membrane technology (5 to 25 psi driving force) that enables cost effective compliance with the Safe Drinking Water Act. It produces a more consistent and higher quality water than is obtainable with conventional media filters. Because of the extremely small pore size, permeate quality is typically superior to that produced by microfiltration. It is this small pore size that makes UF a positive barrier to viruses, bacteria, Giardia cysts and parasites like *Cryptosporidium*. A barrier means that the membrane structure is such that a properly operated system prevents contaminant breakthrough to the product water side, regardless of how long the system is in service, or what feed pressure is used. Since packaged UF systems operate automatically, they can run unattended for long periods of time with little risk of degradation of the water that is sent to distribution. Membrane fouling is customarily indicated and measured by permeate decline at constant operating conditions. Although this method has been widely used for many years, it is inadequate for characterizing fouling development in modem full-scale RO processes. It has been demonstrated that in some full-scale RO processes with highly permeable membranes, the initial development of membrane fouling may not apparently affect the permeate flux. When the permeate flux is noticeably affected, the membrane is so severely fouled that restoration to its original permeability may become impossible. In this study, the permeate flux of a full-scale RO process and its controlling mechanisms are studied with numerical simulations. Particular attention is given to the relationship between permeate flux and the change of membrane resistance. Results show that the filtration coefficient (a collective parameter related to distribution of the membrane resistance) can be a better indicator or measurement of membrane fouling in a full-scale RO process. With this new fouling indicator, the development of membrane fouling in a full-scale RO process under any conditions can be accurately characterized. The advantages of the new characterization method include early and accurate indication of fouling development and quantitative assessment on the effectiveness of membrane cleaning. In the mid decades of the 20th century, industrial world understood that they should not rely solely on crude oil as an energy source; and solar energies, nuclear energies and the like could be better substitution with the current energy sources. Therefore, non-petroleum sources became gradually important and processes such as OCM and GTL (gas to liquid), which are based upon using natural gas in order to produce fuels, are being developed. A decrease in petroleum sources and a great fluctuation in their prices are the principal motivations that in near future, non-petroleum sources be used for the production of petrochemical materials as the feed of petrochemical plants. In the case of separation operations, currently available methods are easy and reliable but they are expensive, energy-intensive and environmental pollutants. Due to the enormous potential of membranes to function as separators for mixtures of materials, there has been resurgence in interest in membrane based processes. Continual efforts to develop economical and efficient membranes for various end uses have resulted in a variety of application oriented membranes. It is clear that in near future, membranes will have an outstanding role in separation industries. Membranes work with very low energies, do not need any heating-cooling systems and do not have any environmental pollution. There is one discussable point in developing these systems and that would be their long term stability in industrial operations. Production of polyolefins is important in petrochemical industry. An important step in the manufacture of olefins is large scale separation of olefin from the corresponding paraffin. New applications of membrane processes continue to emerge, such as membranes for the development of biosensors and molecularly imprinted polymeric membranes for separation of molecules. The very large range of chemical/physical mechanisms, which may occur in a membrane (*e.g.*, permeation, reaction and mixing), appears to be one of the most attractive aspects of membrane processes. For this reason, there is no doubt that new successful applications will continue to appear in the near future. Their industrial success will again depend on their advantages compared to those offered by other competing processes and on their acceptance on the industrial market.

Reverse osmosis technology is presently undergoing rapid growth in the area of municipal and industrial wastewater reuse. Control of membrane fouling due to organic compounds and particles in wastewater is a major expense in the design and operation of these facilities. Current research in the field includs some recent developments in modeling and fouling control. Use of ultraviolet radiation showed promise as an alternate to chlorination for RO feedwater disinfection. Inorganic fouling can be controlled by pretreatment processess including softening and removal of iron and manganese, and by reducing water recovery. Desalination plays an important role in producing pure water from brackish water. Reverse osmosis (RO) is by far the most efficient way to remove colloidal and dissolved silica, which can be found in high concentrations in brackish water. The presence of silica and its ability to foul membranes limits the use of silica bearing waters for desalination and when used, it has many economic penalties. Silica polymerisation affects RO systems in the presence of polyvalent cations and anions. Use of glassware is minimised to avoid the possibility of any contribution by silica leaching into solution. Some commercially available cleaners were capable at partially restoring the flux. Further investigation is underway to examine the effectiveness of new silica specific inlu'bitors.

A methodology for calcium sulfate (gypsum) scale control in nanofiltration of saline waters is being in use. The methodology involves the use of both theoretically and experimentally determined parameters. Pitzer's thermodynamic equations for electrolytcs are used to determine the gypsum scaling potential of the feed water based on its ionic composition, whereas the extent of concentration polarization at the membrane surface is determined from the film model. The model is tested by nanofiltration membrane experiments utilizing model solutions simulating saline agricultural drainage waters. Extractive process research to date focuses on high rate nitrate removal with little consideration of either nitrite formation, electron donor carry over or the implications of fouling. Autotrophic growth conditions are conceivably preferable over their heterotrophic counterparts on the basis of low waste yields and substitution of organic carbon addition, though culmination of their lower adaptability, lower denitrifying rates and associated electron donor safety risks cast doubt on their applicability at full scale. The challenge with the application of heterotrophic electron donors is dose control, since under dosing produces nitrite (a more strictly regulated contaminant) and over dosing leads to secondary contamination. The latter may constitute direct health and sociological concerns, or indirect concerns over trihalomethane formation potential or stimulation of bacterial growth in the distribution network, though reports from a full scale pressure driven process implies dosing control to be achievable. Where efficient sustainable gas transfer can be proven and fouling understood under low flow conditions, opportunities exist for *in-situ* gas transfer technology for groundwater remediation where breakthrough is irrelevant. Extractive and pressure driven configurations hold greater possibilities for mainstream treatment or blending, although further research into donor addition, the origin and control of biologically produced organics and their influence at the membrane surface is required.

Pervaporation modules operating in two-phase flow appear to be effective to reduce polarization effects and to supply the energy required for the separation process. The membrane area, the amount of auxiliary equipment and the related capital and operating cost are reduced as compared with single phase flow expanding the application window of pervaporation. Industrial applications are already viable with the multi-tubular modules available in the market. For applications with hollow fibre membranes, it is necessary to obtain more information on wetting behavior as a function of hydrodynamics and mixture composition. For pervaporation-distillation operations in one single column information on dynamic behavior and multiple steady states are important for optimization and start up. Interesting developments are related with tuning of membrane properties with an external source, the use of catalytic membranes in distillation columns and the use of internal two-phase flow in capillary membranes. Pervaporation has been successfully applied for the separation of mixtures in combination with distillation. In particular, this is because the energy consumption of distillation for the separation of azeotropic mixtures or mixtures with low relative volatility is high. Pervaporation is attractive because the energy required for the separation is strongly reduced due to the selective vaporization of the permeating components only. However, pervaporation involves higher capital cost than distillation processes. Thus, a hybrid process in principle combines the advantages of both separation techniques while the disadvantages can be minimized. The hydrodynamics and flux in a pervaporation module have been studied using CFD and ultrasound computer tomography for several density ratios between the permeating and the non-permeating component. Calculated density and temperature profiles are similar to those reconstructed from measurements. An inversion point has been found on the surface of tubular membranes above which the fluid moves upwards and below which the fluid moves downwards. The inversion point suggests that for the dewatering of mixtures with a low density ratio a multi-tubular pervaporation module with triangular configuration is convenient. For dewatering of mixtures with a high density ratio a squared configuration is preferred. The performance of a two-phase pervaporation module at low liquid flow rates is close to the performance in single phase at turbulent conditions.In addition, pervaporation modules operating with co-current two-phase flow around hollow fibre membranes have been analyzed. Gas void fractions and mass transfer coefficients between the liquid and the membrane surface have been measured in two-phase flow using an electrochemical technique in a transversal unit with hollow fibre membranes. The mass transfer coefficient can be four times higher than the one obtained in single phase. The resulting dimensionless correlation has been included in a rate-based model for designing a pervaporation module for dewatering of an IPA-water mixture. A reduction of the membrane area of about 45% can be obtained as compared with pervaporation modules with an economically optimal number of inter-stage heat exchangers. The lower membrane area required and the avoidance of inter-stage heat exchangers results in a strong reduction of capital cost. Pervaporation continues to evolve as a feasible separation technology for many different applications. As a proven method of separation as low temperatures and pressure, further application development for food processing is likely. Using pervaporation to clean wastewater streams by removing a variety of organic compounds also holds much promise. Since the commercialization of pervaporation for ethanol dehydration launched by GFTin 1980s based on the cross-linked PVA/PAN composite membrane, both the scope of application and the types of the pervaporation membranes were extensively enlarged. It should be noted that ensuring low transport resistance in the permeate side is a critical consideration in pervaporation module design, this is because the efficient evaporation of the permeate molecules in the downstream face of the membrane needs an extremely low absolute pressure, presence of resistance in the permeate channel can very sensitively affect the pervaporation separation process. The tubular membrane is also a feasible module configuration for pervaporation, such as the tubular zeolite membranes. Pervaporation has played an important role in solvent dehydration, and this application can be further extended by integrating pervaporation with other viable liquid-separating technologies, and by finding right materials for dehydrating some caustic solvents (*e.g.*, nitric acid). Higher membrane productivity and selectivity is always a concern for application, and this was conventionally accomplished by operating the pervaporation membranes at higher temperatures. The improved membrane flux can also be achieved by conducting optimized membrane cross-linking, particularly when membranes are used for dehydrating solvents of lower water contents. And the improved separating performance of the membrane can also be anticipated by incorporating various active nanostructures in the polymeric membrane matrix. Membrane technology offers an alternative method where a non-dispersive interface is created between the two phases. Permanent investigations in separation processes led to improvement results in contrast to conventional ones. It is an alternative technology that provides many advantages in contrast to traditional fluid/fluid contactors. They reduce the volume of equipment, and with much more interfacial area the membrane can ensure higher values of the mass transfer coefficient (k). All of these advantages enable development of commercial applications and a number of laboratory investigations prove why the membrane contactor technology is the best way for applying of membrane processes. During the past 15 years removal of water from organic liquids and liquid mixtures by means of pervaporation/vapour permeation has developed into a mature technology. More than 150 industrial plants have been installed around the world, with capacities between 20 kg/h to several tons per hour. Nearly all of these plants are equipped with polymeric membranes, which also are used in the plants for removal of methanol from its azeotropic mixtures. The introduction of ceramic membranes will not only enlarge the area of application of the process to higher temperature and aggressive mixtures, but also help to overcome the prejudice which some engineers still have against the application of polymer membranes. Development of new membrane, probably more inorganic than organic one, may in the future even allow for real organic-organic separation. Pervaporation is an efficient membrane process for liquid separation. The past decades had witnessed substantial progress and exciting breakthroughs in both the fundamental and application aspect of pervaporation has the potential for separating liquid mixtures in terms of the solubility parameter and the kinetic parameter of solvents. The thermodynamic approach of pervaporation, featuring emphasizing membrane/species interactions, though gained great successes in the past decades, is now facing its toughest challenge in the org–org separation. A kinetic era of pervaporation, featuring emphasizing diffusion selectivity, as well as the synergy between the selective diffusion and sorption, is in the making, and this approach will eventually find solutions to the challenging org–org separation. New way for applying membrane processes for purification water in different industries is using membrane contactors. Membrane contactor technology has applications in wastewater treatment, pharmaceuticals, fermentation, semiconductor manufacturing, protein extraction, carbonation of beverages, and VOC removal from waste gas and so forth. There are a few advantages to using membrane contactors in contrast to conventional methods. Some of them are no flooding at high flow rates, no unloading at low flow rates, absence of emulsions, no density difference between fluids required, and high interfacial area. Membrane contactor are designed and operated especially in contactors for removing dissolved oxygen from aqueous solutions by vacuum degassing

Liquid membrane-based extractions are capable of complementing conventional techniques (liquid–liquid and solid-phase extractions) in environmental and biomedical sample preparation. Their attractiveness in sample preparation is based on being able to design a configuration that can suit specific application at no cost. They offer the possibility of determining the concentration of freely dissolved pollutants on a time-weighted average basis, thus reflecting the episodic changes in real time. They are also suitable for exposure risk assessment of metal ions in both environmental and biological samples since the bioavailable fraction with its specific charge is measured. They can also be easily automated and interfaced to other separation techniques. The use of liquid membrane-based extractions is increasingly seen as offering an alternative to conventional sample preparation procedures in analysis of chemical species in both environmental and biological media. Liquid membrane-based extraction techniques are used in variety of analytical applications such as speciation of metal ions in biological fluids, time-weighted average field sampling, and time-weighted average passive field sampling. The area of designing new and much simpler miniaturized configurations, which is enjoying much attention in liquid membrane-based extraction techniques, is also a large area of research. Liquid membrane technology is introduced and is identified as a subset of membrane science. The reactive liquid membranes combines with traditional unit operations such as extraction or absorption with stripping are becoming popular. PLATINUM Liquid Membrane is a solar reflective, biobased urethane roof coating to be applied as a liquid waterproofing membrane over various low slope roofing substrates such as smooth asphalt built up roofs; granular and smooth modified bitumen; single-ply membranes such as TPO, aged Hypalon and aged PVC membranes; EPDM; metal roofing assemblies; Polyurethane. The use of ILs is opening new opportunities in different areas of separation science, with new many applications. Further applications in separations are related to the pharmaceutical, biomedical, environmental and other industries. ILs has been explored in separations for extraction, supported liquid membranes, as additives and as stationary phases in chromatography. The use of ILs as additives in chromatography exhibits great advantages compared to other additives. However, the applications of user designed ILs are not well developed. ILs with different cations and anions must be synthesized to separate various target compounds with different properties. IL-modified materials have proven to be an important new type of stationary phases. These IL stationary phases could be commercialized after some improvements involving the separation selectivity, thermal stability, immobilization bonding stability and so on. The stability of the stationary phase is considered an emerging handicap for commercially applications. Additionally, the range of separation compounds must be broadened from normal chemicals to bioactive compounds. ILs has been advertised as green solvents due to their nonvolatility. However, the environmental benefits of ILs need to be carefully considered. Not all ILs are safe and nontoxic, especially during their synthesis. Therefore, the development of real green ILs for extraction is still a challenge. Ionic liquids (ILs) have been applied in different areas of separation, such as ionic liquid supported membranes, as mobile phase additives and surface-bonded stationary phases in chromatography separations and as the extraction solvent in sample preparations, because they can be composed from various cations and anions that change the properties and phase behavior of liquids. Although the applications of ILs in separations are still in their early stages, the academic interest in ILs is increasing. The liquid membrane can be recycled but the efficiency of extraction decreased. Therefore, to maintain the extraction performance, this recycling liquid membrane should be mixed with fresh liquid membrane with certain ratio before homogenizing process.

## 5.5. ADSORPTION/ BIOSORPTION

Most studies in the biosorption field have been and are still being conducted to pinpoint the best biosorbent and/or alternative adsorbents with high adsorption capacity. During the last decade, however, the modification of biomass to advance its metal removal capability has become more attractive. Despite the availability of a significant number of publications, the real and major applications of biosorption from an industrial perspective are still far from realization due to economic constraints. The use of biosorbent and alternative adsorbents on an industrial scale are growing the heated areas of research. Arrays of non-conventional low cost biomass have been largely assessed for their latent use to remove heavy metals from synthetic or real wastewater. Excellent removal capabilities were apparent for several biomasses. More than a few factors, such as pH, temperature, adsorbent dose, etc. significantly affect the biosorption capacities. On the other hand, utilization of them in industrial-scale applications is still some distance from reality. While most available biomasses have the capability to sequester heavy metals from solutions, not all of them fit as alternative adsorbents in real wastewater treatment plants. There is no doubt that many biosorbents and/or alternative adsorbents as mentioned in a majority of journal articles have a high adsorption capacity to the extent that even some are better than commercially available adsorbents. Looking from this perspective only, it seems that most biosorbents and/or alternative adsorbents have potential for industrial application. Yet, several biomasses that have low binding capacity in nature still widely exist. Their adsorption capacity normally can be improved by pretreatment or modification using physical or chemical methods. Chemically, modification is usually performed by adding some chemicals such as acid alkali or other oxidizing and organic chemicals while in the physical method, pretreatment is facilitated by heat, autoclaving, freeze-drying and boiling. Specifically speaking, chemical activation methods are not favorable, as the main advantage gained by using biomass as a biosorbent and/or alternative adsorbent, *i.e.*, it is environmentally friendly (waste forwaste treatment) and costeffective, is lost. Unused chemicals entail more serious problems and commonly necessitate expensive waste treatment facilities. Looking at microorganism alones, while they are plentiful in nature, their isolation, screening and harvesting on a large scale is complicated and very expensive. Accordingly, it is become impractical to employ microorganisms obtained directly from nature for industrial biosorbent applications, although several research groups tested pure strains of microorganisms for biosorption studies by isolating the microorganisms from a polluted environment and cultivating them in their laboratory. However, large-scale fermentation might be able to supply large amounts of microorganisms for industry.

Immobilization of microorganisms and algae in distinct types of materials to enhance its physicochemical properties for continuous heavy metal biosorption from aqueous solution has been demonstrated by several research groups. Undeniably, this technique has achieved its initial purpose. Still, a large operational cost is obvious, making this technique unrealistic for industrial application. Additionally, the formation of coatings by microorganisms on complex pore structures of active immobilization agents generally reduces the sorption capacity of the immobilization agents. A low cost and an inert immobilization agent would preferably be chosen if this technique can be justified. The ability to recover heavy metals from biosorbents without destroying their sorption capability is also crucial. It is widely known that adsorption mechanisms can be classified as either physical adsorption or chemisorption.

Physical adsorption involves a weak Van der Waals attraction of the adsorbate to the surface, thus the adsorbate can be detached easily by heat treatment. On the other hand, chemical sorption to the adsorbate forms chemical bonds with the surface; hence chemical agents are usually required to break down the chemical bonds in order to detach the adsorbate from the solid. It must be noted that environmental concerns can sometimes override the economic considerations. As a note in most studies, heavy metals can be recovered quite easily using various chemicals, although in several cases, the metal binding capability deteriorates after regeneration. All these arguments converge into one conclusion: more effort is required to implement biomass materials as adsorbents for industrial wastewater treatment. High metal binding capacity is not the only factor to judge its suitability. Instead, after the appropriate biomass is already sorted out, further testing in real wastewater should be conducted, and at the same time, appropriate mathematical models need to be developed. Heavy metal pollution has become one of the most serious environmental problems today. Biosorption, using biomaterials such as bacteria, fungi, yeast and algae, is regarded as a cost-effective biotechnology for the treatment of high volume and low concentration complex wastewaters containing heavy metal(s) in the order of 1 to 100 mg/l. Among the promising biosorbents for heavy metal removal which have been researched during the past decades, *Saccharomyces cerevisiae* has received increasing attention due to the unique nature in spite of its mediocre capacity for metal uptake compared with other fungi. *S. cerevisiae* is widely used in food and beverage production, is easily cultivated using cheap media, is also a by-product in large quantity as a waste of the fermentation industry, and is easily manipulated at molecular level. A vast array of biological materials, especially bacteria, algae, yeasts and fungi have received increasing attention for heavy metal removal and recovery due to their good performance, low cost and large available quantities. The biosorbent, unlike mono functional ion exchange resins, contains variety of functional sites including carboxyl, imidazole, sulphydryl, amino, phosphate, and sulfate, thioether, phenol, carbonyl, and amide and hydroxyl moieties. Biosorbents are cheaper, more effective alternatives for the removal of metallic elements, especially heavy metals from aqueous solution. Molecular biotechnology is a potent tool to elucidate the mechanisms at molecular level, and to construct engineered organisms with higher biosorption capacity and selectivity for the objective metal ions. One trend is to use hybrid technology for pollutants removal, especially using living cells. Another trend is to develop the commercial biosorbents using immobilization technology, and to improve the biosorption process including regeneration/reuse, making the biosorbents just like a kind of ion exchange resin, as well as to exploit the market with great endeavor.

Biosorption of Cu (II) and Cr (III) ions from aqueous solutions by peanut shell biomass were investigated as a function of initial pH, initial biomass concentration and temperature. Results showed that peanut shells biomass is an attractive, alternative low-cost biosorbent for removal of heavy metal ions from aqueous media. There has been an explosion of biosorption-related research in recent years. However, it is doubtful whether such a dramatic rise in published output has significantly improved knowledge of the process, or aided any commercial exploitation, which so often is the prime rationale for such work. Most studies involve characterization of a chosen biomass type in sorbing a given substance from solution, and the effect of physico-chemical parameters in affecting biosorption. Most such studies involve metals, although an increasing number involve organic pollutants. Immobilized and/or granular biomass preparations may overcome the robustness and separation issue, but still do not overcome the specificity problem. It should also be noted that (bio) sorption technology transfers the sorbate from one medium to another, and so also raises questions regarding the safe disposal of loaded biosorbent, sorbate recovery, and regeneration or replacement of the biosorbent. Some of the disadvantages like incomplete metal removal, high reagent and energy requirements, generation of toxic sludge or other waste products that require careful disposal have often been used as the basis for arguments supporting a cost-effective biological approach. In natural systems, bioavailability of nutrients, including essential metals, and pollutants is determined by interactions with environmental components. In terrestrial and aquatic ecosystems, such components include clay and other minerals, humic substances, soil colloidal materials, biogenic debris and exudates, and living organisms. Sorption is one of the most important reactions that influence bioavailability, and therefore biosorption must also have a role to play within the spectrum of sorptive interactions with environmental components. These will include influencing distribution of substances between aqueous solution and particulate matter (including microorganisms) and their transport through environmental compartments and ultimate fate, *e.g.* settling in ocean sediments. Sorption of metals to cells is likely to play a critical role in all microbe–metal–mineral interactions. Interactions with specific groups on the surface of the cell may also enhance or inhibit metal transport and, thus, metal transformations and biomineralization processes.

In natural systems, sorption–desorption processes can occur over wide time scales, ranging from milliseconds to years, with lower reaction rates being due to such factors as diffusion into micropores, the existence of sites with low reactivity and surface nucleation-precipitation. This is often seen in soils, although just as in simple laboratory biosorption systems, biphasic sorption kinetics may be observed with an initial 'rapid' phase being followed by a slower phase occurring over longer time periods, progressing towards equilibrium. The major biosphere compartments, *e.g.* soils, oceans, contain material with high surface area to volume ratios and, of course, microorganisms like bacteria have the highest surface area: volume ratios of any living organism. Microorganisms are major components of the soil environment while biogenic particles dominate detrital phases in the oceans. In fact, several studies have shown that microbial cells, on a specific unit area basis, can exhibit higher sorption values for metals than, *e.g.* clay minerals which are one of the most important metal-sorbing components in soils. It may be concluded that biosorption phenomena have a more significant role in metal/radionuclide speciation, bioavailability and mobility in aquatic and terrestrial environments than has previously been supposed. Further, it should be emphasized that accompanying nucleation and precipitation can lead to biomineral formation. Biosorptive processes may be a component of varying significance in any form of primary or secondary biological treatment process for aqueous waters and process streams including domestic, municipal and industrial wastes, and in some circumstances, solid wastes. Sewage treatment, activated sludge plants, biofilters, biofilm reactors, fixed and suspended film systems, lagoon treatments, stream meanders, nitrification and denitrification treatments, biological phosphate removal processes, wetlands and reed-bed technologies, composting, *in situ* and *ex situ* bioremediation processes all rely on the activities of microorganisms to break down organic substances. Many wastes contain metals as well as organics and therefore biosorption of metals and related materials may also play a part in the overall process. The significance of such a role is difficult to establish but it may be significant in some cases: the problem of metal-laden sewage disposal is well known. In biological treatment processes, a clear advantage is that metabolic activities contribute to the degradation of organic materials.

Biosorption is a ubiquitous property of living or dead biomass and derived products, and is undoubtedly an important process in the environment, and in several conventional waste treatment processes. It has been proposed as a cheap and effective biotechnology for many years, yet has extremely limited industrial exploitation to date, even as an addition to conventional pollutant treatment approaches in hybrid technologies. Biosorption is frequently compared with ion exchange technology and often stated to provide a cheaper alternative. However, as mentioned previously, specificity is a problem while biosorbents also exhibit a shorter life cycle. Common suggestions for future research directions include identification of better and more selective biosorbents, more development of biosorption models and identification of biosorption mechanisms, and further assessments of market size, and costs of development. After so many years of biosorption research, it is debatable whether any more efforts in these directions will result in significant developments or novel contributions to understanding. Despite the apparent advantages of biosorption, it is ironic that many established and novel biotreatment methods for pollutants rely on living cell systems.

It is possible that biosorption may be combined with such systems in the future although at present this seems unlikely. The development of specific metal-binding molecules and/or engineered highly-specific biosorbents was heralded as a promising research direction, although there seems to have been little progress in industrial application. More recently, biosorption has been proposed for the purification and recovery of high-value proteins, steroids, pharmaceuticals, etc. by identifying a system that is highly selective for a given substance. This seems speculative in view of the general non-specificity of biosorption for both inorganic and organic substances, although the application of monoclonal antibodies for selective protein sorption has been suggested. While it can be concluded that the rationale for many biosorption studies is rather weak, especially if based on commercial development and application, the importance of biosorption in the environment and conventional biotreatment processes perhaps suggests further research should be directed in these areas.

Biosorption is being demonstrated as useful alternative to conventional systems for the removal of toxic metals from industrial effluents. The development of the biosorption processes requires further investigation in the direction of modeling, of regeneration of biosorbent material and of testing immobilized raw biomasses with industrial effluents. Due to the extensive research and significant economic benefits of biosorption, some new biosorbent materials are poised for commercial exploitation. Banana peel has proven to be a promising material for the removal of contaminants from olive mill wastewaters. Not only banana peel is an abundant cheap adsorbent, but also it is highly efficient for removing phenolic compounds from olive mill wastewater. All those add more credits to banana peel for removing pollutants from wastewaters.

There has been a high increase in production and utilization of dyes in last few decades resulting in a big threat of pollution. It is worthwhile noting that the removal of dyes can be done by various techniques; however, there exists no such methodology which can successfully remove all types of dyes at low cost. Above lead to the conclusion that for removal of dyes either adsorption alone or a combination of processes involving adsorption can give fruitful results. However, as noticed the main drawback in adsorption is the high cost of adsorbents as activated carbons are not low-cost materials and hence, in spite of their good efficiency and applicability for adsorbing a wide variety of materials, their use can sometimes be restricted due to economic considerations. Besides this, the regeneration is expensive and involves adsorbent loss as well as efficiency. As regulations become stricter, the effectiveness and cost of treatment processes for dye becomes more significant. There is a need for more detailed systematic studies on dye removal process and also some technical improvements in preparing and utilizing adsorbent. As regards to future work, the following recommendations are suggested. There exists a big variety of dyes and their removal is based on various factors which include adsorbate–adsorbent relationships (interactions), size of adsorbate, role of functional groups on both adorbate as well as adsorbent, a more detailed work on these interactions is needed for the studies to correlate and compare. Since sometimes differences are found in adsorption capacities for different adsorbents of the same origin, this indicate that there exists some relationship between adsorptive properties and the molecular structure or the surface groups of the adsorbent, besides the surface area/porosity effect, and needs more detailed critical investigations. This hopefully will help in predicting and selecting adsorbents. If the materials are of different origins then a critical evaluation of materials with a comparison of some similar adsorbent materials should be preferred. Further, detailed systematic studies are needed to find out which component in the precursor is primarily responsible for the development of surface area or micro/meso porosity or is helpful in development of functional groups. If possible, pore size distribution of adsorbent, size of dye molecule/s, functional groups present on surface of adsorbent, initial pH, batch or column conditions, temperature, particle size of adsorbent, the furnace as well as sample holder size (in case activation is performed) should be provided. The adsorbent cost (raw material, final product at lab scale) may also be made available, so that other workers can also compare the materials easily.

Once batch conditions, characterization of adsorbents, adsorptive capacity and the uptake mechanism are investigated work should be done to design and carry out some pilot-plant scale studies to check their feasibility at industrial level. As the dye effluents contain several other pollutants, attention has to be given to adsorption of dyes from mixtures and the LCAs should further be investigated for their efficiency using dye effluents from industries. Research workers should be encouraged to work on real effluents at local level. Since a vast number of the low-cost adsorbents are produced from locally available materials, the researchers utilizing LCA for dye removal can work and promote mantra 'think globally, act locally'. Combination of methodologies/system should also be encouraged since one perfect method/system is practically difficult. Besides these, strategies to minimize dye and related chemical effluent at all levels and designing more environmentally friendly chemicals are also needed. As far as we have seen a wide scope of research in these separation processes, there are still more areas where we can switch over to have more and more benefits from the results on these researches. Biosorption of metal ions and organics from industrial wastewater on full-scale was mainly focused on the biosorbent of peat during the past decades. Some peat on-site wastewater treatment systems for ion removal were operated in Maine, Alaska, Canada, and Ireland. The peat was regarded as utilizable and disposable. With the emphasis on using readily renewable biosorbents, peat-involving systems become much more engineered and much more specific, such as membrane-media extraction process. The process used peat moss capsules and was able to effectively treat As, Cd, Pb, Ni, Se and other metals from several type of wastewater, including electroplating rinsewater, pulp and paper mill discharge, municipal wastewater and acid mine drainage. In order to meet the renewable demand, another alternative is to take a pure component in a load-regeneration cycle system. Peat was regarded as the most successful and the rigorously scrutinized biosorbent in its natural state or in a modified form. Designs for large-scale peat-sorption processes are also available for application. However, the supply of peat is finite, and probably not the best biomaterial resource for commercialization. Thus, the development of other forms of biosorbents is essential. Three types of biomasses, including algae, fungi and waste biomass were suggested as potential biosorbents after predicting the future for biosorption application.

Biosorption application is facing to great challenge, some investigators proposed several suggestion. For the future of biosorption, there are two trends of biosorption development for metal removal. One trend is to use hybrid technology for pollutants removal, especially using living cells. Another trend is to develop good commercial biosorbents just like a kind of ion exchange resin, and to exploit the market with great endeavor. The difficulties existing for biosorption application urge people to consider applying the hybrid technology which comprise of various processes to treat real effluents. Various biotechnology-based processes, such as biosorption, bioreduction and bioprecipitation were suggested. Consequently, application of living cells rather than dead cells for biosorption has gained attention again. The above-mentioned bioprocesses along with other non-biotechnology based processes, for example, chemical precipitation, flotation, electrochemical processes, membrane technology, may also be helpful for treating wastewater in large-scale, even for simultaneous removal of organic substances and heavy metal ions in solution. All these processes may even possibly be realized in a single reactor; hence the corresponding novel reactors should be designed. Biosorption appears to be suitable as secondary or polishing applications for metal removal from dilute waste streams, which would be competitive with ion-exchange resin based on final cost-beneficial analysis, and the greatest use for biosorption may be in modular system for small companies, *e.g.* for specific treatment. Another trend requires the improvement of biomaterials immobilization, as well as the optimization of the parameters of biosorption process and physicochemical conditions, including reuse and recycling. The mechanisms involved in biosorption or metal–microbe interactions should be further studied with great efforts. The sources and type of biosorbent play a major role in determining the overall cost of the biosorbent material. If the biomass needs to be specifically cultured for this purpose, manufacturers will incorporate maintenance and production expenses in the total cost, as well as a commercial fee. These costs can be minimal where certain biomass types such as photo-autotrophic algae can be successfully grown for large-scale commercial use due to their minimal growth requirements (water, sunlight and $CO_2$). Many industrial waste biomass types have been investigated for their biosorptive potential. *S. cerevisiae* from the food and beverage industry and *Candida albicans*, a clinical isolate; the moulds, *R. arrhizus* from the food industry, *P. chrysogenum* from antibiotic manufacturers and *A. niger* from citric acid and industrial enzyme producers; the bacteria, *Bacillus* spp., utilized in amino acid and antibiotic fermentations and *Streptomyces noursei* from the pharmaceutical industry. These potential biosorbents can usually be obtained relatively free of charge from the respective producers since they already present disposal problems to them. The only costs incurred should be those of drying, if required, and transport. These low-cost biosorbents will make the process highly economical and competitive particularly for environmental applications in detoxifying effluents of *e.g.* metal-plating and metal-finishing operations; mining and ore processing operations; metal processing, battery and accumulator manufacturing operations; thermal power generation; nuclear power generation. The researchers of various types of scientific background, from engineering to biochemistry, working together, will make a significant contribution to elucidating the biosorption mechanisms. Interdisciplinary efforts are mandatory and represent quite a challenge. The optimization of specific biosorption process applications has to be done in conjunction with industrial users/clients and requires specific process engineering expertise and a serious development capital commitment. A variety of investigation demonstrated that biosorption is a useful alternative to the conventional systems for the removal of heavy metal ions from aqueous solution. The development of biosorption process requires further investigation in the direction of modeling, of regeneration and immobilization of biosorbents, and of treating the real industrial wastewater.

## 5.6. HYBRID SYSTEMS

For a pressure sensitive azeotropic mixture such as acetonitrile -water, it appears effective to use a relatively low pressure in the distillation column and a higher pressure in the pervaporation unit. As compared to distillation-based processes, a reduction between 25% and 60% in the total separation cost can be achieved. Hybrid distillation processes that combine a pervaporation unit operating in vapor-liquid flow have been analyzed. The pervaporation unit has been integrated with distillation in one single column (DPSU). The pervaporation zone, containing hollow fibre pervaporation membranes, replaces a section of packing or trays in a distillation column. The proposed hybrid system is evaluated for the dewatering of binary mixtures such as ethylene diamine-water and IPA-water and for removal of methanol from methyl-tertbutyl ether (MTBE)-butene-methanol mixture involved in the production of MTBE. The DPSU can overcome the azeotropic composition and the distillation boundaries. The hybrid distillation pervaporation process in a single unit (DPSU) is compared with a conventional hybrid process where the pervaporation unit is externally connected to the distillation column (DPEC). The performance of the DPSU has been calculated using a modified rate based model where the interface between the liquid and the membrane has been included. It has been found that a hybrid DPSU is more convenient for the removal of one component from a multicomponent mixture when the component can not be obtained from the distillation column as a pure distillate or bottom product. Thus, for purification of binary mixtures a DPEC process is more convenient while the DPSU is recommended for the removal of methanol from the multicomponent mixture. For the multicomponent system, the difference in performance between a DPEC and a DPSU lies in the higher methanol concentration that occurs in the pervaporation section in a DPSU. In general, two-phase flow pervaporation units can expand the application window of pervaporation in stand alone and hybrid applications. Industrial applications are already viable using the multi-tubular pervaporation modules available in the market. The development of pervaporation modules with hollow fibre membranes will take some more time and it requires more information about the mechanical resistance and the wetting properties, since the wetted fraction will probably affect more strongly the energy consumption of the separation process than the required membrane area for a specific separation duty. A distillation-pervaporation process in one single column opens the possibility to separate multicomponent mixtures in one unit. However, it is necessary to explore the dynamic behavior of these systems and the existence of multiple steady-states which are important for industrial operation and start up.

# Solved Examples

## DISTILLATION COLUMN

(1) The feed to a binary distillation column has 40 mol % vapor and 60 mol % liquid. Then, the slope of the q-line in the McCabe-Thiele plot is

(A) –1.5 (B) –0.6 (C) 0.6 (D) 1.5

[GATE 2009, Q 37]

**Solution (A)**

(2) In a binary mixture containing components A and B, the relative volatility of A with respect to B is 2.5 when mole fractions are used. The molecular weights of A and B are 78 and 92 respectively. If the compositions are however expressed in mass fractions the relative volatility will then be

(A) 1.18 (B) 2.12 (C) 2.5 (D) 2.95

[GATE 2008, Q 48]

**Solution (C)**

(3) A batch distillation operation is carried out to separate a feed containing 100 moles of a binary mixture of A and B. The mole fraction of A in the feed is 0.7. The distillation progresses until the mole fraction of A in the residue decreases to 0.6. The equilibrium cure in this composition range may be linearized to $y^* = 0.7353\,x + 0.3088$. Here x and y are the mole fractions of the more volatile component A in the liquid and vapor phases respectively. The number of moles of residue is

(A) 73.53 (B) 418.02 (C) 40 (D) 30.24

[GATE 2008, Q 50]

**Solution (B)**

(4) In distillation where q is defined as the moles of liquid flow in the stripping section per mole of feed introduced, for saturated liquid feed –

(A) $q > 1$ (B) $q < 1$ (C) $q = 1$ (D) $q = 0$

[GATE 1995, Q 2(n)]

**Solution (C)**

(5) The reason for preferring packed towers over plate towers in distillation practice is that the packed tower operation gives:

A. Low pressure drop and high hold up
B. High pressure drop and low hold up
C. Low pressure drop and low hold up
D. High pressure drop and high hold up

[GATE 1996, Q 1.12]

**Solution (D)**

(6) Solvent used in extractive distillation

A. Is of low volatility
B. Forms a low-boiling azeotrope
C. Forms a high-boiling azeotrope
D. Does not alter the relative volatility of the original components

[GATE 1999, Q 1.17]

**Solution (A)**

(7) In distillation column sizing calculations by short cut methods, match the following

| | | | |
|---|---|---|---|
| P | Underwood's Equation | 1 | Number of real trays |
| Q | Fenske's Equation | 2 | Column diameter |
| R | Gilliland's Equation | 3 | Minimum number of ideal trays |
| S | Vapour velocity at flooding | 4 | Actual number of ideal trays |
| | | 5 | Minimum reflux ratio |
| | | 6 | Tray efficiency |

(A) P–1, Q–3, R–4, S–6 (B) P–2, Q–5, R–1, S–3
(C) P–5, Q–3, R–6, S–2 (D) P–5, Q–3, R–4, S–2

[GATE 2003, Q 84]

**Solution (C)**

(8) In distillation column design, the McCabe-Thiele procedure is inadequate and a Ponchon-Savarit procedure is needed when –

A. Saturated feed is not used
B. An azeotrope forms
C. The latent heats of vapourisation of the more and less volatile components are greatly different
D. A total condenser is used

[GATE 1998, Q 1.18]

**Solution (D)**

(9) In binary distillation, the separation of the components is easier if the relative volatility ($\alpha$) is –

(A) $\alpha >> 1$ (B) $\alpha << 1$ (C) $\alpha = 1$ (D) none of these

[GATE 1997, Q 1.14]

**Solution (A)**

(10) For a fixed number of ideal stages in a distillation column, as the reflux ratio is increased, the difference in composition between the top and bottom product streams –

(A) Increases (B) Decreases

(C) Remains unaffected (D) Passes through a maximum

[GATE 1998, Q 1.15]

**Solution (A)**

(11) When a multistage tray tower uses a minimum reflux ratio it implies –

A. Infinite trays and maximum reboiler heat load

B. Infinite trays and minimum reboiler heat load

C. Minimum trays and minimum reboiler heat load

D. Minimum trays and maximum reboiler heat load

[GATE 1996, Q 1.13]

**Solution (A)**

(12) An equi-molar mixture of four hydrocarbons (1, 2, 3, 4) is to be separated into high purity individual components using a sequence of simple distillation columns (one overhead and one bottom stream). Four possible schemes are shown below.

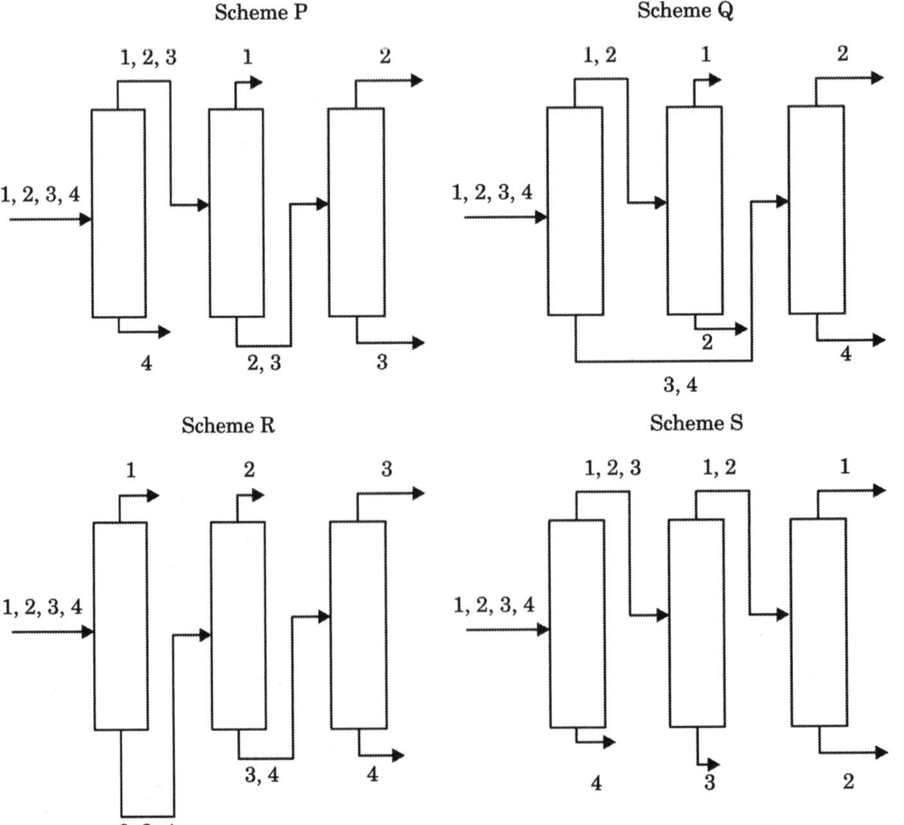

| Component | $K_i$ |
|---|---|
| 1 | 6 |
| 2 | 3 |
| 3 | 2.5 |
| 4 | 1.1 |

Using the $K_i$ (= $y_i^*/x_i$) values given above, the optimal scheme is

(A) P (B) Q (C) R (D) S

[GATE 2009, Q 46]

**Solution (D)**

(13) The interfacial area per unit volume of dispersion, in a gas – liquid contractor, for fractional hold-up of gas = 0.1 and gas bubble diameter = 0.5 mm is given by (in $m^2/m^3$)

(A) 500 (B) 1200 (C) 900 (D) 800

[GATE 2001, Q 2.12]

**Solution (B)**

(14) According to the Fenske equation, what will be the minimum number of plates required in a distillation column to separate an equimolar binary mixture of components A & B into an overhead fraction containing 99 mole % A and a bottoms fraction containing 98 mole % B? [Assume that the relative volatility ($á_{AB}$ = 2) does not change appreciably in the column]

(A) 5 (B) 9 (C) 12 (D) 28

[GATE 2002, Q 2.8]

**Solution (C)**

(15) A distillation column with N plates is being operated under normal conditions. At some point in time, the operation is shifted to total reflux condition (*i.e.*, no product and residue are being withdrawn and feed to the column is stopped). At the new steady state,

A. Composition of vapours and that of liquid do not vary throughout the column
B. Reboiler load and condenser load are minimum
C. The top and bottom compositions are unchanged with and without total reflux
D. The top and bottom compositions correspond to the maximum enrichment achievable.

[GATE 2004, Q 69]

**Solution (D)**

(16) An aqueous solution of methanol is to be distilled in a tray column. High-pressure steam is available as a source of heat. For a given reflux ratio and overhead composition, two options are being explored: (i)

*a reboiler is used,* and *(ii) no reboiler is used* but steam is fed directly to the bottom of the column. As compared to option (i), in option (ii).

A. Less number of trays are required ,
B. Composition of the residue remains unchanged
C. More number of trays are required but the residue composition remains unchanged
D. More number of trays are required and the residue composition is more dilute in methanol.

[GATE 2004, Q 70]

**Solution (A)**

(17) A distillation column at a pilot plant is scaled up by 3 times for industrial use at steady state. After scaling up

A. The number of theoretical trays increases by 3 times
B. The minimum reflux ratio is increased by three times
C. The feed flow rate and product flow rates are increased by three times
D. The feed composition and product compositions are increased by three times

[GATE 2005, Q 4]

**Solution (C)**

(18) For the nth tray (counted from the bottom of a distillation column), the Murphee tray efficiency is given by

(A) $\dfrac{y_{n+1} - y_n}{y_n^* - y_{n-1}}$ (B) $\dfrac{y_n - y_{n-1}}{y_n^* - y_{n-1}}$ (C) $\dfrac{y_{n-1} - y_n}{y_{n+1} - y_n}$ (D) $\dfrac{y_n^* - y_{n-1}}{y_n^* - y_{n+1}}$

[GATE 2000, Q 1.16]

**Solution (B)**

(19) In a tray column, separating a binary mixture, with non-ideal stages, ONE of the following statements in TRUE,

A. Point efficiency can exceed 100%
B. Murphree efficiency cannot exceed 100%
C. Murphree efficiency can exceed 100%
D. Both Murphree and point efficiencies can exceed 100%

[GATE 2005, Q 19]

**Solution (C)**

(20) For a two-phase feed, where 80% of the feed is vaporized under column conditions, the feed line slope in the McCabe-Thiele method for distillation column design, is

(A) –1/4 (B) +1/4 (C) +4 (D) –4

[GATE 2005, Q 62]

**Solution (B)**

(21) In a binary distillation column, if the feed contains 40 mol% vapour, the q line will have a slope of

(A) 1.5 (B) –0.6 (C) –1.5 (D) 0.6

[GATE 1997, Q 2.12]

**Solution (C)**

(22) The McCabe ALL law states that the

A. Molar heats of vaporization of components are nearly equal
B. Linear crystal growth rate depends on the degree of super saturation
C. Linear crystal growth rate does not depend on the crystal size
D. Linear crystal growth rate depends on the crystal size.

[GATE 2000, Q 1.17]

**Solution (D)**

(23) Minimum reflux ratio in a distillation column results in

(A) Optimum number of trays (B) Minimum reboiler size
(C) Maximum condenser size (D) Minimum number of trays

[GATE 2003, Q 18]

**Solution (C)**

(24) Given the following statements listed from P to T, select the correct combination of TRUE statements from the choices that follow this list.

(P) Plate columns are preferred when the operation involves liquids containing suspended solids.

(Q) Packed towers are preferred if the liquids have a large foaming tendency.

(R) The pressure drop through packed towers is more than the pressure drop through plate columns designed for the same duty.

(S) Packed columns are preferred when large temperature changes are involved in distillation operations.

(T) Packed towers are cheaper than plate towers if highly corrosive fluids must be handled.

(A) T, S, P (B) P. Q. T (C) S, R, T (D) R, Q, S

[GATE 2007, Q 64]

**Solution (B)**

The boiling points for pure water and pure toluene are 100°C and 110. 6°C, respectively. Toluene and water are completely immiscible in each other. A well-agitated equimolar mixture of toluene and water is prepared.

(25) The temperature at which the above mixture will exert a pressure of one standard atm is

(A) Less than 100°C (B) 100°C

(C) between 100 and 110°C (D) 110.6°C

[GATE 2004, Q 62]

**Solution (A)**

(26) AT a total pressure of one standard atm exerted by the vapours of water and toluene, the mole fraction of water $x_w$ in the vapour phase satisfies

(A) $0 < x_w < 0.5$ (B) $x_w = 0.5$

(C) $0.5 < x_w < 1.0$ (B) $x_w = 1.0$

[GATE 2004, Q 63]

**Solution (C)**

(27) 100 moles of a binary mixture F containing 60 mol% A (more volatile) and 40 mol% B is treated in a batch distillation still. After 1 hour, 70 moles of the distillate D is collected leaving behind the residue W. Relative volatility á is 2. The governing equation is

$$\log \frac{F\, x_F}{W\, x_W}\ \alpha \log \frac{F\,(1 - x_F)}{W\,(1 - x_W)}$$

The average mole fraction of A in the distillate is

(A) 0.43 (B) 0.61 (C) 0.69 (D) 0.73

[GATE 2006, Q 48]

**Solution (C)**

(28) A saturated vapor is fed to a distillation column at 180 kmol/hr. Both the rectifying and stripping sections of the column operate at 60% of their respective flooding velocities. The flooding velocity of the rectifying section is twice that of the stripping section. The assumptions of constant molar overflow and constant molar vapor density throughout the column are valid. If the boil-up rate is 60 kmol/hr, then the relationship between the diameters of the rectifying section ($d_r$) and the stripping section ($d_s$) is

(A) $d_r = \sqrt{2}\ d_s$ (B) $d_r = \sqrt{3}\ d_s$ (C) $d_r = 2\ d_s$ (D) $d_r = 3\ d_s$

[GATE 2006, Q 65]

**Solution (A)**

(29) In a distillation operation, what is the effect of the temperature of the reflux stream (given below) on the condenser and reboiler loads? Reflux conditions:

(i) Reflux stream is completely liquid and is at its bubble point

(ii) Reflux stream is below its bubble point.

A. Condenser and reboiler loads are the same in both the cases,

B. Reboiler load is the same in both the cases but condenser load is higher in case (ii)

C. Condenser load is the same in both the cases but reboiler load is higher in case (ii)
D. Both condenser and reboiler loads are higher in case (ii) as compared to case (i)

[GATE 2004, Q 64]

**Solution (d)**

(30) In the McCabe-Thiele diagram for binary distillation, vertical feed line represents ____________ feed and horizontal feed line represents ____________ feed.

[GATE 1992, Q 6(b)]

**Solution** (Saturated Liquid & Saturated Vapor)

(31) In small columns local efficiency is __________ Murphee efficiency; in large columns local efficiency is __________ Murphee efficiency.

[GATE 1992, Q 6(d)]

**Solution** (Larger than & Smaller than)

## EXTRACTION (LIQUID-LIQUID)

1. Compound A is extracted from a solution of A + B into a pure solvent S. A co-current unit is used for the liquid–liquid extraction. The inlet rate of the solution containing A is 200 moles of B/hr $m^2$ and the solvent flow rate is 400 moles of S/hr $m^2$. The equilibrium data are represented by $Y = 3X^2$ where Y is in moles of A/moles of B and X is in moles of A/moles of S. The maximum percentage extraction achieved in the unit is

   (A) 25% (B) 50% (C) 70% (D) 90%

   [GATE 2003, Q 68]

   **Solution (B)**

2. An ideal single stage extraction process is used to treat 100 mol/s of an organic feed solution. The solute concentration in this solution is to be reduced from 0.5 mol% to 0.1 mol%. A pure solvent S is used. To reduce the solvent requirement by half for the same separation,
   A. Add one more ideal co-current stage
   B. Use another pure solvent S* whose partition coefficient is twice of S
   C. Use solvent S containing 0.02 mole fraction of the solute,
   D. Double the residence time of the solvent S in the contactor

   [GATE 2006, Q 14]

   **Solution (C)**

3. In a binary liquid solution of components A and B, if component A exhibits positive deviation from Raoult's law, then components B,
   A. Exhibits positive deviation from Raoult's law
   B. Exhibits negative deviation from Raoult's law

C. Obeys Raoult's law

D. May exhibit either positive or negative deviation from Raoult's law.

[GATE 2000, Q 1.4]

**Solution (A)**

4. In a single stage extraction process, 10 kg of pure solvent S (containing no solute A) is mixed with 30 kg of feed F containing A at a mass fraction $X_F = 0.2$. The mixture splits into an extract phase E and a raffinate phase R, containing A at XF = 0.05 and XR = 0.05 respectively. The total mass of the extract phase is (in kg).

   (A) 6.89 (B) 8.89 (C) 10 (D) 8.25

[GATE 2001, Q 2.14]

**Solution (B)**

5. Penetration theory states that the mass transfer coefficient is equal to (where De is diffusivity and t is time)

   (A) $\left(D_{e^t}\right)^{1/2}$ (B) $\left(\dfrac{D_e}{t}\right)^{1/2}$ (C) $\left(\dfrac{4D_e}{\pi t}\right)^{1/2}$ (D) $\left(\dfrac{4D_e}{t}\right)^{1/2}$

[GATE 1999, Q 1.19]

**Solution (C)**

6. In liquid-liquid extraction 10 kg of a solution containing 2 kg of solute C and 8 kg of solvent A is brought into contact with 10 kg of solvent B. Solvents A and B are completely immiscible in each other whereas solute C is soluble in both the solvents. The extraction process attains equilibrium. The equilibrium relationship between the two phases is $Y^* = 0.9X$ where $Y^*$ is kg of C/kg of B and X is kg of C/kg of A. Choose the correct answer

   A. The entire amount of C is transferred to solvent
   B. Less than 2 kg but more than 1 kg of C is transferred to solvent B
   C. Less than 1 kg of C is transferred to B
   D. No amount of C is transferred to B.

[GATE 2004, Q 66]

**Solution (B)**

7. A feed (F) containing a solute is contacted with a solvent (S) in an ideal stage as shown in the diagram below. Only the solute transfers into the solvent. The flow rates of all the streams are shown on a solute free basis and indicated by the subscript S. The compositions of the streams are expressed on a mole ratio basis. The extract leaving the contactor is divided into two equal parts, one part collected as the product (P) and the other stream is recycled to join the solvent. The equilibrium relationship is $Y^* = 2X$.

   The product flow rate (Ps) and composition ($Y_{out}$) are

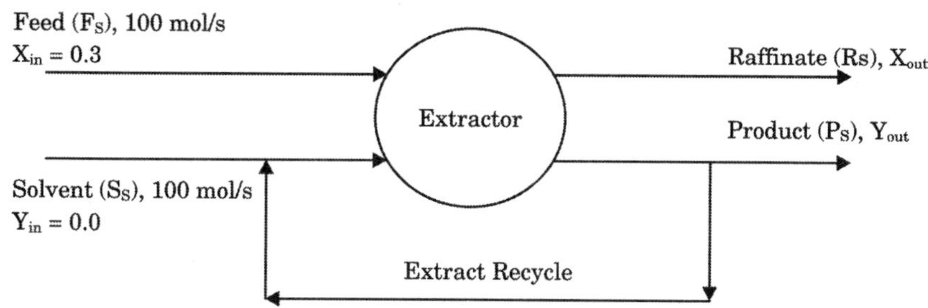


(A) $P_S = 50$ mol/s, $Y_{out} = 0.3$ (B) $P_S = 100$ mol/s, $Y_{out} = 0.2$
(C) $P_S = 200$ mol/s, $Y_{out} = 0.1$ (D) $P_S = 100$ mol/s, $Y_{out} = 0.4$

[GATE 2008, Q 54]

**Solution (B)**

8. ____________ number in mass transfer corresponds to Nusselt number to heat transfer and __________ number to Prandtl number.

[GATE 1992, Q 6(c)]

**Solution** (Sherwood & Schimit)

2. A continuous – contact extraction column is used to extract a solute from an aqueous stream (F) using an organic solvent (S). The distribution coefficient (y/x) is 1.0, where x and y are the mass fractions of solute in raffinate and extract phases, respectively. The height of transfer unit based on the extract phase is 1.0 m. The rest of the data are given in the figure. Assuming that the phase flow rates are constant, find the height of the tower.

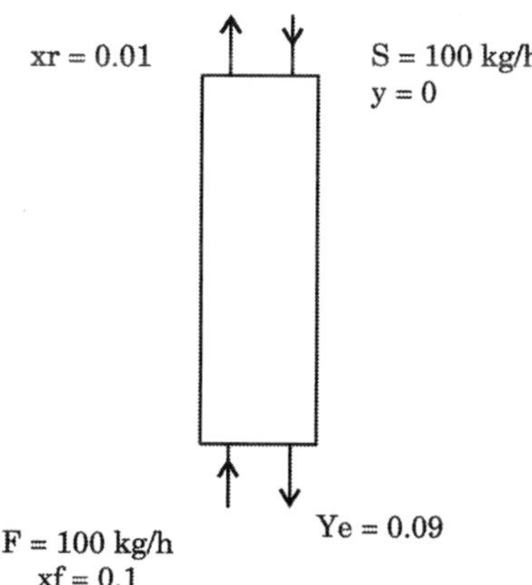


[GATE 2000, Q 10]

**Solution**

$H_{oe}$ = HIV based on extract = m find height of tower

$H_{oe} = 1$

Z = tower height

$N_{oe}$ = no. of transfer unit = $\dfrac{Y - Y_2}{(Y^* - Y)_{LM}}$

$N_{oe} = 9$

$(Y^*\text{-}Y)_1 = 0.01$

$(Y^*\text{-}Y)_2 = 0.01$

Since (Y*-Y) is same throughout $(Y^*\text{-}Y)_1 = (Y^*\text{-}Y)_2 = (Y^*\text{-}Y)_{LM}$ –**Ans**

10. The equilibrium data for the distribution of a solute C between feed solvent A and the extracting solvent B is given below :

| X conc. of C is raffinate kg C/kg A | 0 | 0.1 | 0.2 | 0.3 | 0.4 | 0.45 |
|---|---|---|---|---|---|---|
| Y conc. of C extract kg C/kg B | 0 | 0.096 | 0.17 | 0.232 | 0.275 | 0.28 |

In a counter current contact operation 10 kg/s of the feed containing 20% C in A (w/w) is treated with pure solvent so as to reduce the solute concentration to 5% (w/w). Calculate the solvent rate for operation with 2.5 times the minimum solvent rate.

[GATE 1995, Q 20]

**Solution**

Based on the given X-Y data, it can be considered that solvent A and S are immisicible.

For Stage 1 $Y_2 = \dfrac{A}{S}(X_1 - X_f) + Y_1$

For stage n $Y_{n+1} = \dfrac{A}{S}(X_n - X_{n-1}) + Y_n$

Overall $Y_{n+1} = \dfrac{A}{S}(X_n - X_f) + Y_1$

And slope = $\left(\dfrac{A}{S}\right)_{min} = 0.85$

S = 2.55 min

A/S = 0.34

Solvent rate should be 29.412 kg/s.

## UNSOLVED EXAMPLES

11. Inlet water solution of 100 kg/hr containing 0.010 wt fraction nicotine in water is stripped with a kerosene stream of 200 kg/hr containing 0.0005 wt fraction nicotine in counter current stage tower. Water and kerosene are essentially immscible in each other. Desired concentration of the exit water is 0.0010 wt fraction nicotine. Determine the number of stages

| x | 0.001010 | 0.00246 | 0.00500 | 0.00746 | 0.00988 | 0.0202 |
|---|---|---|---|---|---|---|
| y | 0.000806 | 0.001959 | 0.00454 | 0.00682 | 0.00904 | 0.0185 |

12. Inlet water solution of 1000 kg/hr containing 1.5% nicotine in water is stripped with a kerosene stream of 2000 kg/hr containing 0.05% nicotine

in counter current stage tower. Water and kerosene are essentially immscible in each other. Desired concentration of the exit water is 10% nicotine. Determine the number of stages

| x | 0.001010 | 0.00246 | 0.00500 | 0.00746 | 0.00988 | 0.0202 |
|---|---|---|---|---|---|---|
| y | 0.000806 | 0.001959 | 0.00454 | 0.00682 | 0.00904 | 0.0185 |

13. Feed 200 kg of unknown composition containing water, acetic acid, isopropylether is contacted in single stage with 230 kg of mixture containing 40 wt % AA, 10 wt % water and 50 wt % isopropylether. The resulting raffinate later weighs 320 kg and contains 29.5 wt % AA, 66.5 wt % water and 4 wt % isopropylether. Determine the original composition.
14. Two liquids A and B which have nearly identical boiling point are to be separated by extraction with solvent C. The data below is the equilibrium at 95°C

| | | | | | | | | | | |
|---|---|---|---|---|---|---|---|---|---|---|
| A% | 0 | 1 | 1.8 | 3.7 | 6.2 | 9.2 | 13 | 18.3 | 24.5 | 31.2 |
| B% | 7 | 6.1 | 5.5 | 4.4 | 3.3 | 2.4 | 1.8 | 1.8 | 3.0 | 5.6 |
| C% | 93 | 92.9 | 92.7 | 91.9 | 90.5 | 88.4 | 85.2 | 79.9 | 72.5 | 63.2 |
| A% | 0 | 9 | 14.9 | 25.3 | 35 | 42 | 48.1 | 52.0 | 47.7 | Pliat point |
| B% | 92 | 81.5 | 75 | 63 | 51.5 | 41 | 29.3 | 20 | 12.9 | |
| C% | 8 | 9.3 | 10.1 | 11.7 | 13.5 | 17 | 22.6 | 28 | 40 | |

Determine the number of ideal stages is the operation is

A. Cocurrent

B. Countercurrent.

The feed contain 35% A and 65% B on solvent free basis. The extract should contain 83% A and 17% B and raffinate contains 10% A and 90% B both on solvent free basis.

15. A mixture weighing 1000 kg contains 23.5% acetone and 76.5 % water and is to be extracted by 500 kg MIBK. Determine the amount and concentrations of the extract and raffinate after 4 stages in the countercurrent mode.
16. In the train of mixture settlers, 100 kg/h of 40:60 acetone water solution is to be reduced to 10% acetone by extraction with pure 1,1,2-trichloroethane at 25°C. Find the number of mixer settlers required if the operation is cocurrent.
17. Propyl ether at 600 kg/hr is used to extract aqueous solution of $L_0 = 200$ kg/hr containing 30 wt % acetic acid. It is desired that the exit acid concentration in aqueous phase is 4%. Calculate the number of stages is the column is operated counter currently.
18. An aqueous feed of 1000 kg/h 23.5 wt% acetone and 76.5 wt % water is extracted using pure MIBK at 298 K. The outlet raffinate contains 2.5 wt% acetone. Calculate the number of stages is the column is operated counter currently.

| Composition data (wt %) | | | | | | | | | | | | |
|---|---|---|---|---|---|---|---|---|---|---|---|---|
| MIBK | 98 | 93.2 | 77.3 | 71 | 65.5 | 54.7 | 46.2 | 12.4 | 5.01 | 3.23 | 2.12 | 2.20 |
| Acetone | 0 | 4.6 | 18.95 | 24.4 | 28.9 | 37.6 | 43.2 | 42.7 | 30.9 | 20.9 | 3.73 | 0 |
| Water | 2 | 2.33 | 3.86 | 4.66 | 5.53 | 7.82 | 10.7 | 45.0 | 64.2 | 75.8 | 94.2 | 97.8 |
| **Acetone distribution data (wt %)** | | | | | | | | | | | | |
| Water phase | 2.5 | 5.5 | 7.5 | 10 | 12.5 | 15.5 | 17.5 | 20.0 | 22.5 | 25.0 | 26.0 | |
| MIBK phase | 4.5 | 10 | 13.5 | 17.5 | 21.3 | 25.5 | 28.2 | 31.2 | 34 | 36.5 | 37.5 | |

| Limiting solubility curve | | | | | | | | | | | | |
|---|---|---|---|---|---|---|---|---|---|---|---|---|
| $C_2H_3Cl_3$ wt % | 94.73 | 79.58 | 67.52 | 54.88 | 38.31 | 24.04 | 15.39 | 6.77 | 1.72 | 0.92 | 0.65 | 0.44 |
| Water wt % | 0.26 | 0.76 | 1.44 | 2.98 | 6.84 | 15.37 | 26.28 | 41.35 | 61.11 | 74.54 | 87.63 | 99.56 |
| Acetone wt % | 5.01 | 19.66 | 31.04 | 42.14 | 54.85 | 60.59 | 58.33 | 51.88 | 37.17 | 24.54 | 11.72 | 0.00 |
| Tie lines wt % in water layer | 0.52 | 0.73 | 1.02 | 1.17 | 1.60 | 2.10 | 3.75 | 6.52 | | | | |
| | 93.52 | 82.23 | 72.06 | 67.95 | 62.67 | 57.00 | 50.20 | 41.70 | | | | |
| | 5.96 | 17.04 | 26.92 | 30.88 | 35.73 | 40.90 | 46.05 | 51.78 | | | | |
| Tie lines wt % in $C_2H_3Cl_3$ layer | 90.93 | 73.76 | 59.21 | 53.92 | 47.53 | 40.00 | 33.70 | 26.26 | | | | |
| | 0.32 | 1.10 | 2.27 | 3.11 | 4.26 | 6.05 | 8.90 | 13.40 | | | | |
| | 8.75 | 25.14 | 38.52 | 42.97 | 48.21 | 53.95 | 57.40 | 60.34 | | | | |

| AA | 0 | 0.69 | 1.41 | 2.89 | 6.42 | 13.30 | 25.5 | 36.7 | 44.3 | 46.4 |
|---|---|---|---|---|---|---|---|---|---|---|
| Water | 98.8 | 98.1 | 97.1 | 95.5 | 91.7 | 84.4 | 71.1 | 58.9 | 45.1 | 37.1 |
| IE | 1.2 | 1.2 | 1.5 | 1.6 | 1.9 | 2.3 | 3.4 | 4.4 | 10.6 | 16.5 |
| AA | 0 | 0.18 | 0.37 | 0.79 | 1.93 | 4.82 | 11.4 | 21.6 | 31.1 | 36.2 |
| Water | 0.6 | 0.5 | 0.7 | 0.8 | 1 | 1.9 | 3.9 | 6.9 | 10.8 | 15.1 |
| IE | 99.4 | 99.3 | 98.9 | 98.4 | 97.1 | 93.3 | 84.7 | 71.5 | 58.1 | 48.7 |


| Composition data (wt %) | | | | | | | | | | | | |
|---|---|---|---|---|---|---|---|---|---|---|---|---|
| MIBK | 98 | 93.2 | 77.3 | 71 | 65.5 | 54.7 | 46.2 | 12.4 | 5.01 | 3.23 | 2.12 | 2.20 |
| Acetone | 0 | 4.6 | 18.95 | 24.4 | 28.9 | 37.6 | 43.2 | 42.7 | 30.9 | 20.9 | 3.73 | 0 |
| Water | 2 | 2.33 | 3.86 | 4.66 | 5.53 | 7.82 | 10.7 | 45.0 | 64.2 | 75.8 | 94.2 | 97.8 |
| **Acetone distribution data (wt%)** | | | | | | | | | | | | |
| Water phase | 2.5 | 5.5 | 7.5 | 10 | 12.5 | 15.5 | 17.5 | 20.0 | 22.5 | 25.0 | 26.0 | |
| MIBK phase | 4.5 | 10 | 13.5 | 17.5 | 21.3 | 25.5 | 28.2 | 31.2 | 34 | 36.5 | 37.5 | |

19. 100 kg of cottonseed oil – oleic acid solution containing 25% acid is extracted using 1000 kg with propane. Determine the amount and concentrations of the extract and raffinate after 4 stages in the countercurrent modes

20. Pure water is used to extract acetic acid from 400 kg of a feed solution containing 23 wt % acetic in propylether. Determine the amount and concentrations of the extract and raffinate after 3 stages in the cocurrent mode

| | | | | | | | | | | | | |
|---|---|---|---|---|---|---|---|---|---|---|---|---|
| Cottonseed | 63.5 | 57.2 | 52 | 46.7 | 39.8 | 31 | 26.9 | 21 | 14.2 | 8.3 | 4.5 | 0.8 |
| Oleic acid | 0 | 5.5 | 9 | 13.8 | 18.7 | 26.3 | 29.4 | 32.4 | 37.4 | 39.5 | 41.1 | 43.7 |
| Propane | 36.5 | 37.3 | 39 | 39.5 | 41.5 | 42.7 | 43.7 | 46.6 | 48.4 | 52.2 | 54.4 | 55.5 |
| Cottonseed | 2.30 | 1.95 | 1.78 | 1.50 | 1.36 | 1.20 | 1.10 | 1 | 0.8 | 0.7 | 0.4 | 0.2 |
| Oleic acid | 0 | 0.76 | 1.21 | 1.90 | 2.73 | 3.8 | 4.4 | 5.1 | 6.1 | 7.2 | 6.1 | 5.5 |
| Propane | 97.7 | 97.3 | 97 | 96.6 | 95.9 | 95 | 94.5 | 93.9 | 93.1 | 92.1 | 93.5 | 94.3 |

| | | | | | | | | | | |
|---|---|---|---|---|---|---|---|---|---|---|
| AA | 0 | 0.69 | 1.41 | 2.89 | 6.42 | 13.30 | 25.5 | 36.7 | 44.3 | 46.4 |
| Water | 98.8 | 98.1 | 97.1 | 95.5 | 91.7 | 84.4 | 71.1 | 58.9 | 45.1 | 37.1 |
| IE | 1.2 | 1.2 | 1.5 | 1.6 | 1.9 | 2.3 | 3.4 | 4.4 | 10.6 | 16.5 |
| AA | 0 | 0.18 | 0.37 | 0.79 | 1.93 | 4.82 | 11.4 | 21.6 | 31.1 | 36.2 |
| Water | 0.6 | 0.5 | 0.7 | 0.8 | 1 | 1.9 | 3.9 | 6.9 | 10.8 | 15.1 |
| IE | 99.4 | 99.3 | 98.9 | 98.4 | 97.1 | 93.3 | 84.7 | 71.5 | 58.1 | 48.7 |

## LEACHING

### Numericals

1. Vegetable oil seeds containing 120 gm insoluble solid and 12 gm oil are contacted with 250 gm of organic solvent in a single stage leaching operation. The solvent used is fresh. Determine amount of oil left in the oil seeds after leaching. The equilibrium data can be expressed as $N = -4y + 8$, where N is gm insoluble/(gm solvent + gm oil), y is gm oil/(gm solvent + gm oil) in the seed phase and x is gm oil/(gm solvent + gm oil) in the solvent phase. The tie line data are

| y | 0.26 | 0.28 | 0.31 | 0.34 |
|---|---|---|---|---|
| x | 0.02 | 0.04 | 0.06 | 0.08 |

2. Oil is to be extracted from meal by means of benzene using continuous counter current extractor. The unit is to treat 1000 kg of meal (based on completely exhausted solid) per hour. The untreated meal contains 350 kg of oil and 30 kg of benzene. The fresh solvent is to contain 50 kg of unextracted oil. Experiments carried out under conditions identical with those of the projected battery, show that the solution retained depends on the concentration of the solution as follows

| 'conc. Kg oil/kg solution | 0 | 0.1 | 0.2 | 0.3 | 0.4 | 0.5 | 0.6 | 0.7 |
|---|---|---|---|---|---|---|---|---|
| Solution retained kg/kg solid | 0.5 | 0.505 | 0.515 | 0.530 | 0.550 | 0.571 | 0.595 | 0.620 |

3. 300 kg per hour of halibut liver is to be extracted in a counter current cascade with ether to recover oil. The ether which has been partially purified contains 3% oil. The fresh liver contains 15 % oil and is to be extracted to composition 0.8% oil (on solvent free basis). 200 kg of solvent is to be used.
   1. What percentage of oil entering with the liver is recovered in the extract?
   2. How many equilibrium stages are required?

| Concentration of over flow (kg oild/kg solution) | 0 | 0.1 | 0.2 | 0.3 | 0.4 | 0.5 | 0.6 | 0.67 |
|---|---|---|---|---|---|---|---|---|
| Entrainment (kg solution/kg extracted livers) | 0.28 | 0.34 | 0.40 | 0.47 | 55 | 0.66 | 0.80 | 0.96 |

4. Roasted copper ore containing copper as $CuSO_4$ is to be extracted in a counter current extractor. The feed charge to be treated per hour comprises of 12 tonnes of gangue, 1.5 tonnes of copper sulphate and 0.5 tonne of water. The strong solution produced is to consist of 90% water and 10% copper sulphate by weight. The recovery of copper sulphate is to be 95% of that of the ore. Pure water is to be used as the fresh solvent. After each stage one tonne of inert gangue retains 2 tonnes of water plus the copper sulphate dissolved in that water. Equilibrium is attained in each stage. How many stages are required?

5. 50 tons per day of oil sand (25 mass % oil and 75 mass % sand) is to be extracted with 50 tons per day of naphtha in a continuous counter current extraction battery. The final extract from the battery contains 40 mass % oil and 60 mass % naphtha and the underflow from each unit is expected to consist of 35 mass % solutions and 65 mass % sand. If the overall efficiency of the battery is 60%, how many stages will be required?
6. 2000 kg of waxed paper per day are to be dewaxed in a continuous counter current extraction system which contains a number of ideal stages by using kerosene as the solvent. The waxed paper contains, by weight 25% paraffin wax and 75 % paper pulp. The extracted pulp is put through a dryer to evaporate the kerosene. The pulp, which retains the unextracted wax after evaporation, must not contain over 0.2 kg of wax per 100 kg of wax free pulp. The kerosene used for extraction contains 0.05 kg of wax per 100 kg of wax free kerosene. Experiments show that the pulp retains 2 kg of kerosene per kg of kerosene and wax free pulp as it is transferred from cell to cell. The extract from the battery is to contain 5 kg of wax per 100 kg of wax free kerosene. How many stages are required?
7. It is desired to extract the oil from halibut livers by continuous counter current extraction with ethyl acetate. The quantity of solution retained by the granulated livers is given as

| `conc. Kg oil/kg solution | 0 | 0.1 | 0.2 | 0.3 | 0.4 | 0.5 | 0.6 | 0.65 | 0.7 | 0.72 |
|---|---|---|---|---|---|---|---|---|---|---|
| Solution retained kg solution /kg oil free liver | 0.205 | 0.242 | 0.286 | 0.339 | 0.405 | 0.489 | 0.60 | 0.672 | 0.765 | 0.81 |

The liver contains 0.257 mass fraction oil. If 95 % of the oil is to be extracted and the strong solution obtained from the system is to contain 0.7 mass fraction oil, determine

A. The quantity and composition of discharged solids
B. Kg of oil free ether required per 1000 kg charge of fresh livers
C. The number of ideal stages required.

## EVAPORATION

(1) An ideal flash vaporization is carried out with a binary mixture at constant temperature and pressure. A process upset leads to an increase in the mole fraction of the heavy component in the feed. The flash vessel continuous to operate at the previous temperature and pressure and still produces liquid and vapor. After steady state is re-established,

A. The amount of vapor produced will increase
B. The amount of liquid produced will decrease
C. The new equilibrium compositions of the vapor and liquid products will be different,

D. The new equilibrium compositions of the vapor and liquid products will remain as they were before the upset occurred.

[GATE 2008, Q 49]

**Solution (D)**

(2) If the percent humidity of air (30°C, total pressure 100 kPa) is 24 % and the saturation pressure of water vapor at that temperature is 4 kPa, the percent relative humidity and the absolute humidity of air are

(A) 25.2, 0.0062 (B) 25, 0.0035

(C) 20.7, 0.0055 (D) 18.2, 0.00

[GATE 2007, Q 29]

**Solution (A)**

(3) The Lewis relation for air-water humidification is given by (ky: Mass transfer coefficient of moisture in air; $h_G$ : heat transfer coefficient; $C_S$: heat capacity of vapour gas mixture)

(A) $\frac{h_{G^2}}{k_y C_S} = 1$ (B) $\frac{K_y C_S{}^2}{h_G} = 1$ (C) $\frac{h_G}{k_y C_S} = 1$ (D) $\frac{K_y{}^2 h_G}{C_S} = 1$

[GATE 2001, Q 1.13]

**Solution (C)**

(4) Multiple effect evaporator has capacity to process 4000 kg of solid caustic soda per day when it is concentrating from 10% to 25% solids. The water evaporated in kg per day is,

(A) 800 (B) 24,000
(C) 60,000 (D) 48,000

[GATE 1992, Q 2(d)]

**Solution (B)**

(5) A multiple effect evaporator as compared to a single effect evaporator of the same capacity has

(A) Lower heat transfer area (B) Lower steam economy
(C) Higher steam economy (D) Higher solute concentration in the product

[GATE 1998, Q 1.14]

**Solution (C)**

(6) An evaporator while concentrating an aqueous solution from 10 to 40% solids evaporates 30000 kg of water. The amount of solids handled by the system in kgs is –

(A) 4000 (B) 9000

(C) 4600 (D) 3000

[GATE 1995, Q 2(e)]

**Solution (A)**

(7) The advantage of backward-feed multiple-effect evaporators over forward-feed units is that,

A. Heat sensitive materials can be handled

B. There is no additional cost of pumping

C. Most concentrated liquor is at highest temperature

D. Equal heat transfer coefficients exist in various effects

[GATE 1992, Q 5(d)]

**Solution (C)**

Pure aniline is evaporating through a stagnant air film of 1 mm thickness at 300 K and a total pressure of 100 kPa. The vapour pressure of aniline at 300 K is 0.1 kPa. The total molar concentration under these conditions is 40.1 mol/m$^3$. The diffusivity of aniline in air is $0.74 \times 10^{-5}$ m$^2$/s.

(8) The numerical value of the mass transfer coefficient is $7.4 \times 10^{-3}$. Its units are

(A) m/s (C) mol/(m$^2$ s Pa)

(B) cm/s (D) kmol/(m$^2$ s Pa)

[GATE 2004, Q 19]

**Solution (A)**

(9) The rate of evaporation of aniline is $2.97 \times 10^{-4}$. Its units are

(A) mol/s (C) mol/(cm$^2$ s)

(B) mol/(m$^2$ s) (D) kmol/(m$^2$ s)

[GATE 2004, Q 20]

**Solution (D)**

An air-water vapour mixture has a dry bulb temperature of 60°C and a dew point temperature of 40°C. The total pressure is 101.3 kPa and the vapour pressures of water at 40°C and 60°C are 7.30 kPa and 19.91 kPa, respectively.

(10) The humidity of air sample expressed as kg of water vapour/kg of dry air is

(A) 0.048 (B) 0.079 (C) 0.122 (D) 0.152

[GATE 2004, Q 21]

**Solution (D)**

(11) The wet bulb temperature Tw for the above mixture would be

(A) less than 40°C (B) 40°C

(C) 40°C < Tw < 60°C (D) 60°C

[GATE 2004, Q 22]

**Solution (C)**

Design a forward feed double effect evaporator to produce an aqueous solution with 50% solid from a solution containing 15% solid flowing at

9000 lb/h. The steam at 15.3 psig was used as heating medium and pressure in second evaporator is 7 psi. The solutions shows BPR of 3 and 7°F and the heat transfer coefficients are 450 Btu/lb ft °F and 200 Btu/lb ft °F respectively.

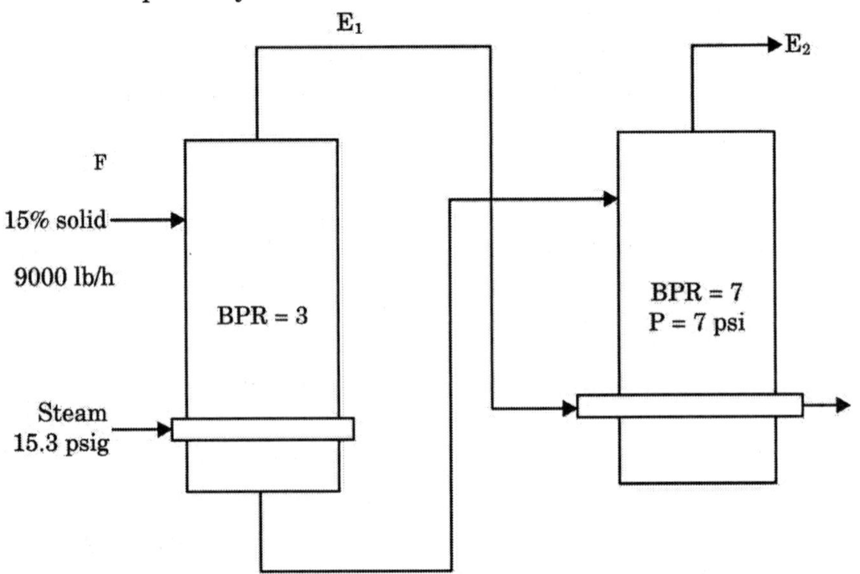


$$F = E_1 + E_2 + P_2$$

$$9000 = E_1 + E_2 + P_2 \quad (1)$$

First effect

$$F_1 = E_1 + P_1$$

Second effect

$$P_1 = P_2 + E_2$$

Component continuity equation

$$Fx_F = P_2 x_{P2} \quad (2)$$

$$P_2 = 9000*0.15/0.5 = 2700$$

From (1)

$$E_1 + E_2 = 9000 - P_2 = 6300$$

Energy balance

$$Fh_f + S\lambda_s = E_1H_1 + P_1H_1 \quad (3)$$

$$P_1h_1 + E_1\lambda_{E1} = E_2H_2 + P_2H_2 \quad (4)$$

$\Delta T$ ovarall determination

Corresponding to 15.3 psig = 15.3 + 14.7 = 30 psia

$$T_s = 250.33\ °F \quad \text{(from steam table)}$$

Corresponding to 7 psia = 176.85 psia (from steam table)

$$\Delta T = 250.33 - 176.85 = 73.48\ °F$$

Total BPR = 10

Hence

$$\Delta T = 73.48 - 10 = 63.48\ °F$$

$$U_1 \Delta T_1 = U_2 \Delta T_2$$

Adding $U_2\,\Delta T_1$ on both sides

$$U_1\,\Delta T_1 + U_2\,\Delta T_1 = U_2\,\Delta T_2 + U_2\,\Delta T_1$$

$$\Delta T_1(U_1 + U_2) = U_2\,\Delta T$$

$$\text{Hence } \Delta T_1 = \frac{U_2}{U_1 + U_2}\,\Delta T = 19.53\ °F$$

Hence $\Delta T_2 = 43.94$ °F

Using eq. (3) and (4)

$$9000 * 118 + s * 945.3 = E_1 * 1156.3 + P_1 * 200$$

$$P_1 * 200 + E_1 * 960.1 = E_2 * 1136.9 + 2700 * 151$$

$$P_1 = F_1 - E_1 = 9000 - E_1$$

Solving simultaneous these three equation results

$E_1 = 3041.73$

$S = 3857.83$

$E_2 = 3258.27$

$P_1 = 5958.27$

Temperature distribution

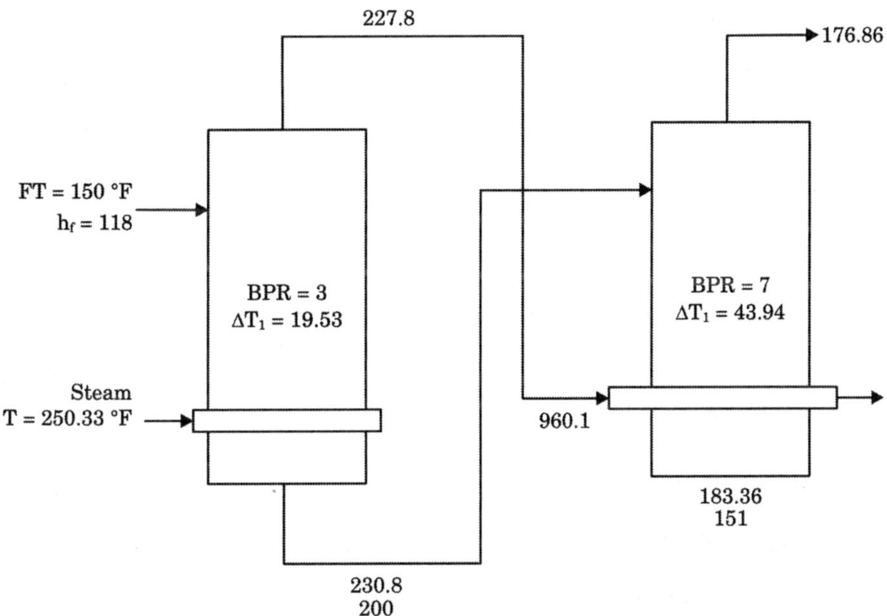


$$A_1 = \frac{S\lambda_s}{U_1\Delta T_1} = 3857.83 * 945.3 / 450 * 19.53 = 415$$

$$A_2 = \frac{E_1\lambda_1}{U_2\Delta T_2} = 3041.73 * 960.1 / 200 * 43.94 = 332.3$$

% age difference = 415 – 332.3/415 * 100 = 20 %

## CRYSTALLIZATION

1. Molecular diffusivity of a liquid –
   A. Increases with temperature
   B. May increase or decrease with temp.
   C. Decreases with temperature
   D. Is independent of temperature

[GATE 1997, Q 1.15]

**Solution (A)**

2. In an interphase mass transfer process, the lesser the solubility of a given solute in a liquid, the higher are the chances that the transfer process will be
   A. Gas phase resistance controlled
   B. Liquid phase resistance controlled
   C. Impossible
   D. Driven by a nonlinear driving force

[GATE 1998, Q 1.17]

**Solution (B)**

3. The surface renewal frequency in Danckwerts' model of mass transfer is given by ($k_L$; mass transfer coefficient, m/s)

   (A) $\sqrt{k_L^2 D_A}$ (B) $k_L^2 D_A$ (C) $\frac{k_L^2}{D_A}$ (D) $\frac{k_L}{D_A^2}$

[GATE 2001, Q 1.11]

**Solution (C)**

4. A supersaturated solution of a sparingly soluble solute, at a concentration of C, is being fed to a crystallizer at volumetric flow rate of V. The solubility of the solute is $C_S$. The output rate of solids from an efficient crystallizer is ___________.

[GATE 1994, Q 2(1)]

**Solution** [(C-$C_s$)V]

## LEACHING & WASHING

1. The mass transfer coefficient for a solid sphere of radius 'a' dissolving in a large volume of quiescent liquid, in which D is the diffusivity of the solute, is

   (A) D/a (B) D/2a
   (C) Roportional to $(D)^{1/2}$ (D) Dependent on the Reynold's number

[GATE 1998, Q 1.16]

**Solution (D)**

2. According to the film theory of mass transfer, the mass transfer coefficient is proportional to

   (A) D (B) $D^2$

   (C) $D^{0.5}$ (D) 1/D

   Where D is the molecular diffusivity.

   [GATE 1997, Q 2.13]

   **Solution (A)**

3. A binary mixture of 100 mm size having densities of 2 gm/cm$^3$ and 4 gm/cm$^3$ is to be classified by lutriation technique using water. Estimate the range of velocities that can do the job and recommend a suitable value.

   [GATE 1995, Q 16]

   **Solution**

   Terminal settling velocity by Newton' s law region

$$u_o = \sqrt{3D_p g \frac{(\rho_p - \rho)}{\rho}}$$
$$u_1 = 0.05425m / s$$
$$u_2 = 0.094m / s$$

   By Strokes Law region

$$u_0 = \frac{D_p^2 g}{18\mu}(\rho_g - \rho_e)$$
$$u_1 = 0.00545m / s$$
$$u_2 = 0.01635m / s$$

   Velocity range is 0.000545 to 0.094 m/s

## DRYING OF SOLID

(1) At equilibrium, the concentration of water is vapor phase (C*) in kg/m$^3$ of air space and the amount of water (m) adsorbed per kg of dry silica gel are related by C* = 0.0667 m. To maintain dry conditions in a room of air space 100 m$^3$ containing 2.2 kg of water vapour initially, 10 kg of dry silica gel is kept in the rom. The fraction of initial water remaining in the air space after a long time (during which the temperature is maintained constant) is

   (A) 0.0 (B) 0.2 (C) 0.4 (D) 1.0

   [GATE 2004, Q 67]

   **Solution (C)**

(2) A 25 cm × 25 cm × 1 cm flat sheet weighing 1.2 kg initially was dried from both sides under constant drying rate conditions. It took 1500 seconds for the weight of the sheet to reduce to 1.05 kg. Another 1 m × 1 m × 1 cm flat sheet of the same materials is to be dried from one side

only. Under the same constant drying rate conditions, the time required for drying (in seconds) from its initial weight of 19.2 kg to 17.6 kg is

(A) 1000 (B) 1500 (C) 2000 (D) 2500

[GATE 2004, Q 68]

**Solution (C)**

(3) At 25°C and 90% relative humidity, water evaporates from the surface of a lake at the rate of 1.0 kg/m$^2$/h. The relative humidity that will lead to an evaporation rate of 3.0 kg/m$^2$/h, with other conditions remaining the same, is

(A) 30% (B) 50% (C) 60% (D) 70%

[GATE 2010, Q 38]

**Solution (D)**

(4) It takes 6 hours to dry a wet solid from 50% moisture content to the critical moisture content of 15%. How much longer will it take to dry the solid to 10% moisture content, under the same drying conditions (the equilibrium moisture content of the solid is 5%).

(A) 15 min (B) 51 min (C) 71 min (D) 94 min

[GATE 2002, Q 2.19]

**Solution (C)**

(5) In a laboratory test run, the rate of drying was found to be $0.5 \times 10^{-3}$ kg/m$^2$ s when the moisture content reduced from 0.4 to 0.1 on a dry basis. The critical moisture content of the material is 0.08 on a dry basis. A tray drier is used to dry 100 kg (dry basis) of the same material under identical conditions. The surface area of the material is 0.04 m$^2$/kg of dry solid. The time required (in seconds) to reduce the moisture content of the solids from 0.3 to 0.2 (dry basis) is

(A) 2000 (B) 4000 (C) 5000 (D) 6000

[GATE 2000, Q 2.15]

**Solution (C)**

(6) 1000 kg of wet solids are to be dried from 60% to 20% moisture (by weight). The mass of moisture removed in kg is –

(A) 520 (B) 200 (C) 400 (D) 500

[GATE 1995, Q 2(f)]

**Solution (C)**

(7) 200 kg of solid (on dry basis) is subjected to a drying process for a period of 5000 s. The drying occurs in the constant rate period with the drying rate as $N_C = 0.5 \times 10^{-3}$ kg/m$^2$s. The initial moisture content of the solid is 0.2 kg moisture/kg dry solid. The interfacial area available for drying is 4 m$^2$/100 kg of dry solid. The moisture content at the end of the drying period is (in kg moisture/kg dry solid).

(A) 0.5 (B) 0.05 (C) 0.1 (D) 0.15

[GATE 2001, Q 2.13]

**Solution (C)**

(8) A solid is being dried in the linear drying rate regime from moisture content $X_0$ to $X_F$. The drying rate is zero at X = 0 and the critical moisture content is the same as the initial moisture, $X_0$. The drying time for M = (LS/LC) is

(A) M ($X_0 - X_F$) (B) M ($X_0/X_F$)

(C) M ln ($X_0/X_F$) (D) $MX_0$ ln ($X_0/X_F$)

where $L_S$: total mass of dry solids, A: total surface area for drying, $R_C$ constant maximum drying rate per unit area, and X moisture content (in mass of water/mass of dry solids)

[GATE 2003, Q 66]

**Solution (D)**

(9) The following plot gives the saturated humidity (He) versus Temperature (T)

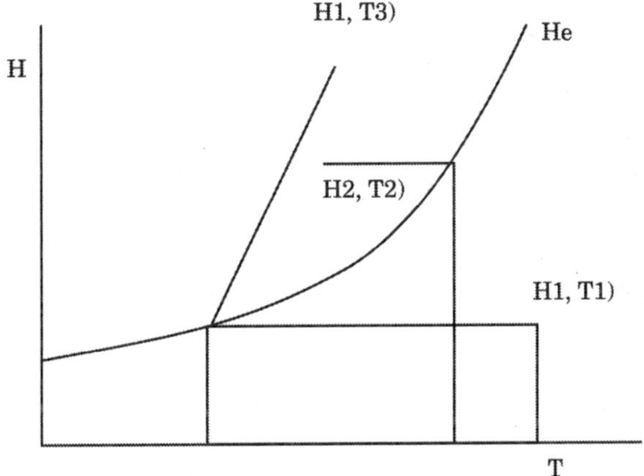


Line joining (HI, TI) and (H2, T2) is the constant enthalpy line. Choose the correct one from among the alternatives A, B, C and D

(a) T1 – Dew Point Temp.; T2 – Dry Bulb Temp.; T3 – Wet Bulb Temp.

(b) T1 – Dew Point Temp.; T2 – Wet Bulb Temp.; T3 – Dry Bulb Temp.

(c) T1 – Wet Bulb Temp.; T2 – Dry Bulb Temp.; T3 – Dew Point Temp.

(d) T1 – Dry Bulb Temp.; T2 –Wet Bulb Temp.; T3 – Dew Point Temp.

[GATE 2003, Q 67]

**Solution (D)**

(10) In drying a solid containing moisture above the critical moisture content, the number of degrees of freedom is ____________.

[GATE 1992, Q 6(a)]

**Solution (Two)**

(11) The equilibrium moisture curve for a solid is shown below:

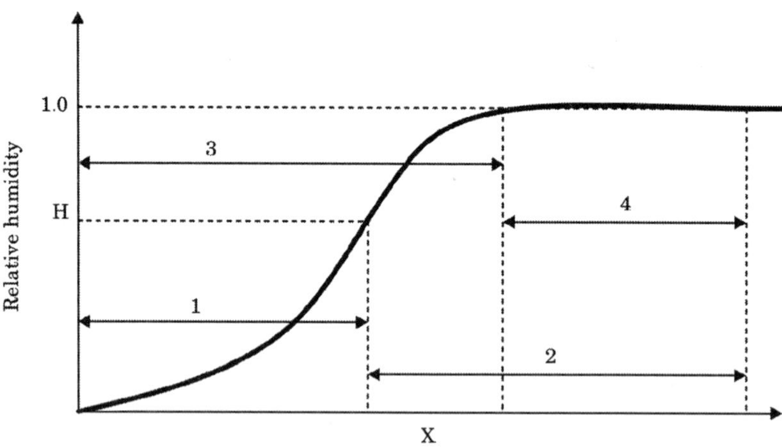


Moisture content (kg water/kg dry solid)

The total moisture content of the solid is X and it is exposed to air of relative humidity H. In the table below, Group I lists the types of moisture, and Group IIrepresents the region in the graph above

| | Group I | Group II |
|---|---|---|
| P. | Equilibrium moisture | 1 |
| Q. | Bound moisture | 2 |
| R. | Unbound moisture | 3 |
| S. | Free moisture | 4 |

Which ONE of the following is the correct match?

(A) P-1, Q-2, R-3, S-4 (B) P-1, Q-3, R-4, S-2
(C) P-1, Q-4, R-2, S-3 (D) P-1, Q-2, R-4, S-3

[GATE 2009, Q 38]

**Solution (B)**

(12) Calculate the time required to reduce the moisture content of a solid from 0.66 to 0.25 kg moisture/kg dry solid. The rate of drying N (kg water evaporated/m$^2$.s) is given as

$$N = 0.0015\,X \text{ for } x < 0.3$$
$$= 0.00045 \text{ for } x > 0.3$$

where X = moisture/kg dry solid. The drying surface is 0.025 m$^2$/kg dry solid.

[GATE 1997, Q 19]

**Solution**

Constant rate period

$$t_c = \frac{W_i - W_c}{R_c A} = 32000 \text{ sec} = 8.89 hr$$

$$mA = \frac{1}{t_c}\left[\frac{W_i - W_c}{W_c - W_e}\right] = 3.75x10^{-5}$$

Falling rate period

$$t_f = \frac{1}{mA}\ln\left[\frac{W_c - W_e}{W_f - W_e}\right] = 47780\,\text{sec} = 13.27\text{hr}$$

Total drying time = 22.16 hr-**Ans**

(13) It is necessary to dry a batch of 160 kg of a wet solid material from 30% to 5% moisture content, under constant rate and failing rate period. The falling rate is assumed to be linear. Calculate the total drying time considering an available drying surface of 1 $m^2$/40 kg of dry solid. A constant drying flulx of $3 \times 10^{-4}$kg/$m^2$ s is given.

Xc the critical moisture content = 0.2 kg moisture/kg solid.

X* the equilibrium moisture content = 0.05 kg moisture/kg solid.

[GATE 1995, Q 19]

**Solution**

Constant rate period

$$t_c = \frac{W_i - W_c}{R_c A} = 19048\,\text{sec}$$

$$mA = \frac{1}{t_c}\left[\frac{W_i - W_c}{W_c - W_e}\right] = 3.5x10^{-5}$$

Falling rate period

$$t_f = \frac{1}{mA}\ln\left[\frac{W_c - W_e}{W_f - W_e}\right] = \infty$$

## NUMERICALS

Q.1. Slab of paper pulp 1m*1m*0.01m is to be dried under constant drying condition from 70% to 30% moisture. The value of equilibrium moisture for the material is 0.4%. If the critical moisture content is 55% and the rate of drying of critical point is 1.4 kg/hr.$m^2$, calculate the drying time. The dry weight of each slab is 2.5 kg. Assume all moisture content on the wet basis.

**Solution**

Considering drying from two big faces A = 1*1*2 = 2$m^2$

$S_s$= dry weight of slab = 2.5 kg

Initial moisture =70%, *i.e.* 0.7(wet basis)

Initial moisture content $X_1$ = 0.7/1-0.7 = 2.303 kg moisture/kg dry solid

Final moisture content $X_2$ = 0.3/1-0.3 = 0.4285

Equilibrium moisture content X* = .004/1-.004 = .004016

Critical moisture content $X_c$ = 0.55/1-0.55 = 1.22

$N_c$ = Constant rate of drying = 1.4 kg/hr.$m^2$

Total time of drying $\theta = \theta_c + \theta_f$

$= S_s/A^* N_c [(X_1-X_C) + (X_C-X^*) \ln\{X_C-X^*/X_2-X^*\}]$

= 2.5/2*1.4 [(2.333-1.22) +(1.222-.004016)ln{1.22-.004016/0.4285-0.004016}]
= 2.136 h.

Q.2. A certain material was dried under constant drying conditions and it was found that 2.5 hrs are required to reduce the free moisture concentration from 30% to 10%. How much longer would it be required to reduce the free moisture to 5%. Assume that no constant rate period is encountered.

**Solution**

Let equilibrium moisture concentration $X^* = 0$

Given $X_1 = 0.3/1\text{-}0.3 = 0.4285$

$X_2 = 0.14/1\text{-}0.1 = 0.1111$

And time required is 2.5 hrs

Since Total time of drying $\theta = \theta_c + \theta_f$

$2.5 = S_s/A^* N_c [(X_C\text{-}X^*) \ln\{X_C\text{-}X^*/X_2\text{-}X^*\}]$

$2.5 = S_s/A^* N_c[(0.4285\text{-}0) \ln\{0.4285/0.1111\}]$

$S_s/A^* N_c = 4.322$

For next condition $X_2 = 0.05/1\text{-}0.05 = 0.05263$

Hence time required $\theta = S_s/A^* N_c [(X_C\text{-}X^*) \ln\{X_C\text{-}X^*/X_2\text{-}X^*\}]$

$= 4.322[(0.4285\text{-}0)\ln\{0.4285/0.05263\}]$

$= 3.785$ Hrs.

Q.3. A batch of 150 kg of wet material is to be dried from 30% to 3% moisture content. The falling rate is assumed to be linear. Calculate the total drying time considering an available drying surface of 2 m²/50 kg dry solid. Drying flux is given as $2.5*10^{-4}$ kg/m².sec. Critical moisture content is 0.25 kg moisture/kg solid and equilibrium moisture content is 0.02.

**Solution**

Given drying surface area $A/S_s = 2$ m²/50 kg

Or $S_s/A = 25$ kg/m²

$X_1 = 30\% = 0.3/1\text{-}0.3 = 0.4285$

$X_2 = 3\% = 0.03/1\text{-}0.03 = 0.0309$

$Xc = 0.25$

$X^* = 0.02$

Given $N_c = 2.5 *10\text{-}4$ kg/m².sec

Total time of drying $\theta = \theta_c + \theta_f$

$= S_s/A^* N_c [(X_1\text{-}X_C) +(X_C\text{-}X^*) \ln\{X_C\text{-}X^*/X_2\text{-}X^*\}]$

$= 25/2.5*10\text{-}4*3600[(0.4285\text{-}0.25)+(0.25\text{-}0.02)\ln\{0.25\text{-}0.02/0.0309\text{-}0.02\}]$

$= 24.44$ Hrs

Q.4. A wet solid is dried from 40% to 10% under constant drying condition in 6 Hrs. if the equilibrium moisture content is 5% and the critical moisture content is 20%,how long will it take to dry 7% moisture under the same condition.

**Solution**

Data given for the first case

Initial moisture $X_1 = 40\% = 0.4/1\text{-}0.4 = 0.667$

Final moisture $X_2 = 10\% = 0.1/1\text{-}0.1 = 0.111$

Critical moisture $X_c = 20\% = 0.2/1\text{-}0.2 = 0.25$

Equilibrium moisture $X^* = 5\% = 0.05/1\text{-}0.05 = 0.05263$

Since Total time of drying $\theta = \theta_c + \theta_f$

$\theta = S_s/A^* N_c [(X_1\text{-}X_C) + (X_C\text{-}X^*) \ln\{X_C\text{-}X^*/X_2\text{-}X^*\}]$

$6 = S_s/A^* N_c[(0.667\text{-}0.25) + (0.25\text{-}0.05263)\ln\{0.25\text{-}0.05263/0.111\text{-}0.05263\}]$

$S_s/A^* N_c = 9.126$

For the second case $X_2 = 7\% = 0.07/1\text{-}.007 = 0.0752$

Total time of drying $\theta = S_s/A^* N_c [(X_1\text{-}X_C) + (X_C\text{-}X^*) \ln\{X_C\text{-}X^*/X_2\text{-}X^*\}]$

$= 9.126^*[(0.667\text{-}0.25) + (0.25\text{-}0.05263)\ln\{0.25\text{-}0.05263/0.0752\text{-}0.05263\}]$

= 7.7 hrs

## ABSORPTION

1. The ratio of the liquid to gas flow rate in a counter-current gas absorption column is increased at otherwise identical conditions. Which **ONE** of the following statements is **TRUE**?

   A. The operating line shifts towards the equilibrium curve

   B. The operating line shifts away from the equilibrium curve

   C. The concentration of the absorbed species increases in the exit liquid stream

   D. The operating line does not shift.

   [GATE 2009, Q 12]

   **Solution (B)**

2. A batch distillation operation is carried out to separate a feed containing 100 moles of a binary mixture of A and B. The mole fraction of A in the feed is 0.7. The distillation progresses until the mole fraction of A in the residue decreases to 0.6. The equilibrium cure in this composition range may be linearized to $y^* = 0.7353\ x + 0.3088$. Here x and y are the mole fractions of the more volatile component A in the liquid and vapor phases respectively. The number of moles of residue is

   (A) 73.53 (B) 418.02 (C) 40 (D) 30.24

   [GATE 2008, Q 51]

   **Solution (A)**

3. The diffusion rate of ammonia from an aq. solution to the gas phase is $_{10}{}^{-3}$ kmol/m$^2$ s. The interface equilibrium pressure of $NH_3$ is 660N/m$^2$ and the concentration of $NH_3$ in gas phase is 5%. If total pressure is 101N/m$^2$, temperature is 295k and diffusivity of $NH_3$ is 0.24 cm$^2$/s, the gas film thickness is ________ mm.

   [GATE 1991, Q 6(i)]

   **Solution (0.00641)**

4. In gas lliquid contact operation the number of ideal stages $N = \frac{x_a - x_b}{x_b - x_b^*}$.
This is true when the stripping factor S is

(A) S > 1 (B) S < 1

(C) S = 1 (D) S = ∞

[GATE 1995, Q 2(o)]

**Solution (A)**

5. For absorbing a sparingly soluble gas in a liquid –

A. Gas side mass transfer coefficient should be increased
B. Liquid side mass transfer coefficient should be increased
C. Liquid side mass transfer coefficient should be decreased
D. Mass transfer coefficient must be kept constant.

[GATE 1995, Q 2(p)]

**Solution (B)**

6. For gas absorption the height of a transfer unit, based on the gas phase, is given by (G: superficial molar gas velocity; L: superficial molar liquid velocity; $F_G$: mass transfer coefficient, mol/m²s; a: interfacial area per unit volume of tower)

(A) $\frac{G}{F_G\, a}$ (B) $\frac{F_G}{G\, a}$ (C) $\frac{G\, a}{F_G}$ (D) $\frac{L}{F_G\, G}$

[GATE 2001, Q 1.12]

**Solution (A)**

7. Hydraulic mean diameter for flow through packed bed of spherical particles of size $D_P$ with porosity ε

(A) $D_P \frac{1}{6} \frac{\varepsilon}{(1-\varepsilon)}$ (B) $D_P \frac{1}{6} \frac{(1-\varepsilon)}{\varepsilon}$ (C) $D_P \frac{2}{3} \frac{(1-\varepsilon)}{\varepsilon}$ (D) $D_P \frac{2}{3} \frac{\varepsilon}{(1-\varepsilon)}$

[GATE 1995, Q 2(i)]

**Solution (D)**

8. In a countercurrent gas absorber, both the operating and equilibrium relations are linear. The inlet liquid composition and the exit gas composition are maintained constant. In order to increase the absorption factor

A. The liquid flow rate should decrease
B. The gas flow rate should increase
C. The slope of the equilibrium line should increase
D. The slope of the equilibrium line should decrease

[GATE 2008, Q 16]

**Solution (D)**

9. Experiments were conducted to determine the flux of a species A in a stagnant medium across a gas-liquid interface. The overall mass transfer coefficient based on the liquid side for dilute systems for the above was estimated to be $4 \times 10^{-3}$ kg.mol/m$^2$ s. The equilibrium data for the sytem is given as y = 2x. The flux across the interface (in kg mol/m$^2$ s) or bulk concentrations of A in gas phase and liquid phase as y = 0.4 and x = 0.01, respectively is

   A. $5.6 \times 10^{-4}$

   B. $8.5 \times 10^{-4}$

   C. $5.6 \times 10^{-3}$

   D. $8.5 \times 10^{-3}$

[GATE 2003, Q 62]

**Solution (B)**

10. $H_2$ S is being absorbed in a gas absorber unit. The height of the transfer unit based on the overall mass transfer coefficient on the gas side is 0.4 m. The equilibrium data are given by y = 1.5 x. The bulk concentration of $H_2$ S has to be reduced from 0.05 to 0.001 mole fraction in the gas side. The height of the tower (in meters) corresponding to an operating line given by y = 5 × + 0.001 is

   (A) 2.0 (B) 1.56 (C) 1.0 (D) 0.56

[GATE 2003, Q 63]

**Solution (A)**

11. The Reynolds Number of the liquid was increased 100 fold for a laminar falling film used for gas-liquid contacting. Assuming penetration theory is applicable, the fold – increase in the mass transfer coefficient (ke) for the same system is

   (A) 100 (B) 10 (C) 5 (D) 1

[GATE 2003, Q 64]

**Solution (B)**

12. The individual mass transfer coefficients (mol/m$^2$ s) for absorption of a solute from a gas mixture into a liquid solvent are $k_L$ =4.5 and $k_G$ =1.5. The slope of the equilibrium line is 3. Which of the following resistance(s) is (are) controlling?

   (A) liquid – side (B) gas – side

   (C) interfacial (D) both liquid and gas-side.

[GATE 2000, Q 2.14]

**Solution (B)**

13. Acetone is to be removed from air in an isothermal dilute absorber using pure water as solvent. The incoming air contains 5 mol% of acetone ($y_m$ = 0.05). The design equation to be used for obtaining the number of trays (N) of the absorber is

$N + 2 = 6 \log\left(\frac{y_{in}}{y_{out}}\right)$. For 98% recovery of acetone, the number of trays required is/are

(A) 1 (B) 8 (C) 9 (D) 10

[GATE 2004, Q 28]

**Solution (C)**

14. Use of Raschig rings in place of crushed stones as packing in packed beds (other things being same) –
    A. Increases pressure drop, increases surface area
    B. Increases pressure drop, decreases surface area
    C. Decreases pressure drop, increases surface area
    D. Decreases pressure drop, decreases surface area

[GATE 1995, Q 2(j)]

**Solution (C)**

15. The carbon dioxide issuing out of a Fermenter contains 0.01 mole fraction of ethanol, which has to be reduced to 0.0001 mole fraction by scrubbing with water in a counter current packed tower. The gas flowrate is 227.3 kmol/h and may be assumed constant throughout the tower. The equilibrium mole fraction of ethanol in the gas phase ($y^*$) is related to that in the liquid (x) as $y^* = 1.07\,x$

    Determine the minimum liquid rate needed, and the number of overall gas-side transfer units needed at 1.5 times the minimum liquid rate. The entering liquid may be assumed to be free of ethanol.

[GATE 1998, Q 19]

**Solution**

Equilibrium relation $y=1.07x$

$L_{min}(X_1\text{-}X_2) = G(Y_1\text{-}Y_2)$

$L_{min} = 361.16$ kmol/hr-**Ans**

$$N_{og} = \int_{y1}^{y2} \frac{dy}{y_e - y} = \ln\left(\frac{y_{e1} - y_1}{y_{e2} - y_2}\right) = 1.87 = 2$$ **-Ans**

16. A counter current plate absorber is to be installed for scrubbing of an air mixture containing 5% ammonia by volume. The scrubber is fed with water containing 0.002 mole $NH_3$ per mole of water. The scrubbing water flows at a rate of 1.0 mole water per mole air. It is necessary to absorb 85 percent of the ammonia present in the gas by operating the absorber at 200C.

    K = (0.80 mol $NH_3$ / mol air) / (mol $NH_3$ / mol $H_2O$)

[GATE 1995, Q 21]

**Solution**

L $(x_1-x_2) = G(y_1-y_2)$
$X_1 = 0.0095$

Equilibrium line is y = 0.8 × or m = 0.8 $\frac{mG}{L} = 0.85$

$$\frac{y_1}{y_2} = \frac{1}{0.85} = 1.1765$$

Now no. of stage

$$N = \frac{1}{\left(1-\frac{mG}{L}\right)} \ln\left[\left(1-\frac{mG}{L}\right)\frac{y_1}{y_2} + \frac{mG}{L}\right] = 0.1734 \text{ –}\textbf{Ans}$$

17. It is desired to absorb acetone from a dilute mixture of acetone/air containing I mol% acetone by contacting it counter currently with pure water in an absorber consisting of two theoretical (ideal) stages. The total inlet gas flow rate is 30 k.mol/hr and that of water is 90 k.mol/hr. Under the operating conditions, the equilibrium relationship for acetone in gas-liquid is y = 2x. Estimate the mol fraction of acetone in the water stream leaving the absorber.

[GATE 1997, Q 18]

**Solution**

Given:- Counter current absorption of acetone from acetone/air mixture inti water.
No. of stage = 2
Equilibrium relation , y=2x
Estimate $x_{1,}$

$$N_{vg} = 2 = \frac{1}{\left(1-\frac{mG}{L}\right)} \ln\left[\left(1-\frac{mG}{L}\right)\frac{y_1}{y_2} + \frac{mG}{L}\right]$$

$$\frac{mG}{L} = 0.667$$

and $y_1 = 0.01$ and $y_2 = 0.001066$
Now L $(x_1-x_2) = G(y_1-y_2)$
$\therefore x_2 = 0.003$-**Ans**

18. Water is used to absorb ammonia from a gas mixture in a single separation stage contactor. The process is schematically represented in the figure below,

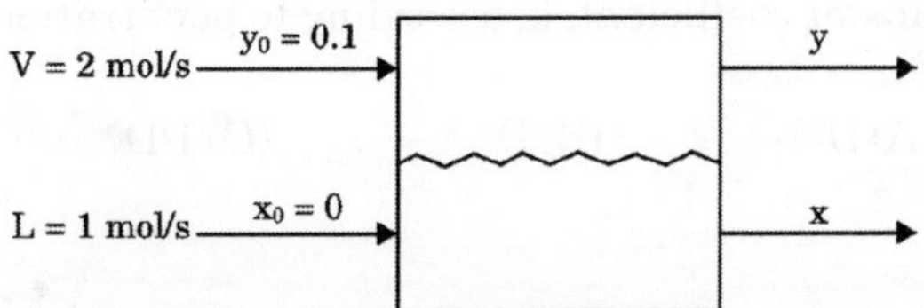

The molar gas and liquid flow rates, and the inlet mole fractions are given in the figure. Both the liquid and the gas phases are well mixed, and the equilibrium relation between y and x is given by $y^* = x$.

If the stage is ideal, then the value of y is

(A) 1/15 (B) 1/10 (C) 2/15 (D) 1/6

[GATE 2010, Q 52]

If the stage efficiency is 50%, then the value of y is

(A) 1/12 (B) 1/6 (C) 1/4 (D) 1/3

[GATE 2010, Q 53]

**Solution**

Mole balance for ammonia

$2 \times 0.1 + 1 \times 0 = 2y^* + 1.x$

Since, $y^* = x$

$0.2 + 0 = 2y^* + y^*$

$Y^* = 0.2/3 = 1/15$ – **Answer**

$NH_3$ transferred (M) from vapor to liquid phase

$M = 1/15$

Stage efficiency = 50%

So, mole transfer = 0.5 x 1/15 = 1/30 mol/s

$M = 2[0.1 - y']$

$1/30 = 2[0.1 - y']$

Or $1/10 - y' = 1/30$

$Y' = 1/12$ – **Answer**

## ADSORPTION

(1) For laminar flow of a fluid through a packed bed of spheres of diameter d, the pressure drop per unit length of bed depends upon the sphere diameter as

(A) d, (B) $d^2$, (C) $d^4$, (D) $1/d^2$,

[GATE 1997, Q 2.6]

**Solution (D)**

(2) Diffusion coefficient of a binary gas mixture at low pressures varies with pressure as,

(A) P (B) $P^2$ (C) 1/P (D) Independent of P

[GATE 1994, Q 1(o)]

**Solution (C)**

(3) Mass transfer coefficient, k, according to penetration theory varies with mass diffusivity as,

(A) $D^{0.5}$ (B) D (C) 1/D (D) $D^{1.5}$

[GATE 1994, Q 1(p)]

**Solution (A)**

(4) A packed gas liquid contractor employs ceramic Intalox saddles (38 mm). The gas flow rate, G = 1.5 kg/s and gas density, PG = 1.5 kg/m³. The liquid flow rate, L = 30 kg/s and liquid density PL = 1000 kg/m³. The column is to be designed for a pressure drop of 42 mm $H_2O$. Norton's correlation for this pressure drop is

$$K = \frac{13.1 G_W^2 F_p \left\{\frac{\mu_L}{\rho_L}\right\}^{0.1}}{\rho_G(\rho_L - \rho_G)}$$

Where K = flow coefficient, $m_L$, liquid viscosity = $10^{-3}$ Ns/m², $G_w$ = gas flow rate per unit cross sectional area, kg/(m²s). Fp = packing factor = 170 m⁻¹. K is given by the equation: K = 0.62 – 0.5 (m – 0.5) for 0.5 £ m £ 1.0 where

$$m = \frac{L}{G}\sqrt{\frac{\rho_G}{\rho_L}}$$

Calculate the diameter of the contactor.

[GATE 2001, Q 19]

**Solution**

$$m = \frac{L}{G}\sqrt{\frac{\rho_G}{\rho_L}} = \frac{30}{1.5}\sqrt{\frac{1.5}{1000}} = 0.7746$$

K = 0.62-0.5 (m-0.5) = 0.4827.

Now

$$K = \frac{13.1 G_W^2 F_p \left\{\frac{\mu_L}{\rho_L}\right\}^{0.1}}{\rho_G(\rho_L - \rho_G)}$$

Cross Section Area = $G_w = 1.1368 \dfrac{Gasflowrate}{CrossSectionArea}$

$$\frac{\pi D^2}{4} = \frac{G}{G_w} = \frac{1.5}{1.1368}$$

D = 1.3 m –**Ans**

# Subject Index

## C



## D



## E

## S



## T

## U



## V



## W



## X



## Y



## Z